Management of Soil Problems

Khan Towhid Osman

Management of Soil Problems

Khan Towhid Osman
Department of Soil Science
University of Chittagong
Chittagong, Bangladesh

ISBN 978-3-319-75525-0   ISBN 978-3-319-75527-4 (eBook)
https://doi.org/10.1007/978-3-319-75527-4

Library of Congress Control Number: 2018935130

© Springer International Publishing AG, part of Springer Nature 2018
This work is subject to copyright. All rights are reserved by the Publisher, whether the whole or part of the material is concerned, specifically the rights of translation, reprinting, reuse of illustrations, recitation, broadcasting, reproduction on microfilms or in any other physical way, and transmission or information storage and retrieval, electronic adaptation, computer software, or by similar or dissimilar methodology now known or hereafter developed.
The use of general descriptive names, registered names, trademarks, service marks, etc. in this publication does not imply, even in the absence of a specific statement, that such names are exempt from the relevant protective laws and regulations and therefore free for general use.
The publisher, the authors and the editors are safe to assume that the advice and information in this book are believed to be true and accurate at the date of publication. Neither the publisher nor the authors or the editors give a warranty, express or implied, with respect to the material contained herein or for any errors or omissions that may have been made. The publisher remains neutral with regard to jurisdictional claims in published maps and institutional affiliations.

Printed on acid-free paper

This Springer imprint is published by the registered company Springer International Publishing AG part of Springer Nature.
The registered company address is: Gewerbestrasse 11, 6330 Cham, Switzerland

For my uncle
A. K. M. Abdur Rashid
Who was there by the grace of Almighty for
me and my family when we were in dire need

Preface

Management of Soil Problems is intended for students of all levels of Soil Science, Agronomy, Horticulture, Forestry, Geography and Environmental Sciences. The soil is a limited non-renewable resource, and the formation of one centimeter depth of fertile surface soil may need several hundreds to thousands of years in nature. Still the area of fertile and productive soils is continuously decreasing for increasing pressures on soils, land-use changes, and soil mismanagement. Moreover, only twelve percent of the global soil area has a few or no limitations to agricultural use; others have some sorts of problems of variable degrees. But, there are many misconceptions about soil use and management, particularly regarding the so-called problem soils and soil problems. Actually, there are a few soils that can be called problem soils in their natural state. If allowed to remain in an undisturbed condition, all soils can function appropriately in their own ecosystems. However, some soils can acquire, during their genesis and evolution under a set or sets of soil-forming factors, certain unique characteristics that may not suit our intended use. These are not the problems of soils themselves but their peculiarities; the problem lies with our dire need to use them for different purposes and in different ways. FAO Soils Portal on management of problem soils states, "Soils are neither good nor bad because the distinction is often based on their intended use. However, many soils have characteristics that make specific management interventions desirable to avoid problems for agricultural production or environmental degradation." The demands of land for agricultural, urban, industrial, and other uses have enormously increased due to an ever-increasing human population. High demands of food, wood, fiber, fuel, and other materials and services are causing continuous land-use changes. More soils having unsuitable properties with regard to agriculture are being brought at present and will be in future under crop production. Some soils offer serious limitations to their agricultural use; traditionally, these soils are known as "problem soils," for which special, innovative, and sustainable management practices should be adopted. Some soils develop problems due to misuse and mismanagement by humans.

Most soil problems are use-oriented. For example, many wetlands are valuable habitats of diverse flora and fauna; they are rich fishing and bird-nesting grounds;

they purify water and recharge aquifers; these uses pose no problem for them, but draining them for cropping offer many problems. I have found beautiful mangrove forests in acid sulfate and potential acid sulfate soils of Chakaria in Bangladesh; rice also thrives there, but using adjacent lands for shrimp culture has been a failure. Drought is a problem for farming, but not for constructing roads and highways. However, the capacity of performing normal ecological functions of some soils has been lost or deteriorated due to human activities with concomitant decline in soil quality; these soils are the degraded soils. Degradation may be reversible and irreversible. If irreversible degradation has occurred, it may need some centuries to return naturally to a state which might be economically and ecologically valuable again. It is very difficult and sometimes uneconomic to restore the original structures or functions of some moderately degraded soils as well. Drained wetlands and reclaimed peat in the north are good examples. In agricultural viewpoint, there are some specific soil types that limit the growth of crop plants, such as shallow soils, sandy soils, saline soils, acid soils, and peat soils. According to Natural Resource Conservation Service of the United States, common soil problems are compaction, crusting, poor drainage, salinity, acidity, and low-organic matter content. However, there are some limitations that can occur in many different soil types. For example, phosphate fixation, which reduces the availability of native and applied phosphorus to plants, can occur in acid soils, acid sulfate soils, calcareous soils, clay soils, metalliferous soils, and so on.

In this book, I have included thirteen chapters on major problems of soil use and their management preceded by an introduction (Chap. 1). In Chap. 1, several soil problems/soil limitations/soil constraints/soil stresses have been introduced and discussed. Soils posing specific problems mainly to agricultural use are discussed in Chaps 2–13 (Chap. 2 Dryland Soils, Chap. 3 Sandy Soils, Chap. 4 Shallow Soils, Chap. 5 Soils with Drainage Limitations, Chap. 6 Expansive Soils, Chap. 7 Peat Soils, Chap. 8 Soils on Steep Slopes, Chap. 9 Poorly Fertile Soils, Chap. 10 Saline and Sodic Soils, Chap. 11 Acid Soils and Acid Sulfate Soils, Chap. 12 Polluted Soils, and Chap. 13 Soils Degraded Due to Use and Misuse). In Chap. 6, the problems of expansive and dispersive soils associated with their engineering use were also discussed. According to the American Society of Civil Engineers, about half of the houses built in the United States each year are situated on unstable (expansive and dispersive) soil, and about half of these will eventually suffer some soil-related damage. Cracking of foundations, walls, driveways, swimming pools, and roads costs millions of dollars each year in repairs.

I reviewed the recent and relevant literature for up-to-date information regarding all the issues discussed above in this book. Despite my efforts, I could not avoid some repetitions. Readers are certainly aware that some management options, such as organic matter addition, conservation farming, mulching, cover crops, and crop rotations, are common for the mitigation of many soil problems. They have appeared in different chapters although their relevance was always kept in mind. Readers should also be aware that management options for soils posing problems to a specific use need extra inputs. Poor farmers of the developing countries of Africa and Asia cannot adopt many effective but costly practices. Management depends on socioeconomic conditions, tradition, farm facility, cropping patters, etc. And some practices may not be profitable as well.

Chittagong, Bangladesh Khan Towhid Osman
November, 2017

Acknowledgements

I am grateful to my colleagues, students, and friends for their appreciation and encouragement. Many people and organizations helped me with photographs, images, and information; I am thankful to them. Particular mention should be made of my student Mueed-Ul-Zahan; he drew many valuable sketches for the book. My colleagues Mr. Md. Enamul Haque and Mr. Md. Imam Hossain offered their assistance during preparation of the manuscript. I am thankful to them and to other colleagues including Dr. Md. Abul Kashem, Dr. Shoffikul Islam, Ms. Ashoka Sarkar, and Mr. Sajol Roy who inspired me in several occasions. My sons Khan Tanvir Osman assisted in searching literature, and Khan Touseef Osman helped me in the finalization of the manuscript. I am grateful to the unknown reviewers who suggested many improvements.

Abbreviations

BSP	Base saturation percentage
CEC	Cation exchange capacity
CHN	Cold hardiness number
CPATSA	Centro de Pesquisa Agropecuaria do Tropico Semi-Arido (Agricultural Research Centre for the Semi-arid Tropics, Brazil)
DECCW	Department of Environment, Climate Change and Water
DLWC	Department of Land and Water Conservation
ECe	Electrical conductivity of saturation extract
EMBRAPA	Empresa Brasileira de Pesquisa Agropecuária (Brazilian Enterprise for Agricultural Research)
ESP	Exchangeable sodium percentage
FAO	Food and Agriculture Organization of the United Nations
FCC	Fertility capability classification
FYM	Farm yard manure
GPCR	Gas phase chemical reduction
GR	Gypsum requirement
GWT	Groundwater table
HRB	Highway Research Board (of USA)
ICIMOD	International Centre for Integrated Mountain Development
ICRISAT	International Crops Research Institute for the Semi-Arid Tropics
IDRAS	International Desert Research of Academia Sinica
ILCA	International Livestock Centre for Africa
IPCC	International Panel of Climate Change
ISRIC	International Soil Reference and Information Centre
KFUPM	King Fahd University of Petroleum and Minerals
LR	Leaching requirement
M ha	Million hectare
MAAT	Mean annual air temperature
MCD	Mechanochemical dehalogentation
NRCS	Natural Resource Conservation Service
NTCHS	National Technical Committee on Hydric Soils

PET	Potential evapo-transpiration
PVAL	Poly(vinylalcohol) copolymer
RSG	Reference Soil Group
RUSLE	Revised universal soil loss equation
SALT	Sloping agricultural land technology
SAR	Sodium adsorption ratio
SHNC	Superabsorbent hydrogel nanocomposite
SSSA	Soil Science Society of America
TDS	Total dissolved solids
UNCCD	United Nations Convention to Combat Desertification
UNEP	United Nations Environmental Program
UNESCO	United Nations Educational, Scientific and Cultural Organization
USCS	Unified Soil Classification System
USDA	United States Department of Agriculture
USEPA	United States Environment Protection Agency
USLE	Universal soil loss equation
VAM	Vesicular-arbuscular mycorrhizae
WRB	World Reference Base

Contents

1 Management of Soil Problems: An Introduction 1
   1.1 'Problem Soils' and Soil Problems 1
   1.2 Soil Constraints to Different Uses 4
   1.3 Land Resource Stress Classes 11
   References 13

2 Dryland Soils 15
   2.1 Drylands of the World 15
   2.2 Global Distribution of Drylands 16
   2.3 Land Use in Drylands 18
   2.4 Soils of the Drylands 21
      2.4.1 Soil Limitations 22
   2.5 Droughts in Drylands 23
   2.6 Dryland Agriculture 24
   2.7 Integrated Soil and Crop Management for Drylands 24
      2.7.1 Principles of Integrated Soil and Crop Management
           for Drylands 25
      2.7.2 Growing Dryland Crops 25
      2.7.3 Diversity in Crops and Cropping Systems 26
      2.7.4 Conservation Tillage 27
      2.7.5 Water Conservation 30
      2.7.6 Supplemental Irrigation 30
      2.7.7 Fertilizer Application 31
      2.7.8 Use of Herbicides 31
      2.7.9 Soil Conservation 31
   References 33

3 Sandy Soils 37
   3.1 Nature of Sandy Soils 37
   3.2 Distribution of Sandy Soils 38
   3.3 Properties of Sandy Soils 38
   3.4 Advantages of Sandy Soils 40
   3.5 Constraints of Sandy Soils 40

	3.6	Management of Sandy Soils	41
		3.6.1 Selecting Suitable Plants	41
		3.6.2 Tillage	41
		3.6.3 Soil Amendments	45
		3.6.4 Fertilizer Application	50
		3.6.5 Irrigation	51
		3.6.6 Cover Crops	53
		3.6.7 Mulching	55
		3.6.8 Stabilization of Sand Dunes	57
	References		60
4	**Shallow Soils**		67
	4.1	Soil Depth Classes	67
		4.1.1 Very Shallow Soils	68
		4.1.2 Shallow Soils	68
		4.1.3 Moderately Deep Soils	69
		4.1.4 Deep Soils	69
	4.2	Properties of Shallow Soils	70
		4.2.1 Shallow Soils on Mountain Slopes	70
		4.2.2 Shallow Soils on Calcareous Materials	71
		4.2.3 Shallow Soils with Root Restrictive Layers	71
		4.2.4 Soils with Shallow Groundwater Table	74
		4.2.5 Shallow Lateritic Soils	75
	4.3	Limitations of Shallow Soils	75
	4.4	Management of Shallow Soils	76
		4.4.1 Selection of Suitable Crops	76
		4.4.2 Management of Shallow Mountain Soils	76
		4.4.3 Management of Shallow Soils with Compacted or Root Restrictive Layers	79
		4.4.4 Management of Soils with Shallow Groundwater Table	80
	References		81
5	**Soils with Drainage Limitations**		83
	5.1	Wetland Soils, Hydric Soils, Poorly Drained Soils	83
	5.2	Criteria of Hydric Soils	87
	5.3	Features of Hydric Soils	88
		5.3.1 Redoximorphic Features	88
		5.3.2 Chemical Transformations in Hydric Soils	90
		5.3.3 Hydric Organic Soils	94
	5.4	Land Use in Hydric Soils	95
		5.4.1 Natural Wetland Ecosystems	95
		5.4.2 Wetland Rice Production Systems	98
	5.5	Plants Suitable for Poorly Drained Soils	101

	5.6	Soils that Need Artificial Drainage	102
		5.6.1 Benefits of Artificial Drainage	103
		5.6.2 Drainage Systems	103
	5.7	Environmental Impact of Agricultural Drainage	108
	5.8	Drainage Water Reuse	109
	5.9	Wet and Cold Soils	110
		References	113
6	**Expansive Soils**		**117**
	6.1	Types and Distribution of Expansive Soils	117
	6.2	Parent Materials of Vertisols	120
	6.3	Properties of Expansive Soils	121
		6.3.1 Morphological Features	121
		6.3.2 Physical Properties	124
		6.3.3 Chemical and Mineralogical Properties	126
		6.3.4 Engineering Problems Associated with Expansive Soils	129
	6.4	Agricultural Uses of Expansive Soils	130
	6.5	Limitations of Expansive Soils to Agricultural Uses	131
	6.6	Integrated Soil and Crop Management for Expansive Soils	132
	6.7	Conservation Tillage in Vertisols	137
	6.8	Amendments in Vertisols	138
		References	139
7	**Peat Soils**		**145**
	7.1	Organic Soils (Histosols, Peat and Muck)	146
	7.2	The Nature, Distribution and Significance of Peatlands	150
	7.3	Peat Soils	153
		7.3.1 Physical Properties	154
		7.3.2 Engineering Properties of Peat Soils	155
		7.3.3 Chemical Properties	155
	7.4	Reclamation and Management of Peat Soils	157
		7.4.1 Peatland Selection for Reclamation	158
		7.4.2 Modification of Peatland for Use	159
		7.4.3 Afforestation in Peatlands	161
		7.4.4 Cultivation of Oil Palm in Drained Peat Soils	162
		7.4.5 Cropping in Naturally Drained Peat Soils	164
		7.4.6 Cropping in Artificially Drained Peat Soils	165
	7.5	Peat Extraction	169
	7.6	Risks Associated with Peatland Use	170
	7.7	Peatland Conservation	172
		References	176

8	**Soils on Steep Slopes**		185
	8.1	Slopes and Steep Slopes	185
	8.2	Slope Failures and Mass Movement	189
	8.3	Factors Affecting Landslides	192
		8.3.1 Geology	193
		8.3.2 Rainfall	193
		8.3.3 Slope Gradient	194
		8.3.4 Soil	194
		8.3.5 Natural Events	195
		8.3.6 Anthropogenic Factors	195
	8.4	Management of Steep Slopes	196
		8.4.1 Mechanical Measures	196
		8.4.2 Agronomic and Agroforestry Measures	203
	8.5	Formation of Gullies	210
	8.6	Gully Control Measures	211
		8.6.1 Rock Check Dams	212
		8.6.2 Bamboo and Rock Structures	212
		8.6.3 Rock and Brush Grade Stabilization	213
		8.6.4 Soilcrete	213
		8.6.5 Aggregate-Filled Geotextiles	213
		8.6.6 Gabions	213
		8.6.7 Land Smoothing or Reshaping	214
		8.6.8 Vegetative Barriers	214
	References		215
9	**Poorly Fertile Soils**		219
	9.1	Soil Fertility and Plant Nutrients	219
	9.2	Poorly Fertile Soils	223
		9.2.1 Processes of Nutrient Depletion	227
	9.3	Management of Poorly Fertile Soils	230
		9.3.1 Organic Fertilizers	231
		9.3.2 Biochar Amendment	235
		9.3.3 Green Manuring	235
		9.3.4 Industrial Fertilizers	236
		9.3.5 Mixed Fertilizers	238
		9.3.6 Liquid and Fluid Fertilizers	239
		9.3.7 Loss of Added Fertilizers	239
		9.3.8 Methods of Fertilizer Application	241
		9.3.9 Cropping Systems in Relation to Soil Fertility Management	242
		9.3.10 Adjustment of Soil pH	247
		9.3.11 Residue Management and Conservation Tillage	247
		9.3.12 Cover Crops	247
		9.3.13 Integrated Crop-Livestock Farming Systems	248
	References		249

10	**Saline and Sodic Soils**		255
	10.1	Characteristics of Saline and Sodic Soils	256
	10.2	Development of Salinity and Sodicity in Soils	259
	10.3	Distribution of Saline Soils	263
	10.4	Field Indicators of Soil Salinity	265
	10.5	Effect of Soil Salinity on Plants	266
	10.6	Effects of Sodicity on Plant Growth	268
	10.7	Reclamation and Management of Saline Soils	269
		10.7.1 Principles of the Management of Saline Soils	270
		10.7.2 Selection of Salt Tolerant Crops	270
		10.7.3 Salt Scraping	273
		10.7.4 Salt Flushing	274
		10.7.5 Leaching	275
		10.7.6 Irrigation	278
		10.7.7 Drainage	282
	10.8	Management of Dryland Salinity	283
	10.9	Reclamation and Management of Sodic Soils	283
		10.9.1 Crop Selection for Sodic Soils	284
		10.9.2 Amendments for Sodic Soils	285
		10.9.3 Phytoremediation of Sodic Soils	290
		10.9.4 Management of Calcareous Saline-Sodic Soils	291
		10.9.5 Fertilizers for Sodic Soils	292
	10.10	Environmental Impact of Saline Soil Reclamation	292
	References		293
11	**Acid Soils and Acid Sulfate Soils**		299
	11.1	The pH Scale, Acidity and Alkalinity	299
	11.2	Soil Reaction, Acid Soils and Acid Sulfate Soils	300
	11.3	Global Extent of Acid Soils	301
	11.4	Measurement of Soil pH	304
	11.5	Development of Soil Acidity	305
	11.6	Buffering Capacity of Soils	309
	11.7	Effects of Acidity on Soil Processes	310
		11.7.1 Solubility and Availability of Chemical Elements	310
		11.7.2 Microbial Processes	315
	11.8	Effect of Soil Acidity on Plants	317
		11.8.1 Phytotoxicity in Acid Soil	318
		11.8.2 Deficiency of Nutrients	320
	11.9	Effects of Soil Acidity on Soil Fauna	321
	11.10	Management of Acid Soils	321
		11.10.1 Selection of Crops for Acid Soils	322
		11.10.2 Liming	322
	11.11	Management of Acid Sulfate Soils	326
		11.11.1 Reclamation of Acid Sulfate Soils	327
		11.11.2 Aquaculture in Acid Sulfate Soils	327

		11.12	Risks of Overliming	328

References.. 328

12 Polluted Soils 333
- 12.1 Soil Pollution 333
- 12.2 Sources of Soil Pollutants 334
  - 12.2.1   Municipal Wastes 334
  - 12.2.2   Sewage Sludge 335
  - 12.2.3   Composts 336
  - 12.2.4   Medical Wastes 337
  - 12.2.5   Veterinary Pharmaceuticals 337
  - 12.2.6   Industrial Wastes 338
  - 12.2.7   Agrochemicals 339
  - 12.2.8   Mining Wastes 344
  - 12.2.9   Atmospheric Deposition 345
  - 12.2.10 Traffic 345
  - 12.2.11 Leaded Petrol and Vehicle Exhausts 346
- 12.3 Nature of Soil Pollutants 346
  - 12.3.1   Organic Pollutants 346
  - 12.3.2   Heavy Metal Pollutants 350
  - 12.3.3   Radionuclides 355
  - 12.3.4   Toxicity of Heavy Metals 356
- 12.4 Prevention of Soil Pollution 359
  - 12.4.1   Waste Management 360
- 12.5 Remediation of Polluted Soils 362
  - 12.5.1   Remediation of Organic Pollutants 362
  - 12.5.2   Remediation of Heavy Metal Pollutants 376
  - 12.5.3   Remediation of Dispersed Radioactive Contaminats 387
- References 390

13 Degraded Soils 409
- 13.1 Soil Use and Misuse 410
- 13.2 Soil Degradation 413
  - 13.2.1   Causes of Soil Degradation 414
  - 13.2.2   Types of Soil Degradation 420
- 13.3 Prevention of Soil Degradation 429
  - 13.3.1   Principles of Soil Degradation Prevention 429
  - 13.3.2   Practices for the Prevention of Soil Degradation 431
- 13.4 Restoration and Rehabilitation of Degraded Soils .. 443
- 13.5 Desertification and Desert Reclamation 447
- References 449

Index 457

Chapter 1
Management of Soil Problems: An Introduction

Abstract In this chapter several problems, constraints, limitations and stresses of soils associated with their different uses are introduced and discussed. Their general characteristics, productivity relations and some management options are indicated.

Keywords Problem soils · Soil problems · Soil limitations · Soil constraints · Land resource stress classes

1.1 'Problem Soils' and Soil Problems

People use soils for a variety of purposes including cropping, pasturing, gardening, forestry, agroforestry, constructing buildings, airports, roads, railways, golf courses, and the like. They always look for the most suitable soils to get the best outcome. For agriculture, humans have cleared forestlands and grasslands that seemed physically most suitable for the purpose. If these soils were fertile, they give good harvests; otherwise, they have to be abandoned, and the most fertile soils are needed for crop production. Desired yield of a crop could be obtained from a soil which is physically, chemically and biologically suitable or productive for that crop. A few soils are naturally very productive; and some soils, which do not have many limitations, can be made productive by appropriate soil and crop management practices. About ten percent of the total soil resources of the world have little or no considerable limitations to cropping, but most soils have some kind of limitations in their capacity or flexibility to grow crops. Some soils can be productive for only a limited number of crops, but offer serious limitations for others. Soils that have serious limitations to agricultural and other land uses are traditionally called problem soils; and they need special management techniques for their profitable and sustainable use. But in reality, hardly any soil can be called problem soils in their natural settings, because, if not disturbed, all soils can perform their respective ecological functions. During their genesis and evolution under a set or sets of soil forming factors, however, some soils can acquire certain unique characteristics that may not suit our intended use. These are their characteristics, not problems. It is the incompatibility of our desired use with the qualities or characteristics of a particular soil that poses the problem. So, most soil problems are use-oriented; for example, many wetlands are rich fishing grounds, there is no problem in them; but draining them for cropping may create many problems. Beautiful mangrove forests can thrive in potential acid sulfate soils; wetland rice can be grown successfully there too, but using them for shrimp culture and arable crops can be unsuccessful and environmentally hazardous. However, the capacity for performing normal ecological functions of some soils has been lost or deteriorated due to human activities, so that soil quality has considerably declined. It is very difficult and sometimes costly to restore them to their original structures or functions. As pressures on soil resources have enormously increased recently due to increased human population, many marginal and unsuitable soils have now been brought under cultivation. These soils deserve special attention to protect them from further deterioration (Eswaran et al. 2001).

Soil limitations to various uses may be physical, such as dryness, wetness, steepness, extreme textures, erosion hazard, compaction, shallowness, shallow groundwater table and flooding; chemical, such as acidity, salinity, sodicity, lack of fertility, phosphate fixation, and pollution; and/or biological, such as reduction in the activities of beneficial soil organisms, for example, earthworms and vesicular-arbuscular mycorrhizae (VAM), or an increase in pathogens or plant parasitic nematodes. Soil characteristics may impose limitations to growth and yield of crops directly through effects on germination, seedling emergence and early growth, or through the content of available nutrients and water. They may also do so through their effects on root development and functioning. Some soil limitations are difficult to alleviate (too cold, too hot, too shallow); some limitations can be alleviated in whole or in part (irrigation in dry areas, drainage of wet soils, fertilizing in poorly fertile soil, liming in acid soil), and alleviation of some limitations widens the range of crop suitability (draining wet soil, irrigating dry soils). Agricultural scientists have described management options for many different soil limitations. For example, Murphy et al. (2004) mentioned the following soil limitations to cropping: soil structure decline, wind erosion, water erosion, mass movement, acidification, soil carbon loss, soil contamination, soil fertility decline, acid sulfate soil, and dryland salinity. The Expert Consultation of the Asian Network on Problem Soils proposed the following 11 categories of problem soils (FAO/AGL 2000).

1. **Cold soils:** Land areas with a 24-hr mean temperature of less than 5 °C during the growing period.
2. **Dry soils:** Desert and semi-desert soils with growing periods which are rainless-dry.
3. **Steep soils:** Soils which have steep slopes in excess of 30 percent.
4. **Shallow soils:** Soils which have depth limitations within 50 cm of the surface caused by the presence of coherent and hard rock or hard pans.
5. **Poorly drained soils:** Soils which are waterlogged and/or flooded for a significant part of the year.
6. **Coarse textured soils (sandy soils):** Soils having coarse texture with less than 18 percent clay and more than 65 percent sand, or have gravel, stones, boulders or rock outcrops in surface.

1.1 'Problem Soils' and Soil Problems

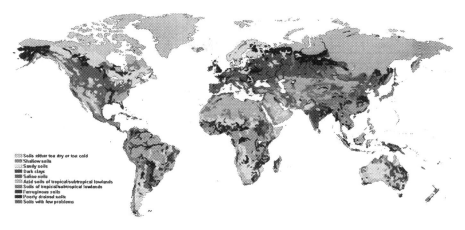

Fig. 1.1 Distribution of some 'problem soils' (Source: http://www.fastonline.org/CD3WD_40/ INPHO/VLIBRARY/U8480E/EN/U8480E0B.HTM)

7. **Heavy cracking clays (Vertisols):** Soils which crack at least 1 cm wide at 50 cm depth at some period in most years.
8. **Poorly fertile soils:** Soils which exhibit deficiencies in plant nutrients.
9. **Saline and sodic soils:** Soils with high salt content and high exchangeable sodium saturation, respectively within 100 cm of the surface.
10. **Acid sulfate soils:** Soils in which sulfidic materials have accumulated under permanently saturated brackish water conditions in general.
11. **Peat soils (Histosols):** Soils in which more than half of the upper 80 cm is composed of organic materials.

There are some other categories of problem soils as well, such as.

Eroded soils: Soils that have seriously suffered from water and wind erosion.
Polluted soils: Soils contaminated with inorganic and organic pollutants.
Compacted soils: Some soils are naturally compacted; others get compacted by the load of heavy farm machineries.

Figure 1.1 shows the global distribution of different so-called problem soils.

Some of the categories of soils listed above have multiple problems. For example, sandy soils are not only low water retentive and poorly fertile, some sandy soils are extremely acidic and very cold. Dry soils are found in arid and semi-arid regions where precipitation is very low. Dry soils are also usually sandy in texture, low in organic matter and poorly fertile. Peat soils are susceptible to subsidence, and many peat soils are submerged. Saline soils may have poor drainage or dryness problems along with salt problems. A soil posing problems to a certain land use may not do so for another land use. For example, some saline soils bear excellent mangrove forests in natural conditions, and soil salinity is not a problem there; but when these areas are cleared for agricultural crop production, they do not give satisfactory yields. Steep soils are not good agricultural lands, but they can support dense forest vegetation if they are left undisturbed. However, soil problems may be natural or acquired. Natural soil limitations usually arise from climate, parent material and topography related situations. Human actions have also created many soil problems including erosion, waterlogging, compaction, fertility depletion, salinization, acidification, pollution, etc.

1.2 Soil Constraints to Different Uses

Using the fertility capability classification (FCC), FAO (2000) listed eight soil constraint classes on the basis of inherent features which offer problems to soil use and management. These soil constraint classes are listed below.

1. **Hydromorphy:** poor soil drainage.
2. **Low cation exchange capacity:** low capacity to retain added nutrients.
3. **Aluminium toxicity:** strong acidity.
4. **High phosphorus fixation:** a high level of ferric oxides in the clay fraction.
5. **Vertic properties:** dark, expanding and contracting ('cracking' and 'expansive') clays.
6. **Salinity and sodicity:** presence of free soluble salts.
7. **Shallowness:** rock or a rock-like horizon close to the soil surface.
8. **Erosion hazard:** a high risk of soil erosion in moderate to steep slopes in association with erosion-prone soils.

According to FAO (2000), the four major constraints, each of which occupies 13–16 percent of the global land area, are in order of extent: erosion hazard, aluminium toxicity, shallowness, and hydromorphy. Four other constraints, each of which covers 2–6 percent of the area, are: salinity and sodicity, low cation exchange capacity (CEC), high phosphorus fixation, and vertic properties.

DECCW (2010) identified several landscape constraints, soil physical constraints, and soil chemical constraints for which special management practices need to be adopted for various uses.

Landscape constraints:	Steep slopes, water erosion hazard, flood hazard, acid sulfate soils, mass movement, wave attack, poor site drainage/waterlogging, general foundation hazard, shallow soils, rock outcrop.
Soil physical constraints:	Shrink-swell potential, low soil strength, low or high permeability, low plant available water holding capacity (water stress), stoniness.
Soil chemical constraints:	Salinity, acid and alkaline soils, sodicity, low fertility/nutrient availability, high phosphorus sorption (low P availability).

DECCW (2010) also identified five intensity classes of soil constraints ranging from very low to very high, and described their management strategies.

1.2 Soil Constraints to Different Uses

Very low
: Very low constraint; low treatment costs; straightforward or no maintenance; associated with negligible financial, environmental or social site costs; very low residual risk.

Low
: Low constraint; associated with minor financial, environmental or social site costs; straightforward or low maintenance; low residual risk.

Moderate
: Moderate constraint; moderate financial, environmental or social costs beyond the standard; frequent maintenance erequired; moderate residual risk; marginally acceptable to society – other factors may intervene.

High
: High constraint; high financial, environmental or social costs beyond the standard; special mitigating measures are required; regular specialist maintenance; moderate to high residual risks and costs.

Very high
: Very high constraint; risks very difficult to control even with site-specific investigation and design; very high financial, environmental or social costs beyond the standard; regular specialist maintenance may be mandatory; high residual risk; generally not acceptable to society.

Gugino et al. (2014) listed the following soil constraints that need specialized management: Physical: low aggregate stability, low available water capacity, high surface density; Biological: low organic matter content, low active carbon, low mineralizable nitrogen, high root rotting; and Chemical: unfavorable pH, low P, K, and minor elements, high salinity, and high sodium content. They put forward some short-term and long-term management suggestions (Table 1.1).

Following are the major soil constraints specific to different regions:

Sub-Saharan Africa: Aluminium toxicity and low cation exchange capacity
North Africa and Near East: Salinity and sodicity
Asia and the Pacific: Aluminium toxicity, hydromorphy, salinity and sodicity
North Asia, east of Urals: Hydromorphy, salinity and sodicity
South and Central America: Aluminium toxicity, high phosphorus fixation, and hydromorphy
North America: Hydromorphy and aluminium toxicity
Europe: Hydromorphy

The global extent of problem soils/soil constraints are given in Table 1.2.

A brief description of some soil problems/constraints are given below:

Stoniness

Stoniness refers to the presence of rock fragments, such as gravel, stone and cobbles in high proportion. These materials do not provide or retain nutrients and moisture. They offer physical resistance to movement of agricultural implements and restrict tillage operations. They reduce the workability of a soil and hinder cultivation.

Table 1.1 Soil constraints and suggested management practices

Suggested management practices		
	Short term or intermittent	Long term
Physical concerns		
Low aggregate stability	Fresh organic materials (shallow-rooted cover/rotation crops, manure, green clippings)	Reduced tillage, surface mulch, rotation with sod crops
Low available water capacity	Stable organic materials (compost, crop residues high in lignin, biochar)	Reduced tillage, rotation with sod crops
High surface density	Limited mechanical soil loosening (e.g. strip tillage, aerators); shallow-rooted cover crops, bio-drilling, fresh organic matter	shallow-rooted cover/rotation crops; avoid traffic on wet soils; controlled traffic
High subsurface density	Targeted deep tillage (zone building, etc.); deep rooted cover crops	Avoid plows/disks that create pans; reduced equipment loads/traffic on wet soils
Biological concerns		
Low organic matter content	Stable organic matter (compost, crop residues high in lignin, biochar); cover and rotation crops	Reduced tillage, rotation with sod crops
Low active carbon	Fresh organic matter (shallow-rooted cover/rotation crops, manure, green clippings)	Reduced tillage, rotation
Low mineralizable N (Low PMN)	N-rich organic matter (leguminous cover crops, manure, green clippings)	Cover crops, manure, rotations with forage legume sod crop, reduced tillage
High root rot rating	Disease-suppressive cover crops, disease breaking rotations	Disease-suppressive cover crops, disease breaking rotations, IPM practices
Chemical concerns		
Unfavorable pH	Liming materials or acidifier (such as sulfur)	Repeated applications based on soil tests
Low P, K and Minor elements	Fertilizer, manure, compost, P-mining cover crops, mycorrhizae promotion	Application of P, K materials based on soil tests; application of organic matter; reduced tillage
High salinity	Subsurface drainage and leaching	Reduced irrigation rates, low-salinity water source, water table management
High sodium content	Gypsum, subsurface drainage, and leaching	Reduced irrigation rates, water table management

Rock outcrop

Rock outcrop impedes working for developments, such as construction of roads, buildings and the cultivation of land for cropping and plantation forestry. It may also restrict recreational uses, particularly sporting fields. Rock outcrop reduces the space for growing plants in a landscape. There is less yield of crop or pasture per hectare of land. It often becomes a significant limitation where it covers over 20 percent of the land surface.

1.2 Soil Constraints to Different Uses

Table 1.2 Total areas of different types of soil problems of the world

Problem soils/soil constraints	Million hectares (M ha)
Hydromorphy (wet soils)[a]	1738.2
Aluminium toxicity[a]	1986.7
Eroded soil (water erosion)[b]	1094.0
Eroded soil (wind erosion)[b]	548.0
Low CEC[a]	615.1
High P-fixation[a]	542.1
Vertisols[c]	335.0
Peat soils[d]	400.0
Acid sulfate soil[e]	17.0
Salt affected soil[f]	950.0
Polluted soils[g]	21.0
Compacted soils[g]	68.0
Degraded soil[g]	1965.0

[a]FAO (2000)
[b]Lal (2001)
[c]Spaargaren (2008)
[d]Bain et al. (2011)
[e]Andriesse and van Mensvoort (2006)
[f]Eswaran et al. (2001)
[g]Oldeman (1994)

Shallowness of soil

Shallowness of soil refers to the presence of a thin soil layer over the bed rock or a hard cemented layer/horizon or a root restrictive layer near the surface. The depth of shallow soils varies from 30 to 50 cm depending on the rooting depth of crops grown. These soils are often stony or gravelly, prone to desiccation, and frequently occur on steeplands. These soils were formerly called Lithosols, Rendzinas and Rankers. Now, FAO (2006) classifies these soils as Leptosols in World Reference Basefor Soil Resources. In Soil Taxonomy (USDA-NRCS 1999), they fall in several Soil Orders.

Low soil strength

Soil strength represents the ability of the soil to support loads. It is related to wind-throwing of trees, bearing capacity of farm animals and machineries, and as foundation of buildings, roads and highways. The rating applied is taken as the most limiting (worst) of the following three attributes: (i) plasticity – highly plastic soils are unsuitable for foundation of buildings and road materials. (ii) low weight bearing strength accounts for low capacity to support loads when wet (iii) organic- peats or peaty soils have low weight-bearing strength and may be subject to contraction. The Unified Soil Classification System (USCS) is a relatively simple classification of soil materials and is widely used for low level engineering purposes (Hicks 2007).

Steep slopes
Land gradient has a major influence on land uses; almost all land-use operations become increasingly difficult as the slope increases, particularly above >20 percent. The greater the slope, the greater is the potential for erosion due to the increased amount and rate of runoff, reducing infiltration and the increased gravitational force on the soil particles. Steeper slopes mean that access is more difficult and cumbersome, especially where heavy machinery is required or heavy loads are being transported. Preparation of site for construction work is more difficult and require greater cut and fill operations.

Erosion
Erosion greatly reduces land-use efficiency. It is more prominent in areas of steep slopes (8–30 percent) and very steep slopes (>30 percent) in conjunction with an abrupt textural contrast in the soil profile, denoting these as having a severe erosion hazard. The total world area having high erosion hazard is somewhat greater than for the other major soil constraints: shallowness and aluminium toxicity.

Mass movement
Mass movement includes landslides, earth slumps and rock falls and is a serious threat to many land uses including grazing, shifting cultivation, forestry and construction of roads and bridges. It may involve the collapse of a slope at one point and the accumulation of the failed material further downslope. Mass movement may demolish houses causing injury or loss of life and damage of infrastructure. Mass movement occurs in several forms, but it occurs basically when wetting soil materials increase the weight in upper slopes beyond the restraining capacity of the underlying substratum. It frequently occurs during periods of rainfall when the weight of the ground material has increased and the internal friction has been reduced. Construction, particularly cuttings into slope bases, may exacerbate this hazard (Rosewell et al. 2007).

Vertic properties
Vertic properties are the features of Vertisols which swell on wetting and shrink on drying. They are also called shrink-swell soils and expansive soils. In India, dark colored Vertisols were formerly called 'black cotton soils'; the dark color of these soils was due to humus complexes with dispersive clays. Vertisols produce wide and deep cracks upon drying. The cause is a high clay content coupled with >50 percent 2:1 lattice (montmorillonitic) clay minerals. The distribution of Vertisols is highly localized, being linked with mafic (basic) rocks, semiarid climates, or both these conditions. The greatest absolute extents are found in India (the lavas of the Deccan), Australia, and Sudan (especially the Gezira zone). Twelve countries, in all the major continental regions, have over 10 percent of their land with vertic properties, Uruguay and India having the highest relative extent (FAO 2006).

Flood hazard
Flooding is a major constraint to many land uses. It can damage or destroy crops, farm houses, infrastructure, and other assets. Flash floods rapidly builds up water flow in narrow and confined valleys, typically in hilly or mountainous areas, while

1.2 Soil Constraints to Different Uses 9

riverine floods are the more extensive inundation of floodplains adjacent to rivers following heavy rains in the catchment, and coastal floods are the inundation of low lying coastal lands by ocean or estuarine waters, resulting from severe storm events and/or unusually high tides (NSW Government 2001).

Hydromorphy

Hydromorphy or wetness in the soil profile for all or part of the year is generally found in flat and low-lying areas with respect to the surrounding land. These lands include alluvial and coastal plains, deltas, and river valleys, including valley floors, peat bogs, marshes and other wetlands. About 20 countries have a substantial proportion (>25 percent) of land affected by hydromorphic constraints, with the highest proportions (>50 percent) in the Falklands, the United Kingdom, and Bangladesh (FAO 2000).

Low cation exchange capacity

Soils that have a low organic matter content, low clay content, clay minerals with low CEC, or all these properties in surface soil possess low cation exchange capacity. These soils have a low inherent fertility and also a low capacity to retain nutrients added as fertilizers. Generally, these include highly sandy soils (Arenosols), and tropical soils dominated by kaolinite clay and sesquioxides. In humid and sub-humid regions there are extensive areas of sandy or highly-weathered soils having low CEC.

Soil acidity and alkalinity

A soil pH value below 7.0 indicates acidity and a value above it denotes alkalinity. Acidity and alkalinity influence the growth of plants and microorganisms, and affect chemical transformation, nutrient availability, and elemental toxicity. At soil pH below 4.5, aluminium and many heavy metals are released into soil solution and nitrogen, phosphorus, potassium and other nutrients becomes virtually unavailable. Different plants have different levels of tolerance to acidity and alkalinity, but most plants thrive at pH levels between 6.0 and 7.5. Levels below 4.5 or above 8.5 present significant limitations for growth of most plants (Fenton and Helyar 2007). Highly acidic or alkaline soils are not suitable for agriculture.

Acid sulfate conditions

Acid sulfate soils offer serious limitations to most uses. They usually develop in estuarine margins inundated with brackish water and supplied with plenty of organic matter. They are a potential constraint to land uses involving excavation or soil disturbance. Potential acid sulfate soils contain pyrite (FeS_2) which reacts with atmospheric oxygen to form sulfuric acid (H_2SO_4) on exposure to air through drainage. It creates extreme acidity that causes the release of toxic materials, including aluminium and heavy metals. Acid sulfate soils may also be highly saline.

Salinity and sodicity

Soil salinity and sodicity occur naturally in semi-arid to arid regions. They are also found in coastal lowlands. The state of salinity is caused by the accumulation of free salts in the profile, and sodicity due to dominance of the exchangeable sodium on colloidal surfaces. Of the 21 countries with over >15 percent of their land affected, 13 lie in a broad belt extending from the African Sahara and its bordering Sahel zone through the Middle East and into Central Asia. Sodicity appears to be even more strongly localized, with six countries (three of them in Central Asia) affected over more than 10 percent of their extent (FAO 2000).

Aluminium toxicity

Soils with aluminium toxicity have the exchange complexes dominated by alumina. This occurs with a soil reaction of pH <5.5, often <5.0. This problem is commonly associated with strongly acid soils. The main cause is intensive weathering and strong leaching under high rainfall. This is predominantly a problem of the humid tropics, although found also in the subhumid tropics and in a few temperate-zone countries with areas of high rainfall, e.g. New Zealand. It affects some 800 M ha in South America and 400 M ha in both Africa and Asia. The most affected eight countries lie largely or entirely in the rain forest zone. More than 50 countries have >25 percent of their territory affected by aluminium toxicity (FAO 2000).

Phosphorus fixation

The problem of phosphorus fixation in soils is generally associated with high content of free iron oxides (Fe_2O_3) in the clay fraction, which insolubilizes phosphate ions and renders it unavailable to plants. It is a dominant feature of strongly acid soils, and hence found often in conjunction with aluminium toxicity. The eight countries with more than 20 percent of land affected by high phosphorus fixation are also affected by aluminium toxicity (FAO 2000). Acidic and high Al containing soils are highly phosphate fixing.

Low soil fertility

Low soil fertility refers to the inadequate supply of available nutrients due to low nutrient availability, imbalance of nutrients, and presence of some kind of toxicity. The low fertility results from strong weathering, low cation exchange capacity, low organic matter content, strong soil acidity, strong phosphate sorption capacity, and strong nutrient leaching or nutrient imbalances (Asher et al. 2002). Low supply of micronutrients causes considerable constraints to food production in the tropics.

Eswaran et al. (2005) recognized some 'edaphic constraints' to food production while discussing problems and prospects of utilizing sandy soils of Asia. These constraints include some intrinsic and induced stresses. Intrinsic stresses result from physical (high susceptibility to erosion; steep slopes, shallow soils; surface crusting and sealing; low water-holding capacity; impeded drainage; low structural stability; root restricting layer; high swell/shrink potential), chemical (nutrient deficiencies; excess of soluble salts – salinity and alkalinity; low base saturation, low pH; aluminum and manganese toxicity; acid sulfate condition; high P and anion retention; calcareous or gypseous condition) and biological conditions (low or high organic matter content; high termite population) of the soils, climatic conditions (low rainfall, high evapotranspiration; excess rainfall, extreme temperature regimes; insufficient length of growing season) and catastrophic events (floods and droughts;

landslides; seismic and volcanic activity). Induced stresses occur due to physical, chemical and biological processes of soil degradation and modification of the landscapes.

1.3 Land Resource Stress Classes

Using soil and climate information, Moncharoen et al. (2001) and Eswaran et al. (2003) defined 25 broad land resource stress classes (Table 1.3) in the global context. Each of these stress classes requires a different level of management for agricultural use. A stress class may have several problems at the same time. For example, areas designated as 'continuous moisture stress', which essentially are the drylands, may have soils with salinity problems and sandy or skeletal soils that have low water holding capacity.

Agriculture in the developed countries is being practiced at present on the most suitable soils according to their intended uses. And, due to steadily rising productivity of these soils, EEC countries and the USA have even been able to reduce their cultivated land area. However, in the developing countries, the trend is almost the reverse. There, although the productivity of the better soils could still be improved substantially, much of the agriculture takes place on soils that are unsuitable or only marginally suitable. In large areas of Asia and Africa, the overall productivity is declining because of soil exhaustion and because areas with soil limitations are being used for cultivation. By drawing attention to these soils and emphasizing the ways their properties affect their reclamation and improvement, we can have a better understanding of the problems encountered and the risks involved when such soils are used for agriculture. More than two billion people depend on the world's arid and semi-arid lands. Preventing land degradation and supporting sustainable development in drylands may have major implications for food security, climate change and human settlement (Foreword by the United Nations Secretary-General, United Nations 2011).

Sources: Moncharoen et al. 2001; Eswaran et al. 2003

Finally, a list of problems of soil associated with various uses is presented below:

Physical problems Steep slopes, Erosion, stoniness, coarse texture, stiff very fine texture, extended dryness, extended wetness, shallow depth, poor soil structure, low infiltration, impermeability, compaction, crusting, root restrictive layers, dispersion, vertic properties, low soil temperature, high soil temperature, shallow groundwater table, poor drainage, etc.

Chemical problems Low organic matter, poor fertility, soil acidity, acid sulfate condition, soil alkalinity, salinity, sodicity, low CEC, low BSP, low buffering capacity, high Al, high P fixation, nitrate leaching, micronutrient deficiency, nutrient toxicity, soil pollution, etc.

Table 1.3 Description of major land resource stresses or conditions

Stress class	Land quality class	Major land stress factor	Criteria for assigning stress
25	IX	Extended periods of moisture stress	Aridic Soil Moisture Regimes (SMR), rocky land, dunes
24	VIII	Extended periods of low temperatures	Gelisols
23	VIII	Steep lands	Slopes greater than 32 percent
22	VII	Shallow soils	Lithic subgroups, root restricting layers <25 cm
21	VII	Salinity/alkalinity	"Salic, halic, natric" categories;
20	VII	High organic matter	Histosols
19	VI	Low water holding capacity	Sandy, gravelly, and skeletal families
18	VI	Low moisture and nutrient status	Spodosols, ferritic, sesquic&oxidic families, aridic subgroups
17	VI	Acid sulfate conditions	"Sulf" great groups and subgroups
16	VI	High P, N, organic compounds	Anionic subgroups, acric great groups, oxidic,
15	VI	Low nutrient holding capacity	Loamy families of Ultisols, Oxisols.
14	V	Excessive nutrient leaching	Soils with udic, perudic SMR, but lacking mollic, umbric, or argillic
13	V	Calcareous, gypseous conditions	With calcic, petrocalcic, gypsic, petrogypsic horizons; carbonatic and gypsic families; exclude Mollisols and Alfisols
12	V	High aluminum	pH <4.5 within 25 cm and Al saturation > 60 percent
11	V	Seasonal moisture stress	Ustic or Xeric suborders but lacking mollic or umbricepipedon, argillic or kandic horizon; exclude Vertisols
10	IV	Impeded drainage	Aquic suborders, 'gloss' great groups
9	IV	High anion exchange capacity	Andisols
8	IV	Low structural stability and/or crusting	Loamy soils and Entisols except Fluvents
7	III	Short growing season due to low temperatures	Cryic or frigid Soil Temperature Regime (STR)
6	III	Minor root restricting layers	Soils with plinthite, fragipan, duripan, densipan, petroferric contact, placic, <100 cm
5	III	Seasonally excess water	Recent terraces, aquic subgroups
4	II	High temperatures	Isohyperthermic and isomegathermic STR excluding Mollisols and Alfisols
3	II	Low organic matter	With ochric epipedon
2	II	High shrink/swell potential	Vertisols, vertic subgroups
1	I	Few constraints	Other soils

Biological problems Low biological activity, low organic matter decomposition, low mineralization, high denitrification, etc.

Some problems are associated with the engineering aspects of soil use. In many states of the USA, building structures are damaged due to construction on expansive (shrink-swell) and collapsing soils. Cracking of foundations, walls, driveways, swimming pools, and roads costs millions of dollars each year in repairs. American Society of Civil Engineers suggests that about half of the houses built in the United States each year are situated on unstable soil, and about half of these will eventually suffer some soil related damage. In the last decade or so, swelling and shrinkage in clay soils have caused losses of up to 3 billion pounds in Britain (Bell and Culshaw 2001). Buildings, roads, farms and human habitations are often damaged by slope failures.

Soil constraints and management of major problems resulting from its use are discussed in different chapters of this book. For convenience and in order to avoid repetition, management of soils with multiple problems is treated in a single chapter. For example, problems of Al toxicity, soil acidity and acid sulfate conditions are discussed in Chapter 11. The readers must be aware that there are some management options common for several soil problems. For example, mulching, cover crops, and conservation tillage are applied in agricultural use of all of dryland soils, sandy soils, shallow soils, etc. Some repetitions could not, therefore, be avoided.

References

Andriesse W, van Mensvoort MEF (2006) Acid sulfate soils: distribution and extent. In: Lal R (ed) Encyclopedia of soil science. Taylor and Francis, Boca Raton

Asher C, Grundon N, Menzies N (2002) How to unravel and solve soil fertility problems. Australian Center for International Agriculture Research (ACIAR), Monograph No 83. Canberra, Australia

Bain CG, Bonn A, Stoneman R, Chapman S, Coupar A, Evans M, Gearey B, Howat M, Joosten H, Keenleyside C, Labadz J, Lindsay R, Littlewood N, Lunt P, Miller CJ, Moxey A, Orr H, Reed M, Smith P, Swales V, Thompson DBA, Thompson PS, Van de Noort R, Wilson JD, Worrall F (2011) IUCN UK Commission of Inquiry on Peatlands. IUCN UK Peatland Programme, Edinburgh

Bell FG, Culshaw MG (2001) Problem soils: a review from a British Perspective. British Geological Survey, Nottingham

DECCW (2010) Soil and land constraint assessment for urban and regional planning. Department of Environment, Climate Change and Water, NSW, Sydney

Eswaran H, Lal R, Reich PF (2001) Land degradation: an overview. In: Bridges EM, Hannam ID, Oldeman LR, Pening de Vries FWT, Scherr SJ, Sompatpanit S (eds) Responses to land degradation. In: Proceeding of 2nd International Conference on Land Degradation and Desertification, Khon Kaen, Thailand. Oxford Press, New Delhi

Eswaran H, Beinroth FH, Reich PF (2003) A global assessment of land quality. In: Wiebe K (ed) Land quality, agricultural productivity, and food security: biophysical processes and economic choices at local, regional, and global levels. Edward Elgar, Northampton

Eswaran H, Vearasilp T, Reich P, Beinroth F (2005) Sandy soils of Asia: a new frontier for agricultural development? Session 1"Global extent of tropical sandy soils and their pedogenesis". In: Proceedings of the conference on Management of Tropical Sandy Soils for Sustainable

Agriculture "A holistic approach for sustainable development of problem soils in the tropics", 27th November – 2nd December 2005, Khon Kaen, Thailand

FAO (2000) Land Resource Potential and Constraints at Regional and Country Levels. World Soil Resources Reports 90. Land and Water Development Division, Food and Agriculture Organization of the United Nations, Rome

FAO (2006) World reference base for soil resources 2006. World Soil Resources Report No. 103, FAO, Rome

FAO/AGL (2000) Agricultural problem soils (FAO/AGL Problem soil database). wwwfaoorg/ag/agl/agll/prosoil/calchtm. Accessed 5 Dec 2011

Fenton G, Helyar KR (2007) Soil chemical properties: soil acidification. In: Charman PEV, Murphy MW (eds) Soils: their properties and management, 3rd edn. Oxford University Press, Melbourne

Gugino BK, Omololu JI, Robert RS, van Es HM, Moebius-Clune BN, Wolfe DW, Thies JE, Abawi GS (2014) Managing soil constraints: Interpreting the Soil Health Test Report. College of Agriculture and Life Sciences, Cornell University

Hicks RW (2007) Soil engineering properties. In: Charman PEV, Murphy MW (eds) Soils: their properties and management, 3rd edn. Oxford University Press, Melbourne

Lal R (2001) Soil degradation by erosion. Land Degrad Dev 12:519–539

Moncharoen P, Vearasilp T, Eswaran, H (2001) Land resource constraints for sustainable agriculture in Thailand. In: Stott DE, Mohtar RH, Steinhardt (eds) Sustaining the Global Farm. In: 10th International Soil Conservation Organization Meeting held May 24-29, 1999 at Perdue University and the USDA-ARS National Soil Erosion Research Laboratory, USA

Murphy BW, Murphy C, Wilson BR, Emery KA, Lawrie J, Bowman G, Lawrie R, Erskine W (2004) A revised land and soil capability classification for New South Wales. In: ISCO 2004 – 13th International Soil Conservation Organization Conference – Brisbane, July 2004, Conserving Soil and Water for Society: Sharing Solutions

NSW Government (2001) Floodplain management manual. NSW Government, Sydney

Oldeman LR (1994) The global extent of land degradation. In: Greenland DJ, Szabolcs I (eds) Land resilience and sustainable land use. CABI, Wallingford

Rosewell CJ, Crouch RJ, Morse RJ (2007) Forms of erosion: water erosion. In: Charman PEV, Murphy MW (eds) Soils: their properties and management, 3rd edn. Oxford University Press, Melbourne

Spaargaren O (2008) Vertisols. In: Chesworth W (ed) Encyclopedia of soil science. Springer Science + Business Media, Dordrect

United Nations (2011) Foreword by the United Nations Secretary-General. Global Drylands: a UN system-wide response. United Nations Environment Management Group. UN, Washington, DC

USDA-NRCS (1999) Soil Taxonomy: a Basic System of Soil Classification for Making and Interpreting Soil Survey. United States Department of Agriculture, Natural Resources Conservation Service. Govt. Printing Office, Washington, DC

Chapter 2
Dryland Soils

Abstract Dryland soils generally occur in arid, semi-arid and dry sub-humid regions with some occasional occurrence in other regions. The basis of definition of aridity or dryness is the ratio of mean annual precipitation to potential evapotranspiration, and this ratio is called the aridity index (AI). Dryland soils occur in regions of AI between 0.05 and 0.65. There are drylands in 40 percent of the world's land area with 38 percent of the world's population. According to an estimate, the total dryland areas of the world are 6310 million hectares (M ha) distributed mainly in Africa (2000 M ha), Asia (2000 M ha), in Oceania (680 M ha), in North America (760 M ha), South America (56 M ha) and in Europe (300 M ha). Drylands are characterized by low, irregular, and unevenly distributed rainfall and high potential evapotranspiration. Frequently there are high or low temperatures and occurrences of drought. Agriculture is difficult there mainly because of inherent soil moisture deficit and scarcity of irrigation water. All these factors make dryland regions prone to desertification hazards. In spite of all these difficulties, however, drylands contribute significantly to the production of cereals, pulses and livestock, particularly in different parts of Argentina, Australia, Canada, the former Soviet Union and the United States of America. Historically livestock rearing has been widespread in the dryland regions. Dryland soils are usually sandy, and may be saline, sodic, calcic, or gypsic in nature. These soils are of low fertility and can support low plant biomass productivity. Salinization, lack of adequate irrigation, rapid leaching and fertility depletion, sparse vegetation, over-grazing and erosion cause widespread soil degradation including desertification. As water is scarce there, human settlements are found around rivers, springs, wells, water catchments, reservoirs and oases.

Keywords Arid regions · Dryland salinity · Dryland crops · Dryland agriculture · Desert reclamation

2.1 Drylands of the World

UNEP (1997) defined and classified drylands on the basis of aridity index (AI) which is the ratio between mean annual precipitation (P) and potential evapotranspiration (PET).

$$\text{Aridity index}, \text{AI} = \frac{P}{PET}$$

Values of AI <1 indicate annual moisture deficit; and as the AI decreases, the moisture deficit increases. According to UNEP (1997), drylands are areas with AI <0.65. There are four subtypes of drylands: (i) hyper-arid – AI <0.05, (ii) arid – AI 0.05 to <0.20, (iii) semi-arid – AI 0.20 to <0.50 and (iv) dry sub-humid – AI 0.50 to <0.65. On this basis, 47.2 percent of the global land areas fall in drylands including 7.5 percent hyper-arid, 12.1 in arid, 17.7 percent semi-arid and 9.9 percent in dry sub-humid subtypes. The UNCCD (2000) considers hyper-arid areas as true deserts and excluded them from drylands. So, according to the UNCCD, drylands include areas of AI between 0.05 and 0.65. On the other hand, FAO (2000) used the length of the growing period (LGP) for annual crops as the basis for defining drylands. Here, the growing period means the period when monthly precipitation exceeds half of the monthly potential evapotranspiration. In hyper-arid areas the monthly precipitation never exceeds half of the monthly potential evapotranspiration; so these areas have no agricultural potential. Arid, semi-arid and dry sub-humid regions have LGP of 1–59, 60–119 and 120–179 days respectively. On this basis, the drylands constitute 45 percent of the world's land area with 7 percent arid, 20 percent semi-arid, and 18 percent dry sub-humid lands. Unfavorable hydrologic conditions, loss of vegetation, degradation of soil quality, and desertification are major challenges of the drylands. According to D'Odorico et al. (2013) these factors combinedly threaten ecosystem services and human life in drylands.

Safriel et al. (2005) suggested that moisture deficiency over long periods has several adverse effects on natural and managed ecosystems. According to Molden and Oweis (2007), the mean annual rainfall in drylands is less than 500 mm, and most of the rain water is lost through evaporation. Whatever rain is fallen, it is erratic, uncertain and unevenly distributed. Drylands frequently suffer from drought and desertification hazards. Drylands are regions of low productivity, low investment and poverty. Still, Luc Gnacadja, Executive Secretary of The UNCCD was optimistic enough when he wrote "Often, when people think of drylands, they associate them with deserts and hostile living conditions, economic hardship and water scarcity. But that is not what drylands are all about. If managed well, the drylands are also fertile and capable of supporting the habitats, crops and livestock that sustain nearly one-third of humanity (Desertification – A Visual Synthesis; Hori et al. 2011)." Hori et al. (2011) reported that about 44 percent of all cultivated lands are distributed in the drylands, and they support 50 percent of the world's livestock (Table 2.1).

2.2 Global Distribution of Drylands

The major areas of drylands are found in Asia, Africa, Australia, Canada, Central America and the USA. Two-thirds of the drylands are distributed in Asia (11 million km^2) and Africa (nearly 13 million km^2) (White et al. 2002; De Pauw 2009). Countries like Australia, China, Russia, the United States and Kazakhastan comprise

2.2 Global Distribution of Drylands

Table 2.1 Characteristics of dryland subtypes of the world

Drylandsubtypes	Characteristics
Arid	Arid lands have aridity indices between 0.05 and <0.20, covering 12.1 percent of the global land area on UNEP basis; and on FAO criteria they have LGP between 1 and 59 days that cover 7 percent of the world's land area, with 4.1 percent percent of global population. Rainfall ranges in winter from 100 to 150 mm and in summer from 200 to 350 mm. The inter-annual variability is from 50 to 100 percent. Natural vegetation is semi-desert with scattered bushes, and small woody, succulent, thorny or leafless shrubs. The area is suitable for only light pastoral use, but rainfed agriculture is not possible.
Semi-arid	On aridity index, semi-arid lands cover 17.7 percent, and on LGP 20 percent of global land area with 14.4 percent of global population. Semi-arid lands have a mean annual rainfall from 300–400 to 700–800 mm in winter rainfall regimes and from 200–250 to 450–500 mm rainfall in summer rainfall regimes. The inter-annual variability ranges from 25 to 50 percent. The semi-arid region is known as the steppe zone and the dominant natural vegetation is grassland with scattered savannahs and scrubs. The semi-arid lands support livestock farming well, especially sheep and goats. Rainfed agriculture is possible in many areas.
Dry sub-humid	On the basis of aridity index, dry sub-humid region constitutes 9.9 percent of global land area and on LGP 18 percent. This region has higher rainfall than the other dryland subtypes. The inter-annual variability in rainfall is less than 25 percent. Several vegetation types are found in this region, such as tropical savannah, maquis and chapparal, steppe, etc. rangelands, and agriculture, both rainfed and irrigated, are the normal land uses. The main agricultural crops of this region are the cereals.

UNEP (1997), UNCCD (2000), FAO (2000), Bantilan et al. (2006), and Hori et al. (2011)

the largest dryland areas. According to Rosegrant et al. (2002) nearly two-thirds of the total cropped areas in India are in the drylands. A major part of Sub-Sahelian West Africa is semi-arid. Dryland hazards affect most of North Africa and the Near East and North Asia, and over half of North America, Sub-Saharan Africa, and Asia and the Pacific. According to FAO, Drylands occupy 90–100 percent of agriculturally productive lands in 36 countries. Six countries including Botswana, Burkina Faso, Iraq, Kazakhastan, Maldova and Turkmenistan have 99 percent of their surface area covered with drylands (Hori et al. 2011). The map in Fig. 2.1 shows the distribution of drylands in the world.

The UNEP (1997) estimated the total drylands of the word to be 6310 M ha distributed in more than 110 countries. The major areas are in Africa (2000 M ha), Asia (2000 M ha), Oceania (680 M ha), North America (760 M ha), South America (56 M ha) and Europe (300 M ha). About 2000 million people live in these drylands and produce cereals and pulses. A major occupation is livestock rearing in bushlands, grasslands and savannahs. Among the dryland sub-types, the arid zones extend about 15 percent of the land surface mainly in Africa and Asia. These continents have about two-thirds of hyper-arid and arid zones in the world. About 18 percent of the land area is climatically semi-arid and occurs in all the continents. Mean annual rainfall is about 500 mm and is highly seasonal. The inter-annual variability varies from 25 to 50 percent. Here, both livestock farming and cropping are vulnerable. Human settlements centre around water reservoirs. The extent and distribution of different dryland sub-types in different regions of the world are given in Table 2.2.

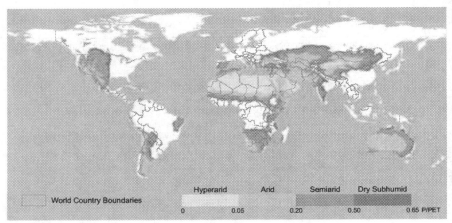

Fig. 2.1 Distribution of drylands in the world (Source: IIASA/FAO 2003)

Table 2.2 Extents of drylands in different regions of the world

Regions	Areas under different dryland categories (M ha)		
	Arid	Semi-arid	Dry sub-humid
Africa	467.60	611.35	219.16
Asia	704.30	727.97	225.51
Oceania	459.50	211.02	38.24
Europe	0.30	94.26	123.47
North & Central America	4.27	130.71	382.09
South America	5.97	122.43	250.21
Total	1641.95	1897.74	1238.68

Source: FAO (2004)

Dietz and Veldhuizen (2004) reported that tropical and subtropical drylands occur in Mexico, southern United States, northern Venezuela, north-eastern Brazil, western Ecuador, Peru, some parts of Chile, southern Bolivia, western Paraguay and northern Argentina of the Americas. They are also found in northern Africa, eastern and southern Africa, and in Israel, Jordan, Syria, Lebanon, south-western Arabia and Yemen, large areas of Iran, and Afghanistan of West Asia. Such drylands also occur in major parts of Pakistan, western and southern India of the Indian subcontinent and in some parts of China, and in large parts of Australia.

2.3 Land Use in Drylands

Bushes, savannahs and grasslands are the types of natural vegetation that may develop in the drylands. Scattered woodlands may also develop near rivers or lakes. However, the types of vegetation in different dryland sub-types are usually

2.3 Land Use in Drylands 19

Fig. 2.2 A landscape of semi-arid region. Scattered bushes of shrubs and bare soil (Image courtesy of dryland-permaculture design)

different. For example, thorny savannahs with annual and perennial grasses are dominant vegetation in the semi-arid region. These areas can be used for grazing or cleared for farming. Scattered bushes and perennial grasses separated by bare soils are commonly observed (Fig. 2.2). On the other hand, broad-leaved savannah woodlands with relatively dense and long trees and perennial grasses commonly develop in the sub-humid region. Large irrigated areas are also found along rivers in this region. These areas are intensively farmed. The United Nations (2011) identify three primary economic functions of drylands: rangelands (65 percent of the global drylands including deserts); rainfed and irrigated farmlands (25 percent), and forests and urban areas (10 percent).

Rangelands occupy 69 percent of the drylands of the developing world (Reid et al. (2004). Rangelands are mostly found in the semi-arid subtype. Extent of rangeland generally increases with increasing aridity. For example, they occupy 34 percent in the dry sub-humid, 54 percent in the semi-arid, and 87 percent in the arid regions. However, livestock densities for rangelands within the dryland subtypes are relatively uniform – 32 to 35 animals per square kilometer of rangeland (Reid et al. 2004; Thornton et al. 2002). Table 2.3 shows areas of land under different land use systems in drylands.

Among drylands, agriculture is practiced mainly in semi-arid and dry sub-humid areas. The main rainfed food crops are sorghum, millet and maize and other crops are groundnuts and cotton for agro-industry and for export. The amount, intensity and distribution of rainfall are so irregular and uncertain that cultivation of other crops is restricted and there is high risk of crop failure due to frequent drought incidence (Dietz and Veldhuizen 2004). The yield of rainfed crops is also low because of scarcity of water and due to low inputs in irrigation and fertilizers. However, animal

Table 2.3 Land use in drylands

Dryland type[a]	Rangelands		Cultivated land		Urban land		Others	
	Area (km²)	%	Area (km²)	%[b]	Area (km²)	%	Area (km²)	%
Dry sub-humid	4,344,897	34	6,096,558	47	457,851	4	1,971,907	16
Semi-arid	12,170,274	54	7,992,020	35	556,515	2	1,871,146	6
Arid	13,629,625	87	1,059,648	7	152,447	1	822,075	5
Hyper-arid	9,497,202	97	55,592	0.6	74,050	1	149,026	2
Total	**39,642,202**	**65**	**15,203,818**	**25**	**1,240,863**	**2**	**4,818,155**	**8**

Source of data: Reid et al. (2004); Thornton et al. (2002)
[a]Based on UNEP classification
[b]Percentage of the dryland type

husbandry is a good opportunity in many dryland areas for adding to food supply and earning cash by producing meat, milk and wool as well as hide. The major animals are cattle, sheep and goat. However, stock density is an important aspect of environmental concern there. Overgrazing might damage the already vulnerable biomass productivity and increase susceptibility to erosion and desertification.

Dryland farming has significant contribution to the production of cereals, pulses and livestock. The major dryland farming areas of North America are distributed in the Prairie regions in Canada, the Great Plains, and different regions of the United States. Production of cereals and livestock in large scale takes place in Argentina, Australia, Canada, the United States and several countries of the former Soviet Union. About 300 million people depend on dryland agriculture for their livelihood. However, poverty is a recurring event in the life of dryland inhabitants because of the low yield and frequent crop failures. Hence, enhancing yields of dryland crop is vital to maintain food security and to improve the livelihoods of the dryland people (Ryan and Spencer 2001).

In the Pacific Northwest of the United States, the mean annual rainfall is low (<300 mm), and Schillinger and Young (2004) observed that winter wheat–summer fallow (one crop in 2 years) is traditionally practiced there in more than 2 million hectares of lands. Thirteen months spanning between harvest of wheat in summer and planting the next wheat crop are kept fallow there to restore water in the soil for the establishment of the subsequent winter wheat crop. Schillinger and Young (2004) also pointed out that fallowing may encourage weed growth and greater number of passes of traditional tillage implements may need to control it. As a result, soil degradation and denudation may occur predominantly through wind erosion and also sometimes by water erosion. Saxton et al. (2000) suggested that wind erosion may cause air quality deterioration in arid and semi-arid regions. According to Bewick (2007), it may also cause pest problems, and reduced crop yields. Farmers, governmental agencies, and conservation groups are now interested in finding options to prevent consequences associated with the traditional wheat-fallow rotation. Intensification of the crop rotation, reduced tillage and no tillage are some possible alternatives.

Ffolliott et al. (2002) stated that widespread and unconfined livestock grazing has been historically practiced in the dryland regions and, according to them, it will remain as the dominant land use type also in the future. Small scale rainfed agriculture is found on some sites, and larger-scale intensive agriculture is practiced mainly in the dry sub-humid regions where water and irrigation facilities are available and profitable. Integration of cropping, livestock farming, forestry, and other production systems is done frequently on the same piece of land, simultaneously, rotationally or spatially.

2.4 Soils of the Drylands

Climate, vegetation and parent materials primarily influence the nature of soil formation in drylands. The major components of climate in the context of soil formation are precipitation, evapotranspiration and temperature. The characteristics of rainfall and evapotranspiration in drylands have been mentioned earlier. Temperature in drylands is usually high, although there are cold dryland areas too. The diurnal variation is generally wide. It has also been mentioned that the natural vegetation types include thorns, bushes, grasses, and savannah, depending on the degree of aridity and moistness. There are short growing season, low growth rate and biomass production, and very low organic matter supply. There are a variety of parent materials including predominantly desert sands, sand dunes, aeolian sands and loess. The rocks and minerals can undergo physical weathering satisfactorily in dryland environments, but the biogeochemical weathering is limited due to scarcity of moisture. Under such conditions there is little clay formation and the soils tend to become coarse textured such as sand, loamy sand and sandy loam. Production, distribution, redistribution and leaching of soluble materials are also restricted by low moisture availability. Where there is enough moisture to initiate some chemical reactions and dissolve some minerals, but not enough to leach soluble salts downward, there is accumulation of salts in the soil profile and the resulting soils are saline. Since there is wide spatial variation in precipitation and evapotranspiration among the dryland sub-types, soils in drylands are also diverse in their origin, structure and physicochemical properties (FAO 2004). Biomass production, deposition, accumulation or decomposition are all low in drylands; so there is little organic matter content in the soils. The natural soil fertility is, therefore, also low (FAO 2008). There is insufficient eluviation for inadequate percolation, and thus there is little scope of developing a B horizon. The soil profiles are usually shallow and of A-C or A-R types. For absence or inadequacy of circulation and redistribution of clay and released salts, there is often accumulation of lime ($CaCO_3$), gypsum ($CaSO_4.2H_2O$) and sodium salts on the soil surface or at a depth close to the surface. Common features of soils of arid and semiarid regions are their high erodibility, susceptibility to seal and crust formation, poor water-holding capacity, structural instability, low content and activity of clay, compaction, and high surface temperatures in summer. Often subsurface horizons of the soil profile, in areas where a little eluviations can occur, accumulate suspended or dissolved minerals, such as silicate clays, calcium carbonate, gypsum and soluble salts. These layers can be cemented by carbonates, gypsum or silica. These cemented layers are called hardpans which restricts the penetration of roots of plants. Soils in the drylands can be saline (accumulation of excess salts), sodic (accumulation of excess exchangeable Na$^+$), gypsic (accumulation of gypsum) and calcareous (accumulation lime). Dryland soils belong to several orders of Soil Taxonomy (Soil Survey Staff 2015) such as Entisols, Aridisols, Mollisols, Alfisols and Vertisols (Ito and Kondo 2000). According to Srinivasarao et al. (2009), these soils have some common characteristics such as coarse to medium texture, low moisture retention, little organic matter content, weak soil profile development, accumulation of salts, or lime or gypsum, and low biological activity. Dryland soils are classified in WRB system (FAO 2006) into Calcisols, Gypsisols, and Leptosols and Steppe soils (FAO 2004).

Dryland salinity develops in regions of low precipitation to evapotranspiration ratio. The level of salinity, however, may be high or low depending on moisture availability for chemical weathering and leaching. Dryland salinity is a serious limitation to agriculture in many countries like Canada, the United States, South Africa, Iran, Afghanistan, Thailand and India. Management of saline soils has been discussed in Chap. 10.

2.4.1 Soil Limitations

Despite climatic constraints, there are many inherent soil limitations to cropping in the drylands. Dryland soils are generally coarse textured, poorly aggregated, excessively drained, low moisture retentive, shallow, sometimes having root restrictive layers, poorly fertile and often with unfavorable chemical conditions. When irrigation and fertilizers are applied in some areas, water and nutrients are easily lost through percolation and leaching. Many dryland soils have surface crusts, compact or cemented subsoil layers. Chemical limitations mainly include salinity and sodicity. Salinity affects crop growth by creating water stress and salt injury. Many of the natural salts become toxic to plants when present in high concentrations. Sodicity destroys soil structure and makes the soil reaction alkaline that reduces availability of some of the micronutrients including Zn and B (Srinivasarao and Vittal 2007). Most dryland soils are deficient in nitrogen and micronutrients. Sahrawat et al. (2007) reported widespread sulfur, boron and zinc deficiencies in dryland soils of India.

Some of the dryland soils have predominance of expanding type of clays. These soils shrink in volume when dried and expand when wet. Deep and wide cracks develop in these soils during the dry season. These cracks interfere with tillage operations. Very hard consistence in the dry season and slaking in the wet season of these soils limits their agricultural use, although their fertility status may be quite good. These soils are classified as the Vertisols and their management has been discussed in detail in Chap. 6.

Soil moisture availability is determined by precipitation, evapotranspiration, soil organic matter content, texture, and amount and type of clay. It influences crop distribution and length of the growing season. The erratic rainfall patterns in the drylands also affect soil moisture availability. At times, soil moisture is deficient when the crop needs it most and the crop plants cannot give satisfactory yield. At some other time, the rainfall intensity may be high to generate run-off. In some areas particularly on steep slopes and with high intensity rainfall, a considerable proportion of rain water is lost by run-off. This is largely a natural phenomenon, but it is worsen by some inappropriate land and crop management practices such as burning of crop residues, excessive tillage, eliminating hedges, etc. Such actions damage soil structure, reduce soil organic matter content, and decrease infiltration.

Many dryland soils are shallow and poorly fertile. Shallow profiles develop due to shallow depth of weathering. Some soils are shallow in the sense that they have root restrictive layers such as hardpans or crusts of lime, gypsum or silica. Soil fertility is low due to coarse texture, low organic matter, absence of enough clay, low CEC, and low retention capacity of added fertilizers.

2.5 Droughts in Drylands

Drought is a frequent, devastating and historic event in the drylands of the world, particularly the arid and semi-arid regions where there are strong temporal and spatial climatic variability and recurring occurrence of extreme dry situations. Drought is one of the most serious constraints to cropping in the drylands and it will remain so in the future. Moreover, some socio-economic and land use including urbanization, increased population and increased demand for water, etc. have exacerbated the vulnerability of vegetation, livestock, wildlife and human populations to drought. However, rangelands may have some degree of resilience and can recover from drought hazards faster, but crops, food supply, and human health suffer the most. According to Msangi (2004), land degradation or desertification due to drought is a great threat to sustainable land management in the arid region. Plants cannot attain their full production potential in arid and semi-arid regions even in the periods of rain, because soil moisture is still low then due to low moisture retention capacity of the soils. Thus, soil moisture availability is a vital factor of the food security of dryland population. Under prevailing rainfall and soil moisture conditions, there seems to have little scope of varying choices for crops and livestock. Innovative soil and crop management techniques need to be adopted to ensure sustainable agricultural production in the drylands which account for about three-quarters of the world's cultivated lands.

Humphreys et al. (2008) suggested that crops may suffer from water deficit even if water is not scarce, but low soil fertility, unsuitable crop variety and inappropriate soil and crop management combinedly prevent the crop plants from fully utilizing whatever soil water is available. They termed it, 'agricultural droughts'. According to the United Nations (2011), total seasonal rainfall in the wetter semi-arid and sub-humid regions often exceeds crop water needs, and if appropriate levels of inputs are used, there is enough scope of improving yields more than double or even quadruple. Lack of inputs in such poorly fertile soils is the major constraint to increasing crop productivity. Many investigators (Hilhorst and Muchena 2000; Morris et al. 2007; Twomlow et al. 2008) mentioned that most poor and smallholder farmers in sub-Saharan Africa do not apply any fertilizer to their crops.

2.6 Dryland Agriculture

Cereals and pulses are the major crops in drylands. Among the cereals wheat and barley are mainly grown as rainfed crops in the Mediterranean and the middle Eastern areas. On the other hand, maize and sorghum are important cereals in sub-Saharan Africa. Irrigated cotton is grown often for export in Egypt, Syria and sub-Saharan Africa. Among pulses, faba bean, chickpea and lentil are the major food crops which supply a large proportion of protein in the diet mainly of the poor inhabitants of drylands. Some oilseeds are also grown. Olive, almond, fig, pistachio, apple, apricot, peach, hazelnut, grape, quince, date palm, cucumber and melon are the major fruits and vegetable crops in different dryland regions. According to Parr et al. (1990), more than 50 percent of the groundnuts, 80 percent of the pearl millet, 90 percent of the chickpeas, and 95 percent of the pigeon peas are produced in the drylands.

Stubble mulch tillage for the wheat-fallow production system has long been a popular practice in drylands of the west-central Great Plains of the USA. This system favors the efficient utilization of the soil moisture reserves and gives relatively stable yields. But for weed control and seedbed preparation, intensive tillage has reduced soil organic matter, deteriorated soil structure, and enhanced risk of wind erosion. However, farmers have adopted reduced tillage, no tillage and some diversification of crop and reduced summer fallow. Now, there can be three crops in 4 years because reduced tillage stores more water in the soil and crops which are present during initial stage of rainfall make efficient use of water as measured by the amount of yield per unit area. Diversification of cropping systems improves and restores soil quality and increases profitability. There can be several diversified dryland cropping systems; the most common systems are winter wheat-corn-fallow, winter wheat-sorghum-fallow, winter wheat-proso millet- fallow, and winter wheat-corn-proso millet-fallow. There are some diversified cropping systems without a fallow period such as wheat-corn-proso millet and wheat-proso millet and continuous proso millet. According to Miller et al. (2002), continuous diversified crop rotations can utilize water and nutrients and sustain crop yields in comparison to monocropping in areas with limited water. For example, pea, a legume that has nitrogen fixing capacity, uses less soil water than spring wheat and barley and retains more water available. It also increases yields of the succeeding crops (Miller et al. 2002; Lenssen et al. 2007a, b; Sainju et al. 2009). Miller et al. (2002) stated that crop diversification could control weeds, diseases, and pests effectively, and Gregory et al. (2002) suggested that it reduces the risks of crop failure, farm inputs, and duration of fallow, and improvement in economic and environmental sustainability.

2.7 Integrated Soil and Crop Management for Drylands

Resource-poor, and subsistence-based farmers or the small-holders are engaged in different activities to survive in these hostile environments of the drylands with multiple risks through diversification, flexibility and adaptability (FAO 2004). According to FAO (2008), an integration of soil, water, crop and nutrient managements is needed for successful dryland farming. The improvement of soil fertility by concerted use of locally available nutrient sources such as manure and compost and water conservation to enhance water-nutrient synergy (Zougmore et al. 2003; Stroosnijder et al. 2012) along with covering the soil by crop residues to reduce erosion (Adimassu et al. 2014; Baptista et al. 2015) may contribute to sustainable dryland farming. Buerkt et al. (2002) reported that the integration of water and nutrient management can be an ecologically sound and economically viable strategy for optimizing dryland crop production.

2.7.1 Principles of Integrated Soil and Crop Management for Drylands

There are some important principles of integrated soil and crop management for drylands. These are (i) selection of suitable crops, (ii) reducing evaporation, (iii) conserving soil moisture, (iv) control of weeds, (v) storing water in reservoirs of watersheds, (vi) harvest of rain or run-off water, (vii) using supplemental irrigation where feasible, and (viii) protection against erosion.

2.7.2 Growing Dryland Crops

Some crop plants can adapt to soil dryness by different mechanisms such as (i) short life cycle - they can germinate and grow during a very short period of available moisture, (ii) deep and extensive root systems enabling them to draw water from a large volume of soil, (iii) succulent leaves and stems for storage of water in their tissues to use in need, and (iv) very narrow leaves or leaves modified to thorns to reduce transpiration. Creswell and Martin (1998) classified crops on the basis of tolerance to water stress. They ranked some crops from 0 to 3; 0 for no tolerance, 1 for slight tolerance, 2 for moderate tolerance and 3 for high tolerance. A list of crops having moderate to high tolerance to water stress (ranks >2) is given below.

Crop types	Crops
Cereals	Pearl millet (*Pennesitum americanum*)
Grain legumes	Pigeon pea (*Cajanus cajan*), lablab bean (*Dolichos lablab*), Mungbean (*Vigna radiate*), Tepary bean (*Phaseolus acutifolius*), mat bean (*Vigna aconitifolius*), Marama bean (*Tylosema esculentum*)
Leafy vegetables	Chaya (*Cnidoscolus chayamansa*), horseradish tree (*Moringa oleifera*), Leucaena (*Leucaena leucocephala*)
Root crops	Cassava (*Manihot esculenta*), African yam bean (*Sphenostylis stenocarpa*)
Fruit trees	Karanda (*Carissa carandus*), dove plum (*Dovyalisa byssinica*), Pomegranite (*Punica granatum*), Cashew (*Anacardium occidentale*), prickly pear (*Opuntia* sps.), date (*Phoenix dactylifera*)
Oil plants	Shea butter (*Butyrospermum paradoxum*)

(continued)

(continued)

Crop types	Crops
Feed legumes	Mesquite (*Prosopsis* sps.), Leucaena (*Leucaena leucacephala*), Apple ring acacia (*Acacia albida*), umbrella thorn (*Acacia tortilis*), Jerusalem thorn (*Parkinsonia aculeate*)
Fiber plants	Henequen (*Agave fourcroydes*), Sisal (*Agave sisalana*)
Timber plant	Umbrella thorn (*Acacia tortilis*)

There are cool-weather crops and warm-weather crops. Wheat, barley, chickpeas, horse beans, lentils, linseed, oats, peas, sugar beets, vetches, etc. are cool weather crops for the semi-arid regions. On the other hand, beans, corn (maize), cotton, cowpeas, groundnuts, millets, pigeon pea, sorghum, sunflowers, sesame, etc. are warm weather crops. Many dryland soils suffer from both scarcity of water and salinity. A list of salt tolerant crops is provided in Chap. 10. However, there are some crops which are tolerant to both salt stress and water stress. Some of such crops are barley, safflower, sugar beet, triticale wheat, etc. These crops can be grown in dryland saline soils, and supplemental irrigation, if feasible, may leach away some salts of the root zone.

Cacti can grow well in the dryland soils of the arid regions and on degraded soils. *Opuntia* species have a great capacity to withstand severe dry conditions. Roots of these plants can reduce wind and water erosion and can grow in degraded areas. According to Nefzaoui (2011), cacti can be used: (i) as forage, (ii) as vegetables (young cladodes), (iii) as fruit, (iv) as processed foods such as concentrated foods, juices, liquors, semi-processed and processed vegetables, food supplements, and (v) for medicinal applications.

2.7.3 Diversity in Crops and Cropping Systems

Dryland farmers often grow several crops together in the same field at the same time to match existing environmental stresses. This type of mixed cropping system is followed to avoid total crop failures; under stresses some of the crops survive and give some yield. All the crops in a mixture are not equally susceptible to extremes of climate, pests, and diseases. The choices of crops are based on their own food requirement, for feed of livestocks, and sometimes for export. Farmers grow sweet potatoes on steep slopes in some areas for providing cover and reducing erosion; and in addition sweet potato can be harvested in the period of food shortages before harvest of maize. Moreover, sweet potatoes can thrive even in periods of drought. The criteria of selecting crops by the farmers differ. Some crops have high yield potentials, some are tolerant to drought and other stresses, some have resistance to pests and diseases, and some can give higher profit.

Sorghum, pearl millet, rice, wheat and maize are some of the most important cereal crops of the semi-arid tropics. Pulses and oilseeds are important cash crops. These crops are grown in different cropping systems including mono-cropping,

2.7 Integrated Soil and Crop Management for Drylands

multiple cropping, intercropping, mixed cropping, mixed row cropping and relay cropping. Intercropping, mixed cropping and relay cropping are different variants of multiple cropping. Multiple cropping is a cropping system in which more than one crop is grown simultaneously or successively in a piece of land within a stipulated period. Mixed cropping is a system when more than one crop is grown simultaneously in the same field in the same time. It can be a regular mixture such as in alternate rows or two rows of one crop separated by one row of the other, and the like. This is intercropping. However, several crops can be grown haphazardly in the same field at the same time. In mixed cropping, crops of different habits such as root and canopy characteristics, demand of water and nutrients, period of maturity, etc. are selected. Dryland farmers prefer haphazard mixing and intercropping because only a short growing season is available there. In dry sub-humid regions, however, some crop succession can be practiced; usually one rainfed and the other with supplemental irrigation. In mixed cropping farmers prefer combinations of cereals and legumes (pulses), because the legumes improve the yields of the cereals by contributing N_2. Pigeon pea may be a good companion crop in an intercrop because it draws water and nutrients from greater depth of the soil by their deep roots, they compete little with the cereals for water and nutrients, they have high N_2 fixing capacity and they produce high biomass residues. Wheat–fallow cropping system is followed in many dryland areas. In some areas, crop rotations include spring barley, lentil, peas, flax, buckwheat, and oilseeds. In crop-livestock integrated farming systems, farmers also include perennial grasses into the rotation. Sometimes a particular planting technique known as skip-row technique is followed. In this method seeds are sown into every other or every third row, but the number of plants is increased to maintain the original plant density. Soil moisture of the blank rows is not depleted at the initial stage of the growing season. As the crop plants grow they expand their roots laterally into the soil of the blank row area. However, herbicides may be needed to kill the weeds in the blank rows.

Keeping a period extending from harvest of one crop to planting the next crop in the field without an established crop is called fallowing. Fallowing is done usually in summer in the drylands. It is a component of the conservation farming which conserves soil moisture, breaks the pest cycles and restore soil fertility.

2.7.4 Conservation Tillage

Tillage refers to the physical manipulation of soil for making a good tilth in the seedbed or root bed. A 'good tilth' means a condition of the prepared soil favorable for seed germination and root proliferation, so that plants can acquire adequate water and nutrients to develop a good crop stand. Tillage has the advantages of (i) loosening and aerating soil, (ii) breaking pans and clods, (iii) facilitating root growth, (iv) exchanging gases between soil and atmosphere, (v) mixing soil and amendments together, and (vi) killing weeds. But conventional tillage characterized by frequent, deep, and intensive operations is often responsible for soil degradation

Fig. 2.3 No till seeding keeping the stubbles of the previous crop in field (Image courtesy of Agriculture and Agri-Food Canada)

through organic matter depletion, impoverishing soil fertility, crusting, compaction, and erosion. Conservation of soil and water is a priority in dryland production systems. Tillage practices should be carefully chosen there because more tillage will cause more soil surface evaporation as the generally wetter subsoil is exposed to the sun and wind. Therefore, conservation tillage is preferred in dryland farming to sustain soil fertility and crop productivity. Different conservation tillage systems can be employed for effective conservation of soil moisture under dryland agriculture. According to Alvarez and Steinbach (2009), conservation tillage systems can effectively reduce soil erosion. Su et al. (2007) stated that conservation tillage systems can improve soil quality and reduce soil loss because all conservation tillage systems involve maintenance of crop residues on soil surface. Zero tillage or no tillage, minimum or reduced tillage, strip tillage, mulch tillage, etc. are some examples of conservation tillage systems.

Seeding without any soil preparation is known as zero or no tillage (Fig. 2.3). In this system, stubbles of the previous crop are left on the field and seeding is done within it. In minimum tillage or reduced tillage, however, only a slit or a small hole is made in the soil, and seeds are sown into it. Limited soil disturbance is done in minimum tillage with a kind of a specific tiller suited to row crops. Narrow rows are tilled with the tiller and the space between the rows is left undisturbed. Minimum tillage may be combined with mulching. Crop residues, leaf litter, straw, wood chips, saw dust, etc. can be used for mulching. Strip-till is also a type of reduced tillage that is being increasingly popular among the dryland farmers. In strip tillage narrow strips are tilled, but the spaces between rows are left untilled and covered with crop residues, stubbles or some other mulch (Fig. 2.4). Here the tilled strips are wider than in minimum tillage. Cotton farmers usually prefer strip tillage. Residue management is an integral part of conservation tillage.

2.7 Integrated Soil and Crop Management for Drylands

Fig. 2.4 Strip tillage along with stubble mulching (Image courtesy of MK Martin Enterprise)

When crop residues or any other material form enough cover or provide shade to protect the soil from raindrop impact and to conserve soil and water it is known as mulch. Mulching can be integrated with tillage, and then it is called mulch tillage. There are several types of mulch tillage such as cover mulch, stubble mulch, dust mulch, etc. Sadegh-Zadeh et al. (2009) suggested that mulching is an efficient method of conserving water in semi-arid and arid regions. Cover mulch refers to materials that are spread over to cover the soil surface. Cover mulches formed by the residues reduce evaporation, increase infiltration, conserve soil moisture, and keep the soil cool in the warm season and warm in the cold season. Residues also increase the content of soil organic matter. Several types of organic and inorganic materials are used for cover mulching. Common organic materials are compost, composted manure, grass clippings, newspaper, straw, shredded leaves, straw, etc. Inorganic materials for cover mulching include gravels, pebbles, plastic sheets, woven cloth, ground rubber tires, etc. However, organic mulches are preferred for dryland farming. Stubble mulches shade the soil surface and reduce evaporation. They also decrease wind erosion by reducing wind velocity at the soil surface. According to Hemmat et al. (2007), stubbles left on the soil surface after harvest form an organic mulch that reduces raindrop impact, impedes run off, reduces evaporation, and increases infiltration. When a very fine textured layer is created on the soil surface by intensive hoeing, it is called 'dust mulch' which breaks the soil capillarity and destroys the continuity of pores. It reduces loss of soil water through evaporation. It is a common practice in India and China.

Low organic matter containing soils of the arid region are relatively easily compacted and crusted so that their infiltration, seedling emergence and root growth are reduced. Some soil may be hard setting or may have root restrictive layers. These soils may need deep tillage and mechanical loosening for decompaction.

These soils tend to recompact unless generous addition of organic residues is done. Many authors (Hemmat and Eskandari 2004, 2006; Mosaddeghi et al. 2009; Shirani et al. 2002) have discussed the effects of different tillage methods but the combined effects of tillage systems and mulching have less been investigated.

2.7.5 Water Conservation

The most severe constraint to agricultural production in the drylands is the scarcity of water during the growing season, and according to Baumhardt and Jones (2002), conserving rain water properly for use during crop growth is also a major challenge in these areas. Water storage in the soil can be optimized if the soil has satisfactory infiltration capacity, permeability and water retention capacity. Sadegh-Zadeh et al. (2011) discussed dryland farming in the West of Iran and suggested that like other dryland areas water is the main constraint to crop production also here. Precipitation increases during winter, but there are high temperatures during growth season of crops in spring; Hemmat and Eskandari (2004) suggested that such conditions also limit crop production in dryland agriculture in many other countries.

2.7.6 Supplemental Irrigation

Addition of small amounts of water to rainfed crops in periods of low rainfall, when plants show signs of water stress or if rainfall cannot provide sufficient moisture for normal plant growth, is known as supplemental irrigation. The objective of supplemental irrigation is to improve and stabilize yields. It is, however, difficult to ascertain when and how much water would be needed for supplemental irrigation well in advance because of the uncertainty of rains. Still, supplemental irrigation is very effective in periods of critical water shortage. It can alleviate the adverse effects of soil water stress, particularly in sensitive growth stages like flowering and grain filling, on the yield of dryland crops. According to some authors (Oweis et al. 2000; Ilbeyi et al. 2006), applying a limited amount of water as supplemental irrigation during the critical crop growth stages can substantially improve yield of crops in drylands. For example, Oweis and Hachum (2012) conducted an experiment on several barley genotypes in an area under a Mediterranean climate with total rainfall of only 186 mm. They applied water to replenish 33, 66, and 100 percent of the soil moisture deficit in the crop root zone. They obtained mean grain yield of 0.26, 1.89, 4.25 and 5.17 t ha^{-1} under rain-fed, 33 percent, 66 percent and 100 percent supplemental irrigation, respectively. Liu et al. (2005) reported the results of supplemental irrigation on wheat and maize in the semi-arid region of China. Run-off water was collected and stored in small tanks. It was used for supplemental irrigation in the dry spells and yield of wheat and maize increased by more than 50 percent. Similarly, on-farm experiments in Kenya and Burkina Faso gave yield increases of 70 to 300 percent of maize and sorghum over conventional farming practices with combined application of supplemental irrigation and fertilizers (Rockstrom et al. 2002).

2.7.7 Fertilizer Application

Dryland soils have inherently low nutrient status and low nutrient retention capacity. So, there is a need of frequent fertilizer application in low doses, particularly in irrigated crops. Since these soils have very little organic matter, they are usually deficient in nitrogen and often in phosphorus especially in sandy and calcareous soils. Nitrogen fertilizers are always necessary, and in sandy and calcareous soils phosphate fertilizer may also be beneficial. According to Alloway (2008) sandy dryland soils are usually low in B, Cu, Fe, Mn, Mo, Ni and Zn. As crops often show signs of Zn and other micronutrient deficiencies, micronutrient fertilizers may be needed in coarse textured soils. Fertilizers in sandy soils should be applied in low and frequent split doses, because applying all the fertilizers at a time may add to salt stress in already salt affected soils and in soils with low moisture content. Applied fertilizers may be lost by leaching as well.

2.7.8 Use of Herbicides

Weeds are a severe problem during the fallow season in dryland agriculture. Killing these weeds by tillage may need many operations and these operations may expose and dry out the soil. According to Regehr and Norwood (2008), tillage operations during the fallow period have made many soils of the Great Plains extremely vulnerable to wind erosion. Peterson and Westfall (2004) reported that using herbicides for weed control reduced significantly the need for tillage and resulted in storage of considerable amounts of precipitation. Use of herbicides, inclusion of summer corn and sorghum in cropping patterns and supplemental irrigation reduced greatly the areas of summer fallowing in drylands of the United States. Derksen et al. (2002) stated that improved herbicide options could eliminate the need for fallow in most areas of the Great Plains. Herbicides have reduced the need of tillage, enhanced the maintenance of crop residue cover on the soil surface, and conserve soil moisture. Integrated no-till+herbicide+residue management may favor intensive cropping in drylands.

2.7.9 Soil Conservation

Conservation tillage, as discussed earlier, is also a soil conservation technique, but some other physical and chemical methods of soil conservation are discussed in this sub-section. Physical soil conservation measures involve reducing degree and length of slope for slowing run-off velocity and enhancing infiltration. Trapping sediments, preventing formation of rills and gullies and restoration of eroded lands are some of the objectives of physical soil conservation methods. In some parts of Kenya `fanyajuu' terracing (Gichuki 2000), cut-off drains and grass strips (Fall 2000) are common soil conservation measures.

Fig. 2.5 Perennial grass barriers for wind erosion control (Image courtesy of USDA-NRCS)

Wind erosion is a major form of soil degradation in the drylands. Chemical soil stabilizing agents including polymers are used, sometimes, to reduce blowing out of the soil particles from the soil matrix. Some of these polymers are coherex, DCA-70, Petroset SB, Polyco 2460, Polyco 2605, and SBR Latex S-2105. The effects of these products on controlling wind erosion are often temporary. On the other hand, vegetative measures such as wheat straw anchored with a rolling disk packer can offer easy and inexpensive stabilization. Wind barriers are popular and effective wind erosion control measures. Wind barriers are usually linear plantings of a single row or multiple rows of trees or shrubs planted perpendicular to the direction of wind. Their effects include (1) reduction of field width, (2) decrease in the distance that wind travels across the field, (3) decreasing wind velocity, and (4) trapping wind-blown and saltated soil. Trees, shrubs, and perennial grasses and their combinations can effectively reduce wind erosion. Annual crops such as corn, sorghum, Sudangrass, and sunflowers can also act as wind barriers. Most barrier systems, however, occupy some space of the field that is not available for producing crops (Fig. 2.5).

Thus, an integration of suitable crops, wise use of soil moisture, soil moisture conservation, harvest of rain and run-off water for supplemental irrigation, conservation tillage, application of fertilizers and organic residues, and soil conservation practices can improve and sustain food security and human health in addition to preventing soil degradation and desertification in the drylands.

Study Questions
1. Define dryland soils. Describe the principal characteristics of dryland soils and list their major limitations to agricultural use.
2. Discuss the distribution and land use of dryland soils.

3. Give a list of crops suitable for dryland farming. Discuss the methods of soil moisture conservation in dryland soils.
4. Describe soil conservation measures for dryland soils.
5. Write notes on: (a) Aridity index, (b) Dryland agriculture, (c) Water conservation in drylands, (d) Limitations of drylands, (e) Supplemental irrigation.

References

Adimassu Z, Mekonmen K, Yirga C, Kessler A (2014) Effect of soil bunds on runoff, soil loss and crop yield in the central highland of Ethiopia. Land Degrad Dev 25(6):554–564

Alloway BJ (2008) Micronutrients and crop production: an introduction. In: Alloway BJ (ed) Micronutrient deficiencies in global crop production. Springer, New York

Alvarez R, Steinbach H (2009) A review of the effects of tillage systems on some soil physical properties, water content, nitrate availability and crops yield in the argentine pampas. Soil Tillage Res 104:1–15

Bantilan MCS, Anand BP, Anupama GV, Deepthi H, Padmaja R (2006) Dryland agriculture: dynamics, challenges and priorities. Research bulletin no. 20. International crops research Institute for the Semi-Arid Tropics Andhra Pradesh, India

Baptista I, Ritsema CJ, Querido A, Ferreira AD, Geissen V (2015) Improving rainwater-use in Cabo Verde drylands by reducing runoff and erosion. Geoderma 237–238:283–297

Baumhardt R, Jones O (2002) Residue management and tillage effects on soil-water storage and grain yield of dryland wheat and sorghum for a clay loam in Texas. Soil Tillage Res 68:71–82

Bewick LS (2007) No-till facultative wheat production in the dryland winter wheat-fallow region of the Pacific northwest. MS thesis, Washington State University, Pullman

Buerkt A, Piepho HP, Bationo A (2002) Multi-site time-trend analysis of soil fertility management on crop production in sub-Saharan West Africa. Exp Agric 38:163–183

Creswell R, Martin FW (1998) Dryland farming: crops and techniques for arid regions. ECHO Technical Note 1–23

D'Odorico P et al (2013) Global desertification: drivers and feedbacks. Adv Water Resour 51:326–344

De Pauw EF (2009) Management of dryland and desert areas. In: Verheye WH (ed) Land use, land cover and soil sciences. National Science Foundation Flanders- Belgium and Geography Department, University of Gent, Belgium

Derksen DA, Anderson RL, Blackshaw RE, Maxwell B (2002) Weed dynamics and management strategies for cropping systems in the northern Great Plains. Agron J 94:174–185

Dietz T, Veldhuizen E (2004) The World's drylands: a classification. In: Dietz AJ, Ruben R, Verhagen A (eds) The impact of climate change on drylands, with a focus on West Africa, Environment and policy series 39, pp 19–26, Springer, Netherlands

Fall A (2000) Makueni District profile: Livestock management, 1989–1998. Drylands research working paper 7. Drylands Research, Crewkerne, UK

FAO (2000) Land Resource Potential and Constraints at Regional and Country Levels. World Soil Resources Reports 90. Land and Water Development Division, Food and Agriculture Organization of the United Nations, Rome

FAO (2004) Carbon sequestration in dryland soils. World Soil Resources Reports 102. Food and Agriculture Organization of the United Nations, Rome

FAO (2006) World reference base for soil resources 2006. A framework for international classification, correlation and communication. FAO–UNESCO–ISRIC. FAO, Rome

FAO (2008) Drylands, people and land use. In Water and Cereals in Drylands. Food and Agriculture Organization of the United Nations and Earthscan, Rome

Ffolliott PF, Brooks KN, Fogel MM (2002) Managing watersheds for sustaining agriculture and natural resource benefits into the future. Quart J Int Agri 41(1/2):23–40

Gichuki FN (2000) Drylands Research Working Paper 4. Makueni District Profile: Soil Management and Conservation 1989–1998. Drylands research, Crewkerne, Somerset, UK

Gregory PJ, Ingram JSI, Anderson R, Betts RA, Brovkin V, Chase TN, Grace PR, Gray AJ, Hamilton N, Hardy TB, Howden SM, Jenkins A, Meybeck M, Olsson M, Ortiz-Montasterio I, Palm CA, Payn TW, Rummukainen M, Schulze RE, Thiem M, Valentin C, Wikinson MJ (2002) Environmental consequences of alternative practices for intensifying crop production. Agri Ecosyst Environ 88(3):279–290

Hemmat A, Eskandari I (2004) Tillage system effects upon productivity of a dryland winter wheat-chickpea rotation in the northwest region of Iran. Soil Tillage Res 78:69–81

Hemmat A, Eskandari I (2006) Dryland winter wheat response to conservation tillage in a continuous cropping system in northwestern Iran. Soil Tillage Res 8(6):99–109

Hemmat A, Ahmadi I, Masoumi A (2007) Water infiltration and clod size distribution as influenced by ploughshare type, soil water content and ploughing depth. Biosyst Eng 97:257–266

Hilhorst T, Muchena F (2000) Nutrients on the move: soil fertility dynamics in African farming systems. International Institute for Environment and Development, London

Hori Y, Stuhlberger C, Simonett O (2011) Desertification – a visual synthesis. United Nations Convention to Combat Desertification (UNCCD); Zoï Environment Network (Zoï). http://www.unccd.int/Lists/SiteDocumentLibrary/Publications/Desertification-EN.pdf

Humphreys E, Peden D, Twomlow S, Rockström J, Oweis T, Huber-Lee A, Harrington L (2008) Improving rainwater productivity: topic 1 synthesis paper. CGIAR Challenge Program on Water and Food, Colombo

IIASA/FAO (2003) Distribution of drylands in the world. Cited from FAO (2008) Drylands, people and land use. In Water and Cereals in Drylands. Food and Agriculture Organization of the United Nations and Earthscan, Rome

Ilbeyi A, Ustun H, Oweis T, Pala M, Benli B (2006) Wheat water productivity in a cool high land environment: effect of early sowing with supplemental irrigation. Agric Water Manag 82:399–410

Ito O, Kondo M (2000) Crop and resource management for improved productivity in Dryland Farming Systems. In: Watanabe K, Komamine A (eds) Proceedings of the 12th Toyota conference: challenge of plant and agricultural sciences to the crisis of biosphere on the Earth in the 21st century. Landes Bioscience, Texas

Lenssen AW, Johnson GD, Carlson GR (2007a) Cropping sequence and tillage system influences annual crop production and water use in semiarid Montana. Field Crop Res 100(1):32–43

Lenssen AW, Waddell JT, Johnson GD, Carlson GR (2007b) Diversified cropping systems in Semiarid Montana: nitrogen use during drought. Soil Tillage Res 94(2):362–375

Liu FM, Wu YQ, Xiao HL, Gao QZ (2005) Rainwater-harvesting agriculture and water-use efficiency in semi-arid regions in Gansu province, China. Outlook Agricul 4(3):159–165

Miller PR, McConkey B, Clayton GW, Brandt SA, Staricka JA, Johnston AM, Lafond GP, Schatz BG, Baltensperger DD, Neill KE (2002) Pulse crop adaptation in the Northern Great Plains. Agron J 94(2):261–272

Molden D, Oweis TY (2007) Pathways for Increasing Water Productivity. In: Molden, D. (ed), Water for Food, Water for Life. London and International Water Management Colombo, p. 279–310

Morris M, Kelly V, Kopicki R, Byerlee D (2007) Fertilizer use in African agriculture: lessons learned and good practice guidelines. Agriculture and Rural Development Division. World Bank, Washington, DC

Mosaddeghi M, Mahboubi A, Safadoust A (2009) Short-term effects of tillage and manure on some soil physical properties and maize root growth in a sandy loam soil in western Iran. Soil Tillage Res 104:173–179

Msangi JP (2004) Drought hazard and desertification management in the drylands of Southern Africa. Environ Monit Assess 99(1–3):75–87

Nefzaoui A (2011) Cactus crop is need of the day for dryland farming in India. International workshop on "cactus crop to improve the rural livelihoods and to adapt to climate change in the arid and semi-arid regions" during 25–26 November 2011 at National Bureau of Plant and Genetic Resources (NBPGR), New Delhi, India (organized by ICARDA, FAO, ICAR)

Oweis T, Hachum A (2012) Supplemental irrigation, a highly efficient water-use practice. ICARDA, Aleppo, pp 2–28

Oweis T, Zhang H, Pala M (2000) Water use efficiency of rainfed and irrigated bread wheat in a Mediterranean environment. Agron J 92:231–238

Parr JE, Stewart BA, Hornick SB, Singh RP (1990) Improving the sustainability of Dryland Farming Systems: a global perspective. Adv Soil Sci 13:1–8

Peterson GA, Westfall DG (2004) Managing precipitation use in sustainable dryland agroecosystems. Ann Appl Biol 144:127–138

Regehr DL, Norwood CA (2008) Benefits of Triazine herbicides in Ecofallow. In: The Triazine herbicides: 50 years revolutionizing agriculture. Elsevier, San Diego

Reid RS, Thornton PK, McCrabb GJ, Kruska RL, Atieno F, Jones PG (2004) Is it possible to mitigate greenhouse gas emissions in pastoral ecosystems of the tropics? Dev Sustain 6:91–109

Rockstrom J, Barron J, Fox P (2002) Rainwater management for increased productivity among small-holder farmers in drought prone environments. Phys Chem Earth 27(11–22):949–959

Rosegrant M, Cai X, Cline S, Nakagawa N (2002) The role of rainfed agriculture in the future of global food production. EPTD discussion paper 90, IFPRI, Washington, DC

Ryan J, Spencer D (2001) Future challenges and opportunities for agricultural land in the semi-arid tropics. ICRISAT, Patancheru

Sadegh-Zadeh F, Seh-Bardan BJ, Samsuri AW, Mohammadi A, Chorom M, Yazdani GA (2009) Saline soil reclamation by means of layered mulch. Arid Land Res Manag 23:127–136

Sadegh-Zadeh F, Wahid SA, Seh-Bardan BJ, Seh-Bardan EJ, Bah A (2011) Alternative management practices for water conservation in dryland farming: a case study in Bijar, Iran. In: Jha MK (ed) Water conservation. InTech, Croatia

Safriel U, Adeel Z, Niemeijer D, Puigdefabregas J et al (2005) Dryland systems. In: Hassan R, Scholes R, Ash N (eds) Ecosystems and human well-being: current state and trends, vol volume 1. Island Press, Washington, DC

Sahrawat KL, Wani SP, Rego TJ, Pardhasaradhi G, Murthy KVS (2007) Widespread deficiencies of sulfur, boron and zinc in dryland soils of the Indian semi-arid tropics. Curr Sci 93(10):1428

Sainju UM, Lenssen AW, Caesar-Ton That T, Evans RG (2009) Dryland crop yields and soil organic matter as influenced by long-term tillage and cropping sequence. Agron J 101(2)

Saxton K, Chandler D, Stetler L, Lamb B, Claiborn C, Lee BH (2000) Wind erosion and fugitive dust fluxes on agricultural lands in the Pacific Northwest. Trans Am Soc Agric Eng 43(3):623–630

Schillinger WF, Young DL (2004) Cropping systems research in the world's dries rainfed wheat region. Agron J 96(4):1182–1187

Shirani H, Hajabbasi M, Afyuni M, Hemmat A (2002) Effects of farmyard manure andtillage systems on soil physical properties and corn yield in central Iran. Soil Tillage Res 68:101–108

Soil Survey Staff (2015) Illustrated Guide to Soil Taxonomy, Version 1.1. U.S. Department of Agriculture, Natural Resources Conservation Service, National Soil Survey Center, Lincoln, Nebraska

Srinivasarao C, Vittal KPR (2007) Emerging nutrient deficiencies in different soil types under rainfed production systems of India. Indian J Ferti 3:37–46

Srinivasarao C, Vittal KPR, Venkateswarlu B, Wani SP, Sahrawat KL, Marimuthu S, Kundu S (2009) Carbon stocks in different soil types under diverse rainfed production systems in tropical India. Commun Soil Sci Plant Anal 40:2338–2356

Stroosnijder L, Moore D, Alharbi A, Argaman E, Biazin B, van den Elsen E (2012) Improving water use efficiency in drylands. Current Opin Environ Sustain 4:1–10

Su Z, Zhang J, Wu W, Cai D, Lv J, Jiang G, Huang J, Gao J, Hartmann R, Gabriels D (2007) Effects of conservation tillage practices on winter wheat water-use efficiency and crop yield on the Loess Plateau, China. Agric Water Manag 87:307–314

Thornton PK, Kruska RL, Henninger N, Krisjanson PM, Reid RS et al (2002) Mapping poverty and livestock in the developing world. ILRI (International Livestock Research Institute), Nairobi

Twomlow S, Rohrbach D, Dimes J, Rusike J, Mupangwa W, Ncube B, Hove L, Moyo M, Mashingaidze N, Mahposa P (2008) Microdosing as a pathway to Africa's Green Revolution: evidence from broad-scale on-farm trials. Nutr Cycl Agroecosyst 88(1):3–15

UNCCD (2000) An introduction to the United Nations Convention to Combat Desertification. http://www.unccd.int

UNEP (1997) World atlas of desertification, 2nd edn. United Nations Environmental Program, Nairobi

United Nations (2011) Global drylands: a UN system-wide response. United Nations Environment Management Group, UN, Washington, DC

White R, Tunstall D, Henninger N (2002) An ecosystem approach to drylands: building support for new development policies. Information policy brief no 1. World resources institute, Washington, DC

Zougmore R, Mando A, Ringersma J, Stroosnijder L (2003) Effect of combined water and nutrient management on runoff and sorghum yield in semiarid Burkina Faso. Soil Use Manag 19:257–264

Chapter 3
Sandy Soils

Abstract Sandy soils contain high proportion of sand particles, more than sixty eight percent by weight in their mechanical composition. They are common in the drylands, but many soils in the humid and subhumid regions are also sandy. They may be moist for variable periods in the low lands and humid regions, and dry in the high lands where they are low moisture retentive. Sandy soils are naturally poorly fertile, and when they are fertilized and irrigated, they cannot retain much water and nutrients. Natural vegetation in sandy soils is usually composed of grasses and forests, and, through innovative and intensive management, they can be profitably used for a variety of cereals and vegetables.

Keywords Psamments · Aerenosols · Low moisture retentive soils · Hydrogel · Biochar · Mulching · Dune stabilization

3.1 Nature of Sandy Soils

Sandy soils are characterized by sand contents more than sixty eight percent and clay contents less than eighteen percent in the first hundred centimeter of the solum (Bruand et al. 2005). In the soil classification system of World Reference Base (FAO 2006), sandy soils may occur in the following Reference Soil Groups: Arenosols, Regosols, Leptosols and Fluvisols; Arenosols being the most common. In the Soil Taxonomy (Soil Survey Staff 1999), most sandy soils are in the suborder Psamments. Podzols (FAO 2006) or Spodosols (Soil Survey Staff 1999) are also sandy throughout the solum with accumulation of humus and alluminium with or without iron and clay in the B horizon. These coarse textured surface soils are strongly acidic in nature and usually have an organic O horizon above the sandy mineral A horizon. Sandy soils may also occur in the orders Alfisols and Ultisols. On the basis of origin, sandy soils can be of three different categories including residual sands, shifting sands and recently deposited sands. Soil Taxonomy considers dunes and shifting sands as non-soils. Sandy soils may occur in arid to humid and in extremely cold to extremely hot climates. Sandy soils are excessively drained and highly leached. They have low organic matter and poor soil fertility. Sandy soils do not hold much water, and, at the same time, applied water is rapidly drained away. They do not hold much nutrient, and, for low nutrient retaining capacity, native and added nutrients are lost quickly by leaching. Sandy soils are used for extensive grazing in the dry zone; and cropping (wheat, millet, barley, pea, etc.) is done in many areas of semi-arid and sub-humid regions. Management of these soils has been discussed in sufficient details in Chap. 2. Properties and management of sandy soils in humid temperate and humid tropical regions will be discussed in this chapter, although many management options are common. In temperate areas, mixed arable cropping with supplemental irrigation is done during dry spells. Sandy soils in temperate region bear extensive natural forests; they are also useful for intensive horticultural crops because they allow mechanical harvesting when other clay-rich soils may be too wet and plastic during times of heavy rainfall. In the perhumid tropics sandy soils are used for field crops with irrigation and fertilizer application. Satisfactory crop yields may be obtained from sandy soils if some general management rules are adopted.

3.2 Distribution of Sandy Soils

The typical sandy soils are the Psamments, which are deep deposits of sand of alluvial or aeolian origin. Psamments consist of a suborder of Entisols in Soil Taxonomy (Soil Survey Staff 1999). They are very sandy in all layers within the particle-size control section. Some form in poorly graded and well sorted sands on shifting or stabilized sand dunes, in cover sands, or in sandy parent materials that were sorted in an earlier geologic cycle. Psamments occur under any climate without permafrost within 100 cm of the soil surface. In the tropical regions, they are subjected to wetting and drying cycles due to the seasonality of rainfall (Bruand et al. 2005). They can have any vegetation and have been on surfaces of virtually any age from recent historic to Pliocene or older. Psamments on old stable surfaces commonly consist of quartz sand. In WRB and FAO/UNESCO Soil Map of the World (FAO 2006), sandy soils are included in several Reference Soil Groups but principally in Arenosols. The total estimated extent of Arenosols is 900 M ha mainly in Western Australia, South America, South Africa, Sahel, and Arabia. Global distribution of sandy soils is shown in Fig. 3.1.

3.3 Properties of Sandy Soils

Physical properties of sandy soils are dominated by their texture. Sandy soils are not very well aggregated; they have weak soil structure or they remain single grained. They have little shrinking or expansion properties due to the low clay content and the high proportion of low activity clays. Sandy soils have small shrinkage tendency. When they dry, sandy soils develop very few thin cracks organized in a loose network. Sandy soils show a wide range of porosities and bulk densities (D_b). In

3.3 Properties of Sandy Soils

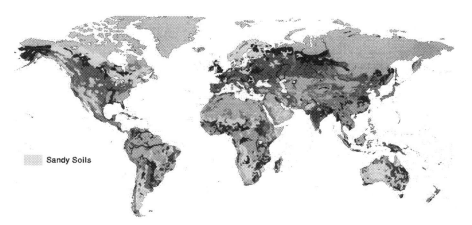

Fig. 3.1 Distribution of sandy soils in the world. (http://www.fastonline.org/CD3WD_40/INPHO/VLIBRARY/U8480E/EN/U8480E0B.HTM)

most sandy soils, bulk density is higher and consequently total porosity is lower than silty and clayey soils. Bruand et al. (2005) reported porosity from 33 percent to 47 percent (bulk density of 1.78 g cm^{-3} and 1.40 g cm^{-3} respectively) in some sandy soils. However, Bortoluzzi (2003) and Lesturgez (2005) observed very high porosity (60 percent; D_b = 1.10 g cm^{-3}) in sandy soils under native vegetation and after recent tillage. Sandy soils retain little water at high water potentials and water content decreases rapidly with the water potential. Kukal and Aggarwal (2002) measured a water content of 0.16 and 0.10 cm^3 cm^{-3} at −33 and −1500 kPa respectively in a sandy loam topsoil (clay content = 10 percent) in India. Sandy soils are highly sensitive to surface crusting (Eldridge and Leys 2003; Janeau et al. 2003; Goossens 2004). Although surface crusts may protect the soil surface from wind and interrill erosion, they favor runoff and rill and gully erosion. Two main types of structural crusts were recognized in sandy soils (Janeau et al. 2003): (i) sieving crusts made of well sorted micro-layers with average infiltrability of approximately 30 mm h^{-1}, (ii) and packing crusts made of sand grains closely packed of with average infiltrability of 10 mm h^{-1}. Unlike other soils, the structure of sandy soils can be easily affected by mechanical compaction over a large range of scales. Usually mechanical compaction preferentially affects large pores resulting from tillage and biological activity. Sandy soils have low structural stability, and a high sensitivity to erosion and crusting.

Sandy soils have low contents of soil organic matter (usually <1 percent) and clay (<18 percent). The clays also commonly have low activity kaolinites. Sandy soils have, therefore, very low cation exchange capacity. Organic matter is relatively rapidly oxidized because of the high aeration and generally warmer temperature. Sandy soils are inherently poorly fertile soils, and added fertilizers are quickly lost through leaching. They have low contents of available nitrogen and micronutrients. Buffering capacity of the sandy soils is low, and they are gradually acidified due to leaching of bases. Due to their dominant mineralogy (generally quartz, kaolinite, iron and aluminium oxides) and their sandy texture, the role of organic matter in the properties of these soils, their potential of productivity and the sustainability of agricultural systems is, thus, fundamental. The control of soil organic matter on chemical (CEC, pH, some cations, such as calcium and magnesium), and physical (porosity, structural stability) properties has often been demonstrated.

3.4 Advantages of Sandy Soils

Despite many constraints, sandy soils have some advantages in their use for cropping. Sandy soils are easy to work with; they are loose and friable and need very little soil working; no-till and mulch tillage suit well; they are well drained, and if not lying on a low land, they are hardly waterlogged. There is no heavy clay to dig through or hard clods to break. Plant roots can penetrate deep and occupy a large soil volume. Sandy soils can be improved without too much effort. Along with peat, vermiculite, manure or compost, sandy soils offer good potting media for nursery seedlings.

3.5 Constraints of Sandy Soils

Sandy soils have a wide range of limitations for agricultural production including nutrient deficiencies, acidity, low water holding capacity and susceptibility to wind erosion on the dunal sands (Hoa et al. 2010). Sandy soils may suffer from water repellency which is caused when hydrophobic 'skins', made from plant waxes and other products from the natural process of plant biodegradation, form around individual sand grains. These waxy skins effectively repel the water from the soil and limit water availability to the crop. Generally, sandy soils are strongly acidic. Their coarse texture, along with low clay and organic matter contents, result in a low water holding capacity and high percolation rate which represent major challenges to agriculture production (Hoa 2008). An average content of only 1.08 percent organic matter was reported from three hundred cultivated sandy soils from Vietnam. Deficiencies of essential nutrients for crop production on sandy soils are often reported. Vinh (2005) conducted an omission experiment which revealed that deficiencies of essential nutrients for crop growth were N > P > K and B > Mo > Zn for red sandy soils and P > N > K and B > Mo > Zn for white sand soils. Sandy soils from more than two hundred cultivated fields in 18 provinces of Vietnam had an average content of 19 mg total Zn kg^{-1} soil which was lower in comparison with the Vietnamese recommendation (200 mg/kg soil) for agricultural soils (Thuy and Ha 2007). Wind erosion occurs more seriously in unprotected sandy soil areas in the dry season due to the lack of soil moisture and strong winds. Wind-blown sand particles spread over cultivated and inhabited areas and create new unstable sand dunes (Hoa et al. 2010). Constraints of sandy soils to agricultural use are: high macroporosity, excessive drainage, low retention of irrigation water, low water holding capacity, high percolation and leaching, high evaporation, low soil organic matter content, low fertility, low retention of added nutrients, high erodibilty and sometimes strong acidity.

3.6 Management of Sandy Soils

The principles of sandy soil management include: choice of suitable plants, improving water holding capacity and ensuring adequate available soil moisture, reducing evaporation, building up of soil organic matter, reducing leaching, integration of water and nutrient management, reducing evaporation, reducing wind erosion, stabilizing dunes, etc. Some soil and crop management practices must be integrated for the best agricultural use of sandy soils. Relevant management practices for sandy soils, particularly under irrigated agriculture, are discussed here.

3.6.1 Selecting Suitable Plants

Some plants thrive well in sandy soils, and some can be grown satisfactorily in them. These plants have some adaptations for sandy soils, such as deep and extensive roots, low consumptive moisture requirement, slow transpiration, tissue water storage and drought tolerance, etc. A list of ornamental, landscape and crop plants known to be suitable for sandy soils is given in Table 3.1.

3.6.2 Tillage

As already mentioned in Chap. 2, tillage is the physical manipulation of soil to optimize conditions for seed germination, seedling establishment and crop growth, and is carried out in a range of cultivation operations, either mechanically or manually. Tillage loosens and aerates soil, facilitates gas exchange and movement of air and water, mixes soil, fertilizers and manures together, kills weeds, breaks up pans and decompacts soil, and enhances root growth. Sandy soils are naturally loose, friable, arable and penetrable, so that the necessity of tillage should be limited there. Still, there can be surface crusts, compaction and presence of pans within the rooting depth, which are needed to be remedied by tillage. Tillage practices should be appropriate to obtain optimum crop yield and to keep the soil in a healthy state. Appropriate tillage practices avoid the degradation of soil properties but maintain crop yields as well as ecosystem stability. Conventional tillage practices may cause organic matter and fertility depletion, and loss of water and erosion in sandy soils, whereas conservation tillage provides very efficient ways of halting degradation and

Table 3.1 Suitable ornamental, landscape and crop plants for sandy soils

Herbs/ground cover		
Common name	Botanical name	General habit
Stonecrop, Gold Moss	*Sedum acre*	Succulent evergreen. Spreads by creeping and rooting along stems. Yellow flowers in spring. Full sun. Many other species of *Sedum* can be used in sandy soils.
Thyme, Common	*Thymus vulgaris*	Evergreen with aromatic foliage and flowers. Full sun. Slightly moist soil required for growth.
Shrubs/ornamental flower plants		
Common name	Botanical name	General habit
Barberry, Japanese	*Berberis thunbergii*	Many thorns. Sun or shade. Profuse, yellow flowers followed by red berries. Many cultivars available.
Bayberry	*Myrica pensylvanica*	Semi-evergreen. Gray berries are fragrant when crushed. Full sun.
Beach Plum	*Prunusmaritima*	Dull purple fruits. Prefers slightly acid soil. Full sun or partial shade.
Blanket flower	*Gaillardia* × *grandiflora*	The blanket flower is a genus of drought-tolerant annual and perennial plants from the sunflower family, native to the North and South America.
Blueberry, Highbush	*Vaccinium corymbosum*	Edible, blue fruits attract birds. Requires acid soil. Red fall color. Full sun or partial shade.
Bush Clover	*Lespedeza thunbergii*	Rosy-purple blooms in late summer. May be killed in winter.
California poppy	*Eschscholzia californica*	*Eschscholzia californica* is a species of flowering plant in the family Papaveraceae, native to the United States and Mexico, and the official state flower of California.
Chokeberry, Red	*Aronia arbutifolia*	White blooms followed by bright red berries in fall. Red fall color. Full sun or partial shade.
Cleome, spider flower	*Cleome hassleriana*	*Cleome* is a genus of flowering plants in the family Cleomaceae. Previously it had been placed in family Capparaceae
Cosmos	*Cosmos bipinnatus*	*Cosmos* is a genus, with the same common name of Cosmos, of about 20–26 species of flowering plants in the family Asteraceae
Firethorn	*Pyracantha coccinea*	Semi-evergreen with bright orange berries. Susceptible to fire blight. Full sun or partial shade.
Gazania	*Gazania* sp.	*Gazania* is a genus of flowering plants in the family Asteraceae, native to Southern Africa.
Shrubs/ornamental flower plants		
Common name	Common name	Common name
Juniper, Creeping	*Juniperus horizontalis*	Many varieties available. Needled evergreen. Full sun.
Juniper, Japanese Garden	*Juniperuschinensis* var. *procumbens*	Dense, needled evergreen. Grows slowly. Full sun.

(continued)

3.6 Management of Sandy Soils

Table 3.1 (continued)

Kerria, Japanese	*Kerria japonica*	Single, yellow flowers in spring. Green twigs through winter. Full sun or partial shade.
Lavender	*Lavandula* sp.	*Lavandula* is a genus of 39 species of flowering plants in the mint family, Lamiaceae. It is native to the Old World.
Penstemon.	*Penstemon* sp.	It is a large genus of North American and East Asian flowering plants.
Potentilla (Shrubby Cinquefoil)	*Potentilla fruticosa*	White to yellow, single flowers through summer. Full sun.
Quince, Flowering	*Chaenomeles speciosa*	White, pink to dark red flowers in spring. Thorns present. Full sun.
Rose, Memorial	*Rosa wichuraiana*	Semi-evergreen foliage. Small, white flowers followed by red fruit. Stems root where they touch, moist ground. Full sun.
Rugosa rose, Japanese rose	*Rosa rugosa*	Rosa rugosa is a species of rose native to eastern Asia, northeastern China, Japan, Korea and southeastern Siberia, where it grows on the coast, often on sand dunes.
Sumac, Fragrant	*Rhus aromatica*	Tall groundcover or shrub. Small, yellow flowers appear before leaves. Fruits turn red in summer. Scarlet fall foliage. Full sun.
Tamarisk	*Tamarix parviflora*	Small, pink flowers in spring. Soft, needle-like foliage. Partial shade.
Weigela	*Weigela florida*	Bright flowers in red, white, or pink. Susceptible to winter die-back. Sun or partial shade.
Yarrow	*Achillea* sp.	Achillea is a genus of about 85 flowering plants in the family Asteraceae. The common name "yarrow" is normally applied to *Achillea millefolium*

Trees

Common name	Botanical name	General habit
Cedar, Red	*Juniperus virginiana*	Needled evergreen. Full sun.
Crape myrtle	*Lagerstroemia indica*	Lagerstroemia indica is a species in the genus Lagerstroemia in the family Lythraceae. From China, Korea and Japan, Lagerstroemia indica is an often multi-stemmed, deciduous tree with a wide spreading, flat topped, open habit when mature.
Crabapples, Flowering	Malus x atrosanguinea, M. floribunda, M. sargentii	Full sun. M. sargentii is the smallest Crabapple available. Attractive flowers, fruits, and red fall color. Choose disease-resistant cultivars.
Hawthorn, Washington	*Crataegus phaenopyrum*	Full sun. Attractive flowers and fruit.
Japanese Pagoda Tree	*Sophora japonica*	Withstands city conditions. Lacks fall color. Attractive flowers in August.
Oak, Pin	*Quercus palustris*	Brilliant red autumn color. Drooping lower branches. Prefers slightly acid soil. Sun.
Oak, White	*Quercus alba*	Slow-growing. Purple-red autumn color. Sun.

(continued)

Table 3.1 (continued)

Common name	Botanical name	General habit
Pine, Austrian	*Pinusnigra*	Grows well on alkaline soils. Good as specimen or mass planting. Sun. Susceptible to Diplodia Tip Blight.
Pine, Pitch	*Pinusrigida*	Scrubby tree good for poor soils. Sun. Susceptible to Diplodia Tip Blight
Pine, Eastern White	*Pinus strobus*	One of the best pines for ornamental use. Sun or partial shade.
Russian Olive	*Elaeagnus angustifolia*	Gray-green foliage and fragrant flowers. Very susceptible to canker. Sun.
Spruce, White	*Piceaglauca*	Tolerates heat and drought well. Sun.
Vines/climbers		
Common name	**Botanical name**	**General habit**
Bittersweet, American	*Celastrus scandens*	Sun to partial shade. Twining vine. Yellow fall foliage with attractive orange berries on female plants(male plants required for pollination). (Invasive)
Grape	*Vitis* spp.	Many cultivars available. Edible fruits. Full sun. Fruit production will not be good on dry sites.
Trumpet Vine	*Campsis radicans*	Clinging vine with orange flowers through summer. Full sun to partial shade.
Virginia Creeper	*Parthenocis susquinquefolia*	Climbing vine. Brilliant red fall foliage. Small, blue fruits attractive to birds. Full sun to partial shade.
Wintercreeper	*Euonymous fortunei*	Semi-evergreen, clinging vine, or small-shrub. Attacked by scale and crown gall. Full sun or shade. Invasive in natural areas.
Crop plants		
Common name	**Botanical name**	**General habit**
Beetroot	*Beta vulgaris*	The beetroot, also known as the table beet, garden beet, red beet, or informally simply as the beet, is one of the many cultivated varieties of beets.
Carrot	*Daucus carota* subsp. *sativus*	The carrot is a root vegetable usually orange in color, though purple, red, white, and yellow varieties exist.
Cucumber	*Cucumis sativus*	The cucumber is a widely cultivated plant in the gourd family Cucurbitaceae. It is a creeping vine which bears cylindrical edible fruits.
Ginger	*Zingiber officinale*	Ginger is the rhizome of the plant, consumed as a delicacy. Medicine or spice. It lends its name to its genus and family Zingiberceae. Other notable members of this plant family are turmeric (*Curcuma longa*) and cardamom (*Elettaria cardamomum*) which can also be grown on sandy soils.
Onion	*Allium cepa*	The onion is the most widely cultivated species of the genus *Allium*. The genus also contains a number of other species variously referred to as onions and cultivated for food. Garlic (*Allium sativum*) also belongs to the genus *Allium*.

(continued)

3.6 Management of Sandy Soils

Table 3.1 (continued)

Peanut or Groundnut	*Arachis hypogaea*	The peanut, or groundnut is a species in the legume family Fabaceae. The peanut was probably first domesticated and cultivated in the valleys of Paraguay.
Potato	*Solanum tuberosum*	The potato is a tuberous crop of the Solanaceae family (also known as the nightshades). The word may refer to the plant itself as well as the edible tuber.
Pumpkins	*Cucurbita* sp.	A pumpkin is a gourd like squash of the genus Cucurbita and the family Cucurbitaceae. It commonly refers to cultivars of any one of the species *Cucurbita pepo, Cucurbita mixa, Cucurbita maxima*, and *Cucurbita moschata*.
Radish	*Raphanus sativus*	The radish is an edible root vegetable of the Brassicaceae family that was domesticated in Europe.
Sweet potato	*Ipomoea batatas*	The sweet potato is a dicotyledonous plant that belongs to the family Convulvulaceae. Its large, sweet-tasting, modified storage tuberous roots are used as vegetables.
Watermelon	*Citrullus lanatus*	Watermelon belongs to the family Cucurbitaceae and is a vine-like (scrambler and trailer) flowering plant which bear large fruits. Sweet when ripe.

Sources: Different web sources including https://www.wariapendi.com.au/hints-tips/plants-for-sandy-soils; http://www.torontomastergardeners.ca/gardeningguides/perennials-for-sandy-soils-a-toronto-master-gardeners-guide

restoring and improving soil productivity. The benefits of various conservation tillage systems including no tillage, minimum tillage or reduced tillage and mulch tillage have been described by many investigators (Mithchel et al. 2009) and also mentioned in Chap. 2 of this book. Details of these systems are not, therefore, included here. Conservation tillage can be employed advantageously for crop production and water and fertility conservation in sandy soils.

3.6.3 Soil Amendments

Several amendments are used for sandy soil management. The chief objectives include: (i) increasing water retention, (ii) increasing nutrient retention by improving surface charges, (iii) decreasing leaching, and (iv) increasing aggregation. According to Lazányi (2005), there are three groups of amendments for sandy soils: (i) organic residues, (ii) farmyard manure and different composts, and (iii) soil modifiers originating from mining industry, e.g. alginite, zeolite, bentonite (Szegi et al. 2008). Natural and synthetic hydroabsorbents and biochar constitute other kinds of amendments.

3.6.3.1 Manuring

Addition of manures increases fertility of sandy soils by slowly recycling nutrients. Inorganic fertilizers, even if added in low installments, are rapidly lost from sandy soils due to excessive natural drainage. On the other hand, manures are not only nutrient carriers which release nutrients slowly and reduces nutrient loss, but they are also soil modifiers which help build up soil organic matter and improve the soil structure and water holding capacity. Sandy soils have low clay contents; their cation exchange capacity depends mainly on organic matter which could also serve as the major controlling factor for water retention (Sam and Vuong 2008). But organic matter content is also low in sandy soils. Hoa et al. (2010) suggested that the systematic use of farmyard manures, crop residues, green manures, and alley cropping could substantially improve the situation.

Enhancing organic matter content has been the key to alleviating the soil moisture and nutrient retention problems in sandy soils. Input of organic residues can be done *in situ* through the adoption of agricultural systems that provide organic input continuously. These systems that produce significant quantity of both aboveground and belowground litter to compensate for the harvestable parts include the incorporation of green manure crops. Inclusion of a green manure legume to the crop rotation can be done. The inputs of organic materials can also be brought in from outside the system, such as application of composts, animal manure, sludges, and some organic industrial by-products. Organic fertilizers contain low concentration of nutrients. In one hectare of field crops, 10–20 t of farm yard manures or composts are needed. Applying continuous inputs of organic residues and minimization of soil disturbances are measures that have been shown to restore soil organic matter in sandy soils. Residues resistant to decomposition, i.e. those that contain large amount of recalcitrant C compounds, such as lignin and polyphenols, and possess high C:N ratios in general, can bring about greater soil organic matter accumulation relative to easily decomposable residues (Vityakon 2005).

Organic fertilizers can improve and maintain soil fertility by their content of almost all the plant nutrients, releasing these nutrients in available forms slowly in soil and counteracting the acidifying effects of inorganic fertilizers. Organic fertilizers can also retain soluble nutrients in soil against leaching by increasing cation exchange capacity (CEC) and base saturation percentage (BSP). Such an example was provided by Bakayoko et al. (2009) who observed that cattle manure increased CEC of their experimental soil from 1.7 to 12.75 cmolc kg^{-1} and the BSP from 47 to 80. They evaluated the effects of cattle and poultry manure on the organic matter and nutrient contents of unsaturated sandy ferrallitic soils under cassava (*Manihot esculenta* Crantz) cultivation. They reported that poultry manure had the highest organic C, N, P, K, Ca contents and the lowest C:N ratio. Increasing effective CEC and nutrient concentrations by amendments of manures or composts was also reported by McClintock and Diop (2005). Improving fertility status of sandy soils with cattle manure was reported by some investigators (Nyamangara et al. 2001; Hao et al. 2004). However, poultry industry has grown greatly throughout the world and broiler houses produce huge quantity of wastes including chicken feces,

3.6 Management of Sandy Soils

urine, bedding materials and feather. These wastes are disposed of regularly to clean the houses. Composting these wastes for several weeks under a cover converts them into a valuable natural product of high fertilizing and soil conditioning value and makes their disposal easier. Another way of increasing organic matter and nitrogen status of sandy soils is green manuring. It is an agronomic practice of a short duration crop being grown densely between two harvestable crops or during a short fallow period in rotations and plowing under the live green crop at a stage when the stems are still slender and succulent. The crop is generally a legume (for example, lucerne, white clover, red clover, vetches, soybean, etc.) with high potential of vegetative growth and nitrogen fixation. Usually, the crop is plowed under at the flowering stage before two or more weeks of planting the next crop so that the residues are decomposed well and the organic matter is intimately mixed with the soil. Some of the fixed nitrogen is made available in the meantime. green manuring contributes to the build-up of organic matter in soils. As has been noted earlier, organic matter improves physical conditions, biological activity, nutrient retention and moisture content of sandy soils. According to FAO (2011), these aspects of green manuring are most significant in sandy soils. Moreover, a green manure crop can act as a cover crop when it occupies the field and can conserve soil against erosion by wind and water.

3.6.3.2 Bentonite Amendments

Sandy soils have a small amount of clay, and the types of clay minerals that are present in them have low activity, small surface area and low charge density. Additionally, sandy soils have very little organic matter content, and hence sandy soils have low cation exchange capacity. As a result, there is high risk of nutrient leaching (Blanchart et al. 2007), high loss of added fertilizers, unsatisfactory growth and yield of crops and reduced fertilizer use efficiency. Amendment of sandy soils with materials containing high activity clays may increase CEC, reduce nutrient loss by leaching and increase nutrient retention by holding nutrient ions on their surfaces. Bentonite is such a natural material usually originating from volcanic substances and contains several clay minerals, predominantly montmorillonite of the smectite group (Pártay et al. 2006). It has high surface area, high CEC and high water holding capacity. Natural bentonites are processed and ground for use for different purposes. Among several types of bentonites, calcium bentonite is usually used to improve sandy soils because of its higher stability (www.natureswayresources.com/nl/44Bentonite.pdf; accessed on 5.7.2016). It stores more water and releases water more easily. Bentonite improves nutrient holding capacity of soils and reduces leaching losses (Noble et al. 2000). So, bentonite can improve fertility and productivity of sandy soils (Noble et al. 2005). Croker et al. (2004) reported that bentonite amendment was found to increase CEC of sandy soils and improve plant biomass in some studies. Gillman (2007) observed that bentonite application increased CEC and cation sorption and reduced NH_4^+ leaching. Bentonite can also retain other cations like Ca^{2+}, K^+ and Mg^{2+} in soil (Berthelsen et al. 2007). Some authors (Usman et al. 2005; Lazányi 2005) suggest that bentonite stimulates biological activity primarily by improving water holding capacity and water content of the soil. However, both positive and negative effects of bentonite, depending on the rate of application, on plant growth were reported by Schnitzler et al. (1994). They observed that mixing bentonite with soil of salad seedlings at 3 percent increased fresh weight of seedlings, but a higher dose depressed growth.

3.6.3.3 Hydroabsorbent Amendments

Hydroabsorbents or hydrogels can play an important role in increasing water availability in sandy soils. The hydrogel also can reduce the effects of salts in the soil matrix. There are generally three classes of hydrogels: natural polymers, semi-synthetic and synthetic polymers. Natural polymers are starch based polysacharides commonly derived from crops, semi-synthetic polymers are initially derived from cellulose and then combined with forms of petrochemicals, and, finally, synthetic hydrogels. Complex hydroabsorbents, such as Terra Cottem, which is a mixture of more than 20 components, assist plant growth processes in a synergetic way (Dewever and Ottevaere 2003). Hydroabsorbents can increase germination of seeds, survival rate of germinated seedlings, and water uptake (Specht and Harvey-Jones 2000). Sarapatka et al. (2004) observed that hydrogels improved the biological and biochemical properties of sandy soils.

Gel-forming polymers were first introduced for agricultural use in the early 1980s. Application of hydrogels has an influence on the improvement of aggregate content and increase in water retention in the soil at all suctions (from 0 to 15 atm), and the available water is significantly increased (El-Hady and Abo-Sedera 2006). The biggest increase of retention capacity is in the range of pF 0–2.2 which may influence retention of gravitational water and creation of unfavorable conditions for the plants (Leciejewski 2009). Hydrogel influences the increase of efficiency of water usage and decrease of irrigation frequency (Sivapalan 2006). El-Hady et al. (2009) stated that the applied Acrylamide hydrogels also had influence on increasing OM, organic carbon, total nitrogen, phosphorus and potassium in the treated soil.

Orikiriza et al. (2009) applied Luquasorb hydrogel at two levels 0.2 and 0.4 percent w/w and grown seedlings of *Eucalyptus grandis, Eucalyptus citriodora, Pinus caribaea, Araucaria cunninghamii, Melia volkensii, Grevillea robusta, Azadirachta indica, Maesopsis eminii and Terminalia superba* in green house. Results suggested that hydrogel amendment enhanced the efficiency of water uptake and utilization of photosynthates of plants grown in soils which have water contents close to field capacity. Djurovic et al. (2011) added a hydrogel (potassium acrylate) in the dosage of 0.5 percent which significantly influenced the reduction in water consumption and resulted in better rooting and development of the barley root system. The use of some novel and efficient crop nutrient-based super absorbent hydrogel nanocomposites (SHNCs) is currently becoming increasingly efficient in improving the crop yield and productivity due to their water retention properties. Shahid et al. (2012)

3.6 Management of Sandy Soils 49

observed that amendment with 0.1, 0.2, 0.3 and 0.4 percent (w/w) of SHNC enhanced the moisture retention significantly at field capacity compared to the untreated soil. Seed germination and seedling growth of wheat was found to be significantly improved with the application of SHNC.

Depending on synthesizing conditions, types and covalent bond densities, hydrogels can absorb water up to 1000 times of their weight. Sodium polyacrylate, which is derived from the polymerization of acrylic acid blended with sodium hydroxide (($C_3H_3NaO_2$)n), is the most widely used hydrogel today. Other commonly used hydrogels are poly (acryl-amide) copolymer, poly (vinylalcohol) copolymer (PVAL), cross-linked poly ethylene-oxide and cross-linked carboxy-methyl cellulose. El-Hady et al. (2009) stated that the application of the polyacrylamide copolymer also resulted in an increase in the content of organic carbon, total nitrogen and accessible forms of nitrogen, phosphorous and potassium. Hydrogels also increase the efficiency of water use and reduce the frequency of irrigation (Sivapalan 2006; Durovic et al. 2012).

3.6.3.4 Biochar

Biochar is a charcoal-like material produced by pyrolysis of biomass (saw dust, grasses, wood chips, straw, corn stover, peanut shells, olive pits, bark, sorghum, and sewage wastes, etc.) by heating in absence of oxygen at a temperature between 400 and 500 °C. When biomass is burnt in the absence of oxygen, pyrolysis occurs and the biomass can be turned into a liquid ('bio-oil'), a gas (syngas) and a high-carbon, fine-grained residue – biochar (Winsley 2007). Biochar is actually a finely ground charcoal with some similarities to activated charcoal. It has a very high surface area which contributes to its applicability in various industrial, environmental and agricultural fields. Biochar increases the capacity of sandy soils to hold nutrients against leaching. Adding fertilizers after biochar amendment is very effective to prevent loss of added fertilizers. It increases organic matter content permanently and improves water holding capacity. Biochar improves soil structure and water retention, enhances nutrient availability, lowers acidity, and reduces the toxicity of aluminium to plant roots and soil microbiota. Modern experimental research demonstrates that biochar application can substantially lift the productivity of crops, such as soybeans, sorghum, potatoes, maize, wheat, peas, oats, rice and cowpeas depending on soil and crop type, biochar concentrations, and nutrient levels, so optimal applications would need to be tailored to local conditions (Winsley 2007). Biochar can reduce nitrogen fertilizer requirements and nitrous oxide emissions (Baum and Weitner 2006). Biochar has a carbon content of about 70–80 percent, which can be permanently sequestered in soil. Biochar may have the potential to increase atmospheric carbon dioxide uptake in the form of glomalin, a major component of humus produced by plant mycorrhizal fungi. Biochar offers an extremely high surface area to support microorganisms that can catalyze processes related to the reduction of nitrogen loss and the increase in nutrient availability.

Mulcahy et al. (2013) observed that 30 percent (v/v) biochar applied in sandy soils around the root zone significantly increased the resistance of tomato (*Lycopersicon esculentum*) seedlings to wilting. Tomato is an important crop which is highly sensitive to moisture stress in the early growth stage. Addition of biochar obtained from fast pyrolysis of hardwood increased the water-holding capacity of sandy soils (Basso et al. 2013). It has been observed that the optimum dose of biochar for crop yield should be determined through trials. Experiments have found that, for some crops and soils, the optimum dose of biochar is around 0.5 Mg ha^{-1}. It can reduce leaching of soluble substances from agricultural soils (Lehmann et al. 2006) by its strong adsorption affinity for soluble nutrients, such as ammonium (Lehmann et al. 2002), nitrate (Mizuta et al. 2004), phosphate, and other ionic solutes (Radovic et al. 2001). This affinity of biochar for ionic solutes can reduce run-off in agricultural watersheds and hypoxia of waterways. Biochar is very stable in the environment; so, it can effectively contribute to carbon sequestration. Charcoal may be stable for several tens of millions of years in the environment. Large accumulations of charred material with residence times in excess of 1000 years have been found in soil profiles (Forbes et al. 2006; Glaser et al. 2001). Charcoal deposits of more than 9500 years have been found in wet tropical forest soils in Guyana (Hammond et al. 2007), and up to 23,000 years old in Costa Rica (Titiz and Sanford 2007).

3.6.4 Fertilizer Application

As nitrate is rapidly leached from sandy soils to the groundwater and poses risk for water pollution, nitrate form of nitrogen fertilizers should be avoided. A nutrient-management plan based on leaching losses and retention ability of the soil should be followed. Urea and ammonium containing fertilizers are converted to nitrate easily in aerobic soils, so these fertilizers are also lost from sandy soils. Nitrogen fertilizers that resist nitrification, such as urea formaldehyde, sulfur-coated urea can be used. Alternatively, application of nitrification inhibitors is used to mitigate nitrogen loss from the ecosystem (Dinnes et al. 2002; Subbarao et al. 2006). Dicyandiamide (DCD) and nitrapyrin (2-chloro-6-(trichloromethyl)pyridine) are the most frequently used commercial nitrification inhibitors in agriculture (Subbarao et al. 2006) and allylthiourea (ATU) has been widely employed in nitrification research. All three inhibitors exert their effects indirectly; while nitrapyrin and ATU are believed to act by chelating copper; the mode of action of DCD is poorly characterized (Subbarao et al. 2006). DCD has been previously suggested to either prevent ammonia uptake or utilization, or act as a copper chelator (Subbarao et al. 2006). Although ammonia oxidizing archaea are numerically, and, in some soils, functionally dominant over their bacterial counterparts (Gubry-Rangin et al. 2010; Stopnišek et al. 2010), investigation of nitrification inhibitors has focused almost exclusively on ammonia oxidizing bacteria. Sandy soils may also be low in other macro- and micro-nutrients. In small scale farms, addition of nutrients mainly as organic manures along with supplemental inorganic fertilizers should be a good practice.

3.6 Management of Sandy Soils

Table 3.2 Rates of nitrogen fertilizers for corn and potato in a sandy soil

Organic matter, percent	Corn		Potato (yield goal, t ha^{-1})		
	Irrigated	Non-irrigated	5–7	7–9	9–12
	Amount to apply, kg ha^{-1} N				
<2.0	225	145	130	170	225
2.0–4.9	180	120	100	140	170
5.0–10.0	145	112	80	112	140
>10.0	90	90	55	85	112

Data converted from: Nitrogen management on sandy soils. http://www.soils.wisc.edu/extension/pubs/A3634.pdf

Too much inorganic fertilizers in sandy soils may cause environmental problems. Fertilizer strategies, such as split application, delayed basal application and varied timing of application to synchronize with plant demand have improved plant production by minimizing nutrient leaching (Sitthaphanit et al. 2009). However, the efficacy of fertilizer strategies may be limited in rainfed areas by high intensity rainfall immediately after fertilizer application (Sitthaphanit et al. 2009).

Fertilizer rates are affected by (i) the crop to be grown, (ii) soil organic matter content, (iii) soil pH and CEC, (iv) soil fertility, (v) yield goal, and (vi) the amount and frequency of irrigation. The nitrogen fertilizer rates that can be given to corn and potato in sandy soils are shown in Table 3.2.

3.6.5 Irrigation

Selection of an irrigation method depends on such factors as natural conditions, crop type, available technology, previous experience with irrigation, labor requirement and cost involvement. Natural conditions for suitability of irrigation method include soil type, slope, climate, water availability and water quality. For example, sandy soils have low water storage capacity and a high infiltration rate; they need frequent but small irrigation applications, particularly if the sandy soil is also shallow. Drip and sprinkler irrigation systems are more suitable than surface irrigation systems. These systems are also more preferable in sloping surfaces. They do not need any land leveling. However, strong winds can disturb the distribution of sprinkler water which may create irregular crop stands, and drip is more preferable to sprinkler. Water application efficiency is generally higher with sprinkler and drip irrigation than surface irrigation, so these methods are preferred when water is in short supply. However, the presence of sediments in irrigation water makes it unsuitable for use in drip and sprinkler systems. The sediments may clog the drip or sprinkler nozzles. If the irrigation water contains dissolved salts, drip irrigation is particularly suitable, as less water is applied to the soil than with surface methods. Sprinkler systems are more efficient than surface irrigation methods in leaching out salts. Surface irrigation can be used for all types of crops. Sprinkler and drip irrigation systems involve high capital investment. They are mostly used for high value cash crops, such as vegetables and fruit trees. Among the surface irrigation systems, furrow irrigation is generally used for row crops, basin irrigation for orchards and border flooding for close growing crops. Surface irrigation methods do not need sophisticated equipments and high degree of technical know-how.

Sprinkler and drip irrigation systems have high water use efficiency but have not become popular enough because their installations need some technical knowledge and their primary installation costs are high. However, some small-scale vegetable farmers have discovered some sort of low cost indigenous drip systems and found them profitable. Water is applied in the root zone where it is most needed. Control on the amount and distribution of water in the field is very efficient in these irrigation systems. Water can be applied at the times and rates of plant demand, and the loss of water is minimized. This is very crucial in arid regions where most sandy soils are found and where water is always scarce. Fertilizers can be dissolved with irrigation water in low concentrations in the central water storage tanks and distributed along with water evenly throughout the field—a system known as 'fertigation'. Loss of added fertilizers can also be reduced and fertilizer use efficiency is improved. So, these systems, although having high cost at the beginning, become profitable in the long run. On the other hand, surface irrigation systems, such as flooding or furrow, need less sophisticated equipment but involve higher amount of water and higher loss of water as well. Surface irrigation systems also involve the risk of leaching loss of nutrients and salinization of soil if drainage is inappropriate and inadequate. Irrigation scheduling is also crucial in sandy soils because sandy soils cannot hold much water and have tendencies of becoming droughty. Therefore, sandy soils need small amount of water at frequent intervals. Late irrigation often reduces growth and the yield of crops and use efficiency of water and agrochemicals.

The drip system has some advantages over sprinkler and surface irrigation systems. Water is applied in drip irrigation in small drops directly at the root zone of the plant. Water is restricted only in a small area around the base of the plant. Roots easily absorb water and little water is lost by evaporation or percolation. There is little loss of nutrients by leaching too. So, water and nutrient use efficiency is the highest in drip irrigation system. Moreover, continuous water dropping dilutes salts in the rhizosphere and minimizes salt injury to plants. In the arid region, many sandy soils are also saline. Tiwari et al. (2003) considered drip irrigation as the most effective way of irrigating for higher yield of crops. Bryla et al. (2003) reported that drip irrigation, based on evapotranspiration, improved yield of faba bean and its water use efficiency. Drip irrigation is mainly a surface distribution system of water, but some investigators also used subsurface drip systems. Lamm and Trooien (2002) suggested that subsurface drip system is more water use efficient in sandy soils. They also found that subsurface drip irrigation reduced the water requirement often more than 50 percent. In a subsurface drip irrigation system developed for peanut, cotton, and corn, Sorensen et al. (2001a, b) observed that there was little difference in pod yield of peanut due to drip tube spacing, the amount of irrigation water applied over several treatments or emitter spacing. Some investigators (Lamm and

3.6 Management of Sandy Soils

Trooien 2005; Neelam and Rajput 2007) attempted to evaluate the appropriate depth of placement of drip lines and found out that it should be shallower than 10 cm under sandy soils for a crop like potato. Grabow et al. (2002) showed that yields of cotton under subsurface drip irrigation were higher than under sprinkler irrigation. Sprinkler irrigation system has also high water use efficiency, but as water travels through the air from the rotating nozzles, some water is evaporated and some tiny water drops are carried away by the wind. Therefore, sprinkler irrigation system involves some loss of water. Thus, an even distribution of water over the crop field may not be achieved by sprinkler in strong wind environments. Now-a-days, irrigation systems are being automatized. Some investigators examined the efficiency of automation of irrigation systems consisting of a soil moisture sensor like tensiometer, a control system and other components. Muñoz-Carpena et al. (2003) reported the use of switching tensiometers in various sandy soil applications. These probes can be installed permanently at suitable points of the crop field and when the soil moisture potential decreases beyond a threshold level, the control system automatically releases water. However, in sandy soils sensing devices may suffer from lack of contact with the soil matrix (Muñoz-Carpena and Dukes 2005).

Irrigation of tomatoes on sandy soils in some areas is primarily done by means of a seepage irrigation system. In this method, perching a water table at 35–50 cm below the ground surface is done and water table is maintained throughout the growing season by supplying water to lateral ditches either through a main supply ditch or, more commonly, through a buried PVC manifold. Spacing of the lateral ditches varies from 3.5 m to around 12 m according to field conditions and grower's production practices. Seepage irrigation can require from 5 to 13 gallons per minute per acre to maintain the water tables. The challenge of a seepage irrigation system is to maintain the water table just below the root zone to provide the crop with enough soil moisture to meet evapotranspiration requirements, while at the same time avoiding root zone saturation, which may negatively impact the crop. Field evaluations of soil moisture content in the crop root zone have shown non-uniform water distribution across the field. The plant rows closer to the irrigation furrows tend to be wetter than the rows in the areas farther from irrigation furrows (center of the beds). A properly managed shallow water table can be a very important source of water to supply crop needs and save water and energy. Sub-irrigation involving controlled water tables has been practiced in north central United States (Fig. 3.2).

3.6.6 Cover Crops

A cover crop is any crop composed of annual, biennial, or perennial herbaceous plants grown in a pure or mixed stand during all or part of the year in order to provide soil cover. When cover crops are grown to conserve soil moisture, and reduce nutrient leaching and erosion following a main crop, they are termed "catch crops." They provide cover during the fallow period between two crops and can reduce nutrient leaching from sandy soils. For example, planting cereal rye following corn

Fig. 3.2 Sprinkler irrigation has a high water use efficiency. (Image courtesy of USDA NRCS)

harvest helps to scavenge residual nitrogen. Legumes, such as clovers, vetches, medics, and field peas, are usually chosen for winter cover crops because they have the added benefits of nitrogen fixation. Legumes, such as cowpeas, soybeans, annual sweet clover, sesbania, guar, crotalaria, or velvet beans may be grown as summer green manure crops. Important non-legumes, such as sorghum, sudangrass, millet, forage sorghum, or buckwheat can be grown as summer cover crops to provide biomass, smother weeds, and improve soil tilth. Figure 3.3 shows that a cover crop is being killed during transplanting tomatoes.

Cover crops are often killed and the residues are left on soil to recycle nutrients, protect against erosion and conserve soil moisture. Killed cover crop residues can improve soil moisture retention properties including reduced evaporation by altering net radiation, reduced surface temperature, and increased overall water infiltration rate as compared to bare, cultivated soils. Soil moisture is affected by cover crops by building soil organic matter, which increases water holding capacity, and intercepts radiation to slow evaporation loss (Lu et al. 2000). Mulched and un-tilled soils can reduce surface evaporation, retain higher moisture and create a favorable environment for root development. Killed cover crop residues left on the surface

3.6 Management of Sandy Soils

Fig. 3.3 Transplanting tomatoes into mechanically killed hairy vetch. cover crop (Image courtesy of USDA NRCS)

have the most impact on the upper 30–40 cm of the soil profile (Bergamaschi and Dalmago 2006). Uchanski and Rios (2012) observed that moisture content at the 30 cm depth of soil ranged from 50 to 60 percent in the killed cover crop treatments throughout most of the season. In comparison, the control and wheat straw ranged from 30 to 50 percent soil moisture content. However, cover crops can cause serious problems if they are not managed properly, they can deplete soil moisture, and as an intercrop they can compete with the cash crop for water, light, and nutrients. In sandy soils, late killing of a winter cover crop may result in moisture deficiency for the main summer crop.

3.6.7 Mulching

Mulching refers to the covering of soil with a variety of organic and inorganic materials and is an essential element in sandy soil management. It reduces evaporation, increases water retention, and suppresses weeds. Mulch keeps the soil cooler in the warm season and warmer in the cold season, reduces run-off and erosion. Mulches can also prevent over-heating of sandy soils by scorching solar radiation. Mulches also keep fruits and vegetables such as strawberries, cucumbers and squash off the soil, keeping them clean, dry and less prone to disease. It provides efficient protection against drought. Figure 3.4 shows straw mulching in a crop.

Many different organic residues, such as straw, leaves, wood chips, saw dust, seed husks, grass clippings, compost and aged manure can be used as organic mulches. Inorganic substances, such as stones, cobbles, brick chips and plastic sheets of variable thickness and color can be used as inorganic mulches. They are effective in particularly exposed and hot positions. Black plastic mulch is most

Fig. 3.4 Straw mulching in a young sweet gourd plant to reduce evaporation loss in a sandy soil. (Image courtesy of late Azad Kamal)

Fig. 3.5 Stubble mulching is effective in reducing wind erosion in sandy soils. (Image courtesy of the Reduced Tillage Linkages)

widely used for two main reasons: early season soil warming and weed control. This mulch alone has enabled producers to extend their growing season and grow many warm season vegetables. Before applying mulch, the soil should be prepared well and thoroughly watered, weeds are removed and the mulch is spread across the soil surface. Figure 3.5 shows stubble mulching used to reduce wind erosion of soil. Figure 3.6 presents plastic mulching for growing strawberry.

Mulching can protect soils against surface crusting in sandy soils. Surface crusting is a common phenomenon in sandy soils of arid and semi-arid regions. Crusting the surface soils impedes seedling emergence (Valenciano et al. 2004; Voortmana et al. 2004), restricts infiltration (Janeau et al. 2003) and favors rill and gully erosion

3.6 Management of Sandy Soils

Fig. 3.6 Strawberry grown on plastic mulch. (Image courtesy of The Noob Farmers)

(Valentin et al. 2005). Much research work on soil crusts on the loess belts in the United States (Ruan et al. 2001) has been done. There the soils are both highly productive and prone to crusting. Recently, investigations revealed high sensitivity of coarse-textured soils to surface crusting in Southern Niger (Valentin et al. 2004), in Northern Burkina Faso (Ribolzi et al. 2003, 2005), in Northeastern Thailand (Hartmann et al. 2002) and in many other parts of the world as Northern China (Shirato et al. 2005), Zimbabwe (Burt et al. 2001), etc. Runoff produced by soil crusts tends to concentrate and form gullies even in sandy soils (Descloitres et al. 2003). Sandy soils are therefore generally eroded not only by sheet but also by gully erosion, even for very gentle slope gradients (Valentin et al. 2005). Mulching of crop residues or straw is generally recommended to protect against crusting by rain drop impact.

3.6.8 Stabilization of Sand Dunes

Flow of wind or water over lands with loose sands, as in deserts and sea or lake shores, create waves of low hills of sands known as dunes. There are different sizes and forms of dunes, and the types depend on such factors as topography, wind direction and velocity, sand particle sizes and the presence of vegetation. There are three main types of dunes depending on their orientation: longitudinal, traverse and parabolic. When dunes are oriented parallel to the prevailing direction of wind they are called longitudinal dunes. On the other hand, traverse dunes are oriented perpendicular to the prevailing wind direction. Parabolic dunes are known for their particular shapes. Dune sands always move; a hill is broken and another is built as if small hills of sands are creeping. They are unstable ecosystems though some dunes have been stabilized by vegetation. Together, active and vegetation-stabilized air-borne sand deposits cover about 6 percent of the global land surface area (Pye and Tsoar 2009).

Dune stabilization becomes a necessity in some circumstances where blown sand damages crops, homesteads, roads and rail roads, etc. Some desertified areas need to be restored to improve food security and livelihood of native population. Stabilizing a dune could be accomplished mechanically, chemically or biologically. Several methods to stabilize dune sand are: (a) transposing, (b) planting, (c) paving, (d) paneling, (e) fencing, and (f) oiling, etc. Successful dune stabilization has been accomplished in large areas of the USA, China, Sahara, West Africa, Australia, New Zealand, Great Britain, Kuwait, Pakistan, India, Madagascar and Brazil (Desert Research Institute (2011). Most dune stabilization research has occurred in desert regions where active dunes have been artificially stabilized to enhance the anthropogenic utility of the soils (Duan et al. 2004; Zhang et al. 2008). Large tracts of sand dunes in the southern Canadian Prairie region have been vegetated over the last 200 years, representing a major transformation of these ecosystems. Currently, less than 1 percent dune fields are mobile and active there, and it is expected that complete stabilization would be achieved soon (Hugenholtz et al. 2010).

Stabilization of dunes is a difficult job because it involves reduction of wind speed, the diversion of wind direction, modification of ground configuration and binding and trapping of shifting sands. However, it can be achieved by several long-term structural and vegetative measures, vegetative measures being more effective and longer lasting. Vegetation establishment in dry loose sands is also difficult, but some grasses, shrubs and trees have been used successfully in many parts of the world including the United States and China. The use of the American beach grass for dune stabilization is common in the United States. Some pines including red pine, white pine and jackpine can be used for the purpose. Other plant species include cottonwood, shrub willow, dogwood, birch, big tooth aspen, etc. The following plant species are suitable for different dune conditions (Osman 2014):

Wet spots of shifting sands: *Hibiscus tiliaceus*, *Populus* spp., *Salix exigua*, *Salix nigra*, *Cornus spp.*
Very mobile strip dunes: *Prosopis juliflora, Aristida pungens*
Deflation zones: *Leptadenia pyrotechnica, Aristida pungens* and *Panicum turgidum*
More stable zones: *Acacia raddiana, A. senegal, Balanitesae gyptiaca, Euphorbia balsamifera* and *Persica salvadora*
Coastal dunes: Halophytes including *Nitraria retusa, Tamarix aphylla, T. senegalensis, Casuarina equisetifolia, Atriplex halimus, A. nummularia* and *Zygophyllum* spp.
Dunes on lake shores: *Populus* spp., *Betula* spp., *Pinus strobus, Quercus alba, Populus grandidentata.*

3.6 Management of Sandy Soils

Fig. 3.7 Straw checkerboard to stabilize sand dune. (Image courtesy of The China Daily)

Initially, the movement of sand is controlled by establishing dune grass. After about two years when the grass cover has not yet become dense, tree seedlings are planted within dune grass usually at a spacing of 2 m × 2 m. Mulching with straw, branches, stalks, plastic film or acrylic fiber and mesh can be done to prevent saltation and to protect seedlings against being blown away by the wind. Structural measures include fences, bunds, and more popularly straw checkerboards. Straw checkerboards (Fig. 3.7) are used in many countries for efficient stabilization of dunes. Rice and wheat straw are used in China to form the checkerboards 10–20 cm high with grid sizes of 1 m × 1 m to 2 m × 2 m. Bleeker et al. (2013) reported that smaller grid sizes were used in localities of strong winds. According to Li et al. (2004), integrating straw checkerboard and vegetation or revegetation in stabilized dunes are effective measures of desert ecosystem rehabilitation.

Some chemical substances, natural and synthetic, have been used with varying success for the stabilization of sand dunes and protection of sandy soils against wind erosion. These substances include lime, bentonite, latexes, resins, bitumins, polymers, etc. Wang et al. (2005) reported that emulsified asphalt, emulsified crude oil, synthetic rubber and synthetic resin were very effective in fixing moving sand. Ibrahim et al. (2002) and Gong et al. (2001) reported the use of various polymers in fixing sands. Some investigators observed the effectiveness of water-based polymer emulsions (Alkhanbashi and Abdalla 2006), polyvinyl alcohol and polyvinyl acetate emulsions (Newman et al. 2005; Han et al. 2007), polyacrylamide (Yang and Zejun 2012), etc. in controlling sand movement. Many other inorganic and organic chemical compounds and complexes have also been used for the protection of sands

(Moghadam et al. 2015). These chemical substances bind single grained sands together by the process of adhesion, and form a crust on the surface. These crusts resist wind erosion of sandy soils. However, integration of chemical sand stabilization with other methods, such as fencing, mulching and vegetation development can be more effective.

Study Questions
1. Give an account of the distribution of sandy soils. Mention the advantages and disadvantages of agricultural use of sandy soils.
2. Make a list of crop plants suitable for sandy soils.
3. Discuss the most efficient way of irrigation and fertilizer management for sandy soils.
4. Explain the feasibility of using inorganic and organic amendments for water and nutrient management in sandy soils.
5. Discuss tillage and mulching in relation to management of sandy soils.

References

AlKhanbashi A, Abdalla SHW (2006) Evaluation of three waterborne polymers as stabilizers for sandy soil. Geotech Geol Eng 24:1603–1625

Bakayoko S, Soro D, Nindjin C, Dao D, Tschannen A, Girardin O, Assa A (2009) Effects of cattle and poultry manures on organic matter content and adsorption complex of a sandy soil under cassava cultivation (*Manihot esculenta*, Crantz). Afr J Environ Sci Technol 3(8):190–197

Basso AS, Miguez FE, Laird DA, Horton R, Westgate M (2013) Assessing potential of biochar for increasing water-holding capacity of sandy soils. GCB Bioenergy 5:132–143

Baum E, Weitner S (2006) Biochar application on soils and cellulosic ethanol production. Clean Air Task Force, Boston

Bergamaschi H, Dalmago GA (2006) Brazil: Can the no-tillage system affect the use of irrigation in tropical and subtropical cropping areas? Food Agriculture Organization. http://www.fao.org/ag/AGP/agpc/doc/publicat/no_tillage_system/no_tillagessystem.htm#_ftn1. Retrieved March 5, 2011

Berthelsen S, Noble AD, Ruaysoongnern S, Huan H, Yi J (2007) Addition of clay based ameliorants to light textured soils to reduce nutrient loss and increase crop productivity. In 'Management of Tropical Sandy Soils for Sustainable Development. Proceedings of the International Conference on the Management of Tropical Sandy Soils, Khon Kaen, Thailand, Nov. 2005', FAO Regional Office for Asia and the Pacific, Bangkok

Blanchart E, Albrecht A, Bernoux M, Brauman A, Chotte JL, Feller C, Gany F, Hien E, Manlay R, Masse D, Sall S, Villenave C (2007) Organic matter and biofunctioning in tropical sandy soils and implications for its management. In: Management of Tropical Sandy Soils for Sustainable Development. Proceedings of the International Conference on the Management of Tropical Sandy Soils, Khon Kaen, Thailand, Nov. 2005. FAO Regional Office for Asia and the Pacific, Bangkok, pp 224–241

Bleeker T, Miceli C, Nieuwsma J, Prather E (2013) Efficacy of sand fences in stabilizing a steep active dune blowout, Castle Park Reserve, Michigan. Dunes Research Report No. 4, Department of Geology, Geography and Environmental Studies, Calvin College, Grand Rapids, Michigan

Bortoluzzi EC (2003) Nature des constituants, propriétéschimiques et physiques des sols. Influence de la gestion de sols sableux au Sud du Brésil. ThèseInstitut National Agronomique, Grignon (INA PG), Paris

References

Bruand A, Hartmann C, Lesturgez G (2005) Physical properties of tropical sandy soils: a large range of behaviours. Session 4 Physical properties of tropical sandy soils. Management of Tropical Sandy Soils for Sustainable Agriculture "A holistic approach for sustainable development of problem soils in the tropics". 27th November – 2nd December 2005, FAO, KhonKaen, Thailand

Bryla DR, Banuelos GS, Mitchell JP (2003) Water requirements of subsurface drip-irrigated faba bean in California. Irrig Sci 22(1):31–37

Burt R, Wilson MA, Kanyanda CW, Spurway JKR, Metzler JD (2001) Properties and effects of management on selected granitic soils in Zimbabwe. Geoderma 101:119–141

Croker J, Poss R, Hartmann C, Bhuthorndharaj S (2004) Effects of recycled bentonite addition on soil properties, plant growth and nutrient uptake in a tropical sandy soil. Plant Soil 267:155–163

Descloitres M, Ribolzi O, Troquer Y (2003) Study of infiltration in a Sahelian gully erosion area using time lapse resistivity mapping. Catena 53(3):229–253

Desert Research Institute (2011) Sand dune stabilization methods – a global tour. Keeler Dunes Meeting August 24, 2011. Great Basin Unified, Air Pollution Control District

Dewever F, Ottevaere D (2003) Terra Cottem in growing media. Proceedings of the international peat symposium in horticulture. Additives in Growth Media, Amsterdam

Dinnes DL, Karlen DL, Jaynes DB, Kaspar TC, Hatfield JL, Colvin TS, Cambardella CA (2002) Nitrogen management strategies to reduce nitrate leaching in tile-drained Midwestern soils. Agron J 94:153–171

Djurovic N, Stricevic R, Pivic R, Petkovic S, Gregoric E (2011) Influence of hydrogel on water Conservation and N uptake by barley irrigated with saline water: a pot study. ICID 21st International Congress on Irrigation and Drainage, 15–23 October 2011, Tehran, Iran

Duan Z, Xiao H, Li X, Dong Z, Wang G (2004) Evolution of soil properties on stabilized sands in the Tengger Desert, China. Geomorphology 59:237–246

Durovic N, Pivic R, Pocuca V (2012) Effects of the application of a hydrogel in different soils. Agric For 53(07, 1–4):25–34

Eldridge DJ, Leys JF (2003) Exploring some relationships between biological soil crusts, soil aggregation and wind erosion. J Arid Environ 53:457–466

El-Hady OA, Abo-Sedera SA (2006) Conditioning effect of composts and acrylamide hydrogels on a sandy calcareous soil. II – Physico-bio-chemical properties of the soil. Int J Agric Biol 8(6):876–884

El-Hady OA, Abd El-Kader AA, Shafi AM (2009) Physico-bio-chemical properties of sandy soil conditioned with acrylamide hydrogels after cucumber plantation. Aust J Basic Appl Sci 3(4):3145–3151

FAO (2006) World reference base for soil resources 2006, a framework for international classification, correlation and communication. FAO–UNESCO–ISRIC. FAO, Rome

FAO (2011) Green manure/cover crops and crop rotation in Conservation Agriculture on small farms. Integrated Crop Management Vol 12-2010, Plant Production and Protection Division, Food and Agriculture Organization of the United Nations, Rome, pp 108

Forbes MS, Raison RJ, Skjemstad OJ (2006) Formation, transformation and transport of black carbon (charcoal) in terrestrial and aquatic ecosystems. Sci Total Environ 370(1):190–206

Gillman GP (2007) Hydrotalcite: leaching-retarded fertilizers for sandy soils. In Management of Tropical Sandy Soils for Sustainable Development. Proceedings of the international conference on the management of tropical sandy soils, KhonKaen, Thailand, Nov. 2005, FAO Regional Office for Asia and the Pacific, Bangkok

Glaser B, Haumaier L, Guggenberger G, Zech W (2001) The 'Terra Preta' phenomenon: a model for sustainable agriculture in the humid tropics. Naturwissenschaften 88:1

Gong FH, He XD, Peng XY (2001) The performance and cost comparison of several sand stabilization methods along the highroads in Tarim Desert comparison study. Deserts China 21(1):45–49

Goossens D (2004) Effect of soil crusting on the emission and transport of wind-eroded sediment: field measurements on loamy sandy soil. Geomorphology 58:145–160

Grabow GL, Huffman RL, Evans RO, Edmisten K, Jordan D (2002) Subsurface drip irrigation research on commodity crops in North Carolina. ASAE Meeting Paper No. 022290. St. Joseph, Mich. ASAE

Gubry-Rangin C, Nicol GW, Prosser JI (2010) Archaea rather than bacteria control nitrification in two agricultural acidic soils. FEMS Microbiol Ecol 74:566–574

Hammond D, Steege H, Van der Borg K (2007) Upland soil charcoal in the wet tropical forests of central Guyana. Biotropica 39(2):153–160

Han Z, Wang T, Dong Z, Hu Y, Yao Z (2007) Chemical stabilization of mobile dune fields along a highway in the Taklimakan Desert of China. J Arid Environ 68:260–270

Hao X, Chang C, Li X (2004) Long-term and residual effects of cattle manure application on distribution of P in soil aggregates. Soil Sci 169:715–728

Hartmann C, Poss R, Janeau JL, Bourdon E, Lesturgez G, Ratanaanupap S (2002) Use of granular material theories to interpret structural changes in sandy soil. In: 17th World Congress of Soil Science, Bangkok, Thailand, 14-20-August 2002. Symposium no. 04, Paper no. 2108

Hoa HTT (2008) Soil characteristics, cropping patterns, and use of organic resources in the coastal sandy area of ThuaThien Hue province, Central Vietnam. PhD Thesis. Universitécatholique de Louvain, Louvain-la-Neuve, Belgium

Hoa HTT, Cong PT, Tam HM, Chen W, Bell R (2010) Sandy soils in South Central Coastal Vietnam: Their origin, constraints and Management. In: 19th World Congress of Soil Science, Soil Solutions for a Changing World, 1–6 August 2010, Brisbane, Australia

Hugenholtz CH, Wolfe SA, Bender D (2010) Declining sand dune activity in the southern Canadian prairies: historical context, controls and ecosystem implications. Aeolian Res 2:71–82

Ibrahim MA, Hamad I, Al-Abdul W et al (2002) Stabilization of dune sand using foamed asphalt. Geotech Test J 25(2):168–176

Janeau JL, Bricquet JP, Planchon O, Valentin C (2003) Soil crusting and infiltration on steep slopes in northern Thailand. Eur J Soil Sci 54:543–554

Kukal SS, Aggarwal GC (2002) Percolation losses of water in relation puddling intensity and depth in a sandy loam rice (*Oryza sativa* L.) field. Agric Water Manag 57:49–59

Lamm FR, Trooien TP (2002) Subsurface drip irrigation for corn production: a review of 10 years of research in Kansas. Irrig Sci 22(3–4):195–200

Lamm FR, Trooien TP (2005) Dripline depth effects on corn production when crop establishment is non-limiting. Appl Eng Agric 21(5):835–840

Lazányi J (2005) Effects of bentonite on the water budget of sandy soil. Technologii de Cultura Pentru Grau Si Porumb Sympozion International. 7-8 iulie Oradea-Romania

Leciejewski P (2009) The effect of hydrogel additives on the water retention curve of sandy soil from forest nursery in Julinek. J Water Land Develop 13a:239–247

Lehmann J, da Silva Jr JP, Rondon M, Cravo MS, Greenwood J, Nehls T, Steiner C, Glaser B (2002) Slash-and-char – a feasible alternative for soil fertility management in the central Amazon?. In: Proceedings of the 17th World Congress of Soil Science, Bangkok, Thailand

Lehmann J, Gaunt J, Rondon M (2006) Biochar sequestration in terrestrial ecosystems a review. Mitig Adapt Strateg Glob Chang 11:403–427

Lesturgez G (2005) Densification des sols sableux sous culture mécanisée. Cas du Nord-EstThaïlandais. Thèse Université Henri Poincaré Nancy, France

Li XR, Xiao HL, Zhang JG, Wang XP (2004) Long-term ecosystem effects of sand-binding vegetation in the Tengger Desert, Northern China. Restor Ecol 12:376–390

Lu YC, Watkins KB, Teasdale JR, Abdul-Baki AA (2000) Cover crops in sustainable food production. Food Rev Intrnl 16:121–157

McClintock NC, Diop AM (2005) Soil fertility management and compost use in Senegal's Peanut Basin. Int J Agric Sust 3(2):79–91

Mithchel JP, Pettygrove DGS, Upadhaya DS, Shreshtha DA, Roy R, Hogan P, Vargas R, Hembree K (2009) Classification of conservation tillage practices in California Irrigated Row Crop Systems. Publication 8364, University of California

Mizuta K, Matsumoto T, Hatate Y, Nishihara K, Nakanishi T (2004) Removal of nitrate nitrogen from drinking water using bamboo powder charcoal. Bioresour Technol 95:255–257

Moghadam BK, Jamili T, Nadian H, Shahbazi E (2015) The influence of sugarcane mulch on sand dune stabilization in Khuzestan, the southwest of Iran. Iran Agric Res 34(2):71–80

Mulcahy DN, Mulcahy D, Dietz D (2013) Biochar soil amendment increases tomato seedling resistance to drought in sandy soils. J Arid Env 88:222–225

Muñoz-Carpena R, Dukes MD (2005) Automatic irrigation based on soil moisture for vegetable crops. Document AE354, Department of Agricultural and Biological Engineering, UF/IFAS Extension, University of Florida, http://edis.ifas.ufl.edu

Muñoz-Carpena R, Bryan H, Klassen W, Dukes MD (2003) Automatic soil moisture–based drip irrigation for improving tomato production. Proc Florida State Horticul Soc 116:80–85

Neelam P, Rajput TBS (2007) Effect of drip tape placement depth and irrigation level on yield of potato. Agric Water Manag 32(3):209–223

Newman JK, Tingle JS, Gill R, McCaffrey T (2005) Stabilization of silty sands using polymer emulsion. Int J Pavemal 4:1–12

Noble AD, Gillman GP, Ruaysoongnern S (2000) A cation exchange index for assessing degradation of acid soil by further acidification under permanent agriculture in the topics. Eur J Soil Sci 51:233–243

Noble AD, Ruaysoongnern S, Berthelsen S, Penning de Vries FWT, Giordano M (2005) Enhancing the agronomic productivity of degraded soils through clay-based interventions. Int J Agric Sustain 3:102–113

Nyamangara J, Gotosa J, Mpofu SE (2001) Cattle manure effects on structural stability and water retention capacity of a granitic sandy soil in Zimbabwe. Soil Till Res 62:157–162

Orikiriza LJB, Agaba H, Tweheyo M, Eilu G, Kabasa JD, Huttermann A (2009) Amending soils with hydrogels increases the biomass of nine tree species under non-water stress conditions. Clean 37(8):615–620

Osman KT (2014) Soil degradation, conservation and remediation. Sprnger Science + Business Media, Dodrecht

Pártay G, Rajkainé VK, Lukács A (2006) Kálium- migrációvizsgálatakáliföldpáttalkezelt-gyökérközegben. Agrokémiaés Talajtan 55(2):395–414

Pye K, Tsoar H (2009) Aeolian sands and sand dunes. Springer, Berlin Heidelburg

Radovic LR, Moreno-Castilla C, Rivera-Utrilla J (2001) Carbon materials as adsorbents in aqueous solutions. In: Radovic LR (ed) Chemistry and physics of carbon. Marcel Dekker, New York

Ribolzi O, Bariac T, Casenave A, Delhoume JP, Ducloux J, Valles V (2003) Hydrochemistry of runoff and subsurface flow within Sahelianmicrodune. Eur J Soil Sci 54:1–1

Ribolzi O, Hermida M, Karambiri H, Delhoume JP, Thiombiano L (2005) Wind processes improve water infiltration in Sahelian sandy rangeland. In: Management of tropical sandy soils for sustainable agriculture: A holistic approach for sustainable development of problem soils in the tropics, 27 November – 2 December 2005, KhonKaen, Thailand; Proceedings, Management of Tropical Sandy Soils for Sustainable Agriculture, Khon Kaen (Thailand), 27 Nov – 2 Dec 2005 / International Union of Soil Sciences, Wageningen (Netherlands); Land Development Department, Bangkok (Thailand); International Water Management Institute, Colombo (Sri Lanka); FAO, Bangkok (Thailand). Regional Office for Asia and the Pacific; KhonKaen University (Thailand). Faculty of Agriculture

Ruan HX, Ahuja LR, Green TR, Benjamin JG (2001) Residue cover and surface-sealing effects on infiltration: numerical simulations for field applications. Soil Sci Soc Am J 65(3):853–861

Sam L, Vuong ND (2008) Real condition of soil – water resources and desert threat, nature catastrophe in the sandy area of Binh Thuan province. Scientific collection in 2008. South Institute of irrigation science. (In Vietnamese, with English abstract)

Sarapatka B, Rak L, Bubenikova I (2004) Effects of hydroabsorbent used on extremely sandy soils on soil biological and biochemical characteristics. EUROSOIL, September, 04 – 12 Freiburg, Germany, CD-Rom

Schnitzler HU, Kalko EKV, Kaipf I, Grinnell AD (1994) Fishing and echolocation behavior of the greater bulldog bat, Noctilioleporinus, in the Æeld. Behav Ecol Sociobiol 35:327–345

Shahid SA, Qidwai AA, Anwar F, Ullah I, Rashid U (2012) Improvement in the water retention characteristics of sandy loam soil using a newly synthesized poly(acrylamide-co-acrylic acid)/AlZnFe2O4 superabsorbent hydrogel nanocomposite material. Molecules 17:9397–9412

Shirato Y, Zhang TH, Ohkuro T, Fujiwara H, Taniyama I (2005) Changes in topographical features and soil properties after exclosure combined with sand-fixing measures in Horqin Sandy Land, Northern China. Soil Sci Plant Nutr 51(1):61–68

Sitthaphanit S, Limpinuntana V, Toomsan B, Panchaban S, Bell RW (2009) Fertiliser strategies for improved nutrient use efficiency on sandy soils in high rainfall regimes. Nutr Cycl Agroecosyst 85:123–139

Sivapalan S (2006) Benefits of treating a sandy soil with a crosslinked-type polyacrylamide. Aus J Expt Agri 46:579–584

Soil Survey Staff (1999) Soil taxonomy: a basic system of soil classification for making and interpreting soil surveys, 2nd edn. Agriculture Handbook No. 436, U.S. Government Printing Office, Washington, DC

Sorensen RB, Wright FS, Butts CL (2001a) Subsurface drip irrigation system designed for research in row crop rotations. Appl Eng Agric 17(2):171–176

Sorensen RB, Wright FS, Butts CL (2001b) Pod yield and kernel size distribution of peanut produced using subsurface drip irrigation. Appl Eng Agric 17(2):165–169

Specht A, Harvey-Jones J (2000) Improving water delivery to the roots of recently transplanted seedling trees: the use of hydrogels to reduce leaf loss and hasten root establishment. For Res 1:117–123

Stopnišek N, Gubry-Rangin C, Höfferle S, Nicol GW, Mandić-Mulec I, Prosser JI (2010) Thaumarchaeal ammonia oxidation in an acidic forest peat soil is not influenced by ammonium amendment. Appl Environ Microbiol 76:7626–7634

Subbarao GO, Ito K, Berry SW, Nakahara K, Ishikawa T, Watanabe T et al (2006) Scope and strategies for regulation of nitrification in agricultural systems – challenges and opportunities. Crit Rev Plant Sci 25:303–335

Szegi T, Czibulya ZS, Makádi M, Szeder B (2008) Szerves szervetlen adalékanyagok hatása a nyírségi homoktalajok talajszerkezeti, nedvességgazdálkodási tulajdonságaira és a terméseredményekre. Talajvédelem, különszám

Thuy LT, Ha PQ (2007) Total Zn content in sandy soils in Vietnam. Vietnam Soil Sci J 28:5

Titiz B, Sanford R (2007) Soil charcoal in old-growth rain forests from sea level to the continental divide. Biotropica 39:673–682

Tiwari KN, Singh A, Mal PK (2003) Effect of drip irrigation on yield of cabbage (Brassica oleracea L. var. capitata) under mulch and non-mulch conditions. Agric Water Manag 58:19–28

Uchanski ME, Rios A (2012) Killed cover crop residue impacts on soil moisture in an irrigated agricultural system. New Mexico J Sci 46:149–168

Usman A, Kuzyakov Y, Stahr K (2005) Effect of clay minerals on immobilization of heavy metals and microbial activity in a sewage sludge contaminated soil. J Soils Sediments 5(4):245–252

Valenciano JB, Casquero PA, Boto J (2004) Influence of sowing techniques and pesticide application on the emergence and the establishment of bean plants (*Phaseolus vulgaris* L.) Agronomie 24(2):113–118

Valentin C, Rajot J-L, Mitja D (2004) Responses of soil crusting, runoff and erosion to fallowing in the sub-humid and semi-arid regions of West Africa. Agric Ecosyst Environ 104:287–302

Valentin C, Poesen J, Li Y (2005) Gully erosion: impacts, factors and control. Catena 63(2-3):129–131

Vinh N (2005) Coastal sandy soils and constraints for crops in Binh Thuan Province, Southern Central Vietnam. In: Proceedings of internaltional management of tropical sandy soils for sustainable agriculture symposium, 28 Nov – 2 Dec 2005. Khon Khaen, Thailand

Vityakon P (2005) Soil organic matter loss and fertility degradation under different agricultural land uses in sandy soils of Northeast Thailand and the use of organic materials of different qualities as a possible restoration measure. In: Proceedings of international management of tropical sandy soils for sustainable agriculture symposium, 28 Nov – 2 Dec 2005. KhonKhaen, Thailand

Voortmana RL, Brouwer J, Albersen PJ (2004) Characterization of spatial soil variability and its effect on millet yield on Sudano-Sahelian cover sands in S.W Niger. Geoderma 121:65–82

References

Wang HJ, Li J, Lu X, Jin Y (2005) A field experimental study of lignin sand stabilizing material (LSSM) extracted from spent-liquor of straw pulping mills. J Env Sci 17(4):650–654

Winsley P (2007) Biochar and bioenergy production for climate change mitigation. New Zealand Sci Rev 64(1):5–10

Yang K, Zejun T (2012) Effectiveness of fly ash and polyacrylamide as a sand-fixing agent for wind erosion control. Water Air Soil Polluton 223:4065–4074

Zhang ZS, Liu LC, Li XR, Zhang JG, He MZ, Tan HJ (2008) Evaporation properties of a revegetated area of the Tengger Desert, Northern China. J Arid Environ 72:964–973

Chapter 4
Shallow Soils

Abstract Shallow soils have less than 50 cm depth of solum. Generally, they have a thin A horizon over the bed rock or the parent material. If there is a B horizon underlying the A horizon the total depth of A and B horizon does not exceed 50 cm. Soils having the solum depth of 50–100 cm are moderately deep soils and those having the solum depth greater than 100 cm are deep soils. Soils of the high mountains and valleys are commonly very shallow and lack significant topsoil. They are highly erodible. Such shallow soils on bedrock were known as Lithosols in earlier soil classification systems. Some very shallow soils might have developed from hard calcareous rocks. These soils were earlier known as Rendzina. Some soils are considered shallow if they have root restrictive layers or shallow groundwater table so that roots cannot penetrate those shallow layers. Natural vegetations of shallow soils include grasslands, bush lands and low forests. Shallow soils have severe limitations to agricultural use. Plant roots remain confined to a small volume of soil that cannot provide adequate anchorage, water and nutrients. Shallow soils with root restrictive layers can, however, be profitably used for cropping under sustainable management.

Keywords Soil depth · Lithosols · Leptosols · Root restrictive layers · Hardpans · Shallow groundwater table

4.1 Soil Depth Classes

The depth of soils refers pedologically to the depth of solum. It includes the distance between the surface of the soil (top of A horizon) and the lower boundary of the B horizon in mineral soil. Some soils do not have a B horizon (e.g. young soils developing on alluvium). In such cases, solum depth is the depth of the A horizon. For soils with an O horizon, the soil surface is the top of the part of the O horizon that is at least slightly decomposed. Fresh leaf or needle fall that has not undergone observable decomposition is excluded from soil. It is often difficult to determine the lower limit of soil, and for many purposes depth of soil is considered to be the rooting depth of plants. Thus, agronomically the depth of soil refers to the vertical distance between the soil surface and the root restricting layer below. According to

Buol et al. (2003), soil depth classes are assigned in all mineral soils, except lithic subgroups that have a root limiting layer present. The depth of soil up to a root restricting layer (virtual absence of root penetration) is known as effective depth of soil because it is the space that allows for the proliferation of plant roots, and for the accommodation of air, water and nutrients. It gives an indication of the soil volume which can be utilized by the plant and which is conducive to moisture retention. Soils are said to be shallow if a root restricting layer lies at a depth within 50 cm. However, in Oxisols, an Order of Soil Taxonomy (Soil Survey Staff 1999), the root restrictive layer lies within a depth of 100 cm. Root restricting layers include duripans, petrocalcic, petrogypsic, and placic horizons, continuous ortstein, densic, lithic, paralithic, and petroferric contacts. Soils having bedrock close to the surface (Lithic subgroup of Soil Taxonomy, Lithosols of Old American Soil Classification and Leptosols in World Reference Base for Soil Resources; FAO 2006) are all shallow soils. The presence of a shallow groundwater table also limits root growth, and hence soils with groundwater table within 50 cm can be considered as shallow soil.

However, soils can be of the following four categories according to effective depth:

Category	Effective depth of soil (cm)
Very Shallow	<25
Shallow	25–50
Moderately deep	50–100
Deep	>100

4.1.1 Very Shallow Soils

Soils of this category have the surface less than 25 cm from a layer below that restricts root development. These soils have often skeletal stony materials or organic debris usually overlying solid or shattered bedrock, located in high elevation exhibiting bare rock outcrops (Fig. 4.1). Very shallow soils with a thin A horizon above a thin C horizon often develops on steep slopes in mountainous regions. Some very shallow soils might have developed from hard calcareous rocks. These soils are not suitable for any economic crop. The usual natural vegetation includes grasses, herbs and shrubs. Low shallow rooted trees may sometimes be found. These soils should be left under natural vegetation without any disturbance. These soils need protection and even grazing should be controlled.

4.1.2 Shallow Soils

Soils of this category have a surface of less than 50 cm from a layer below it, which restricts root development. The root limiting layer lies in a depth between 25 and 50 cm. A "root restrictive layer" is a nearly continuous layer that has

Fig. 4.1 Very shallow soils on mountain slopes (Image courtesy of Prof Gerard Kiely)

one or more physical, chemical, or thermal properties that significantly impede the movement of water and air through the soil or restrict roots or otherwise provide an unfavorable root environment. The root liming layers may include compacted layers, hardpans, lithic materials and a saturated layer of soil groundwater table.

4.1.3 Moderately Deep Soils

Moderately deep soils have a surface between 50 and 100 cm from a layer that restricts root development below.

4.1.4 Deep Soils

These soils lack any root restricting layer; or it is at a depth greater than 100 cm. These soils usually develop on transported parent materials and deeply weathered residual parent materials on level topography and valleys. They are well drained and mostly suited physically for a wide variety of crops. If the soils are fertile, they still have a few limitations to agricultural use. Some deep soils in low-lying lands and in basins are limited in agricultural use by waterlogging.

In the tropical region, shallow soils can be classified into the following categories:

1. Massive laterite lying at 25–50 cm, irrespective of stoniness of overlying soil: *petroferric phase*, which has severe agricultural limitations.
2. Consolidated rock or massive laterite at 50–100 cm: *shallow phase*, which has moderate agricultural limitations.
3. Depth of rock >50 cm, but upper 50 cm dominated (>40 percent by volume) by stones, boulders or gravel consisting mainly of rock fragments: *stony phase*, in which mechanized cultivation is not possible.
4. Depth of rock >50 cm, but upper 50 cm dominated (>50 percent by volume) by hard lateritic concretions: *petric* phase.

Based on agricultural limitations, the following categories can also be identified:

1. Consolidated rock (including ferruginized rock) or massive laterite at <25 cm, irrespective of stoniness of overlying soil: *lithosol*, which has severe agricultural limitations.
2. Consolidated rock lying at 25–50 cm, irrespective of stoniness of overlying soil: *lithic phase*, having severe agricultural limitation.

4.2 Properties of Shallow Soils

4.2.1 Shallow Soils on Mountain Slopes

Soils of high mountains and valleys are commonly shallow and lack significant topsoil. These soils are excessively surface drained, and run-off water continually removes the loose surface materials on steep slopes in humid regions. The residual soils are usually gritty and stony. These soils may have exposed C, thin A over C and truncated BC profiles. There are frequent landslides on slopes of hillsides. These soils were earlier classified as Lithosols, and are now classified as Entisols and Inceptisols in Soil Taxonomy and Leptosols in WRB (World Reference Base for Soil Resources; FAO 2006). These shallow soils store limited amounts of water and nutrients for their shallow depths and continuous erosion. These soils are hard to till and almost equally hard to amend. These soils are suitable for natural adapted plants and traditional herbs. These lands are used for grazing, but overgrazing may degrade them seriously. Some native trees are found to grow there, but their growth is generally stunted. Limited aesthetic and home gardening can be done with generous additions of organic matter, and low stock livestock farming is feasible. However, there are some deep soils rich in organic matter and nutrients in broad valleys of mountainous regions. These soils support dense forests (also there are low growth coniferous forests on shallow soils of steep slopes at high altitude) and are often degraded by shifting cultivation.

4.2 Properties of Shallow Soils

Walker et al. (2010) mentioned the following disadvantages and advantages of shallow soils:

Disadvantages of shallow soils

Disadvantages of shallow soils include reduction of permeability and storage of water, affecting the suitability of soils for cropping, decreasing diversity of land use, and limit the placement of wells, septic systems, foundations, agricultural uses, roads, and utilities.

Advantages of shallow soils

Shallow soils with bedrock outcrops on steep slopes often offer spectacular views, making them tempting sites for recreational developments and homes. They are ideal sites for natural recreation areas, such as hiking trails, forest preserves, and open spaces.

4.2.2 Shallow Soils on Calcareous Materials

Shallow soils often develop on shallowly weathered carbonate rocks, such as limestone, dolomite, marl and chalk. They are generally fine textured soils with neutral to alkaline reaction. They can be found in a variety of landforms ranging from plains to steep slopes. Native vegetation ranges from grasslands to forests. There are two prominent types of shallow soils on calcareous materials called *terra rossas* (red) and *rendzinas* (black) in Old American Soil Classification System. The terra rossas (Fig. 4.2) are variable in texture, but the rendzinas (Fig. 4.3) are generally well-structured clay soils. The rendzinas are shallow soils usually less than 50 cm deep, derived from parent material containing over 40–50 percent carbonates. The surface horizon is dark in color with a moderately strong structure and neutral to alkaline reaction. A calcareous (B) horizon may be present. These soils also belonged to Lithosols, and they are now classified as Entisols and Inceptisols in Soil Taxonomy and Leptosols in WRB. Where they are sufficiently deep, rendzinas are suitable for tillage and pasture, but in many places lack of soil depth precludes tillage. These soils are excellent for winter pasturage. Rendzinas have been extensively drained and developed, and are now mostly devoted to pastures in Australia and elsewhere. They can be deficient in phosphorus and the trace elements, such as copper, zinc and manganese. Terra rossas are well-drained shallow soils and are often stony or intruded by outcropping limestone. Their usefulness is then limited.

4.2.3 Shallow Soils with Root Restrictive Layers

Hard rock on surface, or bed rocks and compact layers on or close to the soil surface restrict root penetration and permeability of water. Crop plants suffer in such soils from deficiency of water and nutrients.

Fig. 4.2 Terra rossa soil (Image courtesy of Dr. Stephen Hallett © Soil-Net.com, Cranfield University, 2013)

Fig. 4.3 Rendzina soil (Image courtesy of Dr. Stephen Hallett © Soil-Net.com, Cranfield University)

4.2 Properties of Shallow Soils

There is also inadequate aeration and root extension. Growth of plants is stunted and the yield is reduced considerably.

Plants show the following symptoms which can indicate the shallow rooting depth:

Stunted, uneven crops: Poor growth of crops due to inadequate absorption of water and nutrients.

Yellow leaves: Yellow leaves and other signs of nutrient deficiencies may be caused by poor rooting systems.

Rapid wilting: Crops may wilt quickly during dry periods as the surface layers of the soil dry out, and storage of moisture is low.

Distorted roots: Roots of crop plants may be distorted. Waterlogging due to poor drainage often develops and cause root damage.

The presence of a root restrictive or dense layer can be examined by pushing a device known as penetrometer into the soil and observing the pressure needed to penetrate. Soils with higher bulk density are generally more compacted, root restricting and impermeable to water. Hardpan soils and other shallow soils are either naturally vegetated or pasteurized. There are two types of hardpans – pedogenic and anthropogenic. Pedogenic hardpans develop by soil physical, chemical and biological processes conditioned by climatic, hydrological and lithological variables. The development and occurrence of hardpans in certain soils have been attributed to such soil factors as (1) iron and aluminum oxides, (2) amount and type of clay (3) dispersed organic matter, (4) soluble aluminum, (5) colloidal silica, (6) close-packing of soil particles. Additionally, applied pressure on the soil surface by agricultural implements is a major factor in hardpan formation in cultivated soils (Chap. 13). Hardpans create major widespread problems for crop production worldwide. Traffic or soil genetic processes can produce horizons with high density or cemented soil particles (Hamza and Anderson 2005); these horizons have elevated penetration resistances that limit root growth and reduce water and airflow. Soil Survey Staff (1999) recognized two major forms of pedogenic hardpan, namely duripans and fragipans. Duripans (Fig. 4.4) are hard, subsurface horizons, cemented by silica or other materials, such as iron oxides or calcium carbonate to the extent that fragments of the air-dry material does not slake after prolonged soaking in water or in HCl. Fragipans are seemingly cemented, with hard to very hard consistence when dry and moderate to weak brittleness when moist. A dry fragment slakes or fractures when placed in water (NSRI 2009). Silcrete is another form of widely distributed hardpan. Silcrete is a brittle, intensely indurated rock comprising primarily quartz grains cemented with siliceous allophane. It occurs at 1.5–7 m deep and is often several meters thick and extremely hard, requiring excavation (Bennett et al. 2005). Silcretes or siliceous duripans are indurated products of surficial and penesurficial (near-surface) silica accumulation. Fragipan formation may involve the following processes individually and in combination: close packing of the particles, clay bridging and amorphous silicate bonding.

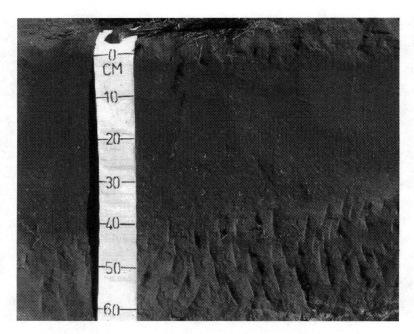

Fig. 4.4 Dark red duripan at 40 cm depth of a South African soil (Image courtesy of ISRIC)

4.2.4 Soils with Shallow Groundwater Table

Here, a shallow groundwater table refers to the presence of a groundwater table (or saturated soil zone) close to the surface that completely depletes oxygen for root respiration. A shallow groundwater table can affect plant growth in various ways. When groundwater table approaches the surface, root growth decreases and eventually stops for want of air. Decay of roots of mesophytic plants in poorly drained soils is a common occurrence in humid areas. Poor root growth in the anaerobic soil limits water and nutrient uptake and reduces plant production. Aeration problem generally occurs in soils where the groundwater tables rise to about 100 cm from surface, depending on soil texture, porosity and plant adaptation. Overall, shallow groundwater influences 22–32 percent of global land area including ~15 percent as groundwater-fed surface water features and 7–17 percent with the water table or its capillary fringe within plant rooting depths (Fan et al. 2013). However, presence of a water table at a safe depth of soil can contribute to the water uptake and evapotranspiration of plants. Capillary rise of water to the root zone takes place from the groundwater table meeting a significant proportion of plant's water demand (Singh et al. 2006). Actually, plant available water includes not only the water stored in the root zone, but also the water moving up from below the root zone (Logsdon et al. 2008).

4.2.5 *Shallow Lateritic Soils*

Early experiences indicated that oil palms grown on shallow lateritic soils in Peninsular Malaysia came into bearing two years later and three times less compared to deep soils. Increasing the fertilizer rates only partially alleviated the constraint and improved yield. Productivity also seemed to improve with palm age. The main problems with shallow lateritic soils are low effective soil volume, poor nutrient status and water holding capacity. Most shallow lateritic soils have low CEC and high P fixing capacity. The main approaches to obtain satisfactory oil palms on shallow lateritic soils are to improve soil fertility and implement soil and water management adroitly. The types and compactness of the laterites also play a major role on the degree of severity of limitations. For example, the less compact and subangular laterites pose only moderate limitation to oil palms compared to very serious limitation in some other shallow lateritic soils.

4.3 Limitations of Shallow Soils

Shallow soils have severe limitations particularly to agricultural use. Plant roots remain confined to a small volume of soil that cannot provide adequate anchorage, water and nutrients. Nutrient and water absorption capacity of the restricted roots is also low. Therefore, plants on otherwise fertile soils may show nutrient deficiency. Growth of plants even in favorable climatic and management situations is unsatisfactory and the plants tend to remain stunted. Yields of crops are low. Shallow soils cannot store enough water to support the plants and the soils are droughty in the dry season. Hardpans are impervious to both roots and water. Soils with shallow hardpans have poor drainage and plants often suffer from poor aeration in monsoon and under irrigation there. Hardpans can be broken with great effort and cost, but these pans usually redevelop unless the environmental conditions are unaltered and the soils are not managed sustainably. Shallow soils on steep slopes are highly susceptible to erosion particularly when the natural vegetation is disturbed by grazing, biomass collection and shifting cultivation. Regrowing the vegetative cover by natural fallow remains to be the only option for conservation of such soils. Shallow soils need higher quantities of irrigation and fertilizers but at low and frequent installments. Management of shallow soils involves extra cost of subsoiling, deep tillage and amendments. However, under sustainable management shallow soils on level topography can be used profitably for cropping.

Shallow soils are often left under natural vegetation, and used for growing grasses as well as forages or for forestry. Forest trees with thick roots may penetrate the compacted layer or the hardpan at shallow depths. Some investigators (Romanya and Vallejo 2004; Dovcʻiak et al. 2003) have, however, shown that soil thickness can limit forest productivity. Water-holding capacity and nutrient conditions may be crucial for tree seedlings and understory shrubs that often have roots concentrated at shallow soil depths (Royce and Barbour 2001; Rose et al. 2002). In mixed-conifer forests of lower density stands dominated by shrubs or Jeffrey pine (*Pinus jeffreyi*) are considered indicative of shallow and exposed mineral soils (North et al. 2002). Meyer et al. (2007) studied soils and weathered bedrock zones to observe patterns of stand composition and regeneration within a single Sierra Nevada mixed-conifer forest stand where mineral soil is thin and weathered bedrock is relatively thick. They found that patches of shrubs and low-density Jeffrey pine may occur on apparently shallow soil conditions.

4.4 Management of Shallow Soils

4.4.1 Selection of Suitable Crops

Properties of soil profoundly influence the rooting behavior of plants. These properties include texture, structure, porosity, compaction, moisture and depth. Restricting soil layers (hardpan, plowpan, etc.), or even abrupt changes in soil texture may affect the rooting depth of crop plants. Plants also widely differ in their requirements of rooting depth in soil. Herbs and shrubs need shallower depth than trees. As soil properties impact the amount of stored moisture and nutrients, the plant's rooting characteristics determine how much of the soil moisture and nutrients can be accessed by the plant. A shallow rooted crop (peas, celery, lettuce, radish, spinach) has a better chance to thrive in a shallow soil. If shallow soils have to be farmed, careful selection of crops, coupled with frequent watering and fertilizing, is required. We all know that a deep-rooted crop has greater access to soil moisture than a shallow-rooted crop does, but in shallow soils the roots are usually malformed and fail to support the plants. Institute of Natural Resources (2004) recommended growing the following crops in shallow soils of Ethekwini, South Africa: bananas (irrigated), cabbage, carrot, cowpeas, dry beans, lucerne (irrigated), maize (rainfed or irrigated), potatoes, sorghum, soybeans (rainfed or irrigated), star grass (rainfed or irrigated), sugarcane (irrigated), sunflower and tomatoes (rainfed or irrigated). McFarlane (2011) considered the following fruit trees to be suitable for shallow soils: Banana, Babaco, Dragon fruit, Grumichama, Guava, Loquat, Papaya, Pepino, Pomegranate, Tamarillo, Yellow pitaya.

A list of plants including shrubs, perennials and trees suitable for shallow soils on chalk is given in Table 4.1.

4.4.2 Management of Shallow Mountain Soils

Mountains are an important source of water, energy and biological diversity. They provide minerals, forest products and agricultural products and recreation. Mountains represent diverse ecosystems with varied soil, climate and vegetation

Table 4.1 Some shrubs, perennials and trees suitable for shallow soils on chalk

Shrubs			
Botanical name	**Local name**	**Botanical name**	**Local name**
Aucuba japonica	Japanese Aucuba	*Juniperus communis*	Juniper
Berberis spp.	Barberry	*Laurus spp.*	Sweet bay
Bergenia spp.	Bergenia	*Lonicera spp.*	Honeysuckle
Buddleia davidii	Butterfly Bush	*Mahonia spp.*	Mahonia
Buxus spp.	Boxwood	*Myrtus spp.*	Myrtle
Caryopteris spp.	Blue Beard	*Hibiscus syriacus*	Hardy
Ceanothus spp.	Wild Lilac	*Osmanthus spp.*	Sweet Olive
Ceratostigma spp.	Hardy Plumbago	*Philadelphus spp.*	Mock Orange
Cistus spp.	Rockrose	*Phlomis fruiticosa*	Jerusalem Sage
Cornus mas	Cornelian Cherry	*Potentilla spp.*	Cinquefoil
Cotinus coggygria	Smoke Tree	*Prunus laurocerasus*	English Laurel
Cotoneaster spp.	Cotoneaster	*Rhus spp.*	Sumac
Deutzia spp.	Deutzia	*Rosa (some spp.)*	Rose
Euonymus spp.	Euonymus	*Rosmarinus spp.*	Rosemary
Euphorbia spp.	Euphorbia	*Ruscus spp.*	Butcher's bloom
Forsythia spp.	Forsythia	*Santolina spp.*	Santolina
Fremontia californica	Fremontodendron	*Saracococca spp.*	Sweet Box
Fuchsia spp.	Fuchsia	*Spiraea spp.*	Spirea
Halimium spp.	Halimium	*Symphoricarpos spp.*	Snowberry
Hebe spp.	Hebe	*Syringa spp.*	Lilac
Helianthemum spp.	Sun Rose	*Thuja plicata*	Western Red Cedar
Hibiscus syriacus	Hardy	*Viburnum spp.*	Viburnum
Indigofera spp.	Indigo	*Weigela spp.*	Weigela
Perennials			
Acanthus spp.	Bear's Breeches	*Helictotrichon spp.*	Blue Oat
Alyssum spp.	Alyssum	*Ixia spp.*	African Corn Lily
Arabis spp.	Wall Cress	*Knautia spp*	Field Scabiosa
Armeria spp.	Thrift	*Kniphofia spp.*	Red Hot Poker
Aster spp.	Aster	*Lamium spp.*	Dead Nettle
Ballota spp.	Ballota	*Linaria purpurea*	Toadflax
Bergenia ciliata	Bergenia	*Lychnis spp.*	Campion
Calamintha spp.	Calamint	*Meconopsis spp.*	Welsh Poppy
Campanula	Bell Flower	*Nepeta spp.*	Catmint
Centaurea spp.	Centaurea	*Origanum spp.*	Oregano
Cheiranthus	Wall Flower	*Paeonia suffruticosa*	Tree Peony
Coreopsis spp.	Coreopsis	*Papaver orientale*	Oriental Poppy
Cosmos spp.	Cosmos	*Perovskia atriplicifolia*	Russian Sage
Dianthus spp.	Pinks	*Phlox spp*	Phlox
Dicentra spp.	Bleeding Heart	*Platycodon spp.*	Balloon Flower
Doronicum spp.	Leopard's Bane	*Primula veris*	Cowslip Primrose
Dryopteris spp.	Wood Fern	*Pulsatilla vulgaris*	Pasque Flower

(continued)

Table 4.1 (continued)

Echinops ritro	Globe Thistle	Raoulia spp.	Raoulia
Erodium spp.	Erodium	Salvia spp.	Sage
Eryngium spp.	Sea Holly	Saxifraga spp.	Saxifrage
Erysimum spp.	Wallflower	Scabiosa spp.	Pincushion Flower
Euphorbia spp.	Euphorbia	Sedum spp.	Stonecrop
Festuca spp.	Fescue	Sidalcea spp.	Mallow
Filipedula spp.	Dropwort	Stachys spp.	Stachys
Foeniculum spp.	Common Fennel	Stipa spp.	Feather Grass
Francoa spp.	Maiden's Wreath	Thymus spp.	Thyme
Gaillardia spp.	Blanket Flower	Verbascum spp.	Mullein
Geranium cinereum	Cranesbill	Vinca spp.	Periwinkle
Gladiolus spp.	Gladiolus	Zauschneria spp.	California Fuchsia
Helichrysum spp.	Strawflower		
Trees			
Aesculus spp.	Horse Chesnut	Laurus spp.	Sweet Bay
Arbutus unedo	Strawberry Tree	Malus spp.	Crabapple
Cercis spp.	Red Bud	Morus spp.	Mulberry
Fagus sylvatica	Beech	Pinus nigra	Australian Pine
Fraxinus spp.	Weeping Ash	Prunus serrulata	Flowering Cherry
		Sambucus spp.	Elderberry

Source: Scarborough Gardens 33 El Pueblo Road, Scotts Valley, CA 95066 http://www.scarboroughgardens.com/online_docs/alkaline.pdf

types within a particular mountain range. For example, there are several climatic systems, such as tropical, subtropical, temperate and alpine at different elevations of the same Himalayan range. Although mountain environments are essential to the survival of the global ecosystem, mountain ecosystems are, however, rapidly changing. They are susceptible to accelerated soil erosion, landslides and rapid loss of habitat and genetic diversity and highly in need of conservation. However, shallow mountain soils represent land capability class VI and soils of this class have severe limitations that make them generally unsuited to cultivation and limit their use largely to pasture or range, woodland, or wildlife food or cover. Limitations include steep slope, severe erosion hazard, stoniness, shallow rooting, etc. These soils are better conserved under undisturbed natural vegetation where a steady state may be reached among the environmental components including soil, landform, vegetation and climate. Local small scale recreational activity and controlled ecotourism can take place.

However, some successful vineyards have been established on shallow soils in the Santa Cruz Mountains. These soils are intensively managed and sustainably farmed. Often soil and compost or other organic residues are mixed together and used as fill on the shallow native soil. Raised beds may be prepared to increase the rooting depth. Farmers employ organic amendments, irrigation and drainage as needed. Every farm operation is done manually and no heavy equipment ever enters the vineyards. Every vineyard is harvested by hand in small bins, and every lot is meticulously hand-sorted to assure mature, luscious and clean fruit. These steep slopes, shallow soils, and the coastal range environment are challenging and demand monitoring of the environmental consequences on the long term basis.

Mountain soils are frequently used for recreational purposes including picnic, tourism, ecotourism, hiking, trailing, etc. These activities often degrade shallow mountain soils. Mountain climbers often use the same trail repeatedly and these trails are constructed mainly on relatively gentle slopes. These trails accumulate huge wastes, including water bottles, papers, packages, burnt cigars, bottles of beer, etc. Trail providers are very reluctant in cleaning these trails. Cattle trailing on mountain slopes is an important economic activity in some regions. To minimize the negative effects of mountain grazing on shallow soils, a deferred grazing system has been developed. It is a process by which the cattle enter the foothills of a pasture system in the spring and migrate to the high country and then drift down the other side to a low pasture in the fall. In the following year, the cattle use the opposite low pasture first and come out through the other pasture in the fall. It allows the grass a recovery time.

Beautiful summer resorts are constructed on high altitudes for tourists on valleys, gentle slopes or summits of hills. Deep green lawns and colorful ornamental gardens are found in the front and backyard. Small trees can also be grown. These systems are intensively managed with regular organic and inorganic amendments, watering and draining.

4.4.3 *Management of Shallow Soils with Compacted or Root Restrictive Layers*

Surface crusts, hard-setting and seals in the surface soil can be managed with normal tillage operations and sustained with manures and cover crops. Sometimes anticrusting amendments are used to prevent surface crusting. Gypsum is often used in soils to prevent crusting or for improving crusted soils. Application of gypsum makes the soil loose and soft. Recently, some synthetic soil conditioners are being tried against soil crusting. Improving soil structure and reducing crust formation by using organic polymers (for example, PAM-Poly Acrylamide) has been under intensive investigation for many years. Because polymers increase the stability of the soil structure, they reduce the tendency of soils to form seals, thereby preventing the decline in infiltration rates, reducing runoff and soil losses. These polymers do not interact directly with the soil matrices but form aqueous gels and act as water reservoirs for the plant-soil system. Tillage is commonly used to remediate hard-layer problems by breaking it physically. Shallow hard layers (<5 cm) can be broken up with tines or cultivators that disrupt the surface soil. Deeper hard layers (>15 cm) can be broken up with shanks of various sizes at desired depths. Shanks are pulled through the soil at the depth of the hard layer shattering it and decreasing its resistance to root growth (Busscher 2008). Soils often reconsolidate leading to reduced water/and airflow, reduced root growth, and lower crop yields (Hakansson and Lipiec 2000). So, repeated and frequent tillage is needed often seasonally or annually. Frequent tillage can be expensive because it often requires large tractors (14–20 kg weight per shank), 20–40 min ha^{-1} of labor, and 20–25 L ha^{-1} of fuel (Busscher 2008). Dense compact subsoils need concerted management efforts. Decompaction by deep tillage should be combined with alternating shallow and deep rooted crops in rotations. Additions of organic residues, mulching and maintaining tap-rooted forage cover crops help build soil structure and prevent recompaction. Tillage is the main cause of accelerated erosion in agricultural soils. Shallow soils suffer more from erosion because erosion may sometimes remove the entire part or a substantial part of the loose material over the bedrock. Management of soil compacted by human actions will be discussed in Chapter 13 on degraded soils.

Slit tillage is a process of cutting slits usually 3-mm-wide through the hardpan with a thin blade mounted on a shallow subsoil shank. It can increase root growth through slits of subsurface hardpans. Subsoiling is usually repeated annually but slit tillage may be an alternative to subsoiling that does not need to be repeated annually. Slit tillage is suitable for row crops. There is a wide variability in depth and thickness of hardpan layers within a field. Several workers (Raper et al. 2000; Gorucu et al. 2001) showed that the depth of root-restricting layer varies greatly from field to field and also within the field. Applying uniform depth tillage over the entire field may be either too shallow or too deep and can be costly. Site-specific tillage involving decompaction where it is needed only for crop growth, could reduce subsoiling cost. Variable depth tillage or site-specific tillage can be implemented either with (1) a pre-tillage map technology, or (2) a real-time sensor. The map would then be used in the site-specific tillage equipment control system to control subsoiling location and depth.

Thus, there are several methods of modifying the soil profile, particularly of shallow soils for crop production. Plowing can completely disrupt the plowpan; and its effects are usually longer lasting than chiseling. However, plowpans may be reformed quickly in some soils; even when complete disruption is done. Deep plowing or trenching may improve fragipan soils if the tillage extends through the fragipan layer and sufficiently mixes soil material from above or below the fragipan. Duripan soils may be modified by deep ripping, slip plowing, trenching, or backhoeing; but power requirements and cost are often prohibitive.

4.4.4 Management of Soils with Shallow Groundwater Table

Maintaining a groundwater table (GWT) at a desirable depth of soil is a critical decision because a shallow groundwater table has both important advantages and disadvantages. In which way the GWT will influence crop production depends soil (capillary pull), distribution of effective plant roots, and groundwater salinity. In arid and semi-arid regions, a shallow groundwater table may contribute significantly to the crop water use. Shallow groundwater table may meet up a considerable portion of the crop water requirement provided that the salinity of groundwater remains below 4 dS m^{-1} at least at a maximum depth of 2.2 m soil (Ayars et al. 2006). So, successful management of shallow groundwater will contribute to a reduction in irrigation water requirements.

If the groundwater table is so shallow that it depletes oxygen for most of the effective roots of crops and if the groundwater is saline, lowering the groundwater table becomes the preferred option. Some crops including cucumbers, onions, lettuce, egg plants, potatoes, melons, strawberries, and spinach can be grown in soils with GWT near 1 m depth. Groundwater table may rise to 20–30 cm at times. Too shallow a GWT is not a problem for paddy (*Oryza sativa*) and *Arum* sp. but for many others drainage is necessary. Two different strategies are now applied for management of the groundwater table: (i) re-use of drainage water for irrigation and (ii) draining away the excess water. Here, a short account of both the methods is provided. Its details will be discussed in Chap. 6 on Hydric Soils.

The removal of excess water from soil is known as drainage. Some soils, sloping or coarse textured porous, are naturally well drained. In some soils, water stagnates on the surface for a considerable period so that cultivation is hampered. In some other soils, the subsoil remains saturated by water for a long time and plant roots starve for oxygen. These soils need artificial drainage. There are several surface and subsurface drainage systems. The choice of the system, the depth of drains, distance between drains, frequency of drains, etc. depends on the crop type, soil type, water volume, etc. Drainage systems will be discussed systematically in Chap. 5. Here, it needs to be noted that the surface ditch drains or underground pipe drains may all serve the purpose if carefully designed and monitored.

Study Questions
1. Define shallow soils. Classify soils on the basis of solum depth. Mention the properties of soils of different depth classes.
2. Distinguish between Lithosols and Rendzina. Describe their general features. How can they be managed for specific purposes?
3. What do you mean by shallow root-restrictive layers? How can soils with shallow root restrictive layers be used for cropping?
4. Define groundwater. Explain shallow groundwater table. What are the problems associated with shallow groundwater table? How can shallow groundwater be managed?
5. Write notes on (a) Hardpan (b) Deep tillage, (c) Subsoiling and, (d) duripan.

References

Ayars JE, Christen EW, Soppe RWO, Meyer WS (2006) Resource potential of shallow groundwater for crop water use – a review. Irrig Sci 24:147–160

Bennett DL, Speed RJ, Goodreid A, Taylor P (2005) Silcrete hardpan in the north-eastern wheatbelt: hydrological implications for oil mallees. Report 297, Department of Agriculture and Food, Western Australia

Buol SW, Southard RJ, Graham RC, McDaniel PA (2003) Soil genesis and classification, 5th edn. Iowa State Press – Blackwell, Ames

Busscher WJ (2008) Hardpan soils: management. In: Chesworth W (ed) Encyclopedia of soil science. Springer, Dordrecht

Dovc̆iak M, Reich PB, Frelich LE (2003) Seed rain, safe sites, competing vegetation, and soil resources spatially structure white pine regeneration and recruitment. Can J For Res 33:1892–1904

Fan Y, Li H, Macho GM (2013) Global patterns of groundwater table depth. Science 339(6122):940–943

FAO (2006) World reference base for soil resources 2006: a framework for international classification, correlation and communication. World Soil Resources Reports 103, Food and Agriculture Organization of the United Nations, Rome

Gorucu S, Khalilian A, Han YJ, Dodd RB, Wolak FJ, Keskin M (2001) Variable depth tillage based on geo-referenced soil compaction data in costal plain region of South Carolina. ASAE Paper No. 01-1016, ASAE, 2950 St Joseph, MI, 49085–9659

Hakansson I, Lipiec J (2000) A review of the usefulness of relative bulk density values in studies of soil structure and compaction. Soil Tillage Res 53:71–85

Hamza MA, Anderson WK (2005) Soil compaction in cropping systems: a review of the nature, causes and possible solutions. Soil Tillage Res 82:121–145

Institute of Natural Resources (2004) Ethekwini Agricultural Status Quo. A report prepared for Ethekwini Municipality, South Africa

Logsdon SD, Ramirez GH, Hatfield JL, Sauer TJ, Prueger JH, Schilling KE (2008) Soil water and shallow groundwater relations in an agricultural Hillslope. Soil Sci Soc Am J 73(5):1461–1468

McFarlane A (2011) Organic fruit growing: your complete guide to producing beautiful fruit all year round. ABC Books, Australia

Meyer MD, North MP, Gray AN, Zald HSJ (2007) Influence of soil thickness on stand characteristics in a Sierra Nevada mixed-conifer forest. Plant Soil 294:113–123

North M, Oakley B, Chen J, Erickson H, Gray A, Izzo A, Johnson D, Ma S, Marra J, Meyer M, Purcell K, Rambo T, Rizzo D, Roath B, Schowalter T (2002) Vegetation and ecological characteristics of mixed conifer and red fir forests at Teakettle Experimental Forest. USDA Forest Service General Technical Report PSW-GTR-186

NSRI (2009) Soil classification system of England and Wales. National Soil Resources Institute, Cranfield University, UK

Raper RL, Schwab EB, Dabney SM (2000) Spatial variation of the depth of root-restricting layers in Northern Mississippi soils. Second International Conference Geospatial Information in Agriculture and Forestry, Lake Buena Vista, FL

Romanya J, Vallejo VR (2004) Productivity of *Pinus radiata* plantations in Spain in response to climate and soil. For Ecol Manag 195:177–189

Rose KL, Graham RC, Parker DR (2002) Water source utilization by *Pinus jeffreyi* and *Arctostaphylos patula* on thin soils over bedrock. Oecologia 134:46–54

Royce EB, Barbour MJ (2001) Mediterranean climate effects. I. Conifer water use across a Sierra Nevada ecotone. Am J Bot 88:911–918

Singh R, Kundu DK, Tripathi VK (2006) Contribution of upward flux from shallow ground water table to crop water use in major soil groups of Orissa. Jour Agric Physics 6(1):1–6

Soil Survey Staff (1999) Soil Taxonomy: a basic system of soil classification for making and interpreting soil surveys, 2nd edn. Natural Resources Conservation Service, US Department of Agriculture Handbook 436, Washington, DC

Walker J, Hoxsie E, Groffman P (2010) The soils of Dutchess County, NY. Natural Resource Inventory of Dutchess County, New York

Chapter 5
Soils with Drainage Limitations

Abstract Several soil types have inadequate natural drainage that severely limits their profitable or intended use. These soils have different names and are distributed in all regions of the world. Wetland soils, submerged soils, waterlogged soils, hydric soils, poorly drained soils, etc. all have drainage limitations, and in addition they have some common and specific characteristics. They have aquic soil moisture regimes characterized by a reducing state that is virtually devoid of dissolved oxygen due to prolonged saturation with water. Some soils of northern Europe and North America are wet and cold. An account of these soils – their properties, land-use problems and management – is presented in this chapter.

Keywords Wetlands · Poorly drained soils · Hydric soils · Redoximorphic features · Wetland rehabilitation · Wetland rice · Flooding · Artificial drainage · Cold and wet soils

5.1 Wetland Soils, Hydric Soils, Poorly Drained Soils

Soils with drainage limitations include wetland soils, hydric soils, poorly drained soils, etc. All these soils have a common characteristic; they have aquic soil moisture regime. Soil Survey Staff (2010) defined an aquic soil moisture regime as the virtual absence of dissolved oxygen in soil due to prolonged saturation with water so that soil conditions in the growing season are dominated by the reducing state. Such soils include intermittently flooded rice fields, shallowly or deeply flooded submerged soils, subaqueous soils and swamp forests (Figs. 5.1, 5.2, 5.3 and 5.4).

Some differences exist between the concepts of waterlogged and wetland soils, although natural drainage in both of them is poor. Waterlogged soils undergo prolonged saturation with water in the growing season, but wetland soils remain under water most part of the year. It is said that wetland ecosystems are transitional lands between terrestrial and aquatic systems where the land is shallowly or deeply flooded. Wetlands are characterized by any one of the following: (i) the vegetation is predominantly hydrophytic, (ii) the substrate is predominantly waterlogged hydric soil, and (iii) the substrate is sediment or rock either saturated with water or covered by shallow water during the greater part of the year. Wetlands include marshes, fens,

© Springer International Publishing AG, part of Springer Nature 2018
K. T. Osman, *Management of Soil Problems*,
https://doi.org/10.1007/978-3-319-75527-4_5

Fig. 5.1 A seasonally waterlogged land in Bangladesh. Rice is the main land use without irrigation in the monsoon and with irrigation in the dry season. (Image courtesy of Khan Tanjid Osman)

Fig. 5.2 A shallowly flooded land in Matagorda County, Texas, USA. (Image courtesy of Ducks Unlimited)

5.1 Wetland Soils, Hydric Soils, Poorly Drained Soils

Fig. 5.3 Dinga Pota Haor in Bangladesh. A wetland where water level fluctuates with seasons; some tiny islands rise above water in the dry season. Some rooted vegetation is noticed in the inner-left side of the photograph. (Image courtesy of Khan Mohammad Rabbi)

Fig. 5.4 Open swamp forest dominated by swamp tupelo (*Nyssa biflora*) and common spatterdock (*Nuphar advena*) at Seashore State Park, City of Virginia. (Image courtesy of Gary P. Fleming)

peatlands, or lands that remain under water. The water that covers the land may be (i) natural or artificial, (ii) permanent or temporary, (iii) static or flowing, and (iv) fresh, brackish, or marine. According to Finlayson and Moser (1991), the depth of water does not exceed 6 meters at low tide. Recurrent and sustained inundation or saturation at or near the surface and soil properties reflective of these conditions are the minimum requirement of the wetlands. In general, diagnostic features of wetlands include the presence of hydric soils and hydrophytic vegetation.

According to Mitsch and Gosselink (2000), wetlands cover about seven percent of the Earth's land area. Authors like Richardson and Vepraskas (2001) and Reddy and DeLaune (2008) presented detail accounts of flooded and wetland soils. Richardson and Vepraskas (2001) reported that wetland soils or hydric soils had the following major characteristics: (i) water saturation (ii) anoxic situation, (iii) organic matter accumulation,(iv) gleying, (v) mottling, and iron/ manganese segregation, (vi) oxidizing root channels and soil pore linings, and (vii) a reduced soil matrix. The synonymous and interchangeable use of the terms – wetland soils, submerged soils, flooded soils, and poorly drained soils – may create some confusion and misconception. To the common people, wetlands are lands under water but ecologists consider wetlands as distinct ecosystems with permanent submergence and hydrophytic vegetation. Flooding and poor drainage may, however, be seasonal events.

Hydric soils include all soils that are submerged, seasonally flooded or saturated and produce an anaerobic condition due to expulsion of oxygen from soil pores and depletion of dissolved oxygen in soil water. The National Technical Committee on Hydric Soils (NTCHS 1985) defined hydric soils as undrained soils that are saturated, flooded, or ponded for a long period in the growing season and have developed anaerobic conditions that favor the growth and regeneration of hydrophytic vegetation. The Natural Resources Conservation Service, USDA modified this definition as: hydric soils are soils formed under conditions of saturation, flooding, or ponding long enough during the growing season to develop anaerobic conditions in the upper part (Federal Register 1994). In this definition the `growing season refers to the portion of the year when soil temperatures remain above 5 °C at 50 cm depth of soil. Thus, hydric soils mainly include wetland and submerged soils. Hydric soils should also include very poorly drained and poorly drained soils. Here the main characteristics of very poorly drained and poorly drained soils as defined by Soil Survey Staff (1993) are mentioned.

Very Poorly Drained Soils In very poorly drained soils the water table is at or near the soil surface for most of the growing season. The soil is saturated with water persistently or permanently and if not artificially drained, mesophytic plants cannot usually grow. The soil is generally situated in a depression or in a low lying level land.

Poorly Drained Soils Poorly drained soils are wet at a shallow depth usually covering the root zone seasonally or for a long period of the growing season. The soil is saturated at shallow or very shallow depth for a large part of the growing season and if the soil is not artificially drained, most mesophytic plants cannot grow. The soil, however, is not continuously wet directly below plow depth.

5.2 Criteria of Hydric Soils 87

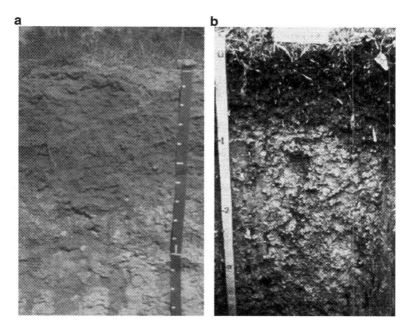

Fig. 5.5 Two soil profiles; (a) non-hydric, (b) hydric with prominent gleying and mottling. (Image courtesy of USDA-NRCS)

According to Egbuchua and Ojeifo (2007), hydric soils occur in (i) depressions, low lying areas, catchments (ii) in humid areas where there is high precipitation and accumulation of excess water, and (iii) places where the groundwater table rises to the root zone. Most hydric soils are found to develop in marshes, swamps, bogs, inland valleys, depressions, coastal plains, tidal flats, mudflats, estuaries and alluvial or marine deposits (Akamigbo 2001). Anaerobic and reducing conditions due to inundation or prolonged saturation produce distinct morphological features in hydric soils. Fig. 5.5 shows contrasting features of a non-hydric and a hydric soil.

5.2 Criteria of Hydric Soils

Soil formation under prolonged wetness gives hydric soils some unique properties. These properties are used as criteria for differentiating between hydric and non-hydric soils (Federal Register 1995). The criteria of hydric soils are discussed below.

1. Hydric soils include all soils in the suborder Histels except Folistels. Histels belong to the order Gelisols of the Soil Taxonomy. They are the soils of the cold regions. Folistels are a great group of Histels that are cold and may remain saturated with water for a short period of time, say less than 30 days, but the other great groups remain saturated with water for most of the year, and they can be

sometimes frosted. Histels have organic horizons similar to Histosols, but they encounter permafrost in the arctic and low arctic regions.
2. Hydric soils include all Histosols except Folists. Folists are a suborder of Histosols formed under upland forests on litter materials deposited over time. All other Histosols develop under wetland conditions by the accumulation of hydrophytic vegetation over a long time. Due to lack of oxygen, the decomposition of organic matter is very slow and thick organic horizons develop. They are permafrost-free soils with accumulation of organic soil materials which are characterized by the presence of 30 percent or more organic matter, if the mineral fraction contains 60 percent or more clay, or 20 percent organic matter if the mineral fraction has no clay; or a proportional intermediate organic matter for intermediate content of clay (Soil Survey Staff 2010).
3. Hydric soils also include soils of different suborders, great groups and subgroups that have an aquic soil moisture regime, Albolls suborder of the order Mollisols, Historthels great group, Histoturbels great group, Pachic subgroups, or Cumulic subgroups that are: (a) poorly drained or very poorly drained with water table at the surface in the growing season, and (b) poorly drained soils with water table within 15 cm if the permeability is low.
4. Hydric soils include those soils that are ponded or flooded for very long duration during the growing season.

Prolonged wetness of soils develops some features in soil that can be easily observed in the field. These are the indicators of hydric soils. Artificial flooding of land for a very long period also give soils hydric features, and some soils drained artificially may also retain some hydric soil properties such as redoximorphic features for some period after draining.

5.3 Features of Hydric Soils

5.3.1 Redoximorphic Features

When a land is waterlogged by rainfall or flooding, all soil pores are filled with water and soil air is excluded. Under conditions of high soil permeability and deep groundwater table, excess water is drained away almost immediately after rains. If the soil is on a sloping surface, flooding does not usually occur, and excess rain water is removed by surface drainage or runoff. If waterlogging does not persist, soil air is renewed again by mass flow and diffusion. On the other hand, if the soil is in a basin, the land is low-lying, permeability of the soil is low and if the groundwater table is at or near the soil surface, waterlogging persists for a long period of time. Lack of renewal of soil air, consumption of residual oxygen by roots and microorganisms and depletion of dissolved oxygen in soil water produce a reducing regime or anaerobiosis in the soil. Because of this prolonged anaerobiosis reducing chemical reactions predominate in the soil and result in distinct morphological features.

5.3 Features of Hydric Soils

Some oxidized zones are also observed in the soil profile as revealed by red or brown coloring, mottles of oxidized Fe and Mn, concretions and as pore lining (Vepraskas 1992). These reduction and oxidation or redox reactions are carried out by the soil bacteria through transfer of electrons from one element to another. Oxidation takes place by removing electrons and increasing valence state (e.g. Fe^{2+} to Fe^{3+}), and reduction occurs due to addition of electrons and decrease in valence state (Fe^{3+} to Fe^{2+}). Organic matter provides most electrons for redox reactions in soil. Decomposition of organic matter is an oxidation process that provides electrons by bacterial respiration. Craft (2001) suggested that microbial oxidation of organic matter in wetlands creates a reducing biogeochemical environment, and according to Verpraskas (2001), it favors redistribution of iron and manganese oxides to produce redoximorphic features. Redox reactions in waterlogged soils are, therefore, biogeochemical reactions.

The development of redoximorphic features in hydric soils may be shown as:

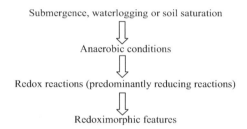

Since molecular oxygen is completely depleted after continued flooding, facultative and obligate anaerobic microorganisms use some oxidized compounds such as NO_3^-, MnO_2, $Fe(OH)_3$, SO_4^{2-} and CO_2 as electron acceptors for respiration and reduce them (Pezeshki and DeLaune 2012). The typical series of reduction reactions include reduction of NO_3^- to N_2, Mn^{4+} to Mn^{2+}, Fe^{3+} to Fe^{2+}, SO_4^{2-} to H_2S, S^{2+} or HS^- and reduction of CO_2 to CH_4. These biochemical reduction reactions are carried out respectively by denitrifying bacteria, Mn-reducing bacteria, Fe-reducing bacteria, sulfate reducing bacteria and methane bacteria.

As a result of submergence for a long period, the redox potential (Eh) is lowered (Pezeshki and DeLaune 2012) and the soil reaction reaches to almost neutrality (pH around 7). The above processes arise from the gradual reduction of oxygen, nitrate, and sulfate as waterlogging continues. On the other hand, carbon dioxide, ammonium, and sulfide gradually accumulate, and solubility of iron as well as manganese increases due to their reduction and lowering of valence states. After draining waterlogged soils and if the groundwater table is lowered, oxidation reactions prevail in the aerated zone of the soil. It usually happens in soils with fluctuating groundwater table in seasonally flooded soils. Substances that were reduced in the anaerobic conditions are oxidized again, for example Fe^{2+} to Fe^{3+} and Mn^{2+} to Mn^{4+}. As a result of such alternate oxidation and reduction, iron/ manganese mottling develops in the B horizon in seasonally flooded soils with fluctuating groundwater table. According to Gray (2010), redoximorphic features are indicative of prolonged soil moisture saturation. The most common features that develop with the period of waterlogging include (i) formation of Histosols (except folists) in permanently submerged soils, (ii) formation of a histic epipedon or O horizon, (iii) presence of sulfidic material or H_2S that gives odor of rotten eggs, (iv) gleying due to presence of reduced iron that gives low chroma (< 1) soil color within 30 cm of soil surface, and (v) presence of red, yellow or orange mottles in the mineral soil matrix. Some soils that have low amounts of soluble organic matter, high pH, cold temperatures, and low amounts of Fe may not exhibit redoximorphic features.

5.3.2 Chemical Transformations in Hydric Soils

Many environmentally significant biogeochemical transformations occur in hydric soils. These transformations lead to changes in pH and Eh, bring about gains and losses of nutrients, and cause accumulation and emission of greenhouse gases. These transformations are related to quality of soil, water and air in regions of vast wetland areas of all geographical regions. Some important transformations are briefly discussed below.

The pH

Oxidation-reduction reactions in soils subjected to prolonged waterlogging may change soil pH by 2 to 3 units. Usually soil pH tends to shift towards neutrality (approximately 7, but not essentially 7) after several weeks of flooding (Vepraskas and Faulkner 2000). Thus, pH of acid soils increases and of alkaline soils decreases. This change depends on the nature of redox reactions and the content of soil organic matter. In acid soils containing low organic matter, soil pH remains below 6.5 even after very long period of submergence (Ponnamperuma 1972). The increase in pH of waterlogged soils is due to the consumption of protons during reduction of mainly Fe and Mn in the form of hydroxides and carbonates as shown in the following reactions (Reddy and Delaune 2008).

$$Fe(OH)_3 + e + H^+ = Fe^{2+} + 3H_2O$$

$$MnO_2 + 2e + 4H^+ = Mn^{2+} + 2H_2O$$

On the other hand, formation of carbonic acid by dissolution of CO_2 in water, and dissociation of carbonic acid into HCO_3^- and H^+ ions cause the decline in pH of alkaline soils. Reddy and Delaune (2008) also concluded that acid or alkaline soils tend to be buffered around neutrality by the products of oxidation-reduction reactions in soils under submergence. Upon draining, hydric soils usually return to their original pH values. For example, acid sulfate soils (*Sulfaquents* according to Soil Taxonomy) under wet conditions reach pH values near 7 due to reduction of sulfate to sulfide, but when drained and dried their pH may fall beyond 3.

The Eh

A series of reactions take place upon flooding a soil and results in reducing the redox potential, Eh (Pezeshki and DeLaune 2012). Flooding cuts off the pathway of gas exchange between soil and atmosphere, and the limited oxygen in soil pores is depleted rapidly by roots, microorganisms, and chemical reductants in soil (Ponnamperuma 1972). Thus, a predominantly reducing regime is established and reactions include denitrification, reduction of iron, manganese and sulfate, and as a result there is a change in soil pH and Eh (Gambrell et al. 1991). The redox potential (Eh) refers to the tendency of a substance to accept or donate electrons. Therefore, Eh indicates the soil chemical environment in which oxidation or reduction reactions are likely to predominate. The Eh of soil can be relatively easily measured in the laboratory and in the field. The Eh values in soils vary widely between -300 to $+700$ mV. Well drained soils have Eh values $> +400$ mV, while in waterlogged soils Eh values are $< +350$ mV (Mitsch and Gosselink 2007). The permanently wet soils may have Eh value as low as -300 mV. The presence of oxygen or other oxidizing agents and the pH of soil determine its Eh, which is measured generally in millivolts (mV). The Eh decreases gradually from $+400$ to -300 mV in waterlogged soils depending on the intensity of reduction of oxidized substances. Generally, when oxygen is not present in soil, Eh values are below $+350$ mV (Pezeshki, 2001). At the initial stage of waterlogging, some O_2 may remain as entrapped air in soil pores, but it is used up rapidly by aerobic microorganisms at Eh values between $+380$ and $+320$ mV, and below this range of Eh there is a virtual absence of O_2 in the soil (Pezeshki 2001). Anoxia develops at further lower levels of Eh, and then there are some microorganisms that can use other electron acceptors than oxygen. Reduction of nitrate (NO_3^-) follows and Eh remains in the range of $+280$ to $+220$ mV. After disappearance of nitrate, Eh falls down, and at this level, reduction of Mn^{4+} commences. Similarly, reduction of Fe^{3+}, SO_4^{2-} and CO_2 occurs sequentially, and the Eh value may fall below -300 mV. According to Inglett et al. (2004), Eh is an index that is most commonly used to measure the degree of wetness or intensity of anaerobic condition in the soil.

Carbon

Organic matter decomposition in hydric soils occurs very slowly due to the absence of molecular oxygen and is carried out by anaerobic bacteria. Carbohydrates are converted to pyruvic acid in anaerobic systems; a portion of pyruvic acid is consumed by microorganisms for building their cell substances. Some pyruvic acid undergoes fermentation and other chemical changes depending on pH, Eh, osmotic pressure, available election acceptors, and microbial population of the soil. These changes include reduction to lactic acid, decarboxylation to CO_2 and acetaldehyde, formation of acetic acid, formic acid, oxaloacetic acid, and butyric acid. These products of anaerobic transformations undergo further biochemical reactions including reduction of acetaldehyde to ethanol, decomposition of ethanol to CH_4, and CO_2, decomposition of several organic acids to CO_2 and H_2, reduction of oxaloacetic acid to succinic acid, and reduction of several acids to alcohol. Ultimately, the decomposition of carbohydrates in anaerobic soils produces CO_2, fatty acids and CH_4. Initially, aerobes and facultative anaerobes cause anaerobic respiration, and finally decomposition of acids and reduction of CO_2 to CH_4 are brought about by strict anaerobes.

Nitrogen

Nitrogen is present in soil mainly in the organic forms, such as proteins and peptides. Organic nitrogen compounds are biologically converted to ammonium though a process known as ammonification. It occurs simultaneously with the decomposition of organic matter. Since the end products of organic matter decomposition under aerobic condition are inorganic substances including CO_2, H_2O, NH_4^+ and several bases, the process is called mineralization. Mineralization and ammonification occur more rapidly under aerobic conditions. However, Vepraskas and Faulkner (2001) suggested that ammonification can occur both aerobically or anaerobically, but it occurs faster in the upper oxidized zone of submerged soils than in the reduced subsoil. Ammonium may be absorbed by roots of plants, and in aerobic condition, it may be oxidized to nitrites and nitrates by the process known as nitrification. Nitrification can occur in two steps – first ammonium is converted to nitrite (NO_2^-), and second – nitrite to nitrate (NO_3^-). Bacteria of the genera *Nitrosomonas, Nitrosococcus,* and *Nitrobacter* carry out these transformations. As the substrate (NH_4^+ and O_2) concentration is low in submerged soil, the rate of nitrification is also low. However, the fate of nitrates in the aerated zone includes (i) absorption by plant roots, (ii) diffusion to the underlying reduced layer and denitrification there, and (iii) leaching to greater depths. Ammonia can be derived from anaerobic deamination of amino acids, degradation of purines, and hydrolysis of urea in the reduced soil where ammonium accumulates due to lack of O_2 which restricts its nitrification. If conditions are favorable, all mineralizable nitrogen in a soil is converted to ammonia within 2 weeks of submergence. Higher ammonium concentration in the reduced layer and lower in upper aerated zone creates a gradient which causes a passive flow to the upper layer of ammonium which can again undergo nitrification and other transformations (Scholz 2011). Nitrate is unstable in the reduced soil layer and it undergoes denitrification which is a process of biological conversion of nitrate to N_2 by a large number of bacteria and fungi. Denitrification causes loss of a substantial amount of nitrogen from waterlogged soils.

Phosphorus

Sources of phosphorus in hydric soils (submerged, saturated and seasonally flooded, poorly drained) include parent materials, sediments and effluents, dissolved phosphates in runoff water, surface flow, groundwater flow, organic residues and fertilizers. Circulation of phosphorus occurs in hydric soils through soil, water and organisms. Phosphorus is an element of constant valence state of 5, and therefore, it does not play any role in the redox systems. However, solubility of phosphorus is greatly influenced by the redox potential. Lowering of the Eh of soil by submergence favors solubilization of iron and manganese – a process that releases sorbed phosphates by these elements. Like non-hydric soils, hydric soils also have three major forms of phosphorus: organic P, fixed mineral P, and soluble inorganic P. Soluble inorganic P can exist as three anions $H_2PO_4^-$, HPO_4^{2-} and PO_4^{3-} in soil; and their prevalence occurs at pH values of 2 to 7, 8 to 12, and >13 respectively (Vepraskas and Faulkner 2001). Phosphorus chemically bound to oxides and hydroxides of Al^{3+}, Fe^{3+}, Ca^{2+} or Mg^{2+} is called fixed inorganic P. A major proportion of phosphorus in hydric soils is found in organic matter, and plant and animal residues in soil. This organic P is present in such compounds as inositol phosphates, phospholipids, and nucleic acids. Organic soils contain more than 90 percent of their P content bound with organic compounds, and mineral soils have the largest proportion of their P content bound to minerals containing Al, Fe, Ca, or Mg. The mineralization of organic P is carried out by heterotrophic microorganisms. As P mineralization is a very slow process in hydric soils due to low O_2, most P remains in the organic form in hydric soils. Water in wetlands can be loaded with phosphates from urban or agricultural wastes, fertilizer residues from croplands, and effluents from industries, and may undergo eutrophication. Hydrophytic plants, floating or rooted, absorb a significant amount of phosphorus from the soil and water. However, most phosphorus taken up by plants is incorporated into the soil/sediment again as dead plant materials, and accumulates in wetlands indefinitely. Scholz (2006) suggested that harvest of macrphytes at the end of the growing season can be a strategy of removing excess load of phosphorus from wetlands. This practice may reduce phosphorus levels in the upper portion of bottom soil or sediment causing phosphorus movement into deeper layers where P sorption capacity increases along with a lower desorption rate (Scholz 2011).

Sulfur
Sulfur undergoes several transformations in wetland soils, including, mineralization of organic sulfur, oxidation of elemental sulfur and sulfides, reduction of sulfates, and immobilization. These transformations are also biogeochemical in nature. As in case of ammonification, S-mineralization is also generally faster in the oxidized layer. Dissimilation of the amino acids cysteine, cystine, and methionine produces H_2S and thiol and fatty acids. Methyl, butyl and isobutyl thiols cause the putrefying odor of cyanobacteria. H_2S gives the odor of rotten eggs. The reduction of SO_4^{2-} in wetlands by the bacteria of the genus *Desulfovibrio* produces H_2S. *Desulfovibrio* uses sulfate as the terminal electron acceptor. Hydrogen sulfide may react with metals to form insoluble metal sulfides. It can also act as hydrogen donor to photosynthetic green and purple bacteria. Hydrogen sulfide can be oxidized chemically and biochemically at the borderline separating the upper oxidized and lower reduced zone. Sulfate reduction is a very rapid process in neutral to alkaline submerged soils. For example, the whole amount of sulfates may be reduced within some weeks of submergence of the soil. Sea water supplies a plenty of SO_4^{2-} ions to tidal and coastal wetlands. Sulfate is reduced to H_2S in wetland systems usually in the anaerobic subsoil, although the reduction process is very slow in acid soils. Hydrogen sulfide may also be produced by the mineralization of organic matter. Some H_2S may be lost from the soil through volatilization and some may diffuse to the upper aerobic layer where it is oxidized first to elemental sulfur and then to sulfate by a group of chemoautotrophic bacteria (for example, of the genus *Thiobacillus*). Some species of *Thiobacillus* gain energy from the oxidation of hydrogen sulfide to sulfur and, some species of this genus, from sulfur to sulfate. Sulfate ions may diffuse to the anaerobic zone and may be reduced there again. Thus, a H_2S-SO_4^{2-} circulation system develops between the oxidized and the reduced zone of the submerged soils.

Pyrite (FeS_2) may accumulate in some coastal wetland soils; upon draining, pyrite may be oxidized to H_2SO_4 and may give the soils a very acid reaction (pH 3 or less). These soils are known as acid sulfate soils (Sulfaquents).

5.3.3 Hydric Organic Soils

It has been mention in section 5.2 that all organic soils (Histosols) except some upland organic forest soils (Folists) of temperate region are hydric soils. Other soils that have an organic O horizon (histic epipedon) and remain submerged or saturated for long periods are hydric soils. Organic soils or organic horizons are characterized by the presence of organic soil materials (40 cm or more of the upper 80 cm in Histosols; at least 20 cm thick in histic epipedon or surface horizon (Soil Survey Staff 2003, USDA-NRCS 2010). Organic soil materials have 12 percent (by weight) or more organic carbon if the soil is saturated with water and if the mineral part of the soil contains no clay. If the soil is saturated and has 60 percent or more clay in the mineral part, it must contain at least 18 percent organic carbon to be organic soil material. If the mineral part of the soil contains intermediate clay between 0 and 60 percent it must contain proportionate intermediate content of organic matter (Soil Survey Staff 2003). Hydric organic soils are found in marshes, peatlands, bogs, swamps, fens and mires. Peat generally refers to an accumulated layer of organic residues in wetland conditions under various stages of decomposition. Peats have high content of fiber. An expansive area of peat is called a peatland which usually forms in depressions, slopes, and raised bogs. According to Rydin and Jeglum (2006), most wetlands have deep peat accumulation because of low decomposability of the submerged organic residues. Some peat forming wetlands have a little inflows or outflows. They are called bogs which are acidic in character and support acidophillic mosses such as *Sphagnum*. Wetlands that support woody vegetation are called swamps (Fig. 5.4 above). According to Collins and Kuehl (2001), the ecosystems in which vast expanse of waterlogged peat accumulation has occurred in raised areas are called mires in Europe. Organic formations are generally divided in two types – peat and muck. Peat contains fibric organic materials; these are plant remains the origin of which can be recognized. These organic materials can be partially decomposed or can remain almost undecomposed. On the other hand, mucks are organic materials in which the original plant remains cannot be recognized because it is well decomposed. Muck is well mixed with mineral matter as well, and is usually darker in color than peat. Roth (2009) stated that the upper layer of peat soils is almost completely made up of organic materials the individual plant parts of which can be identified easily even after centuries. The O horizons have usually three subordinate horizons such as Oa (sapric), Oe (hemic), and Oi (fibric), on the basis of the degree of decomposition. Sapric material includes organic materials in which less than one sixth is recognizable; fibric materials are at an early stage of decomposition with prominent fibers and two thirds of the materials are easily identifiable. Hemic materials are intergrades between sapric and fibric organic materials (FAO 2001).

5.4 Land Use in Hydric Soils

5.4.1 Natural Wetland Ecosystems

Natural wetland ecosystems develop from complex interactions of hydrological, geomorphological, climatic, biotic and edaphic factors under natural conditions. Most wetlands have accumulation of organic residues or organic soil materials in the bottom, but there are also many natural wetlands with non-soil substratum such as sediments, sands, rocks, corals, etc. However, there are hydric soils under the water in most wetlands. As natural wetlands are mostly situated in basins, depressions, lakes, ponds, etc., many of them including bogs, swamps, marshes, fens and peatlands are not cultivable. Their ecological functions are all very important in environmental quality context because they act in storage and filtration of water, recharge of groundwater, settling of sediments and removal of pollutants. They have significant biodiversity of flora and fauna. Wetland vegetation includes reeds, sedges, swamp forests, mangrove forests, etc. Wetlands are rich habitats of aquatic animals including fishes, snails, frogs, snakes, crocodiles, etc. Wetlands are breeding grounds of birds and sources of water for wildlife. Accumulated organic matter in wetlands, particularly peatlands, acts as important reservoirs of carbon. Natural wetlands are disturbed in many areas for different purposes. Drainage, land filling, peat harvest and destruction of vegetation are the main types of human disturbances of wetlands. Such disturbances create the risk of rapid decomposition of organic matter and emission of greenhouse gases. In view of the tremendous ecological importance and economic output of wetlands along with hazards of their degradation, the Ramsar Convention held in Iran in 1971 participated by governments, NGOs and wetland experts agreed to restrain from any kind of disturbance of the existing wetlands. Under undisturbed conditions, these wetlands keep a dynamic equilibrium with their environment. Their soils are not problem soils themselves; draining them for urban, industrial and agricultural developments and over-exploiting their resources create the problems.

5.4.1.1 Wetland Degradation and Wetland Rehabilitation

Many wetlands have been degraded by human actions. Even 50 years ago, people considered wetlands as wastelands and obstacles to industrial development. Swamps and marshes were the most preferable sites for development of farm lands, urban structures and industries without considering the environmental impacts of draining them. One estimate suggests that 50 percent of the wetlands of the United States were drained by 1970. After the Ramsar Convention on wetlands in 1971, the ecological and economic significance of the existing wetlands and the importance of improvement and restoration of degraded wetlands have rightly been conceived. Kirk (2004) highlighted the huge practical importance of wetlands in global element cycles and food production, their unique biogeochemistry and as centers of biodiversity. Wetlands cleanse the environment; environmental scientists consider them as kidneys of nature.

Many wetlands have suffered from biophysical disruption including biodiversity reduction, de-vegetation, change of hydrological conditions, alteration of geomorphology and replacement of ecological settings. The anthropogenic activities behind all these changes include obstructing or diverting water channels, draining, earth-filling, ridging, excavation, extracting peat, developing crop lands and constructing industrial or urban infrastructures. The causes of wetland degradation also include eutrophication, chemical contamination of surface and groundwater, and activities of undesirable terrestrial and aquatic organisms.

Vast wetland areas have undergone irreversible changes; their environmental settings have been so altered almost permanently that their restoration is not practical. However, there are many wetlands that have been degraded by continuous disruption of recharge and outflow of water that reduced area and depth of water, by loss of natural biodiversity, and by altering ecological functions, including geochemical transformations. These degraded wetlands need immediate rehabilitation; further degradation may lead to a stage when their restoration would be difficult and economically prohibitive. In some wetlands deposition of bottom sediments has changed hydrological conditions, reduced water depth and quality, diminished natural fertility and shifted the ecosystem dynamics. According to Liu (2008), the situation will continue to deteriorate if interventions are not made and then outcomes will be worse. The main objectives of rehabilitation and restoration of wetlands are: (i) improvement of water and sediment quality, and (ii) maintenance, and restoration of biodiversity. Here, rehabilitation refers to the improvement and recovery of natural functions and processes but not essentially aiming at reaching to the pre-disturbance state. On the other hand, restoration typically emphasizes returning a degraded ecosystem to its original condition. Rehabilitation and restoration of degraded wetlands and creation of new habitats are needed to replace losses that have occurred. The Department of Environment and Climate Change (2008) suggested that rehabilitation should improve wetland health and functions to such an extent that is feasible. Kentula (2002) mentioned that numerous losses are incurred when a wetland is damaged or destroyed. According to this author, rehabilitation and restoration of degraded wetlands extend their benefits to their surrounding ecosystems as well.

Fig. 5.6 shows a view of rehabilitation of degraded wetland in Australia.

Indicators of wetland dysfunction and degradation include:

(i) **Poor water quality:** dirty water, green/brown water, low dissolved oxygen
(ii) **Hydrology and geomorphology:** sedimentation, infilling, reduced depth of water, reduced area of submergence, raising of tiny islands,
(iii) **Change of biological community dynamics:** devegetation, loss of indigenous plants; prolific growth of environmental weeds, loss of native fauna, and invasion of alien species,
(iv) **Pollution:** litter, refuse, chemical contamination, gross pollutants, and
(v) **Aesthetics:** loss of visual amenity.

According to the Wetland Care Australia (2008), the techniques of rehabilitation of wetlands include: restoring hydrology, preventing soil erosion and sedimentation, managing weeds, revegetation preferably with native species, restoring habitats, managing grazing, controlling feral animals, preventing pollution, improving water

5.4 Land Use in Hydric Soils

Fig. 5.6 Revegetation for rehabilitation of Bibra Lake. Photograph taken in September 2010. (Image courtesy of Linda Metz, City of Cockburn, Western Australia)

quality, improving aquatic biodiversity, etc. However, specific restoration techniques will be needed for different habitats such as floodplains, mangroves, sea grasses, salt marshes, arctic wetlands, peatlands, freshwater marshes and swamp forests on a regional or a large watershed level (Erwin 2009). Wetland rehabilitation measures are taken to reverse or halt the decline of the health of the wetland ecosystems and to improve and regain the lost wetland services (WRC 2009). Interventions are made to reinstate hydrological, ecological and geomorphological conditions by such actions as (i) constructing physical structures to prevent erosion, trap sediments and rewet drained wetland areas, (ii) landscaping to reinstate diminished water quality, (iii) plugging of artificial drainage channels, (iv) addressing inappropriate agricultural practices and other sources of sediments and pollutants, (vi) revegetation and bioengineering, (vii) removing invasive weeds, etc.

Rehabilitation and restoration of ecosystems are challenging jobs. As any change in environmental settings may have major unforeseen and long lasting impacts, proper assessment of baseline information and designing by a team of experts must be carried out before the work is taken in hand. All the stakeholders including the beneficiaries must be consulted. People's participation is a necessary element in any rehabilitation work. Many rehabilitation efforts failed because local people did not accept the changes. In many instances, wetland restoration efforts do not perform as planned for several reasons, including poorly assumed performance criteria, inappropriate designs, inadequate baseline data, unsuitable site, lack of adaptability to situations emerging during progress of restoration operations.

For further reading interested readers may consult *Wetland Drainage, Restoration and Repair* by Thomas R. Biebighauser, The University Press of Kentucky, 2007; *Applied Wetlands Science and Technology* by Kent, Donald M. CRC Press 2001; *Wetlands: Ecology, Conservation and Restoration*, Edited by Raymond E. Russo, Nova Science Publishers, Inc., New York, 2008.

5.4.1.2 Artificial drainage of Peatlands

Peatlands have long been artificially drained for agricultural, horticultural, and forestry purposes, and for harvesting peat as a source of energy. Peatlands bring about environmental buffering by sequestering huge amount of carbon that is not oxidized to any significant extent. Artificial drainage of peatlands causes their rapid oxidation and release of CO_2 and other greenhouse gases. Thus, the sink of carbon is converted to a source of carbon. Peatlands are biologically diverse ecosystems that store carbon, and provide water resources. When peatlands are drained several environmental problems can arise that cannot be easily mitigated. Artificial drainage and burning the vegetation of peatlands strongly influence habitat conditions and ecological diversity. Holden et al. (2004) emphasized the need for rehabilitation and restoration of degraded peatlands. Ramchunder et al. (2009) observed that drainage and burning of peat in peatlands of the United Kingdom result in changing runoff regimes, enhanced organic matter decomposition, and changing biogeochemical cycling particularly of C, N and P. Land subsidence is a common occurrence in drained peatlands. Drainage and development of peatlands, environmental consequences of peatland use and land use change and management of peat soils have been treated separately in chapter 7.

5.4.2 Wetland Rice Production Systems

Some high yielding modern paddy rice (*Oryza sativa* L.) varieties can be grown all the year round in saturated and shallowly flooded soils. More than half of the world's population use rice as their staple food. Although rice can be grown also in well drained soils, most rice is grown in soils with drainage limitations, including permanently but shallowly flooded soils, seasonally flooded soils, and poorly drained soils. More than 95 percent of world's rice production comes from wetland soils. According to FAO (2004), rice provides dietary energy for people of 17 countries in Asia and the pacific, 9 countries in North and South America and 8 countries in Africa. Rice is grown without irrigation in monsoon if there is enough rainfall that is distributed well during the growing season. Rice is grown usually with irrigation in the dry season. Rice areas are classified into some types such as rainfed upland, rainfed shallow water lowland, rainfed deep water lowland and irrigated areas in South and Southeast Asia. Three rice crops can be grown in a year with supplemental irrigation in many countries, The International Rice Research Institute and various national rice research institutes have released several high yielding dwarf rice varieties that can be harvested within 90–110 days after transplanting, and as some of these varieties are day neutral, they can be grown in any season, rainfed or irrigated as the moisture conditions permit. Wetland rice fields are usually situated on floodplains, basins, deltas, estuaries and coastal plains. All types of hydric soils from sandy loam to heavy clay can be used for growing rice but heavy clay soils on river valleys are better suited. If not naturally flooded the land is inundated usually

5.4 Land Use in Hydric Soils

by border flooding, plowed, puddled, leveled and transplanted. A 5 to 15 cm depth of water is maintained after transplanting seedlings usually throughout the growing period to control weeds. However, some hydrophytic grasses and herbs can grow; they are usually eradicated manually. Wetland rice can be classified as follows: (i) **irrigated rice** – in areas with sufficient supply of water during the growing season, (ii) **lowland rice** – rainfed lowland rice areas vary considerably in rainfall characteristics, depth of flooding, depth of standing water during monsoon, frequency of flooding, etc. There are five categories such as shallow and favorable rainfed lowland, shallow and drought prone rainfed lowland, shallow and submergence prone rainfed lowland, medium deeply waterlogged rainfed lowland rice areas, and lowland rice areas that remain stagnant for 2–5 months, (iii) **deep water rice** – these areas are inundated with water for a depth of more than 50 cm; deep water rice can be integrated with aquaculture, and (iv) **tidal wetland rice** – this type is in found in coastal and estuarine areas.

5.4.2.1 Properties of Rice Soils

Land preparation for traditional rice cultivation is usually done by plowing and harrowing under flooded conditions. This practice causes puddling which reduces percolation and prevents loss of water and nutrients. The benefits of puddling also include better resource utilization, maintaining yield stability and high productivity by retaining water and nutrients as well as suppressing weeds (Surendra et al. 2001). Puddling decreases hydraulic conductivity of the soil and diminishes the amount of water needed to maintain saturation. Puddling can increase plant available water within the root zone by changing the pore size distribution. However, puddling destroys aggregates of the surface soil and compresses the subsoil. As a result, the bulk density of the subsoil is increased and a compacted, anthropogenic horizon known as plow sole or plow pan develops. Such physical transformations by wet cultivation are influenced by the intrinsic soil characteristics and the intensity and type of tillage (Singh and Ladha 2004). In rice-based crop rotations including a non-rice crop such as wheat, puddling for rice can hamper the performance of the subsequent non-rice crop. Seed germination rates are reduced in post-paddy cloddy soils due to inadequate seed–soil contact (Rahmianna et al. 2000). Tillage operations for rice may cause subsurface compaction which can induce drought susceptibility for the dry season crops by limiting root penetration to deeper soil (McDonald et al. 2006).

Flooded rice soils exhibit all the chemical features of hydric soils, including the depletion of O_2 with concomitant increase in CO_2 and establishment of a reducing regime. There is a tendency of an increase in pH of acid soils and a decrease in pH of alkaline soils gradually with the period of flooding, and soil pH reaches to about neutrality. The redox potential is lowered and NO_3^-, Fe^{3+}, Mn^{4+}, and SO_4^{2-} are reduced to NH_4^+, Fe^{2+}, Mn^{2+}, and S^{2-} respectively. Flooding enhances availability of P, K, Si, Mo, Cu, and Co and reduces that of N, S, and Zn (Zhou et al. 2014). The concentration of Fe^{2+} and Mn^{2+} increases due to reduction of Fe^{3+} and Mn^{4+} under anaerobic conditions. Nitrogen is lost from rice soils through denitrification and ammonia volatilization. Organic matter is decomposed anaerobically with the production of CH_4 and C_2H_4. However, the rate of production of CH_4 is dependent on other soil properties. Mitra et al. (2002) incubated a large number of rice soils from different locations in the Philippines anaerobically for 100 days to determine CH_4 production potentials and to examine its relationship with other soil properties. There were wide variations in the magnitude of CH_4 production and this variation was related with total N, soil texture, CEC, available K and active Fe content.

The above characteristics are common to almost all submerged soils whether rice is grown or not. Some rice soils, however, are seasonally flooded or have fluctuating groundwater table. The uppermost part of such soils passes through alternate oxidation-reduction cycles and is characterized by pronounced redoximorphic features in the morphology. In rice soils and in some other hydric soils that have hydrophytic vegetation with arenchyma connective tissue, oxygenation by their internal ventilation system takes place in the root zone soil. The unique capability of rice roots to oxidize the rhizosphere soil is due to two separate mechanisms – oxygen release and enzymatic oxidation. This is evidenced by the red coloration of iron oxide precipitates around rice roots. Many hydrophytic plants have internal aeration system which is crucial for their growth in waterlogged soils. Large volumes of aerenchyma connective tissue in rice roots assist in the diffusion of O_2 within the roots and to the soil. According to Colmer (2003a) rice roots also contain a barrier against radial O_2 loss from the basal zones. These mechanisms are responsible for longitudinal diffusion of O_2 towards the root tip and elongation of roots into anoxic part of the soil (Colmer et al. 2006). According to Colmer (2003b), rice is adapted to waterlogging primarily due to these two key features. Dissolved O_2 in floodwater also contributes to the oxidation of a thin layer of soil in the interface between water and soil. Below this layer, the soil is completely reduced and the subsoil becomes dark gray due to the presence of reduced iron and manganese and their complexes with organic matter. Some of the reduced substances move towards the surface soil by diffusion and mass flow and are oxidized again and precipitated. After harvest of the rice crop, the soil becomes unsaturated and aerated, and the part of the soil profile above the water table is oxidized again and it attains a highly mottled appearance.

Continuous rice cultivation is practiced in many irrigated areas. FAO (2004) reported that yields are declining under intensive rice monoculture in tropical Asia and Africa due to imbalanced application of fertilizers. Here, farmers usually apply more nitrogen and less phosphorus and potassium; imbalanced fertilizer application has created deficiency of other nutrient elements in soil for intensive HYV rice production. Sulfur deficiency has been reported from Bangladesh, Brazil, Indonesia, India, Myanmar, Nigeria, The Philippines and Thailand (Singh 2004). Continuous cropping, higher nutrient demands of HYVs, lack of micronutrient fertilizers, decreased use of organic fertilizers, removal of crop residues, etc. have created deficiency of a number of micronutrients. Singh (2004) mentioned deficiencies of Fe, Mn, Zn, B, and Mo in many rice growing soils, including calcareous soils, Vertisols, and Inceptisols. Zinc deficiency is the most widespread in countries of the Indo-Gangetic Plains, including Bangladesh, India, Nepal and Pakistan (Nayar et al. 2001). Coarse textured, calcareous, alkaline or sodic coarse textured soils with high pH and low organic matter are generally low in available zinc.

5.5 Plants Suitable for Poorly Drained Soils

Plants adapted to wetland habitats are generally called hydrophytes or hydrophytic plants. However, some upland plants such as *Lagerstroemia speciosa* (L.) can tolerate prolonged waterlogging, and some wetland plants such as *Barringtonia acutangula* (L.) thrive well in upland habitats. Most plants adapted to wetland conditions are reeds, sedges, swamp forest species, mangroves, etc. A list of trees and shrubs which are suitable for wetland landscaping, and wetland rehabilitation is given below:

Trees

Common Name	Botannical Name
Amur Maple	*Acer ginnala*
Hedge Maple	*Acer campestre*
Norway Maple	*Acer platanoides*
European Hornbeam	*Carpinus betulus*
Cockspur Hawthorne	*Crateagus crusgalli*
Leyland Cypress *x*	*Cupressocyparis leylandii*
Hardy Rubber Tree	*Eucommia ulmoides*
Ginkgo	*Ginkgo biloba*
Thornless Honey locust	*Gleditsia triacanthos*
Crape myrtle	*Lagerstomeia indica*
Crabapple	*Malus spp.*
Metasequoia	*Metasequoia glyptostroboides*
Norway Spruce	*Picea abies*
White Spruce	*Picea glauca*
Callery Pear	*Pyrus calleryana*
Sawtooth Oak	*Quercus acutissima*
Schumard Oak	*Quercus schumardii*
Japanese Pagoda tree	*Sophora japonica*

Some other tree species such as Red maple (*Acer rubrum*), River birch (*Betula nigra*), Hackberry (*Celtis occidentalis*), Green ash (*Fraxinus pennsylvanicum*), Sweet gum (*Liquidambar styraciflua*), Black gum (*Nyssa sylvatica*), Willow oak (*Quercus phellos*), Bald cypress (*Taxodium distichum*), Lacebark elm (*Ulmus parvifolia*), etc. can withstand waterlogging for relatively short duration Sources: www.shelbycountytn.gov/DocumentView.aspx?DID=1099 http://homeguides.sfgate.com/evergreen-trees-poorly-drained-soils-22335.html

Shrubs

Common Name	Botanical Name
Florida Anise	*Illicium floridanum*
Sarccocca	*Sarcococa hookerana*
Inkberry holly	*Ilex glabra*
Yaupon holly	*Ilex vomitoria*
Winterberry holly	*Ilex verticillata*
Chinese witchhazel	*Hamamelis virginiana*

(continued)

(continued)

Common Name	Botanical Name
Butterfly Bush	*Buddleia davidii*
Sweet shrub	*Calycanthus floridus*
Summer sweet clethra	*Clethra alnifolia*
Crape myrtle	*Lagerstroemia indica*
Mockorange	*Philadelphus coronarius*
Arborvitae	*Thuja spp.*

Sources: www.aces.edu/Marion/files/wetsoilshrub.pdf www.bartlett.com/resources/Shrub-Species-for-Poorly-Drained-Soil.pdf

5.6 Soils that Need Artificial Drainage

Soil moisture saturation depends on climate, seasons, geomorphology, soil types, etc. Some soils become waterlogged for a very long period; some others may be flooded temporarily. Presence of excess water at some time of the year is crucial for crop production and some other land use. Most arable crops suffer from poor aeration as a result of prolonged waterlogging for lack of adequate natural internal drainage. Stresses arise from low O_2, high CO_2, methane, H_2S, excess and often toxic Fe^{2+} and Mn^{2+}, and deficiency of micronutrients. The net effects are reduced growth and yield of crops. Excess soil water that causes waterlogging in croplands comes from high precipitation in humid regions, deposition of runoff water in depressions, surplus irrigation, high groundwater table and artesian pressure. Irrigation induced waterlogging may cause soil salinity in absence of adequate drainage. For prolonged waterlogging, preparation of crop field for sowing or transplanting in time, and choice of crops are restricted. Poor drainage hampers seed germination and seedling establishment of crops due to low oxygen supply. Saturated soils restrict root development and root activity of seedlings. Crop plants, unless adapted to wet soil environments, cannot perform their normal physiological functions and suffer from many physiological, nutritional and pathological disorders. Most mesophytic plants are not able to withstand waterlogging for more than a couple of days.

Soil conditions that make artificial drainage in agricultural lands a necessity include those with high groundwater table, low water permeability, dense soil layers that restrict water movement, depressions accumulating water from upper watershed, high clay contents that do not allow to develop many macropores for percolation of water and soils with excess water in the root zone. Fine textured upland soils of the humid region may also need artificial drainage because they can become waterlogged due to intensive rains during monsoon. Irrigated soils may need artificial drainage if they have low permeability. Since plant root systems can develop and extend properly in aerated soils and help the plants to use fertilizers and water

5.6 Soils that Need Artificial Drainage

more efficiently, artificial drainage of waterlogged soils improve yields of crops. Waterlogged soils are suitable for only a narrow range of crops; draining these soils widens the scope of multiple choices for crops. Well-drained soils offer greater flexibility of selecting desired crops.

5.6.1 Benefits of Artificial Drainage

Farmers undertake artificial drainage for protecting crop, increasing crop yield, and making conditions of the field suitable for tillage, planting and harvesting right in time. The benefits of artificial drainage include improving tilth and favoring seed germination and root development, reclaiming soil salinity, improving soil workability, reducing compaction and improving soil structure, porosity and permeability, and modifying soil temperature. Drainage favors gas exchange in soil and removes some toxins. Artificial drainage increases the scope of diversifying crops. Drainage extends growing period and harvesting length, and increases growth and yield. Farmers can include their desirable crops in rotations in well drained soils. High value crops can be introduced with lower risk of disease and pests. The principal objective of a drainage system is to lower the groundwater table (Fig. 5.7).

5.6.2 Drainage Systems

There are two major systems of artificial drainage for agricultural lands – surface and subsurface drainage systems. Both these systems are practiced worldwide, although developing surface drainage systems is relatively easy and cheap, and does not require a high level of technology. Farmers themselves can develop a surface

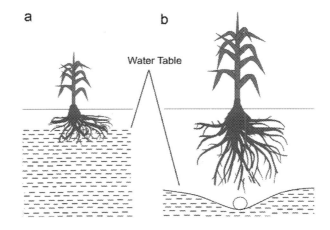

Fig. 5.7 Lowering of the groundwater table by artificial drainage

drainage system, but some land is lost in this system and ditches may hamper tillage operations. On the other hand, establishing a subsurface drainage system is costlier and requires technical assistance. According to Sheler (2013), a combination of these two methods is used for most agricultural drainage. However, the suitability of drainage systems depends on the type of soil, area of land, volume of water to be removed, farm facility, water disposing facility, etc. Surface drainage systems are preferable in impermeable soils; installing subsurface system would require a large number of tiles and that might not be economically feasible.

All drainage systems for agricultural fields have two major components – field drains and main drains. Field drains remove excess water from the field and lower the water table, while the main drains collect water from field drains, transport, and dispose water through an outlet. The surface systems may be of two types – regular systems and checked systems. The regular surface drainage systems remove excess water as soon as it comes through rainfall or irrigation, and gravity drives the water away. They can be divided into – bedding systems for flat lands, and graded systems for sloping lands. These systems usually have ridges and furrows. The subsurface drainage systems consist of horizontal or slightly sloping channels made at a suitable depth of the soil by buried pipes or moles. There may be another type of drainage known as vertical drainage consisting of a series of wells.

5.6.2.1 Surface Drainage

Surface Ditches
Surface ditch system is the most commonly practiced agricultural drainage system taken by farmers for ease of construction, operation, monitoring and maintenance. It is most suitable in soils at a flat topography with slow infiltration and low permeability. If the soil surface is not flat and regular, it must be leveled for draining soil in this system. The system is also suitable in soils with compact layers close to the surface. Channels and ditches distributed strategically or systematically over a carefully leveled land remove the excess water (Fig. 5.8).
There is a main drain, a number of field drains, and several field laterals in this surface drainage system. The main drain drives the water away through the outlet. The main drain is usually a grass waterway that reduces erosion. Field drains are shallow, graded channels that collect water within a field. Surface drainage systems also include diversion drains and interceptor drains. Interceptor drains are generally located at the bases of hills to prevent runoff water from reaching to the crop fields and causing flash floods. Surface drains can be arranged parallel or at random. The parallel system is the more preferable surface drainage system for flat and poorly drained soils. The spacing of the parallel ditches depends on slope of the land, permeability of the soil and the amount of water to be removed. If drainage limitation is not severe, raising the seed bed in the middle and gradually sloping towards the margins where shallow drains are dug to remove the excess water can improve waterlogging problem. This system of surface drainage is called raised bed system.

5.6 Soils that Need Artificial Drainage

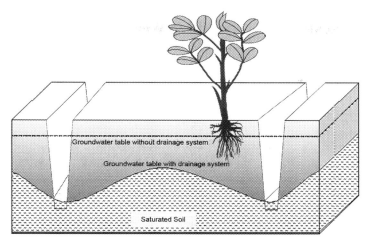

Fig. 5.8 Surface ditches are capable of lowering groundwater table. (Redrawn after FAO Corporate Document Repository, Chap. 6. Drainage; FAO 1997)

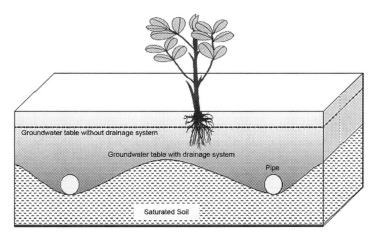

Fig. 5.9 Subsurface drainage systems lower the groundwater table

5.6.2.2 Subsurface Drainage

Subsurface drainage systems are more sophisticated systems because they do not involve any loss of land in unproductive purposes. Either tiles or perforated pipes and moles are placed at a suitable depth and at suitable intervals to lower the ground water table (Fig. 5.9). Installing such systems, however, involve high cost and need expert technical assistance. The depth and frequency of interval at which the pipes need to be placed depend on the type of soil and the depth of the groundwater table. The subsurface drainage systems have the advantage of draining soil to a greater depth than surface drainage systems. In many medium textured soils, subsurface drains placed at a depth of about 1 m and 25 to 30 m apart may be suitable for most crops. If a soil has adequate permeability, the drains need not be placed very closely. Then, the required number of pipes would be fewer and less cost would be involved. Moreover, if the soil is sufficiently productive, the investment could be justifiable. Subsurface systems have an outlet or main, and field laterals. Tile drainage and mole drainage are the major systems of subsurface drainage.

Tile drainage System
The name of the tile drainage system came from earthen or ceramic tiles which were earlier placed end to end at a suitable depth of the soil for draining agricultural lands. Now-a-days, earthen or ceramic tiles are replaced with perforated synthetic pipes and plastic tubing which are placed at a depth of the soil and at intervals depending on the soil and soil moisture characteristics. The factors that determine the size, depth, interval and arrangement of underground tiles or pipes include soil texture, porosity, permeability, compaction, depth of groundwater table, presence of root restrictive layers, rooting characteristics, amount of water to remove, topography, etc. According to Franzmeier et al. (2001), the slower the permeability of the soil is, the more closely must the tile lines be placed to lower the water table in a reasonable time. For installation of a subsurface pipe drainage system, a specific plan is prepared on the consideration of crop and soil types as well as site topography by experts, including engineers or experienced tile installers (USDA-NRCS 2002). A pipe drain system is composed of lateral, sub-main and main line piping. Laterals collect excess water from the soil and dispose it through the main line which acts as the outlet (Fig. 5.10).

Mole drainage System
Mole drainage is not very popular a system of subsurface drainage because the unlined channels of 40–50 mm in diameter at a spacing of 2–6 m may collapse if the soil is loose. However, if maintained well it can give satisfactory results when placed at a depth of 200–700 mm in a soil that has clay subsoil. The channel is constructed with a ripper blade having a cylindrical foot, often with an expander which helps compact the channel wall. Mole drains are used when a heavy clay subsoil prevents downward drainage. Mole drains perform well in soils containing at least 35 percent clay which enables the soil to hold together after pulling the mole through the soil. A series of fine fissures or cracks are created during pulling the mole, and water passes through these cracks in the soil into the mole channels and ultimately through the outlet.

The arrangement of tiles, pipes or mole lines may be of four types – random, herringbone, parallel, and double-main (Fig. 5.11). Any of these arrangements or their combinations can be used depending on the topography of the land. A random system is recommended for rolling land. For large wet areas, the submains and lateral drains for each area may be placed in a herringbone pattern for better drainage performance. In herringbone system, the main or submain drain is often placed in a narrow depression or on the major slope of the land. The parallel and the herringbone systems are combined in the double-main system which is used where a natural watercourse divides the field.

5.6 Soils that Need Artificial Drainage

Fig. 5.10 An outlet of a tile drainage system (Image courtesy of USDA-NRCS)

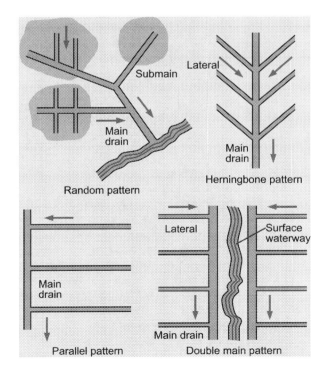

Fig. 5.11 Arrangement of subsurface pipes in field. (Image courtesy of Dr. Richard Cooke of Illinois University)

5.6.2.3 Vertical Drains

Vertical drains are 1 to 2 m deep vertical holes of 2 to 3 cm diameter and are constructed for horticultural plants. These holes also act as the planting holes for seedlings of trees or shrubs. A soil bucket auger or post hole digger is used to excavate the hole. Coarse gravels are filled in the holes to enhance percolation through these holes and to prevent the pooling in the root zone of the plants. A large number of such holes distributed throughout the entire horticultural field can provide satisfactory drainage and aeration so that the plant roots do not suffer from O_2 stress and harmful effects of toxins. Sometimes tube wells are used to withdraw groundwater and to lower the groundwater table. These tube wells are used for enhancing removal of excess water and for controlling soil salinity in some parts of the world. It is also a kind of vertical drainage. The water extracted from the groundwater can be used for irrigation and other purposes if the water is not saline.

5.7 Environmental Impact of Agricultural Drainage

Some investigators (Walker et al. 2000, Mallin 2009) reported the adverse effects of agricultural practices on the environmental components, particularly hydrology, soil erosion and water quality. Installation of artificial drainage systems brings about a change in natural preferential flow paths and significantly alters the hydrologic responses of these landscapes (Sheler 2013). Additions of drain water repeatedly can change the streamflow. For continual removal of soil water through artificial drainage in relatively less humid areas may permanently lower the groundwater table. This can reduce the groundwater supply and make the landscape drier. As a result, the soil become susceptible to drought and wind erosion (Kulhavy et al. 2006). The most noticeable negative effects in the receiving water include poor water quality due to enrichment of nutrients and eutrophication, loading of contaminants including pesticides and their residues, algal bloom and excess plant growth, and fish kills. Excess nitrogen and phosphorus are mainly responsible for eutrophication of surface water. Nitrates and soluble phosphorous move with the drainage water and are transported to the drainage outlets (Anon 2003). Nitrates and some pesticides are dissolved in water and enter into the tiles. Nitrate-N concentration in drainage water is related to cropping system, type, rate and time and frequency of nitrogen application, placement of nitrogen and tillage. Drainage water also contains sediments, nutrients, including nitrogen, phosphorus, and micronutrients, pesticide residues, etc. It has been shown in several occasions that tile drainage outflow contains elevated levels of nitrate, pesticides and pathogens originating from agrochemicals and manure. The discharge of contaminated water to a water body causes its degradation. Many investigators (Dolezal and Kvıtek 2004, Kvıtek et al. 2006, Lexa 2006, Lexa et al. 2006) have observed that tile drainage water containing high nitrate content is discharged directly into streams in some areas of the Czech Republic making the stream water quality unacceptable. Kulhavy et al. (2007) proposed that creation of artificial wetlands for drainage water discharge instead of disposing it directly into the streams may be one of the corrective measures. Sediments would be deposited in the constructed wetlands, and biogeochemical cycling will gradually remediate the toxins.

5.8 Drainage Water Reuse

The drainage water reuse is defined as the use of water in excess of evapotranspiration in addition to canal tail losses (Abdel-Azim and Allam 2005). One option of drainage water reuse is to mix a part of the drainage water with canal water so that the water use efficiency is improved. Usually water of the main drain is mixed with the water of the main canal. The factors affecting drainage water reuse include drainage water availability, crop sensitivity and tolerance to drainage water salinity, and the leaching requirement. However, assuming that the drains should be free of pollution the major suitability parameter of drainage water reuse is its salinity. Usually, drainage water is more saline than canal water. Drainage water, particularly collected in tile drainage system, can be slightly to highly saline and sometimes it can also be sodic depending on the salt and sodium level of the soil and intensity of leaching. There are examples that drainage water can have high concentrations of B, Mo, and Se which limit its reuse (Dudle et al. 2008). Drainage water with EC from 4 to 30 dS m^{-1} (saline) and SAR from 10 to 40 (sodic) can be used to irrigate crops which are moderately sensitive, moderately tolerant and tolerant to salinity (Oster and Grattan 2002). According to these authors, irrigation with saline and sodic drainage water can be sustainable only when particular attention is given to crop selection, preventing further soil salinization, and management so that soil permeability to water and air can be maintained. Sugar beet, sugar cane, dates, cotton and barley are among the most salt tolerant crops that can be irrigated with saline drainage water. Feasibility of using saline drainage water to irrigate a number of salt-tolerant forage crops, including Bermudagrass (*Cynodon dactylon*), salt grass (*Distichlis stricta*), silt grass (*Paspalum vaginatum*), alkali sacaton (*Sporobolus airoides*), Jose wheatgrass (*Agropyron elongatum*), cordgrass (*Spartina alterniflora*), creeping wild rye (*Leymus triticoides*), alfalfa (*Medicago sativa*), perla (*Phalaris tuberosa*), etc. has been tested. Irrigation with drainage water can be sequential and cyclic. The application of drainage water collected from one or more fields within a farm to irrigate salt tolerant crops on another is known as sequential reuse. In sequential reuse, salt concentration gradually increases and the drainage water remains suitable at a stage for application only to halophytes (salt loving plants). On the other hand, application of non-saline irrigation (for example, Ec = 0.5 dS m^{-1}) water during germination to crop establishment and saline water (for example EC = 8 dS m^{-1}) in the following growth period is done in the cyclic system. Using almost 50 percent of water requirement as saline drainage water in cyclic system to salt-sensitive and salt-tolerant crops was demonstrated to be sustainable (Rhodes 1987, Ayars et al. 1993). According to FAO (1997), Egypt has an extensive drainage water reuse program. Over 4000 million m^3 of agricultural drainage water, produced in the upper Nile Delta, is reused to supplement irrigation water requirements in the lower delta.

5.9 Wet and Cold Soils

Many poorly drained soils known as wet and cold soils occur in temperate and colder regions of northern Europe and North America. Some of these soils are seasonally flooded and have low temperatures, but others are permanently wet and cold. All types of hydric soils, including saturated, ponded and submerged soils are there. They are sporadically and spatially distributed in the frosted and permafrost areas in the boreal and boreal-tundra transition zones. There are agricultural lands, marshes, swamps, fens, peatlands, etc. in these areas. Schimel et al. (1996) conducted a study of drainage situations in agricultural lands in these areas and observed that high water content in soils at the onset of freezing reduced infiltration of snow melt. Freezing influences nutrient cycling and retention during these periods. Saturated and cold agricultural soils in these regions are used for growing corn, soybean, canola, vegetables, legumes, etc. in the spring. However, these crops have to face a number of problems. Low temperatures and poor aeration hamper seed germination, seedling emergence, root growth, microbial activity, nutrient cycling and nutrient uptake. Many vegetable crops fail to develop good stands due to low aeration and low temperature. Low soil temperature delays the emergence of vegetable seedlings and increases the risk of yield losses due to diseases and pests. Many warm season vegetable seeds such as snap beans, sweet corn, cucumbers, etc. take about 2 weeks to emerge at temperature of 15.6 °C. When temperature rises to about 21 °C, seedling emergence takes only 1 week. The length of time needed for seedling emergence has a bearing on the crop stand. The greater the time of sitting of the seed in soil, the greater is the risk of stand losses due to rots, diseases and insects. Brainard (2009) observed damping-off in carrots in wet and cold soils. Warming the soil to improve germination of seed and growth of plants is done by some farmers with the use of plastic mulches, raised beds, subsurface insulation, and row covers.

USDA-NRCS (2009) suggested that major alterations of soil properties for use can impact the whole soil ecosystem if conservation practices are not taken. They advised to reduce physical and chemical disturbance and to use cover crops if these soils are cultivated. Physical disturbance caused by tillage may deplete organic matter, increase soil crusting and reduce infiltration. It also increases concrete frost by reducing average pore space, and disturbs microbial function. Use of pesticides and herbicides can alter chemical and biological conditions of the soil. These agrochemicals can be harmful to soil fungi, bacteria, and earthworms which are critical for increasing infiltration and nutrient cycling. It has been shown that residual herbicides can result in more corn injury in cool and wet soils (Anon 2009). Cover crops can be of multiple benefits in such soils. Cold tolerant cover crops such as winter wheat and rye can utilize excess soil moisture and enhance planting date. Cover crops encourage root extension and earthworm activity. They increase nutrient cycling and microbial activity in soil (Osborne et al. 2002). USDA-NRCS (2009) advised to refrain from disturbing wet soils; tillage can make wet soils compact. Reduced tillage and building-up of organic matter are good options for the management of cold and wet soils. Organic matter can be built up if disturbance by tillage

5.9 Wet and Cold Soils

is reduced and crop diversity is increased. For example, Riley et al. (2009) reported that reduced tillage in Northern Europe is used often for winter crops because the quality of seedbed matters little for seed germination during rainy weather in autumn. If alfalfa, clover or grass are included in crop rotation the chance of soil compaction is much reduced because deep roots of alfalfa and clover improve soil porosity and increases soil organic matter.

Several types of poorly drained areas, including bogs, fens, marshes, shallow lakes and other wetlands are found in the boreal forest region and its surroundings. There, the winters are long and the summers are short. About 6 months of winter have mean temperatures below freezing and short summers have 50 to 100 frost-free days. The temperature varies widely between the lows of winter and the highs of summer. The bogs in glacial depressions are covered at some areas with a spongy mat of Sphagnum moss over water. Tundra plant species such as cotton grass and shrubs of the heath family grow on this mat. Trees like black spruce and larch grow around the edge of these cold wetlands. Fig. 5.12 shows a bog area in New York. There is a body of water partly covered by Sphagnum moss. The area surrounding the water body is poorly drained where black spruce and tamarack are found to grow.

The combined effects of cool climates and abundant moisture retard organic matter decomposition rates and often result in the formation of organic soils (Histosols) by the accumulation of organic matter in bogs. They are open hydrological systems receiving little discharge of water from groundwater aquifers. The main source of water is precipitation in these bogs. The runoff is almost normal in bogs, and it can add small amounts of water to regional groundwater systems by recharge These

Fig. 5.12 A bog in the Adirondacks of New York which includes the black spruce (*Picea marina*) and tamarack (*Larix laricina*) forest in the background. (Image courtesy of Donald J. Leopold)

Fig. 5.13 The open areas of Bear Meadows, a poor fen in Central Pennsylvania, that support a lush growth of leatherleaf (*Chamaedaphne calyculata*) and highbush blueberry (*Vaccinium corymbosum*). (Image courtesy of David J Walsh of the US Forest Service)

bogs are poor in nutrients and acidic in reaction so that they have poor to average productivity. Trees produce only very small amount of biomass, but moss productivity is very high. Under such conditions of coldness, wetness, low nutrient content and acidity, there is the development of a complex of vegetation dominated by such adaptable species as black spruce, tamarack, Atlantic white cedar, Northern white cedar, alder, sphagnum moss, sedges and heaths, including highbush blueberry, cranberry and leatherleaf. However, plant communities in fen are more diverse than that of a bog. Heaths are more abundant in bogs, and sedges are more plentiful in fens. Fig. 5.13 shows a poor fen in Central Pennsylvania, USA.

Study Questions
1. Give an account of soils with drainage limitations. Do the wetlands pose any actual problems? "Problems lie with the attitude of using wetlands." Explain.
2. Describe the characteristics of hydric soils. Mention the differences between soils that are permanently submerged and those that are seasonally flooded.
3. Make a list of plants suitable for poorly drained soils. Mention the benefits of artificial drainage of soil. Note soil characteristics that may give a soil good natural drainage.
4. What are the advantages and demerits of surface soil drainage? Describe how surface drainage is used to remove excess water from crop field.
5. What do you mean by wet and cold soils? Give an account of wet and cold soils of northern Europe and North America.

References

Abdel-Azim R, Allam MN (2005) Agricultural drainage water reuse in Egypt: strategic issues and mitigation measures. In: Hamdy A, El Gamal F, Lamaddalena N, Bogliotti C, Guelloubi R (eds) Non-conventional water use: WASAMED project Bari : CIHEAM / EU DG research options Méditerranéennes : Série B. Etudes et Recherches, No 53, pp 105–117

Akamigbo FOR (2001) Survey, classification and land use potentials of wetlands. Proceedings of the 27th annual conference of the Soil Science Society of Nigeria. 5-9th Nov 2001, Calabar, Nigeria

Anonymous (2003) Agricultural drainage water management systems for improving water quality and increasing crop production. Developed by the ADMS Task Force consisting of representatives from industry, universities, USDA-ARS, and USDA-NRCS. Coordination in development provided by William Boyd and Patrick Willey of NRCS. http://www.nrcs.usda.gov/technical/efotg/

Anonymous (2009) Cool, wet soils can result in more corn injury from residual herbicides. http://extension.missouri.edu/explore/agguides/pests /ipm 1007.htm. Published 28 April 2009. Accessed 5 Nov 2015

Ayars JE, Hutmacher RB, Schoneman RA, Vail SA, Pflaum T (1993) Long-term use of saline water for irrigation. Irrig Sci 14:27–34

Brainard D (2009) Cold, wet soils and vegetable seed emergence. Michigan State University Extension, Department of Horticulture, MSU, Michigan. http://www.msue.msu.edu

Collins ME, Kuehl RJ (2001) Organic matter accumulation and organic soils. In: Richardson JL, Vepraskas MJ (eds) Wetland soils, genesis, hydrology, landscapes, and classification. Lewis Publishers, Boca Raton

Colmer TD (2003a) Aerenchyma and an inducible barrier to radial oxygen loss facilitate root aeration in upland, paddy and deep-water rice (*Oryza sativa* L.) Ann Bot 91:301–309

Colmer TD (2003b) Long-distance transport of gases in plants: a perspective on internal aeration and radial oxygen loss from roots. Plant Cell & Env 26:17–36

Colmer TD, Cox MCH, Voesenek LACJ (2006) Root aeration in rice (Oryza sativa): evaluation of oxygen, carbon dioxide, and ethylene as possible regulators of root acclimatizations. New Phytol 170:767–778

Craft CB (2001) Biology of wetland soils. In: Richardson JL, Vepraskas MJ (eds) Wetland soils genesis, hydrology, landscapes, and classification. Lewis Publishers, Boca Raton, Florida

Department of Environment and Climate Change (2008) Saltwater wetlands rehabilitation manual. Department of Environment and Climate Change, NSW, Sydney South, p 1232

Dolezal F, Kvıtek T (2004) The role of recharge zones, discharge zones, springs and tile drainage systems in peneplains of central European highlands with regard to water quality generation processes. Phys Chem Earth 29(11–12):775–785

Dudley LM, Ben-Gal A, Lazarovitch N (2008) Drainage water reuse: biological, physical, and technological considerations for system management. J Environ Qual 37(5 Suppl):S25–S35

Egbuchua CN, Ojeifo IM (2007) Pedogenetic characterization of some wetland soils in Delta State. PhD Thesis, Delta State University, Abraka, Nigeria

Erwin KL (2009) Wetlands and global climate change: the role of wetland restoration in a changing world. Wetl Ecol Manag 17:71

FAO (1997) Management of agricultural drainage water quality. FAO Corporate Document Repository. www.fao.org/docrep/W7224E/W7224E00.htm

FAO (2001) Lecture notes on the major soils of the world. World soil resources reports 94. Food and Agriculture Organization of the United Nations, Rome, p 334

FAO (2004) Rice is life. Food and Agricultural Organization of the United Nations, FAO Rome

Federal Register (1994) Changes in hydric soils of the United States, vol 59. US Govt Print Office, Washington, DC

Federal Register (1995) Hydric soils of the United States. US Govt Print Office, Washington, DC

Finlayson M, Moser M (eds) (1991) Wetlands. Facts on File Limited, Oxlord

Franzmeier DP, Hosteter WD, R. E. Roeske (2001) Drainage and wet soil management: drainage recommendations for Indiana soils. AY-300, Purdue Extension. https://engineering.purdue.edu/SafeWater/Drainage/AY300.pdf

Gambrell RP, DeLaune RD, Patrick WH (1991) Redox processes in soils following oxygen depletion. In: Jackson MB, Davies DD, Lambers H (eds) Plant life under oxygen deprivation: ecology, physiology, and biochemistry. SPB Academic Publishing BV, The Hague, pp 101–117

Gray AL (2010) Redomorphic features induced by organic amendments and simulated wetland hydrology. MS Thesis, University of Maryland.

Holden J, Chapman PJ, Labadz JC (2004) Artificial drainage of peatlands: hydrological and hydrochemical process and wetland restoration. Prog Phys Geogr 28(1):95–123

Inglett PW, Reddy KR, Corstanje (2004) Anaerobic Soils. In: Hillel D et al (eds) Encyclopedia of soil in the environment. Academic Press, New York, pp 72–77

Kentula ME (2002) Restoration, creation, and recovery of wetlands. National Water Summary on Wetland Resources, United States Geological Survey Water Supply Paper 2425. https://water.usgs.gov/nwsum/WSP2425/restoration.html

Kirk G (2004) The biogeochemistry of submerged soils. John Wiley and Sons Ltd. Chichester, UK

Kulhavy Z, Dolezal F, Fucik P, Kulhavy F, Kvitek T, Muzikar R, Soukup M, Svihlas V (2007) Management of agricultural drainage systems in the Czech Republic. Irrig Drain 56:S141–S149

Kulhavy Z, Soukup M, Dolezal F, Cmelık M (2006) Agricultural land drainage: rationalisation of its exploitation, maintenance and repairs. Project QF3095 output. Research Institute for Soil and Water Conservation, Prague (in Czech)

Kvıtek T, Zajıcek A, Pursova K, Zlabek P, Bystricky V, Ondr P (2006) Verification of the effect of grassland and ploughland extent on the nitrate load of surface and subsurface waters, as a background for action programmes measures. Final report QF4062. Research Institute for Soil and Water Conservation, Prague (in Czech)

Lexa M (2006) Nitrate concentrations evaluation in small streams of the Zelivka river basin and analysis of catchments of these streams. PhD thesis, Charles University, Faculty of Science, Prague (in Czech)

Lexa M, Kvıtek T, Hejzlar J, Fucık P (2006) Relation among tile drainage systems, waterlogged soils covered by grassland and nitrate in brooks of the Svihov (Zelivka) drinking water reservoir basin. Vodnı hospodarstvı (8):246–250 (in Czech)

Liu JD (2008) Potential rehabilitation of rugezi highland wetlands. Environmental Education Media Project (EEMP). eempc.org/storage/EEMP_RugeziHighlandWetlands.pdf

Mallin MA (2009) Effect of human land development on water quality. Handbook of water purity and quality,pp 67–94

McDonald AJ, Riha SJ, Duxbury JM, Steenhuis TS, Lauren JG (2006) Soil physical responses to novel rice cultural practices in the rice–wheat system: comparative evidence from a swelling soil in Nepal. Soil Till Res 86:163–175

Mitra S, Wassmann R, Jain MC, Pathak H (2002) Properties of rice soils affecting methane production potentials: 1. Temporal patterns and diagnostic procedures. Nutr Cycl Agroecosyst 64(1):169–182

Mitsch WJ, Gosselink JG (2000) Wetlands, 3rd edn. John Wiley & Sons, New York

Mitsch WJ, Gosselink JG (2007) Wetlands, 4th edn. John Wiley & Sons, Inc, Hoboken, p 582

Nayar VK, Arora CL, Kataki PK (2001) Management of Soil Micronutrient Deficiencies in the Rice-wheat cropping system. In: Kataki PK (ed) The Rice-wheat cropping Systems of South Asia: efficient production management. Food Products Press, New York

NTCHS (1985) Hydric soils of the United States. National Technical Committee for Hydric Soils, USDA. Soil Conservation Service, Washington DC

Osborne SL, Riedell WE, Schumacher TE, Humburg DS (2002) Use of cover crops to increase corn emergence and field trafficability. Progress Report #SOIL PR 02–39, Ag Exp Stn, Plt Sci SDSU. Brookings, SD

Oster JD, Grattan SR (2002) Drainage water reuse. Irrig Drain Syst 16:297–310

Pezeshki SR (2001) Wetland plant responses to soil flooding. Environ Exp Bot 46:299–312

References

Pezeshki SR, DeLaune RD (2012) Soil oxidation-reduction in wetlands and its impact on plant functioning. Biology 1:196–221

Ponnamperuma FN (1972) The chemistry of submerged soil. Adv Agron 24:29–96

Rahmianna AA, Adisarwanto T, Kirchoff G, So HB (2000) Crop establishment of legumes in rainfed lowland rice-based cropping systems. Soil Till Res 56(1–2):67–82

Ramchunder SJ, Brown LE, Holden J (2009) Environmental effects of drainage, drain-blocking and prescribed vegetation burning in UK upland peatlands. Prog Phys Geogr 33(1):49–79

Reddy KR, Delaune RD (2008) Biogeochemistry of wetlands: science and applications. CRC Press, Boca Raton

Rhoades JD (1987) Use of saline water for irrigation. Water Quality Bulletin 12:14–20

Richardson JL, Vepraskas MJ (2001) Wetland soils, genesis, hydrology, landscapes, and classification. Lewis Publishers, Boca Raton

Riley H, Børresen T, Lindemark PO (2009) Recent yield results and trends over time with conservation tillage on clay loam and silt loam soils in southeast Norway. Acta Agriculturae Scandinavica, Section B: Plant Soil Science 59(4):362–372

Roth RA (2009) Freshwater aquatic biomes. Greenwood Press, Westport, London

Rydin H, Jeglum JK (2006) The biology of peatlands. Oxford University Press, Oxford

Schimel JP, Kieland K, Chapin FS (1996) Nutrient availability and uptake by tundra plants. In: Tenhunen JD (ed) Landscape function: implications for ecosystem response to disturbance: a case study of Arctic tundra. Springer-Verlag, New York

Scholz M (2006) Wetland systems to control urban runoff. Elsevier, Amsterdam

Scholz M (2011) Wetland systems: storm water management control. Springer, London

Sheler RJ (2013) The impact of agricultural drainage systems on hydrologic responses. MS thesis, University of Iowa. http://ir.uiowa.edu/etd/2630

Singh MV (2004) Micronutrient deficiencies in Indian soils and field usable practices for their correction. IFA International Conference on Micronutrients, Feb. 23–24, 2004, at New Delhi

Singh Y, Ladha JK (2004) Principles and practices of tillage systems for rice–wheat cropping systems in the indo-Gangetic Plains of India: some experiences. In: Lal R (ed) Sustainable agriculture and the international Rice–wheat system. Marcel Dekker, New York, pp 167–208

Soil Survey Staff (1993) Soil survey manual. USDA–SCS. Agric hand book. US Govt Print Office, Washington, DC, p 18

Soil Survey Staff (2003) Keys to soil taxonomy, 9th edn. USDA-NRCS, Washington, DC

Soil Survey Staff (2010) Keys to soil taxonomy, 11th edn. USDA-NRCS, Washington, DC

Surendra S, Sharma SN, Prasad R (2001) The effect of seeding and tillage methods on productivity of rice–wheat cropping system. Soil Till Res 61(3–4):125–131

USDA-NRCS (2010) In: Vasilas LM, Hurt GW, Noble CV (eds) Field indicators of hydric soils in the United States, version 7.0. United States Department of Agriculture, Natural Resources Conservation Service in cooperation with the National Technical Committee for Hydric Soils, Washington, DC

USDA-NRCS (2002) NRCS conservation practice standards and specifications for sub-surface drains, code 606. Available on the internet at: http://efotg.nrcs.usda.gov/references/public/WI/606.pdf, Accessed 8 Nov 2009

USDA-NRCS (2009) Soil quality: managing cool, wet soils. Soil quality – agronomy technical note No. 20. United States Department of Agriculture and Natural Resources Conservation Service, Auburn

Vepraskas MJ (1992) Redoxmiorphic features for identifying aquic conditions. NC Agric Res Serv, Tech Bull 301, Raleigh

Vepraskas MJ (2001) Morpholocial features of seasonally reduced soils. In: Richardson JL, Vepraskas MJ (eds) Wetland soils: genesis, hydrology, landscapes, and classification. Lewis Publishers, Boca Raton

Vepraskas MJ, Faulkner SP (2000) Redox chemistry of hydric soils. In: Richardson JL, Vepraskas MJ (eds) Wetland soils: genesis, hydrology, landscapes and classification. CRC Press, Boca Raton

Vepraskas MJ, Faulkner SP (2001) Redox chemistry of hydric soils. In: Richardson JL, Vepraskas MJ (eds) Wetland soils, genesis, hydrology, landscapes, and classification. Lewis Publishers, Boca Raton

Walker JT, Aneja VP, Dickey DA (2000) Atmospheric transport and wet deposition of ammonium in North Carolina. Atmos Environ 34(20):3407–3418

Wetland Care Australia (2008) Wetland rehabilitation guidelines for the great barrier reef catchment. Queensland Wetlands Programme/Govt of Australia, Ballina

WRC (2009) Wetlands: tools for effective management and rehabilitation of wetlands. Water Research Commission, South Africa

Zhou W, Teng-Fei L, Chen Y, Westby AP, Ren WJ (2014) Soil physicochemical and biological properties of Paddy-upland rotation: a review. Sci World J 2014:856352

Chapter 6
Expansive Soils

Abstract Soils that contain a large amount of clay – at least more than 30 percent, a large proportion of fine clay in the clay fraction, and the clay fraction generally dominated by 2:1 expanding type of smectitic clay, chiefly montmorillonite, expand in volume when wetted and shrink when dried; they shrink so severely that deep and wide cracks, through which soil materials can slide downward, develop in the dry season. These clay soils are known as expansive soils, shrink-swell soils, cracking soils, or vertic soils. Some clay soils contain high proportion of exchangeable sodium in colloidal surfaces. They remain dispersed and are called dispersive clay soils. Their consistence – very sticky when wet and very hard when dry, their cracks, and their contraction and expansion in volume with changes in soil moisture offer severe limitations to their agricultural and engineering uses. Unique morphological features such as slickensides in the middle of the profile and circular or polygonal landscape features known as gilgai often develop on the surface soil due to their alternate swelling and shrinking behavior. These soils are classified in the Vertisols order of Soil Taxonomy and Vertisol Reference Soil Group of World Reference Base for Soil Resources. These soils were earlier called Regur, Gilgai, Margalite, Tirs, Black Cotton Soils, etc. The major areas of Vertisols are found in Australia, India, Sudan, Chad and Ethiopia. For their profitable and sustainable agricultural use potential, the International Crops Research Institute for the Semi-Arid Tropics (ICRISAT), International Livestock Centre for Africa (ILCA), and the Agricultural Research Centre for the Semi-arid Tropics (CPATSA) have been developing innovative management packages including broad bed and furrow system, reduced tillage systems and their modifications.

Keywords Expansive clays · Vertisols · Shrink-swell soils · Cracking soils · Gilgai · Pedoturbation · Slickensides · Broad bed and furrow · Broad bed maker

6.1 Types and Distribution of Expansive Soils

Expansive soils are clay soils that expand in volume when wetted. They usually contain more than 30 percent clay to a minimum depth of 50 cm, and in majority of the cases, the dominant clay is the swelling type 2:1 smectites (chiefly montmorillonite). These soils are alternatively known as shrink-swell soils because they are contracted as well when dried. When they are dried, they shrink so greatly that very wide and deep cracks develop on the surface to depths often extending more than a meter downward. They are thus called deeply and widely cracking soils, or simply cracking soils. Due to alternate swelling and shrinking, polygonal (or circular) mounds often develop in many of such shrink-swell soils as a distinct landscape feature. These micro-relief features are called "gilgai". Expanding clay soils that contain high proportion of exchangeable Na^+ ions become dispersed, and are then called dispersive soils or dispersive clay soils. Expansive and dispersive clay soils have usually a dark appearance due to dispersed clay-humus complexes and reduced manganese compounds.

In Soil Taxonomy (Soil Survey Staff 1999), these soils are classified in the order Vertisols. These soils are also grouped in the Reference Soil Group Vertisol of the World Reference Base for Soil Resources (FAO 2006). The name of Vertisols is derived from the Latin *vertere* meaning to invert (Dengiz et al. 2012) because soil materials are washed in (inverted) through the cracks downward. The soil order name itself implies the behavior of the soils, such as cracking and the presence of slickensides (a slickenside is a surface of the cracks produced in soils containing a high proportion of swelling clays. Pedogenic slickensides are convex-concave slip surfaces that form during expansion/contraction in expansive clay soils or Vertisols. Slickensides are found in association with other pedogenic features, such as clay-skinned peds, *in-situ* calcareous nodules, and root impressions). However, before recognizing Vertisols as an order of the new system of soil classification in 1960 by the Soil Survey Staff of USDA, these soils were called by different names in different parts of the world. At least some 50 local names could be identified in different regions of the world; the most familiar ones were: Regur (India), Adobe (USA, Philippines), Gilgai (Australia), Margalite (Indonesia), Tirs (Morocco), Black Clays, Black Cracking Clays, Black Cotton Soils, Dark Clay Soils (India, East and South Africa), Dian Pere (West Africa), Firki (Nigeria), Makande (Malawi), Mbuga (Tanzania), Mourcis (Mali), Badobes, Teen Suda (Sudan).

According to Soil Survey Staff (1999), Vertisols are mineral soils that have all of the following characteristics:

1. A layer 25 cm or thicker, with an upper boundary within 100 cm of the mineral soil surface, that has either slickensides close enough to intersect or wedge-shaped structural units that have their long axes tilted 10–60 degrees from the horizontal; and
2. A weighted average of 30 percent or more clay in the fine earth fraction either between the mineral soil surface and a depth of 18 cm or in an Ap horizon, whichever is thicker, and 30 percent or more clay in the fine-earth fraction of all horizons between a depth of 18 cm and either a depth of 50 cm or a densic, lithic, or paralithic contact, a duripan, or a petrocalcic horizon if shallower; and
3. Cracks that open and close periodically.

6.1 Types and Distribution of Expansive Soils

The order Vertisols have six sub-orders:

Aquerts Vertisols with aquic soil moisture regime for most years and show redoximorphic features are grouped as Aquerts. Because of the high clay content, soil permeability is slow and moisture saturation prevails for a large part of the year. Under wet soil moisture conditions, iron and manganese are mobilized and reduced. The manganese may be partly responsible for the dark color of the soil profile.

Cryerts They have a cryic soil temperature regime. Cryerts are most extensive in the grassland and forest-grassland transitions zones of the Canadian Prairies and at similar latitudes in Russia. These soils are not included in Vertisols of FAO Classification.

Xererts They have a thermic, mesic, or frigid soil temperature regime. They show cracks that are open at least 60 consecutive days during the summer, but are closed at least 60 consecutive days during winter. Xererts are most common in the eastern Mediterranean and some parts of California.

Torrerts They have cracks that are closed for less than 60 consecutive days when the soil temperature at 50 cm is above 8 °C. These soils are not extensive in the USA, and occur mostly in west Texas, New Mexico, Arizona, and South Dakota, but are the most extensive suborder of Vertisols in Australia.

Usterts They have cracks that are open for at least 90 cumulative days per year. Globally, this suborder is the most extensive of the Vertisols order, encompassing the Vertisols of the tropics and monsoonal climates in Australia, India, and Africa.

Uderts They have cracks that are open less than 90 cumulative days per year and less than 60 consecutive days during the summer. In some areas, cracks open only in drought years. Uderts are of small extent globally, being most abundant in Uruguay and eastern Argentina, but also found in parts of Queensland and the "Black Belt" of Mississipi and Alabama.

FAO (2006) defined Vertisols specifically as "soils having, after the upper 18 cm have been mixed, 30 percent or more clay in all horizons to a depth of at least 50 cm; developing cracks from the soil surface downward which at some period in most years (unless the soil is irrigated) are at least 1 cm wide and extend to a depth of 50 cm: having intersecting slickensides or wedge-shaped or parallel-piped structural aggregates at some depth between 25 and 100 cm from the surface, with or without gilgai". In the Australian Soil Classification system (CSIRO 2010), soils that consist of more than 35 percent clay throughout the solum, crack at some time in most years, and contain slickensides and/or wedge-shaped peds are recognized as Vertisols. According to FAO (2006) estimate, Vertisols cover 335 M ha worldwide. About 150 M ha is potential cropland. In the tropics, there are some 200 M ha; 25 percent of this is considered to be useful land. Most vertisols occur in the semi-arid

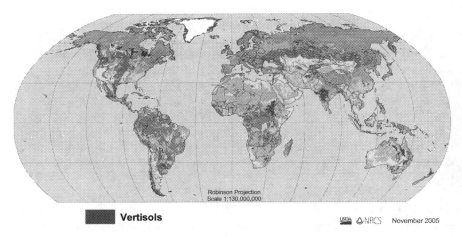

Fig. 6.1 Global distribution of Vertisols

tropics with an average annual rainfall of 500–1000 mm, but Vertisols are also found in the wet tropics, for example Trinidad. The largest Vertisols areas include South Africa, Sudan, India, Ethiopia, Australia, the southwest of the United States of America, Uruguay, Paraguay and Argentina. Two main types of Vertisols can be distinguished: lithomorphic Vertisols and topomorphic Vertisols. Lithomorphic Vertisols are formed on various parent rocks whose weathering generates base-rich environments favourable for smectite synthesis, while topomorphic Vertisols are formed mainly in low landscape positions which favor the accumulation of bases (FAO 2006). The global distribution of Vertisols is shown in Fig. 6.1 (map).

6.2 Parent Materials of Vertisols

Vertisols develop on a variety of parent materials derived from igneous and sedimentary rocks. Vertisols can develop on igneous rocks including basalt, dolerite, ash, tuff and andesite in different regions all over the world. Vertisols have also developed from Rhyolites which are composed of volcanic glass, quartz crystals, orthoclases, biotites and hornblendes as accessories. Vertisols in Central Mexico are derived from basalts, which are abundant in the lowlands. In coastal area of the Gulf of Mexico, Vertisols developed from sedimentary parent materials, mainly from limestone. Sotelo-Ruitz et al. (2013) studied the mineralogical properties of some Vertisols developed on parent materials of igneous and sedimentary origins. Dominant minerals in the sand fraction of soils of igneous origin consisted of volcanic glass (47 percent), quartz (31 percent), and feldspars (22 percent). The clay fraction was dominated by amorphous materials, smectites, vermiculites, illites, and cristobalites. In contrast, the sand fraction in the soils of sedimentary origin was composed of calcite (64 percent), quartz (34 percent), and feldspars (2 percent). Smectites, vermiculites, quartzes, and feldspars composed the clay fraction. The parent material of the soils on igneous rock was rhyolite, while the sedimentary soils were derived from limestone and sediments with high calcium carbonate contents. Özsoy and Aksoy (2007) studied 11 different soil profiles formed on the neogene aged calcareous marl parent materials. Among these soils the Vertisols were deep, dark colored with strong wedge-shaped structure, high in CEC and base saturation with calcium and magnesium occupying more than 90 percent of the exchange sites. Aydinalp (2010) reported the development of Vertisols in different parent materials in northwestern Turkey. These soils occur on flat to gently sloping plains of the region. Clay content is high in the studied sites. The high cation exchange capacity and CEC/clay ratios suggest montmorillonitic and mixed mineralogy of the clay fraction. Calcium was the most dominant extractable cation followed by magnesium. Dengiz et al. (2012) studied the morphological and physico-chemical characteristics of Vertisols formed on the alluvial delta Bafra Plain located in the central Black Sea region of Turkey. These Vertisols were all dark in color in surface soil; they were heavy clayey soils with hardpan (high bulk density and a high compaction) formation under top soil. Vertisols may form residually from weathered limestone or basalt. These soils are generally developed from parent materials that are rich in alkaline earth cations (Ca and Mg). The weathering of these rocks produces smectite type clays. Heidari et al. (2008) studied some Vertisols with diverse parent materials and climates from western Iran. The Vertisols of Fars Province, Lorestan Province and Kermanshah Province have formed on limestone or calcareous sediments. In Ardebil Province, Vertisols developed on volcanic sediments. Pierre et al. (2015) studied the mineralogical properties of some Vertisols of the Logone Valley in Cameroon. Clay minerals are dominated by smectites associated with some amount of kaolinite and illite. Dominant primary minerals are quartz and feldspars. These soils have high contents of SiO_2 (61.07–77.78 percent), moderate content of Al_2O_3 (7.08–15.54 percent) and low amount of Fe_2O_3 (1.78–6.92 percent).

6.3 Properties of Expansive Soils

6.3.1 Morphological Features

The principal morphological features of Vertisols include deep and wide cracks (Fig. 6.2) in the dry season, pedoturbation, minimal horizon differentiation, and unique subsurface features called slickensides. These characteristics in soil develop due to the presence of high clay content (usually >30 percent), by the activity of expanding type of clay (usually smectite, but other clay types may also contribute), and alternate saturation and desaturation of soil with moisture in different seasons.

Fig. 6.2 Deep and wide cracks in expansive soils (Vertisols). (Photo courtesy of Professor Paul McDaniel, University of Idaho)

These soils exhibit minimal horizon differentiation as a result of pedoturbation. They are also very plastic and sticky when wet. For high clay contents and their high surface area, the soil volume changes with the variation in moisture content, i.e. swells when wet and shrinks when dried. The shrink–swell process can generate pressures that cause vertical movement (heaving) on the order of 10–20 cm (Miller and Bragg 2007). Extreme heaves of 45–90 cm have also been found. The physical movement of Vertisols commonly results in the formation of surface mound and depression micro-relief features called gilgai (Miller et al. 2010). Gilgai (Fig. 6.3) is an Australian aborigine term meaning "little water hole".

Tamfuh et al. (2011) studied the morphological properties of some Vertisols of the Sudano-Sahelian Region of North Cameroon. They observed that with a depth of about 2–2.5 m above the water table, these soils show four main horizons from bottom to top: a dark grey horizon with hydromorphic patches (B3g), dark grey horizon (B21), dark grey horizon with slickensides (B1) and a surficial grey humiferous horizon (A1) with desiccation cracks. Also, they show a heavy clayey texture, very massive structure, high bulk density, very low porosity and a high compacity. The microfabric of the soils is marked by abundant plasmas, isotic at the surface but birefringent at depth, with numerous stress cutans.

Gilgai is composed of mounds and depressions; the bottom part of the depression is known as microlow. The microlow can retain water during rainfall and many hydrophytic and mesophytic plants may grow there. The surrounding top convex ridge-like part is called microhigh. The microhigh is usually drier, and generally xerophytic plants grow on microhighs. The area between the lower level of microlow to the upper part of the microhigh is called microslope (Miller et al. 2005). The microlow is concave, the microhigh is convex and the microslope is slightly sloping. The microlow is about 2–5 m deep. Miller and Bragg (2007) reported that in forested Vertisols trees tend to grow on microhighs, and mixed forbes and grasses occupy the

6.3 Properties of Expansive Soils

Fig. 6.3 Gilgai in Laewest clay, Calhoun County, USA. (Image courtesy of USDA)

microlows. Özsoy and Aksoy (2007) observed deep, dark colored Vertisols with strong wedge-shaped structure to develop on calcareous marl parent material.

Extensive shrinking produces wide (>1 cm) and deep (>50 cm) cracks that split and merge periodically. Soil materials are washed downward through the cracks and produce slickensides. Shrinking and swelling followed by pedoturbation creates wedge-shaped structural aggregates that are tilted with an angle from the horizon. Shrink-swell processes in soils are related to total clay content, fine clay content and minerals. Smectite and 'mixed layer' clays comprise an important proportion of the clay fraction in most Vertisols (Özsoy and Aksoy 2007). The shrink-swell phenomenon is responsible for the genesis and behavior of the Vertisols. Expressions of this phenomenon are linear and normal gilgai, cylic horizons, surface cracks and slickensides. Sotelo-Ruitz et al. (2013) suggested that the presence of smectites is responsible for morphological variations in Vertisols. Dengiz et al. (2012) observed prominent slickensides at the middle part of the profiles and a poor differentiation of horizons in Vertisols. The degree and frequency of changes in moisture content of the soil are perhaps the most important parameters that control cracking intensity. Vertisols are typically developed on alluvial material in flat inland areas. A dry layer of fine granules are usually found at the surface of Vertisols in varying thicknesses from several millimeters to a few centimeters. During grazing in the dry season or at the begging of rains, this layer slides into the cracks. Surface soil and sub-surface soil are mixed in this way, a process known as 'churning' or pedoturbation (Deckers et al. 2001).Cracks cause an increased loss of soil moisture with depth, through evaporation from the crack surface, even though this loss may be significantly reduced under fully established crops. Cracks are also the reason for a considerable increase in the irrigation water requirement at the time of the first irrigation after dry season. Cracks may also be the source of tunnel erosion in the semi-arid regions, especially in events of heavy irrigation or high rainfalls. Soil cracks may cause physical damage to crop roots (Elias et al. 2001). Pierre et al. (2015) studied morphological, physical and chemical properties of some Vertisols of the Logone Valley in Cameroon. These Vertisols are characterized by dark color, clayey texture, massive structure, deep and open superficial desiccation cracks and micro-reliefs (gilgai).

6.3.2 Physical Properties

Vertisols have high clay contents, ranging from 40 to 60 percent in most cases, but it may reach to even 80 percent in some instances. As outlined in Soil Taxonomy, clay content in Vertisols generally increases downward to the subsoil. The clay content remains over 35 percent throughout the profile to a minimum depth of 50 cm. Vertisols can store huge water in the root zone. Moisture content generally determines the physical behavior of Vertisols. These soils become very sticky upon wetting, and they are not workable for tillage under wet conditions. Again, they become very hard when dry. Therefore, cultivation of Vertisols under too wet and too dry conditions makes the soil puddled and cloddy respectively. Under such conditions achieving a good tilth cannot be expected. Tillage and seedbed preparation in Vertisols are possible only within a very narrow range of moisture contents. Trafficking in very wet conditions of Vertisols causes structural damage and compaction. On the other hand, tilling is also difficult in very hard soil conditions and it results in seedbeds with large clods. Vertisols swell when wet and shrink when dry. The extent of shrinking and swelling depends on the amount and type of clay, moisture conditions, landscape positions and vegetation type. Hydration and dehydration cause alternate swelling and shrinking and also depend on mineralogical, chemical, and physicochemical properties of soil (Ben-Hur et al. 2009; Lado and Ben-Hur 2004). The presence of high amount of smectite clay, high exchangeable sodium percentage (ESP) and a low electrolyte concentration in the soil solution causes greater swelling.

Dinka (2011) studied shrik-swell dynamics of Vertisol catenae under different land uses in order to:(1) determine if variability in soil cracking on a Vertisol catena, having the same soil and land cover, could be explained by the shrink-swell potential of the soil and changes in soil water content; (2) characterize the temporal and spatial variability of the shrinkage of a Vertisol under different land uses; and (3) determine the relationship between specific volume and water content of soils, particularly between saturation and field capacity. Maximum soil subsidence was 120 mm in the grazed pasture, 75 mm in the native prairie, and 76 mm in the row cropped field. Shrinkage of the whole soil was not equidimensional, and the study generally indicated more horizontal shrinkage than vertical shrinkage. He suggested that a soil layer can subside up to 4 percent while drying from saturation to field capacity. Wide and deep cracks have the capacity to enhance rapid flow of water and nutrients into the subsoil, affecting the hydrology of the soils (Bandyopadhyay et al. 2003). Shrink-swell properties of Vertisols spatially vary with soil properties, microclimate, topography, vegetation, cropping patterns, and soil management practices (Vaught et al. 2006). Soil properties important to shrink-swell that vary in space include clay content, clay mineralogy, and water holding capacity (Azam et al. 2000). High concentrations of clay in a soil, mainly fine clay fraction, result in high specific surface area that helps store water. As a result, the surface area of the fine clay and the bulk volume of the soil increase.

Swelling of clay particles increases the content of small, water-retaining pores at the expense of larger water-conducting pores. On the other hand, clay dispersion is an irreversible process, in which quasi crystals or domains (regions of parallel alignment of individual alumino-silicate lamellae in smectite minerals) break apart and disperse because of mutual-repulsion forces. Dispersion of soil clay occurs instantaneously once the electrolyte concentration of the soil solution falls below a threshold value, termed the flocculation value, and the dispersed clay particles may migrate and plug water-conducting pores, causing a reduction in saturated hydraulic conductivity (K_s) of soil. Clay dispersion is influenced by soil chemistry and mineralogy, and is enhanced primarily by a low electrolyte concentration in the soil solution and high ESP (exchangeable sodium percentage) of the soil (Lado et al. 2004; Laird 2006).

Soils usually contain more clay in lower landscape positions (Dinka and Lascano 2012). Large amounts of water may be present there because of higher clay content and surface or subsurface flow of water. According to Jovanov et al. (2012), very high retention of moisture in Vertisols depends on high clay content particularly montmorillonite, and on organic matter accumulation and pedoturbation. They observed that field capacity in Vertisoils of some regions of the Republic of Macedonia ranged from 22.47 to 40.47 percent by weight, but plants could not absorb much because the wilting point was also very high ranging from 13.55 to 24.68 percent by weight. The difference between field capacity and wilting point is considered as available water which was, on an average only 12.32 percent by weight in soils of their study. However, Vertisols have low hydraulic conductivity and infiltration rate under wet conditions and a high bulk density when dry (Tekluet al. 2004). Structural units in some Vertisols slake easily when wetted, and the soil surface becomes muddy and very sticky. Tewka et al. (2013) conducted a field study to assess the physico-chemical properties of the natural and cultivated soils of Savannah Vertisols in Ethiopia. They collected soil samples from four soil pedons, two from each fallow and cultivated soil. The soils contained about 70 percent clay. Bulk density ranged from 1.25 to 1.40 g cm^{-3} in the fallow soils and from 1.47 to 1.52 g cm^{-3} in the cultivated soils. Total porosity was about 43 percent in fallow soils and 40 percent in cultivated soils with the moisture content ranging from 65 to 80 percent. Their results indicated that cultivation of Vertisols caused considerable compaction of the soil.

In some Vertisols areas in Ethiopia, there are native *Acacia seyal* and *Balanites aegyptiaca* savannah vegetation. Many such lands have been cleared for agricultural crop production in the semi-arid Sahel regions. Shabtai et al. (2014) studied the effects of the changes in land uses on the structure and saturated hydraulic conductivity (K_s) of a Vertisol under sodic conditions. Exchangeable sodium percentage (ESP) increased with soil depth, from 2 percent in the 0–15 cm layer to 8.1–10.6 percent in the 90–120 cm layer. Swelling and dispersion was more pronounced in the subsoil than in the topsoil due to the higher ESP values. In contrast, the topsoil was more sensitive to slaking forces than the subsoil, probably due to increased particle cohesion in the subsoil. This led to lower K_s values of the top soils under fast than slow prewetting. The steady-state K_s values under slow prewetting and leaching with deionized water were significantly higher in the savannah-woodland soil than in the cultivated soils, down to 120 cm depth. These differences in K_s values were associated with higher swelling values in the cultivated soils than in the savannah-woodland soil. Vertisols generally have low hydraulic conductivity, low infiltration rate and high moisture retention capacity ranging from 60 to 70 percent at field capacity because of their high clay content (Zewudie 2000). Marta (2012) analyzed some physical properties of a large number (n = 126, 0–150 cm depth) of Vertisol soil samples of Hungary. He obtained mean values of sand, silt and clay of 15.5, 39.0 and 45.5 percent respectively. Mean bulk density was 1.4 g cm^{-3} and coefficient of linear extensibility (COLE) was 0.21. Asiedu et al. (2000) studied infiltration and sorptivity on the Accra Plains of Ghana under four different Vertisol management technologies including cambered bed, the Ethiopian bed, the ridge, and the flat bed. The initial values of both cumulative infiltration and infiltration rate were the highest in the cambered bed followed by the ridge, the Ethiopian bed, and the flat bed in the decreasing order. The terminal infiltration rates were quite similar for all the landforms and were about 0.05 ms^{-1}. Field-measured sorptivity followed the order: cambered bed > ridge> Ethiopian bed > flat bed. Liu et al. (2010) suggested that if under a condition of no cracks, soil porosity of the entire soil column is 0.45m^3 m^{-3} and soil moisture is 0.1m^3 m^{-3}, the soil porosity of the top soil layer may increase to 0.6m^3 m^{-3} when cracks are open (while soil moisture is still assumed to be 0.1m^3 m^{-3}). The increase in soil porosity increases the proportion of air and decreases the fraction of soil within a unit volume.

6.3.3 Chemical and Mineralogical Properties

Vertisols occurring in India, Australia, Sudan, Ethiopia and other parts of Africa generally have a soil pH ranging between 7.5 and 8.5 in the soil profile due to the presence of high $CaCO_3$ and high contents of exchangeable bases, especially calcium and magnesium. Some Vertisols in tropical areas under irrigation or in depressions may have a soil pH as high as 9.5 due to sodium saturation in the exchange complexes. Alkaline soil pH is conducive to volatilization loss of native and applied ammonia. Most Vertisols are calcareous with either almost uniform distribution throughout the profile or increasing downward. Gypsum has been found to occur in

6.3 Properties of Expansive Soils

the sub-surface of the Vertisol profiles in relatively arid areas indicating the lack of leaching of the slightly soluble gypsum. Dengiz et al. (2012) observed that physico-chemically, the Vertisols on the alluvial plains in the central Black Sea region of Turkey were slightly basic to very basic, non-saline and poor in organic matter. These soils had high cation exchange capacity and total exchangeable bases, and very high base saturation percentage. Özsoy and Aksoy (2007) studied physico-chemical properties of some Vertisols developed on calcareous marl parent materials. These soils were high in CEC and base saturation with calcium and magnesium occupying more than 90 percent of the exchange site, low organic material but sufficient fertility. The agricultural potential of the soils were, however, limited due to high clay and $CaCO_3$ contents of sub-surface horizons and a hard pan formation due to inappropriate soil tilling.

Although the dark color of the Vertisols could be suggestive of high organic matter content, it was found that most of the black cotton soils of India rarely have organic matter exceeding 1.0 percent. Jahknwa and Ray (2014) analyzed some chemical properties of Vertisols of Guyuk area of Nigeria. Soil pH and CEC are generally high in soils of the study site. They attributed these to the high clay content of the soil. Soil properties that exhibited very low values include soil organic matter and available phosphorus.

Pierre et al. (2015) studied chemical properties of some Vertisols of the Logone Valley in Cameroon. These soils are neutral to slightly alkaline in reaction (pH 6.4–7.4), with low organic matter content, and an average CEC ~ 22.8 $cmol_c$ kg^{-1}. Their exchangeable cations are dominated by Ca and K. Tewka et al. (2013) determined some chemical properties in natural and cultivated sites of the soils of Savannah Vertisols of Ethiopia. Soil pH in both soils was near neutral (6.1–7.2) with basic cations (K^+, Ca^{2+} and Mg^{2+}) dominating the exchange sites. High concentrations of Na^+ ions were recorded at lower depths in both fallow and cultivated soils, indicating that the soils are potentially saline-sodic. Giday et al. (2015) found considerable variations in organic matter (0.05–4.39 percent), available P (0.86–22.50 mg kg^{-1}) and total N (0.03–0.23 percent) contents in Vertisols of Southern Tigray, Ethiopia. The soils were slightly acidic to moderately alkaline in reaction (pH 6.5–8.20). The soil exchange complex was mainly dominated by Ca and Mg where the order of occurrence was Ca > Mg > K > Na. The CEC values were very high ranging from 41.42 to 50.37 $cmol_c$ kg^{-1}. Marta (2012) analyzed chemical properties of a large number (n = 126, 0–150 cm depth) of Vertisol soil samples of Hungary. He obtained the following mean values of SOM, $CaCO_3$, pH (H_2O), and CEC: 1.5 percent, 3.2 percent, 7.5 and 35.6 $cmol_c$ kg^{-1} respectively. Exchangeable Ca^{2+}, Mg^{2+}, Na^+ and K^+ represented 72.9, 21.8, 3.1 and 2.3 percent of total exchangeable bases respectively. The mean organic matter content is relatively high; 1.5 percent in the 0–150 cm thick segment of the soils; the mean SOM value stays above 2.4 percent in the upper 40 cm; it is more than 1 percent in the upper 80 cm, and more than 0.5 percent at the depth of 150 cm too. The high mean CEC values are related to the high clay and SOM contents. Tamfuh et al. (2011) studied physico-chemical properties of some Vertisols of the Sudano-Sahelian

Region of North Cameroon. These soils were high in cation exchange capacity (26–42 $cmol_c$ kg^{-1}) and total sum of bases (74.30 and 94.23 $cmol_c$ kg^{-1}), high base saturation, low organic carbon and a very high C/N ratio. Geochemically, Si and Al are the dominant elements, characterized by a Si/Al ratio range of 2.27 and 2.94. According to this rate, 2:1 clay minerals, namely smectite, are predominant and their presence confirms the shrink-swell behavior of the soils. Based on their smectite content, these soils present numerous interesting economic potentials in the chemical industry, pharmaceutics, agronomy and environmental protection (Nguetnkam 2004; Woumfo et al. 2006).

It was a general belief and there were plenty of evidences as well that the shrink-swell behavior of Vertisols is the manifestation of the activities of smectites (chiefly montmorillonite), the 2:1 expanding types of clay in the clay fraction. However, some investigations have shown that expansive layer silicates are not the only clay minerals present in Vertisols. Some studies revealed that it is the proportion of fine clay, regardless of the clay type, together with the wetting and drying cycle in the soil that can produce a high shrink-swell potential (Heidari et al. 2008). Heidari et al. (2008) observed the dominance of palygorskite-chlorite in the clay fraction of some Vertisols in Iran. They concluded that the inter particle pore size that is controlled by the size of primary particles, regardless of its nature, contributes to the shrink-swell potential in soils of their study. Despite the large body of information available today, showing that smectitic clays are by far the most dominant clay minerals (Shirsath al. 2000), these soils may be dominated by other minerals (Heidari et al. 2008). According to Thomas et al. (2000), a combination of physical, chemical, and mineralogical properties can best explain the shrink-swell behavior of soils. No single property can accurately predict shrink-swell potential for all soils. Pierre et al. (2015) observed the dominance of smectites and some amount of kaolinite and illite in the mineralogical properties of some Vertisols of the Logone Valley in Cameroon. Dominant primary minerals are quartz and feldspars. These soils have high contents of SiO_2 (61.07–77.78 percent), moderate content of Al_2O_3 (7.08–15.54 percent) and low amount of Fe_2O_3 (1.78–6.92 percent). Mixed mineralogy has also been revealed in many studies (Shirsath al. 2000). Fassil (2009) studied the relationships of major physico-chemical properties of some Vertisols of northern highlands of Ethiopia. He observed that Si contents ranged from 79.8 to 87.5 g Si kg^{-1} in the cultivated Vertisols of Adigudom, from 97.7 to 115.2 g Si kg^{-1} in Axum, from 113.7 to 117.2 g Si kg^{-1} in Maychew, from 130.0 to 133.9 g Si kg^{-1} in Shire and from 137.3 to 166.3 g Si kg^{-1} in Wukro. The highest concentration was found in areas where the sand content was the greatest.

Vertisols have a satisfactory level of fertility because of favorable pH, high cation exchange capacity and high base saturation percentage. Özsoy and Aksoy (2007) suggested that despite their high fertility under irrigated conditions, Vertisols are often undesirable for agricultural and some engineering purposes due to their high clay contents, puddling under wet and clodding dry conditions, easy slacking and the shrink-swell behavior, deep cracks and compactions.

6.3.4 Engineering Problems Associated with Expansive Soils

Expansive soils offer serious problems to the construction of foundations for roads and buildings. Civil engineers consider expansive soils as potential natural hazard. Expansion of soil can cause extensive damage to structures worldwide if appropriate measures are not taken (Bose 2012). Wide cracks in the wall, distortion of floor, heaving of beds in canal, and rutting of roads are the usual types of damages in expansive soils (Christodoulias 2015). The shrink-swell movement of the soil underneath causes these damages.

The damages due to expansive soils are sometimes minor maintenance issues but often they are much worse, causing major structural distress. In the United States, 10 percent of the 250,000 new houses built on expansive soils each year experience significant damage, some beyond repair (Lucian 2006; Al-Zoubi 2008). Many highway agencies, private organizations and researchers are studying the remedial measures because considerable land areas are covered with such soils in many countries (Radhakrishnan et al. 2014). The problem is of enormous financial proportions and is also a global phenomenon. Australia, Argentina, Canada, China, Cuba, Ethiopia, Great Britain, India, Israel, Kenya, Mexico, Myanmar, Spain and the United States are some of the countries which need to cope with expansive soils.

Expansive soils are responsible for a significant hazard to foundations for light buildings. They can damage foundations by uplifting as they swell with wetting. Swelling soils lift up and crack lightly-loaded, continuous strip footings, and frequently cause distress in floor slabs. The high shrink-swell potential and low bearing strength of fine-textured soils contribute to the failure of structures made of concrete and other non-flexible building materials. The stability and functionality of building foundations (basements), streets and sidewalks are impaired through the internal movement within the soil medium. Embankments and earth dams are thus susceptible to failure through internal slippage along planes of weakness, especially when saturated (Brierley et al. 2011). Expansive soils affect the lightweight structures very severely by high swelling pressure under wet conditions. Damages due to soil swelling or shrinking may be reduced by reducing the swelling pressure of the soil. Another measure can be to build structures resistant to damage from soil expansion.

Lime is commonly used to stabilize shrink-swell soils (Eisazadeh et al. 2012). Other chemical substances that can be used for the purpose include KCl, $CaCl_2$ and $FeCl_3$ instead of lime for their higher solubility in water (Prasada Raju 2001). Cokca (2001) reported that addition of $CaCl_2$ and KOH reduced swelling of the expansive soils. Flyash can also act as a stabilizer (Phanikumar and Sharma 2004). Flyash is produced by coal-fired power plants during the combustion of coal (Hasan 2012). Radhakrishnan et al. (2014) recommended flyash and aluminum chloride ($AlCl_3$) to increase the strength of expansive soils.

Clay particles in some Vertisols can be highly dispersed because of sodium saturation of the exchange complexes. They are dispersive clay soils. Serious engineering problems are caused by these dispersive clay soils when used in hydraulic structures, embankment dams, or other structures such as roadway embankments.

Some natural clay soils disperse or deflocculate in the presence of relatively pure water and are, therefore, highly susceptible to erosion and piping (Zorluer et al. 2010). The amount and type of clay, pH, organic matter, temperature, water content, thixotropy and type and concentration of ions in the pore and eroding fluids are the factors that affect the critical shear stress required to initiate erosion (Umesha et al. 2011). So, dispersive clay soil erodes in the presence of flowing water starting in a drying crack, settlement crack, hydraulic fracture crack, or other channels of high permeability in a soil mass. Dispersive clays have a high proportion of exchangeable sodium whereas other non-dispersive clays may have a predominance of calcium, potassium, and magnesium cations in exchange positions and in the pore water.

6.4 Agricultural Uses of Expansive Soils

Vertisols of the arid areas, are generally under natural grassland and associated vegetation and are used for mainly cattle raising. Where irrigation can be applied, these soils can be used to grow a variety of tropical crops. The predominant crops in African Vertisols are barleys, faba beans, wheat, field peas, oats, lentils, linseed, niger seed, chickpeas, sorghum and rough pea. In Nigeria, many Vertisol soils are used for grazing and for growing a wide range of food crops, such as maize, yams and vegetables. These soils are primarily used in India for raising cereals (sorghum and millets), pulses (pigeon pea, chickpea, mung beans, lentil, etc.), oilseeds (groundnut, mustard, sesame, etc.), and commercially important crops like cotton, chilies, soybeans, sunflower, safflower, etc. In large tracts of Central India, wheat is also grown on the Vertisols. When irrigation is available, crops such as cotton, wheat, sorghum and rice can be grown. Vertisols are especially suitable for rice because they are almost impermeable when saturated.

Yield of crops under traditional management in Vertisols are quite low. Average yields of crops are 846 kg ha^{-1} for barley, 1295 kg ha^{-1} for faba bean, 964 kg ha^{-1} for wheat, 846 kg ha^{-1} for field peas, and 300 kg ha^{-1} for niger seed (Gryseels and Anderson 1983). Rainfed farming is very difficult because of the narrow range of soil moisture when they can be worked on. Vertisols in Australia are highly regarded because they are among the few soils that are not acutely deficient in available phosphorus. The potential productivity of Vertisols may be high because of nearly neutral or slightly alkaline pH, high CEC, high base saturation percentage, high water holding capacity, etc. Unproductive Vertisols can be made productive through improved drainage, fertilizer and adopting suitable crop management practices. Both yield and soil quality can be improved by improved management. Large Vertisol areas are still uncultivated, and some are intensively grazed. Agricultural use of Vertisols in Africa ranges from grazing, firewood production and charcoal burning through small holder post-rainy season crop production (millet, sorghum, cotton), to small-scale (rice), and large-scale irrigated crop production (cotton, wheat, barley, sorghum, and sugarcane).

Most Vertisols in Africa are not cultivated, and are used as grazing grounds. These areas carry large number of domesticated and wild animals. Here, Vertisols can be transformed into productive crop land if their management addresses their inherent physical problems. For example, improvements in tillage quality will lead to higher crop yields on clay soils compared to light soils. Heavy clay soils are very difficult to till by hand; their tillage for cropping tends to be either mechanized or animal powered. Crop/ livestock interactions are strong there, where animal power is used for soil cultivation and where more livestock can be fed on the basis of the enhanced production of crop residues and by-products. More manure and composts could be available too. In small farm holdings, animal power, crop diversity, residue management, and utilization of soil moisture for crop production have benefitted farmers. Most of the Vertisols in the highlands of Ethiopia suffer from excess water and poor workability. They are underutilized, and are largely used for dry season grazing. Only 25 percent of the 7.6 million hectares Vertisols in the highlands are cultivated. The common crops grown on Vertisols are tef (*Eragros tistef*), wheat (*Triticum spp.*), barley (*Hordeum vulgare*), faba bean (*Vicia faba*), field pea (*Pisum sativum*), grass pea (*Lathyrus sativus*), chickpea (*Cicer arietinum*), lentils (*Lens culinaris*), lineseed (*Linum usitaissium*), noug (*Guizotia abyssinica*) and fenugreek (*Trigonella foenum-graecum*). But the yields of these crops are quite low (Ayele 2004).

6.5 Limitations of Expansive Soils to Agricultural Uses

Agricultural management problems of Vertisols include extreme stickiness of the soils when wet and their intractability when dry and the lack of appropriate tillage implements. If tilled in wet condition, the soil is puddled, compacted and crusted; and if tilled in dry condition, which require much energy, large clods are produced. Having a satisfactory tilth in the seedbed in Vertisols under both wet and dry conditions is hardly possible. The soil is workable within a narrow range of soil moisture that can reach for a short span of time between rains. Soil preparation should be done at this time during initial substantial rains or in the post rainy season. The soils can remain saturated for a large part of the rainy season; often in flat lands they are flooded and are highly eroded in slopes. In India, Vertisols are particularly subject to soil loss by water erosion under the traditional systems of bare-fallowing during the rainy season. Losses are promoted by the combination of intense storms and lack of plant cover. In most Vertisols, infiltration and permeability are both very low and internal drainage is impeded. Artificial subsurface drainage systems do not work well. Nutrient deficiencies, especially micro-nutrient deficiencies, sodicity in irrigated fields and soil erosion, are other problems of Vertisols. Sodic Vertisols have dual problems – sodicity (high pH, ammonia volatilization, P and micro-nutrient deficiency) and inappropriate soil consistence. In Nigeria many Vertisols soils are non-saline, but are mildly to strongly sodic, especially the subsoils. They are high in basic cationic nutrients, but are generally low in organic matter, N, P and Cu. The potential productivity of some Vertisols soils is probably quite high, but a number of management problems must be solved before this potential can be fully exploited.

6.6 Integrated Soil and Crop Management for Expansive Soils

A good number of investigations had been carried out about the Vertisols of Ethiopia, particularly in soils of the high lands (ILCA 1990; Astatke and Jabbar 2001; Astatke et al. 2002; Teklu and Gezahegn 2003), regarding their culvation for crop production. According to Astatke and Jabbar (2001) and Teklu and Gezahegn (2003), farmers prepare lands early with the moisture obtained during early short rains and keep the land fallow for 2–3 months. They occasionally till the land and plant seedlings immediately after the rainy season so that the crop thrives on residual moisture. Farmers use five to nine cultivations prior to planting seedlings for making a fine seed bed and controlling weed. Such intensive cultivation disrupts aggregates, increases aeration, and mixes organic residues with soil resulting in higher soil organic matter decomposition and loss. Astatke et al. (2002) suggested that over tillage increases erosion. This is an acute environmental problem of the Ethiopian highlands. Waterlogging is always a problem in Vertisols in the rainy season, so improving surface drainage can advance planting date and increasing crop growing period. Making broad beds and furrows (BBF) is a strategy followed for centuries by women farmers of the Inewari area in the Central Highlands of Ethiopia. International Livestock Centre for Africa has adopted and modified BBF to fit to the smallholder systems in Ethiopia (ILCA 1990).

Improved management practices have been designed to overcome the problems of Vertisols. These practices include drainage of excess water through safe waterways, choice of suitable crops incorporating legumes in rotations, monitoring soil moisture and preparing seedbed in the optimum soil moisture content (nearer but less than field capacity) between or after rains, supplemental irrigation to avoid water stress, dry seeding, double cropping in semi-arid tropics, fallowing during monsoon or cultivating wetland crops (such as rice) in humid conditions, etc. Subsurface drainage is not usually feasible in Vertisols for slow permeability after closing the cracks by wetting. So, special attention has been given to improved surface drainage systems, including cambered beds, ridges, furrows, bunds, and broad banks in Ghana, India, Indonesia, Trinidad, USA and Venezuela.

For sustainable agriculture in Vertisol areas of India, the International Crops Research Institute for the Semi-Arid Tropics (ICRISAT) has developed, almost after a decade of research, some management practices known as Vertisol Technology. In regions of Vertisols where rainfall is dependable but the lands are left fallow in the rainy season, ICRISAT adopted a series of improved practices to increase agricultural productivity. Joshi et al. (2002) reported a system of management where micro-watersheds of 1.5–3 ha size were taken as units for land and water management and agronomic practices. The bed-furrow (ridge-furrow) cultivation system was followed to conserve moisture and to facilitate draining runoff water in a controlled manner.

Important elements of ICRISAT farming system include (i) growing the same crop in both rainy and post-rainy seasons, (ii) using improved crop varieties and improved cropping systems that may include solo, sequential and intercropping systems as the farmers find suitable, and (iii) using appropriate fertilizers. The basic elements of the ICRISAT system are: (i) adoption of a combination of broad bed and furrow (BBF) system and grassed waterways to avoid waterlogging and disposing of excess water safely, (ii) plowing the land roughly after the previous crop is harvested and some moisture is left in the soil, (iii) completion of seedbed preparation after the first pre-monsoon rain, and (iv) appropriate seed and fertilizer placement.

The ICRISAT has modernized the old concept of broad bed and furrow system which encourages controlled surface drainage by forming the soil surface into beds. An old version of this system called "rigg and furrow" was used in the medieval times in Britain for improving pastures; it had been used in the past in North America and in Central Africa. A variation known as the camber-bed system was used in Kenya. ICRISAT recommends broad beds about 100 cm wide, separated by sunken furrows about 50 cm wide with the preference of slope along the furrow between 0.4 and 0.8 percent. Two, three, or four rows of the crop can be grown on the broad bed, and the bed width and crop geometry can be changed if needed (Fig. 6.4). A view of a crop field finished with broad bed making is shown in Fig. 6. 5.

Broad beds can be made on a gentle grade by ox-drawn wheeled broad bed makers. Very simple and cheap broad bed makers are used in Ethiopia.

Broad bed and furrow system work best in deep Vertisols with dependable rainfall averaging 750 mm or more. It has not been as productive in areas of less dependable rainfall, or on Alfisols or shallower black soils although, in the latter case, more productivity is achieved than with traditional farming methods. Sometimes, maize can be planted on the beds and rice in the furrows (Fig. 6.6). Waterlogging will not affect maize in beds, and rice will thrive well in furrows.

Improved management systems were able to increase crop productivity and enhance soil quality in Vertisols of India where average annual rainfall is 800 mm, the average minimum temperature is 19 °C and maximum temperature is 32 °C. Rainfall is variable spatially and temporally and also occurs in torrential downpours. Such erratic rainfall results in spells of excess moisture and drought during the crop growing period. The improved system consisted of a broad-bed and furrow landform treatment. The beds were 1.2 m wide with a 0.3 m furrow prepared at 0.4 ± 0.6 percent gradient using a bullock-drawn bed-maker mounted on a tropicultor. The land was cultivated soon after the harvesting of the post-rainy season crop and, after unseasonal rains, the beds were formed again. Field traffic was confined to the furrows. Excess rainfall drained along the furrows and discharged into grassed waterways. Seeds of high-yielding varieties of pigeon pea, sorghum and maize were dry-sown on the bed with variable spacing for different species. Sorghum and pigeon pea together recorded an average grain yield of 4.7 t ha^{-1} year.$^{-1}$ compared with the 0.9 t ha^{-1} year.$^{-1}$ average yield of sole sorghum in the traditional system (Wani et al. 2003). It appears that the control of soil moisture is the key to sustainable management of Vertisols. Behera et al. (2006) observed that irrigation schedules and frequencies at certain crop growth stages improved yield of wheat (*Triticum aestivum*) and durum wheat (*Triticum durum*) on Vertisols in Central India.

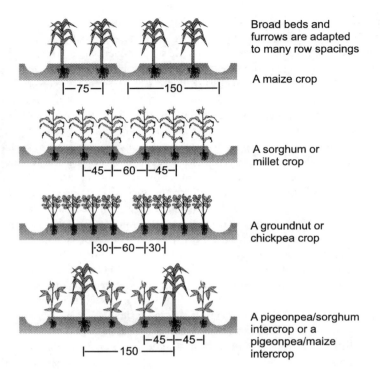

Fig. 6.4 Broad bed and furrows for different cropping systems. (Adapted from FAO Corporate Document Repository; soil and water conservation in semi-arid areas)

Fig.6.5 A section of a crop field prepared with broad bed and furrow. (Image courtesy of FAO)

6.6 Integrated Soil and Crop Management for Expansive Soils

Fig. 6.6 Sketch showing double-row maize in broad bed and rice in furrow

Rajput et al. (2009) described the benefits of community based crop management systems in Vertisols of Central India. They observed that the raised-sunken bed system (RSBS) of land treatment enhanced *in-situ* rainwater conservation and minimized soil erosion and nutrient losses. Grain yields of wheat (*Triticum aestivum*) and chickpeas (*Cicer arietinum*) were higher in this system than in the flatbed system (FBS) of planting. Soybean (*Glycine max*) yield increased nearly 100 percent with the ridge-furrow system (RFS) and about 55 percent in broad-bed and furrow system (BBFS) compared with the FBS. The adoption of integrated nutrient management based on soil testing increased soybean and wheat yields by 71 percent over farmers' practice at Narsinghpur compared with about 100 percent for soybean and 187 percent for wheat at Hoshangabad. The intercropping of soybean with pigeon pea (*Cajanus cajan*) in 4:2 ratio produced higher net return and benefit-cost ratio (3.3:1) than either of the monocropping systems. In this area aquaculture in the ponded water in the bunded field is done during monsoon and growing of wheat or chickpeas is done in the winter season. This system was also profitable.

The Joint Vertisol Project has developed a package composed of the following elements to better utilize Vertisols in the highlands of Ethiopia (JVP 2000):

- A broad bed maker (BBM) by modifying local *mareshas* (wooden plough) to drain excess water from Vertisols plots to allow early planting compared to current practice;
- Wheat variety suitable for early planting on Vertisols;
- Seed rate and fertilizer rate for optimal yield;
- Planting dates for optimal plant growth and yield;
- Weed and pest management recommendations.

Among these, the Broad Bed and Furrow based implement (BBM) is the main element of Vertisols technology. The other components are improved varieties or management practices that can be used along with BBM or traditional practices that could resist waterlogging problems and give better yields.

A field experiment was carried out for 6 years between 1998 and 2003 at CaffeeDoonsa in the central highlands of Ethiopia to evaluate alternative land preparation methods on the performance of wheat (*Triticum durum* Desf.), lentil (*Lens culinaries* Medik L) and tef (*Eragrostis tef* L) grown in rotation (Teklu et al. 2006). Four land preparation methods (broad bed and furrow, green manure, ridge and furrow and reduced tillage) were arranged in a randomized complete block design. Broad bed and furrow (BBF) significantly increased the grain yield of lentils by 59 percent (from 1029 to 1632 kg ha^{-1}) as compared to the control. On the other hand, reduced tillage (RT) resulted in the highest grain yield of wheat (1862 kg ha^{-1}) and tef (1378 kg ha^{-1}) as compared to 1698 kg ha^{-1} of wheat and 1274 kg ha^{-1} of tef for the control although the increase was not statistically significant. A gross margin analysis showed that BBF is the most profitable option for lentil with 65 percent increase in total gross margin. On the other hand, RT resulted in 11 and 8 percent increase in gross margin of wheat and tef respectively as compared to the control. Best combinations of crop and land preparation methods were: lentil sown on broad bed and furrow, and wheat and tef sown after reduced tillage.

Kebede and Bekelle (2008) conducted a field experiment to observe effectiveness of flat seedbed, traditional drainage system, and broad bed and furrow with 100 cm (BBF-100 cm) and broadbed and furrow with 80 cm(BBF- 80 cm) on the yield of wheat. Results revealed that BBF-100 cm, BBF-80 cm and traditional drainage system significantly increased the grain yield of wheat by 51.4 percent, 41.6 percent and 11.2 percent compared to the control respectively. Bhaambe et al. (2001) conducted field experiments to study the effect of sub-surface drain spacing and crop residue incorporation on reclamation of salt-affected Vertisols under soybean-wheat cropping system. Sub-surface drains at 25, 50 and 75 m spacing installed at 1.3 m depth with corrugated PVC perforated pipe efficiently drained out excess water and significantly increased productivity of soybean and wheat crops. They also improved soil physical properties, including bulk density, infiltration rate, hydraulic conductivity, and decreased soil salinity. Crop residues (sugarcane trash @ 5 t ha^{-1} or green manuring with dhaincha – *Sesbania bispinosa*) significantly increased crop productivity and reduced salinity of salt affected Vertisols.

Integrating BBF system with minimum tillage can be a better option for the management of Vertisols. This could be implemented first by constructing broad bed and furrows with an animal drawn broad bed maker (BBM), and then the broad beds could be maintained for several cropping seasons with the minimum tillage practice. In making the land for subsequent seasons with the same practice, BBF will have to be rehabilitated. Retaining the BBFs for repeated use with minimum tillage and row seeding rather than broadcasting conserves soil and increases nutrient use efficiency. It reduces seed rates by the placement of seed uniformly at optimum soil depth and also reduces fertilizer rate by improving nutrient uptake. Crop residues are retained on the soil surface as mulch and the soils get permanent soil cover so as to reduce the extent of land degradation and promote sustainable natural resource management.

6.7 Conservation Tillage in Vertisols

Proper water management is a major component of Vertisol management. Rapid changes in moisture status lead to limitations in use while slow changes could allow for longer periods of soil properties which favor plant growth. FAO (2008) reported that the elimination of runoff can result in waterlogged conditions in Vertisols. On the other hand, reduced tillage and residue management can promote infiltration, improve structure, prevent surface sealing, and decrease evaporational losses in Vertisols. Management techniques for safely redirecting runoff include use of grassed waterways. On-farm tillage research in Ethiopian highland Vertisol area demonstrated that the minimum tillage on participatory basis could be an effective intervention for soil conservation due to the early-vegetative cover of the soil. Application of ash on Vertisols at Chefe Donsa significantly increased grain and straw yields of wheat (Astatke et al. 2004). Results from a factorial experiment including the factors like no-tillage, minimum tillage, full tillage and conventional tillage on growth of soybean under dry farming conditions in the arid or semi-arid region of Iran indicated that root growth and grain yield increased significantly under no-tillage than the other tillage systems (Sani 2013). However, Duiker and Myers (2002) suggested that soils with very low infiltration rates including soils with high concentrations of expansive clays are not likely to show reduced runoff and may experience decreased yields with no-till. An experiment on farmer's field in Northern Ethiopia was conducted to evaluate the short term changes in soil quality of a Vertisol due to the implementation of conservation agriculture practices and to assess their effect on runoff and soil loss, crop yield and yield components of tef (*Eragrostis tef*). The treatments were permanent bed, reduced tillage, and conventional tillage. Soil organic matter was significantly higher in permanent bed than conventional tillage and reduced tillage. A long-term tillage experiment has been carried out (2005–2009) on a Vertisol to observe changes in runoff, soil loss and crop yield due to conservation agriculture in the sub-humid DoguaTembien district of the Northern highlands of Ethiopia (Ugent et al. 2011). The tillage treatments were (i) permanent raised bed (PB) in a furrow and bed system with 30 percent standing crop residue retention and no-tillage on top of the bed, (ii) reduced tillage, locally called terwah (TER), with plowing once at sowing with 30 percent standing crop residue retention and contour furrows made at 1.5 m distance interval, and (iii) conventional tillage (CT) with a minimum of 3 tillage operations and the removal of crop residues. Crops planted during the 5 years were wheat, grass pea, wheat and barley sown together, and grass pea. Glyphosate was sprayed starting from the third year (2007) at 2 Lha^{-1} before planting to control pre-emergent weed in PB and TER. Runoff and soil loss were measured in plastic sheet lined collector trenches, which were located at the lower end of each plot. Significantly different ($p < 0.05$) soil losses of 12.7, 16.2 and 27.3 t ha^{-1} y^{-1} were recorded for PB, TER and CT respectively. Overall, the permanent raised bed and reduced tillage systems significantly reduced sediment loss and runoff and increased crop yield.

Tillage is done to make a seedbed or rootbed with good tilth, mix soil and crop residues to facilitate mineralization and release of nutrients, to make the soil porous, and above all to eradicate the weeds. The main disadvantages of tillage are deterioration of soil aggregation, loss of organic matter and enhanced erosion. If herbicides are used to control the weeds, intensity of tillage may be reduced considerably. In the reduced tillage system, only the portion of the land that is used for seeding or transplanting is tilled keeping the remaining land undisturbed. Moreover, a considerable proportion of the crop residues are retained on soil surface. The crop residues act as cover that conserves moisture and protects the soil against water and wind erosion.

6.8 Amendments in Vertisols

Soil quality of Vertisols could be improved by the application of organic waste products as amendments. Results of an incubation experiment under controlled temperature conditions, 30° C, on a Vertisol with 12 organic amendments resulted in a significant increase in soil-exchangeable K and Na over control. Some of the organic wastes, viz. cotton gin trash (10 Mg ha^{-1}; mega gram per hectare = t ha^{-1}), cattle manure (10 Mg ha^{-1}), biosolids (10 Mg ha^{-1}) and composted chicken manure (3 Mg ha^{-1}) have value as a source of nutrients to soil and hence showed potential to improve Vertisol properties (Ghosh et al. 2010a). Ghosh et al. (2011) conducted another incubation experiment using five organic amendments at various rates and observed their effects on properties of a Vertisol. Cotton gin trash, cattle manure, biosolids (dry weight basis 7.5–120 Mg ha^{-1}), chicken manure (dry weight basis 2.25–36 Mg ha^{-1}) and a liquefied vermicast (60–960 L ha^{-1}) modified soil chemical, physical and microbiological properties: higher light fraction of organic matter, higher N and P content and higher soil microbial activity. In Australia, the surface and subsurface soils of the majority of cotton growing regions are sodic. Application of organic amendments can be an option to stabilize the structure of these sodic Vertisols. To evaluate the possibility, Ghosh et al. (2010b) conducted an incubation experiment with soils of three different sodicity levels, i.e. nonsodic (ESP < 6), moderately sodic (ESP 6–15), and strongly sodic (ESP > 15), and incubated separately with cotton gin trash (60 Mg ha^{-1}), cattle manure (60 Mg ha^{-1}) and composted chicken manure (18 Mg ha^{-1}), keeping an unamended control. The organic amendments improved the physical properties of both Vertisols by decreasing clay dispersion. In the field experiment conducted by Balemi (2012) on the effect of farmyard manure (FYM) and inorganic nitrogen (N) and phosphorus (P) fertilizers on the growth and tuber yield of potato *(Solanum tuberosum* L.), the treatments consisted of a factorial combination of 4 levels of FYM (0, 10, 20 and 30 Mg ha^{-1}) and 3 levels of inorganic NP fertilizers (0, 33.3 percent, 66.6 percent recommended rates) in a randomized complete block design with 3 replications. Results demonstrated that the application of 20 or 30 Mg ha^{-1} FYM + 66.6 percent of the recommended inorganic NP fertilizers significantly increased total tuber yield over the application of the full dose of inorganic NP fertilizers without FYM in Vertisols. Tolessa and Friesen (2001) reported that the application of 25 percent recommended inorganic NP fertilizers + enriched FYM resulted in the highest marginal rate of return in maize indicating that the integrated approach can enable to save up to 75 percent of commercial fertilizers. Likewise, Bayu et al.(2006) also reported the possibility of saving up to 50 percent of the recommended NP fertilizers due to amendment with 5–15 Mg ha^{-1} FYM to sorghum crop. Gypsum amendment is sometime used in Vertisols at a rate of 1.5–3.0 Mg ha^{-1}. Gypsum amendment reduces water dispersible clay, ESP, pH, exchangeable Na, and Mg and improves hydraulic conductivity and exchangeable Ca and Ca/Mg ratio. However, the favorable changes that occur in soil physical and chemical properties are mostly found in the surface soil.

Study Questions
1. Explain the following terms: expansive soils, swell-shrink soils, cracking soils, and dispersive soils. Describe the unique morphological features of Vertisols.
2. Give an account of the physical properties of expansive soils. Justify that the physical properties of the Vertisols are responsible for their limitations to agricultural use.
3. Describe the chemical and mineralogical properties of Vertisols. Explain that the physical behavior of the Vertisols is the manifestation of their clay mineralogy.
4. Discuss the agricultural use of expansive soils. Write a note on engineering problems of shrink-swell soil.
5. Discuss briefly how expansive soils can be managed for better crop yield.

References

Al-Zoubi MS (2008) Swell characteristics of natural and treated compacted clays. Jordan. J Civ Eng 2(1):53–62
Asiedu EK, Ahenkorah Y, Bonsu M, Oteng JW (2000) Infiltration and sorptivity studies on some landform technologies for managing Vertisols. Ghana J Agric Sci 33:147–152
Astatke A, Jabbar M (2001) Low-cost animal-drawn implements for vertisol management and strategies for land-use intensification. In: Syers JK, Penning de Vries F, Nyamudeza P (eds) The sustainable management of Vertisols. IBSRAM Proceedings 20. CAB (Commonwealth Agricultural Bureau) International, Slough, Wallingford, pp 189–201
Astatke A, Jabbar M, Saleem MA, Teklu E (2002) Development and testing of low-cost animal-drawn minimum tillage implements: experience on Vertisols in Ethiopia. Agric Mech Asia Afr Lat Am 33(2):9–14
Astatke A, Mamo T, Peden D, Diedhiou M (2004) Participatory on-farm conservation tillage trial in the Ethiopian highland Vertisols: the impact of potassium application on crop yields. Exp Agric 40(3):369–379
Aydinalp C (2010) Some important properties and classification of Vertisols under Mediterranean climate. Afr J Agric Res 5(6):449–452
Ayele G (2004) Technological innovation, adoption and the management of vertisol resources in the highland Ethiopia. Part I. Natural resources, agriculture and food security issues. wmich.edu/~asefa/.../Papers/2001percent20papers/PaperI8.pdf

Azam S, Abduljauwad S, Al-Shayea N, Al-Amoudi OSB (2000) Effects of calcium sulfate on swelling potential of expansive clay. In: Vaught R, Brye KR, Miller DM (eds) Relationships among coefficient of linear extensibility and clay fractions in expansive, stoney soils. Soil Sci Soc Am J 70:1983–1990

Balemi T (2012) Effect of integrated use of cattle manure and inorganic fertilizers on tuber yield of potato in Ethiopia. J Soil Sci Plant Nutr 12(2):257–265

Bandyopadhyay KK, Mohanty M, Painuli DK, Misra AK, Hati KM, Mandal KG, Ghosh PK, Chaudhary RS, Acharya CL (2003) Influence of tillage practices and nutrient management on crack parameters in a vertisol of central India. Soil Tillage Res 71:133–142

Bayu W, Rethman NFG, Hammes PS, Alemu G (2006) Effect of farmyard manure and inorganic fertilizers on sorghum growth, yield and nitrogen use in semi-arid areas of Ethiopia. J Plant Nutr 29:391–407

Behera UK, Pandey ND, Varma PK (2006) Management of Vertisols with limited water availability for improving the productivity of durum and aestivum wheats 18th World Congress of Soil Science, July 9–15, 2006, Philadelphia, Pennsylvania, USA

Ben-Hur M, Yolcu G, Uysal H, Lado M, Paz A (2009) Soil structure changes: aggregate size and soil texture effects on hydraulic conductivity under different saline and sodic conditions. Aust J Soil Res 47:688–696

Bhaambe PR, Shelke DK, Jadhav GS, Vaishnava VG, Oza SR (2001) Management of salt-affected Vertisols with sub-surface drainage and crop residue incorporation under soybean-wheat cropping system. J Indian Soc Soil Sci 49(1):24–29

Bose B (2012) Geo-engineering properties of expansive soil stabilized with fly ash. EJGE 17:1339–1353

Brierley JA, Stonehouse HB, Mermut AR (2011) Vertisolic soils of Canada: genesis, distribution, and classification. Can J Soil Sci 91:903–916

Christodoulias J (2015) Engineering properties and shrinkage limit of swelling soils in Greece. J Earth Sci Clim Change 6:279

Cokca E (2001) Use of class C fly ashes for the stabilization of an expansive soil. J Geotech Geoenviron 127:568–573

CSIRO (2010) Australian soil and land survey field handbook, 3rd edn, Australian soil and land survey handbooks Series 1. National Committee on Soil and Terrain. CSIRO Publishing, Collingwood

Deckers J, Spaargaren O, Nachtergaele F (2001) Vertisols: genesis, properties and soilscape management for sustainable development. FAO, Rome

Dengiz O, Saglam M, Sarioglu FE, Saygin F, Atasoy C (2012) Morphological and physicochemical characteristics and classification of vertisol developed on deltaic plain. Open J Soil Sci 2(1):20–27

Dinka TM (2011) Shrink-swell dynamics of vertisol catenae under different land uses. Ph D Thesis, Office of Graduate Studies, Texas A & M University, USA

Dinka TM, Lascano RJ (2012) Challenges and limitations in studying the shrink-swell and crack dynamics of vertisol soils. Open J Soil Sci 2:82–90

Duiker SJ, Myers JC (2002) Better soils with the NO-Till system. based on the original document. Better soils, better yields. Conservation Technology Information Center 1220 Potter Drive, Suite 170, West Lafayette, Indiana

Eisazadeh A et al (2012) Solid-state NMR and FTIR studies of lime stabilized montmorillonitic and lateritic clays. J Appl Clay Sci 67-68:5–10

Elias EA, Salih AA, Alaily A (2001) Cracking patterns in the Vertisols of the Sudan Gezira at the end of dry season. Int Agrophys 15:151–155

FAO (2006) World reference base for soil resources 2006. World Soil Resources Reports No 103, FAO, Rome

FAO (2008) Water and cereal in drylands. Earthscan, London /Virginia

Fassil K (2009) Silicon status and its relationship with major Physico-chemical properties of Vertisols of northern highlands of Ethiopia. MEJS 1(1):74–81

Ghosh S, Lockwood P, Daniel H, Hulugalle N, King K, Kristiansen P (2011) Changes in vertisol properties as affected by organic amendments application rates. Soil Use Manag 27(2):195–204

Ghosh S, Lockwood P, Daniel H, King K, Hulugalle N, Kristiansen P (2010a) Short-term effects of organic amendments on properties of a vertisol. Waste Manage Res 28(12):1087–1095

Ghosh S, Lockwood P, Hulugalle N, Daniel H, Kristiansen P, Dodd K (2010b) Changes in properties of Sodic Australian Vertisols with application of organic waste products. Soil Sci Soc Am J 74(1):153–160

Giday O, Gibrekidan H, Berhe T (2015) Soil fertility characterization in Vertisols of southern Tigray, Ethiopia. Adv Plants Agric Res 2(1):00034

Gryseels G, Anderson FM (1983) Research on farm and livestock productivity in the central Ethiopian highlands: initial results, 1977–80. Research Report No. 4. ILCA (International Livestock Centre for Africa), Addis Ababa, Ethiopia

Hasan HA (2012) Effect of fly ash on geotechnical properties of expansive soil. J Eng Dev 16(2):1813–7822

Heidari A, Mahmoodi S, Roozitalab MH, Mermut AR (2008) Diversity of clay minerals in the Vertisols of three different climatic regions in Western Iran. J Agric Sci Technol 10:269–284

ILCA (International Livestock Centre for Africa) (1990) Annual research report. ILCA, Addis Ababa

Jahknwa JC, Ray HH (2014) Analysis of the chemical properties of Vertisols in Kerau, Guyuk area of Adamawa state, Nigeria. IOSR J Agric Vet Sci 7(1):80–89

Joshi PK, Shiyani RL, Bantilan MCS, Pathak P, Nageswara Rao GD (2002) Impact of vertisol technology in India. Impact Series no. 10. International Crops Research Institute for the Semi-Arid Tropics, Patancheru 502 324, Andhra Pradesh, India

Jovanov D, Sijakova-Ivanova T, Ilievski M, Ivanova V (2012) Moisture retention characteristics in the Vertisols of the Stip, Probistip and Sv. Nikole region. Agric Conspec Sci 77(2):69–75

JVP (2000) The joint Vertisol management project: summary of achievements and lessons learnt. Addis Ababa. http://www.fao.org/Wairdocs/ILRI/x5456E/x5456e0b.htm

Kebede F, Bekelle E (2008) Tillage effect on soil moisture storage and wheat yield on the Vertisols of north central highlands of Ethiopia. Ethiop J Environ Stud Manage 1(2):49–55

Lado M, Ben-Hur M (2004) Soil mineralogy effects on seal formation, runoff and soil loss. Appl Clay Sci 24:209–224

Lado M, Paz A, Ben-Hur M (2004) Organic matter and aggregate-size interactions in saturated hydraulic conductivity. Soil Sci Soc Am J 68:234–242

Laird DA (2006) Influence of layer charge on swelling of smectites. Appl Clay Sci 34:74–87

Liu YY, Evans JP, McCabe MF, de Jeu RAM, van Dijk AIJM, Su H (2010) Influence of cracking clays on satellite estimated and model simulated soil moisture. Hydrol Earth Syst Sci 14:979–990

Lucian C (2006) Geotechnical aspects of buildings on expansive soils in Kibaha. Tanzania: preliminary study. Licentiate Thesis. Division of Soil and Rock Mechanics Department of Civil and Architectural Engineering, Royal Institute of Technology Stockholm, Sweden

Marta F (2012) Development of the classification of high swelling clay content soils of Hungary based on diagnostic approach. Ph D Thesis, School of Environmental Sciences, Szent István University, Hungary

Miller WL, Bragg AL (2007) Soil characterization and hydrological monitoring project, Brazoria County, Texas, bottomland hardwood Vertisols. USDA-NRCS, Temple

Miller WL, Kishne AS, Morgan CLS (2005) Using seasonal crack patterns to evaluate the criteria for ustic and udic moisture regimes for Vertisols in the Texas Gulf Coast Prairie. In: 2005 Annual Meetings Abstracts. ASA, CSSA, and SSSA, Madison

Miller WL, ASz K, Morgan CLS (2010) Vertisol morphology, classification and seasonal cracking patterns in the Texas Gulf Coast prairie. Soil Surv Horiz 51:10–16

Nguetnkam JP (2004) Clays of Vertisols and fersiallitic soils of far North Cameroon: genesis, crystallochemical and Textural properties, typology and application in vegetable oil discolouration. Ph D Thesis, University of Yaounde, Cameroon

Özsoy G, Aksoy E (2007) Characterization, classification and agricultural usage of vertisols developed on neogen aged calcareous marl parent materials. J Biol Environ Sci 1:5–10

Phanikumar BR, Sharma RS (2004) Effect of flyash on the engineeering properties of expansive soil. J Geotech Geoenviron 130(7):764–767

Pierre TJ, Pierre NJ, Achile BM, Djakba BS, Lucien BD (2015) Morphological, physico chemical, mineralogical and geochemical properties of Vertisols used in bricks production in the Logone Valley (Cameroon, Central Africa). Int Res J Geol Min 5(2):20–30

Prasada Raju GVR (2001) Evaluation of flexible pavement performance with reinforced and chemical stabilization of expansive soil sub grades. Ph D Thesis, Kakitiya University, AP, India

Radhakrishnan G, Kumar MA, Prasad Raju GVR (2014) Swelling properties of expansive soils treated with chemicals and Flyash. Am J Eng Res 3(4):245–250

Rajput RP, Kauraw DL, Bhatnagar RK, Bhavsar M, Velayutham M, Lal R (2009) Sustainable Management of Vertisols in Central India. J Crop Improv 23:19–135

Sani B (2013) Till-system management technology (TSMT) in soybean farming at Iran. Adv Environ Biol 7(3):458–461

Shabtai IA, Shenker M, Edeto WL, Warburg A, Ben-Hur M (2014) Effects of land use on structure and hydraulic properties of Vertisols containing a sodic horizon in northern Ethiopia. Soil Tillage Res 136:19–27

Shirsath SK, Bhattacharyya T, Pal DK (2000) Minimum threshold value of smectite for vertic properties. Aust J Soil Res 38:189–201

Soil Survey Staff (1999) Soil taxonomy 2nd edn. United States Department of Agriculture. Govt. Printing Office, Washington DC

Sotelo –Ruiz ED, Gutiérrez –Castorena MC, Cruz-Bello GM, Ortiz–Solorio CA (2013) Physical, chemical, and mineralogical characterization of Vertisols to determine their parent material. Interciencia 38(7):357–362

Tamfuh PA, Woumfo ED, Bitom D, Daniel Njopwouo D (2011) Petrological, Physico-chemical and mechanical characterization of the topomorphic Vertisols from the Sudano-Sahelian region of North Cameroon. Open Geol J 5:33–55

Teklu E, Assefa G, Stahr K (2004) Land preparation methods efficiency on the highland Vertisols of Ethiopia. Irrig Drain 53(1):69–75

Teklu E, Gezahegn A (2003) Indigenous knowledge and practices for soil and water management in East Wollega, Ethiopia. In: Wollny C, Deininger A, Bhandari N, Maass B, Manig W, Muuss U, Brodbeck F, Howe I (eds) Technological and Institutional Innovations for Sustainable Rural Development. Deutscher Tropentag 2003. Göttingen, October 8–10, 2003. www.tropentag.de

Teklu E, Stahr K, Gaiser T (2006) Soil tillage and crop productivity on a vertisol in Ethiopian highlands. Soil Tillage Res 85:200–211

Tekwa J, Garjila YA, Bashir AA (2013) An assessment of the physico-chemical properties and fertility potentials of fallow and cultivated Vertisols in Numan area, Adamawa state-Nigeria. Sch J Agric Sci 3(3):104–109

Thomas PJ, Baker JC, Zelazny LW (2000) An expansive soil index for predicting shrink–swell potential. Soil Sci Soc Am J 64:268–274

Tolessa D, Friensen DK (2001) Effect of Enriching Farmyard Manure with Mineral fertilizer on grain yield of Maize at Bako, Western Ethiopia. Seventh Eastern and Southern Africa Regional Maize Conference. 11th–15th February, Nairobi, Kenya

Ugent WTA, Ugent WC, Ugent JN, Govaerts B, Getnet F, Bauer H, Raes D, Gebrehiwot K, Yohannes T, Deckers J (2011) Effects of resource-conserving tillage in the Ethiopian highlands, a sustainable option for soil and water management and crop productivity: a case study from Dogua Tembien. IAG/AIG Regional Conference 2011: Geomorphology for human adaptation to changing tropical environments held in Addis 18–22 February 2011, Addis Ababa, Ethiopia

Umesha TS, Dinesh SV, Sivapullaiah PV (2011) Characterization of dispersive soils. Mater Sci Appl 2:629–633

Vaught R, Brye KR, Miller DM (2006) Relationships among coefficient of linear extensibility and clay fractions in expansive, stoney soils. Soil Sci Soc Am J 70:1983–1990

References

Wani SP, Pathak P, Jangawad LS, Eswaran H, Singh P (2003) Improved management of Vertisols in the semiarid tropics for increased productivity and soil carbon sequestration. Soil Use Manag 19:217–222

Woumfo ED, Elimbi A, Pancder G, Nyada RN, Njopwouo D (2006) Physico-chemical and mineralogical characterization of the Vertisols from Garoua (North Cameroon). J Ann Chim Sci Mat 1:75–90

Zewudie E (2000) Study on the physical, chemical and mineralogical characteristics of some Vertisols of Ethiopia. In: Chekol W, Mersha E (eds) Proceedings of the 5th Conference of the Ethiopian Society of Soil Science, Addis Ababa, Ethiopia

Zorluer I, Icaga Y, Yurtcu S, Tosun H (2010) Application of a fuzzy rulebased method for the determination of clay dispersibility. Geoderma 160:189–196

Chapter 7
Peat Soils

Abstract Peat soils are the most dominant type of organic soils developed through centuries under wetland conditions by the accumulation of partially decomposed and undecomposed plant residues. The other type of organic soil is muck which also develops by the accumulation of organic soil materials, but in this type, materials are relatively well decomposed, and the sources of materials are not identifiable. Saturation or submergence of the substratratum and the complete absence of free oxygen cause very slow anaerobic decomposition of organic matter so that deep organic soils or Histosols can evolve. However, a vast expanse of peat soil is called a peatland. More than half of the global wetlands are composed of peatlands; they cover 3 percent of the land and freshwater surface of the earth. Peat soils develop in several wetland types, including mires (bogs, fens), swamps, marshes, and pocosins. Peat soils occur in all regions, but they are more widespread in the temperate and cold zones of the Northern Hemisphere. There are 12.2 M ha (million hectare) peatlands in Africa, 23.5 M ha in Asia and the Far East, 7.4 M ha in Latin America, 4.1 M ha in Australia, 117.8 M ha in North America and 75.0 M ha in Europe. Peatland vegetation includes *Sphagnum* mosses, rushes and sedges, bog cotton, ling heather, bog rosemary, bog asphodel and sundew. There are also forested peatlands in Europe (Alder forests) and in lowland humid tropical areas of Southeast Asia (fresh water swamp forests and mangroves). Peat soils are characterized by high water table, absence of oxygen, reducing condition, low bulk density and bearing capacity, soft spongy substratum, low fertility, and usually high acidity.

Keywords Mires · Swamps · Marshes · Fen · Organic soils · Peat · Muck · Histosols · Peatlands · Peatland use · Peat extraction · Peatland reclamation · Peatland restoration

Peatlands are reservoirs of huge amount of water and organic carbon. The most important ecosystem services they provide include water regulation, biodiversity conservation, and carbon sequestration. Many peat soils are being used for a variety of land uses including forestry, horticulture, pasturing, vegetable and cereal growing, grazing, and peat harvest for energy after draining, and obviously disrupting their hydrology and lithology. Most of the drained and reclaimed peatlands have undergone degradation, some irreversibly and some are even abandoned as wastelands. These extremely degraded peatlands have lost their capacity of ecological and agrological functions and have become sources of greenhouse gases. There are still some other degraded peatlands that can be rehabilitated and restored to their normal functions, if not necessarily to their original conditions. Disturbance of peatlands should be immediately halted and intact peatlands must be conserved. Most of the peatlands are still in a natural state, and ecosystem experts advice to "keep the wet peatlands wet".

7.1 Organic Soils (Histosols, Peat and Muck)

Organic soils develop by the accumulation of organic residues due to slow anaerobic decomposition because of long-term water saturation, oxygen depletion, acidity and/or low temperature. Organic soils belong to Histels and Histosols; Histosols is an order of Soil Taxonomy (Soil Survey Staff 1999) or a Reference Soil Group of World Reference Base for Soil Resources (FAO 2006). Histels is a suborder of the order Gelisols. Histosols order includes soils that have organic soil materials from surface to a depth that varies due to the type of the substratum on which they rest. Histosols do not encounter permafrost, while Histels are soils of the cold zone having permafrost. Histosols except soils of the suborder Folist develop by the accumulation of organic residues under submerged and reduced conditions prevailing for a very long duration, for example centuries. Absence of oxygen causes very slow decomposition of organic residues and favors their deposition and accumulation. Folists are organic soils that develop in upland moderately well-drained environments by the accumulations of folic materials as thick formations (>10 cm over lithic contact and >40 cm over mineral soil). Folic materials include organic residues of forest vegetation, including leaf litter, branches, roots, and other materials. These soils are classified in Folist suborder of Histosols in Soil Taxcording and Folic Histosol great group in World Reference Base for Soil Resources. These soils are formed under cool, moist, humid environments, and are distributed throughout Canada in forest, heath, and alpine, and most prominently in the Pacific Maritime Ecozone (Fox and Tarnocai 2011).

Soil Survey Staff (2010) defines organic soil materials as: (a) if the materials are saturated for less than 30 cumulative days per year in normal years, they must contain more than 20 percent (by weight) organic matter to be organic soil materials, (b) if the materials remain saturated with water for 30 cumulative days or more, they must contain: (i) more than 12 percent by weight organic carbon if the mineral fraction does not have any clay, (ii) more than 18 percent by weight organic carbon if the mineral contains 60 percent or more clay, and (iii) 12+ (clay percentage multiplied by 0.1) percent by weight organic carbon if the mineral fraction has intermediate clay. So, organic soil materials are mixtures of organic and mineral matter at varying proportions. However, the following types of organic soil materials are generally recognized (Soil Survey Staff 2010; Buol et al. 2011): (i) peat which contains almost undecomposed organic material with more than three fourth volumes of fibers identifiable. It is the most dominant type of organic soil material under wetland conditions, (ii) muck which is well-decomposed, black organic soil material with less than one sixth volume of fibers identifiable, (iii) mucky peat which is intermediate between muck and peat, (iv) humilluvic materials which are found sometimes at the contact between organic and mineral soil in cold regions, and (v) limnic materials which form as deposits in a lake. Soil Survey Staff (1999) included all soils without permafrost that have organic soil materials in half or more of the upper 80 cm of the soil profile into the order Histosols. This order has four suborders – Folists, Fibrists, Hemists and Saprists. Histosols also include soils that have organic soil materials of any thickness (usually >10 cm) resting on rocks, or any densic and lithic substratum. According to World Reference Base for Soil Resources (FAO 2006), Histosols include soils that have a histic (a horizon that occurs at a shallow depth and consists of poorly aerated organic materials), or a folic (a horizon that occurs at a shallow depth and consists of well-aerated organic materials) horizon with a depth of at least 10 cm overlying rock, or of 40 cm or more on mineral soil material.

Folists

Folists include well drained upland soils consisting primarily of O-horizons derived from folic materials (leaf litter, twigs and branches) that rest on rock or on weathered or fragmental materials where the interstices are filled with organic material. Folist soils are saturated with water for less than 30 cumulative days during normal years.

Fibrists

Fibrists include soils which remain saturated with water for 30 or more cumulative days with so slow decomposition of organic matter that the botanic origin of the materials can be readily determined. More than 3/4 of the soil consists of fibers. Fibrists have a very low bulk density (< 0.1 g cm^{-3}) and the water level remains near the surface most of the time.

Hemists

Hemists are wet organic soils consisting of moderately decomposed organic matter the botanic origin of which cannot be easily determined. Hemists have fiber content between 1/6 and 2/3 parts of the organic material and bulk density between 0.1 and 0.2 g cm^{-3}. The groundwater level remains high except in the case of artificial drainage.

Saprists

Saprists are wet Histosols with well decomposed organic material the origin of which is not identifiable. Fiber content in these soils is <1/6th apart of organic material, and the bulk density is greater than 0.2 g cm^{-3}. These soils develop mostly under conditions of fluctuating groundwater table, or if the soils have passed through cycles of aeration. Fibric and hemic materials may decompose to form sapric materials if drained and subjected to aerobic decomposition.

Histels

Histels, a suborder of the order Gelisols, have 80 percent (by volume) or more organic materials from the soil surface to a depth of 50 cm, or to a restricting layer such as densic, lithic, or paralithic contacts. Histels are similar to Histosols except having the presence of permafrost within the upper 100 cm. Huang et al. (2011) stated that Histels are mostly distributed in the Boreal, Subarctic and Low Arctic regions.

Most organic soils are found in areas commonly known as mires, bogs, fens, swamps, marshes, and pocosins. Mire is an area of wet, soggy, and muddy ground. There are two types of mires: bogs (Fig. 7.1) and fens (Fig. 7.2). Bogs are domed-shaped landforms situated at higher than surrounding landscapes. The source of water in bogs is the rainfall. On the other hand, fens are situated in flat lands or depressions. The sources of water for fens are both rainfall and surface water. Helmut (2006) reported that a bog is acidic in reaction and poor in nutrients, while a fen may be either acidic or alkaline, and either poor or rich in nutrients. A swamp (Fig. 7.3) has vegetation dominated by forest trees, while mire is dominated by grass and mosses. A marsh (Fig. 7.4) is different from a mire by its characteristics of water, nutrients and distribution. According to Richardson and Huvane (2008), marshes are rich in nutrients, water is stagnant or slow-moving, and the source of water is mainly groundwater. Marsh plants are generally submerged or floating-leaved. Pocosins are bogs on coastal plains and have evergreen shrub vegetation. Like bogs, pocosins have plenty of *Sphagnum* moss and the water and bottom soil are acidic in reaction and poor in nutrients. The source of water is precipitation. Peat soils are of two types on the basis of their formation and sources of water – ombrogenous peat and topogenous peat. The ombrogenous peat is formed in the centre of a dome shaped landscape and is fed with rain water only. As rain water does not contain much nutrient, ombrogenous peat is low in fertility and high in acidity. The

Fig. 7.1 A blanket bog area. (Blanket bog on Dartmoor, Image courtesy of Dartmoor National Park)

7.1 Organic Soils (Histosols, Peat and Muck)

Fig. 7.2 Namekagon Fen. (Photo courtesy of Matt Bushman of USDA Forest Service)

Fig. 7.3 Ratargul Swamp Forest of Bangladesh. (Image courtesy of Rajib Hassan)

topogenous peat is formed in basins getting water both from runoff and rain water. Runoff water carries considerable nutrients from the surrounding landscape (Mutert et al. 1999; Prasetyo and Suharta 2011). Blanket bog is an area of peatland, forming under conditions of high rainfall and a low evapotranspiration, so that peat develops both in wet hollows and over large expanses of undulating ground (Keddy 2010).

Fig. 7.4 Everglades marsh. (Image courtesy of South Florida Water Management District)

7.2 The Nature, Distribution and Significance of Peatlands

Peatlands or mires develop over a long period of time, for example >10,000 years, by the deposition and accumulation of undecomposed and partly decomposed residues of hydrophytic vegetation, including mosses, sedges, reeds, grasses, trees, etc. mixed with some mineral soil materials. As mentioned earlier, permanent water saturation and consequent anaerobic condition do not allow organic materials to undergo rapid decomposition, and favor their accumulation as a fibrous organic substratum below the water level, and it is known as peat. Vast expanse of peat in the landscape is known as peatland, and according to Schumann and Joosten (2008), peatlands are classified into bogs and fens. . However, not all wetlands have peat accumulations. Some wetlands have rocks, sediments and mineral soils in the substratum. Peatlands represent more than half of global wetlands; they cover about 400 million hectares and represent 3% of the total land surface area of the earth (Joosten and Clarke 2002). However, FAO/UNESCO (1974) gives a lower estimate of total peatland areas of the world. Although peatlands occupy a relatively small area, peatlands provide such important ecosystem services as regulation of water, conservation of biodiversity, and sequestration and storage of carbon. Peats and other organic soils contain 30 percent of the world's soil carbon storage (Joosten et al. 2012).

Peatland ecosystems are distributed in arctic, boreal, temperate, or tropical climates (Charman 2002), and range in altitude from sea level to high alpine conditions at elevations between 1000 and 4000 m asl (Novoa-Munoz et al. 2008). According to Joosten and Clarke (2002), about 80 percent peatlands of the world occur in the boreal region. Most abundant in peatlands are Central and Northern

7.2 The Nature, Distribution and Significance of Peatlands

Canada, Alaska, Northern Finland and Siberia. In the middle latitudes peat mainly occurs in mild maritime climates, and especially in mountaneous regions. There are some peatlands in the tropics, mainly as freshwater swamp forests and mangrove forests in South Asia and in mountaneous areas of Africa and South Africa (Prasetyo and Suharta 2011). Fraser and Keddy (2005) pointed out that the West Siberian Lowland, the Hudson Bay Lowland, and the Mackenzie River Valley contain some of the world's largest peatland areas. Vast natural open peatlands occur in permafrost areas of Russia and Canada, the Everglades in the United States, and the high mountains of the Andes and the Himalayas. According to Lappalainen (1996), peatlands are very unevenly distributed in the world with 43.5 percent of total peatland area in North America, 28 percent in Asia, and 24 percent in Europe. The share of peatland area in the northern hemisphere is 95.3 percent and the remaining 4.7 percent in southern hemisphere. There are 12.2 M ha peatlands in Africa, 23.5 M ha in Asia and the Far East, 7.4 M ha in Latin America, 4.1 M ha in Australia, 117.8 M ha in North America and 75.0 M ha in Europe (FAO-UNESCO Soil Map of the World; FAO-UNESCO 1974; readers should be aware that statistics may vary according to different sources). It has already been mentioned that there are about 400 M ha peatlands in the world according to Joosten and Clarke 2002. Table 7.1 presents major distribution of peatlands in different countries according to IPCC.

Peatland ecosystems have evolved over a long time in a landscape by the interaction of unique sets of hydrology, lithology, vegetation, geomorphology and climate.

Table 7.1 Distribution of Global Peatland Area

Country	Peatland area (ha)
Finland	10,000,000
Canada	129,500,000
Republic of Ireland	1,178,798
Sweden	1,500,000
Northern Ireland	166,960
Scotland	821,381
Iceland	1000,000
Norway	3,000,000
Wales	158,770
USSR	71,500,000
The Netherlands	250,000
Germany	1,618,000
Poland	1,500,000
USA	7,510,000
England	361,690
Austria	22,000
Denmark	60,000
Switzerland	55,000
Hungary	100,000

Source: IPCC web site: http://www.ipcc.i

Peatland plants are adapted to conditions of high water and low oxygen content, of toxic elements and low availability of plant nutrients. Most plants have acquired habitat-specific morphological, anatomical and physiological adaptations. The typical peatland plants are *Sphagnum* mosses, rushes and sedges, bog cotton, bog rosemary, bog asphodel, ling heather, and sundew. There are some rare and protected animals in peatlands, including common frog, Irish hare, otter, hen harrier, white fronted goose, peregrine falcon, golden plover and merlin. According to Schumann and Joosten (2008), peatland organisms have adaptations to (i) prolonged water saturation, absence of molecular oxygen, tolerance to toxicity of Fe^{2+}, Mn^{2+} and S^{2-} in the root zone, (ii) continuous peat accumulation and rising water levels, (iii) spongy substratum that can make trees lodge, (iv) limited nutrient supply, (v) high acidity, (vi) tolerance to toxic organic substances produced by anaerobic decomposition of organic matter. However, for such extreme conditions, peatlands have fewer number of plant species relative to mineral soils within the same biographic region. Schumann and Joosten (2008) noted that many mire species are strongly specialized to peatland environments and not found in other habitats.

Unless disturbed, peatland ecosystems usually reach to a dynamic equilibrium with their vegetation, lithology, geomorphology, hydrology and climate, and if any one component of the system is disrupted by anthropogenic or natural events, there is concomitant disturbance or rearrangement in the whole system. This phenomenon has crucial ecological significance because peatlands are huge reservoirs of water and organic carbon. The organic carbon pool in peatlands represent more than one-third of the global soil carbon, and is almost equal to the global atmospheric carbon pool (Charman 2002). According to Tanneberger and Wichtmann (2011), peatlands have been accumulating peat for millennia and about 550 Gt of carbon has now been sequestered in peatlands. Peatlands are also a major source of methane and dissolved organic carbon (Freeman et al. 2004). So, peatlands play a fundamental role in climate change. These huge carbon sequesters remain under threats of degradation by human actions, including peat extraction, drainage and cropping, afforestation and horticultural plantations, grazing, dumping and burning. These disturbances can lead to the emission of huge amount of carbon dioxide and methane to the atmosphere, and can add to the risk of global warming and permafrost melting in the peatlands of the cold region. According to MacDonald et al. (2006), atmospheric warming can cause thawing of peat bog in Western Siberia leading to the release of billions of tons of methane gas into the atmosphere. According to Forster et al. (2007), methane is a more potent greenhouse gas than carbon dioxide (CO_2), having 25 times the global warming potential of CO_2. Tarnocai et al. (2009) suggests that the Arctic peatlands store about 277 Pg of organic carbon (equivalent to one third of the atmospheric CO_2. IPCC (2007) estimated that there would be a considerably high temperature rise (4–8.1 °C higher annual surface air temperatures) in the high-Arctic regions at the end of this century.

According to Murdiyarso et al. (2010), massive disturbances such as drainage, deforestation and burning cause immediate and long-term changes in hydrological conditions of peatlands and enhance C fluxes to the atmosphere. Rapid deforestation rates in tropical wetlands are of concern for large amounts of greenhouse gas emissions (Langner et al. 2007). Miettinen et al. (2011) suggested an annual deforestation rate of 2.2 percent in peat swamp forests of Southeast Asia between 2000 and 2010. Miettinen and Liew (2010) observed that half of the peat swamp forest area in peninsular Malaysia, Borneo and Sumatra had been converted to other land uses between 1990 and 2008. Rewetting drained peatlands after peat extraction causes the reduction of CO_2 emissions due to peat oxidation (IPCC 2006). Rewetting can also cause an increase in CH_4 emissions which may partly offset the reduction in CO_2 emissions, but Wilson et al. (2008) suggested that rewetting is likely to reduce the global warming potential. According to IPCC (2006), there may be on-site emissions of CO_2 due to peat extraction and off-site emission arising from the use of peat for fuel or for horticultural purposes. Couwenberg (2011) assumed that all C in peat used in horticulture is released within 1 year.

7.3 Peat Soils

Peat soils are formed in wetland environments where anaerobic conditions slow down decomposition of dead plant materials which are deposited under water and accumulate over time. Grozav and Rogobete (2010) mentioned different types of peat soils, including moss peat of boreal tundra, reed or sedge peat and forest peat of the temperate zone, and the swamp forest peat of humid tropics. On the basis of the source of the water, peat soils can be Eutrophic and Oligotrophic. If water is mainly supplied by the ground water then peat becomes eutrophic. Eutrophic peats develop in depressions or small lakes. Oligotrophic peats develop in places where rainwater is the source of water. Since rainwater contains a few nutrients, oligotrophic peats have a low fertility and are acidic. Small plants tolerant of low fertility such as *Sphagnum* moss generally grow there. Peats can be classified into some other categories such as sedimentary peat (residues of water lilies, pond weeds, and planktons mixed together), fibrous peat (residues of sedges, mosses, *Sphagnum*, reeds, cat-tails), and woody peat (residues of trees and shrubs). Peat may be deep (> 3 m depth) and shallow (< 2 m) (Abdullah et al. 2007). Generally, deep peat has lower fertility and poor anchorage compared to shallow peat due to the lower bulk density and nutrient content than shallow peat. Bulk density was also higher in saprist (0.28 g cm^{-3}) than fibrist peat 0.20 g cm^{-3}). According to Holman (2009), there are four types of peat soil: (i) deep peat – peat soils thicker than 100 cm from surface, (ii) thin peat – peat soils thinner than 100 cm from surface, (iii) peat at depth –peat layers of more than 30 cm thickness embedded within the soil horizons of generally alluvial mineral material, and (iv) remnant peat –peat soils that have been left after serious erosion.

7.3.1 Physical Properties

Peat soils vary considerably in their physical properties, including water retention capacity, water storage capacity, water availability, hydraulic conductivity, bulk density, porosity, swelling and shrinkage, etc. due to degree of decomposition, sources of materials, bedding, content of mineral matter and degree of mixing, etc. All these factors may affect water storage capacity of peat soils (Silva et al. 2009a). Peat soils have high water holding capacities, and peats behave like "sponge" by storing large volumes of water in wet conditions. According to Brandyk et al. (2003), water content in peat soils of the Northern hemisphere decreases with the increase in degree of decomposition. Peat structure depends naturally on the parent material, and the subsequent mixing and decomposition. As new layers of organic residues are continually deposited, the peat is buried and the porosity decreases. The proportion of large pores decreases and makes the peat less permeable which can be evident from increased bulk density (Kellner 2003). Campos et al. (2011) collected 90 soil samples at a distance of 20 m from 12 transects over an area of 81.7 ha. They determined rubbed fiber content, bulk density, mineral material, organic matter, maximum water holding capacity and moisture content. They observed that maximum water holding capacity and moisture content were the highest in layers with minimum decomposition. The stages of decomposition in von Post scale (Embrapa 2006) were sapric in 8, hemic in 3 and fibric in 1 transect. Mean bulk density was 0.49 g cm^{-3}. Generally, the surface layers were fibric, the middle layers were hemic and the deeper layers were sapric. According to Silva et al. (2009b) maximum water holding capacity was higher in upper peat layers. Mean bulk density, particle density and porosity of peat at varying stages of decomposition are shown in Table 7.2.

Hydraulic conductivity generally decreases with increasing degree of decomposition. For example, Letts et al. (2000) reported the hydraulic conductivity values of fibric, hemic and saprist peats as 2.8×10^{-4} m s^{-1}, 2.0×10^{-6} m s^{-1} and 1.0×10^{-7} m s^{-1} respectively. Asapo and Coles (2012) obtained higher moisture content (86 percent), organic matter (91 percent in saprist peat than fibrist peat having (82 percent), and 84 percent respectively).

Table 7.2 Mean bulk density, particle density and porosity of peat at varying stages of decomposition

Decomposition stage (von Post).	Bulk density (g cm^{-3})	Particle density (g cm^{-3})	Porosity (percent)
H1	0.05	1.4	96
H4	0.10	1.4	93
H6	0.15	1.4	89
H9	0.20	1.4	86

Adapted from Kellner (2003); Note that bulk density increased and porosity decreased with the advancement of decomposition stage, but particle density remained the same

7.3.2 Engineering Properties of Peat Soils

Peat soils have high permeability, porosity ratio, compressibility and consolidation settlement, bulk density, bearing capacity and shear strength, and relatively low plasticity (Akol 2012; Rahgozar and Saberian 2016). Low bearing capacity is a constraint to construction of embankment, highway, building or any other high load structures on peat soils (Islam and Hashim 2009). Munro (2004) reported that peat consists mainly of decomposing plant fragments and water with little measurable bearing strength. Peat has certain other characteristics that require special consideration such as low shear strength, high spatial variability and increased decomposability under altered environments (Long and Boylan 2012). If peat soils have to be used for engineering purposes, proper ground improvement work is need before starting those works. Ground improvement to enhance soil strength can be done earth displacement and replacement, preloading, stone columns, piles, etc. (Edil 2003). Civil engineers consider peat soils as extremely soft and problematic soils (Kalantari 2010). According to Kalantari (2013), the most useful indices for civil engineers on peat soils are water content, loss on ignition and organic content, fiber content, grain size distribution, density and specific gravity, and Atterberg limits.

7.3.3 Chemical Properties

The chemical composition of peat is complex and widely variable. Properties that are common to all peats are that they are heterogeneous poly-dispersed system consisting organic complexes, soluble organic compounds, low and high molecular weight, hydrophobic sols and hydrophilic colloidal systems. According to Malawska et al. (2006), peat deposits of fens, transition bogs and raised bogs have large differences in their chemical characteristics; peat from distant geographical locations can have similar chemical properties as well. On the other hand, peat within the mire complex may have wide differences in chemical characteristics. Even if the botanical origin of peat materials are similar, chemical attributes can be vary dissimilar, so that vegetation characteristics cannot be very dependable indidices of the chemical properties of peat. Peats contain some organic compounds that are soluble in water, ether or alcohol in addition to such compounds as cellulose, hemicelluloses, lignin and their derivatives, and proteins. The pH of most organic soils is low due to the presence of organic acids, the exchangeable hydrogen and aluminium, iron sulfide and other oxidizable sulfur compounds. Asapo and Coles (2012) observed pH of saprist and fibrist *Sphagnum* peat soils of a natural peat bog of Canada to be 4.2. However, organic soils of rich or extremely rich fens can have neutral to slightly alkaline reaction. Wetland mineral soils also have pH values near neutrality.

Decomposition of organic materials in peat under aerated condition produces CO_2 as the terminal carbon compound; under anaerobic situation and in the reduced subsoil below the thin oxidized surface soil, the terminal compound is CH_4 with some alcohols and acids as unstable intermediate products. Some oxidation of organic matter takes place in soil below the aerated zone at the expense of reduction of nitrate, oxides of metals and sulfur but as their concentrations are generally low, they are completely reduced within a short period after the soil becomes anaerobic, and therefore, their importance is also low (Kellner 2003). As decomposition of organic residues proceeds, hemicellulose, pectins and gums tend to predominate in peat materials because of their low decomsability. Proteins are also rapidly lost; only some amino acids are retained in humic complexes of peat materials. The humic compounds can be fractionated in to humic and fulvic acids; these compounds have a great influence on chemical behavior of peat by their very large surface area, high ion exchange capacity, high sorption capacity, a large colloidial affinity, and a remarkable water absorbing capacity. According to Fredriksson (2002), humified peat becomes a jelly like material by absorbing a huge amount of water. Glycerol and fatty acids produced from lipids along with waxes and steroids tend to remain unaltered in peat. Dissolved organic matter (DOM) in as important constituent of peat; it is composed mainly of fulvic acids, long-chain fatty acids and esters. Droppo (2000) defined DOM as organic substances having a particle size less than 0.45 µm.

Peat soils can sorb large quantities of cations, particularly metal catons, from solution by ion-exchange, surface adsorption and complexation, and large variations are found in this capacity probably depending on peat types and their preparation for analysis and concerned metals (Brown et al. 2000). The dominating exchange sites of peat compounds are the Carboxyl (–COOH) and phenolic hydroxyl (-OH) groups are the dominant sites of exchange. Brown et al. (2000) reported that the presence of monovalent cations such as Na^+ reduced the sorption of metals. (Peat colloids saturated with Ca^{2+}, Mg^{2+} and Fe^{2+} ions can enhance the sorption of heavy metals. Metal sorption on peat is an interesting phenomenon having practical significance in treating wastewater (Brown et al. 2000; Ringqvist et al. 2002). Oxidation-reduction of inorganic substances such as iron, manganese and sulfur along with CO_2-CH_4 transformations regulate the redox potential in peat soils in the under oxidized, lower reduced and the transition zones. Usually, there is a vertical stratification of redox potential with depth undrained peat soil decreasing from aerated surface soil to strongly reduced zone in the profile.

According to Boguta et al. (2011) potassium is generally very low in peat soils because potassium forms soluble complexes or humates. A large amount of potassium is lost from peat due to the high mobility of humates of monovalent metals. For this reason, potassium deficiency occurs in organic soils (Sigua et al. 2006; Brandyk et al. 2008). On the other hand, calcium humates have low solubility and mobility. Among multivalent ions, aluminium or iron have higher affinity to humic acids than manganese, which has great mobility in organic and in low pH soils. Kalembasa and Pakuła (2008), manganese is present as Mn^{2+} in acidic organic soils, except in the surface aerated layer where it can be oxidized and enriched.

The elemental composition of some eutrophic and oligotrophic peats are shown in Table 7.3.

7.4 Reclamation and Management of Peat Soils

Table 7.3 Elemental composition of some peat soils

Element.	Range (percent oven dry)	Average (percent)	
		Eutrophic Peats	Oligotrophic Peats
Aluminium	0.01–5.0	0.5	0.1
Boron	0.00001–0.1	0.01	0.0001
Calcium	0.01–6.0	2.0	0.3
Carbon	12.0–60.0	48.0	52.0
Chlorine	0.001–5.0	0.10	0.01
Cobalt	0.00–0.0003	0.0001	0.00003
Copper	0.0003–0.01	0.001	0.0005
Iron	0.02–3.0	0.5	0.1
Lead	0.00–0.04	0.005	0.001
Magnesium	0.01–1.5	0.3	0.06
Manganese	0.0001–0.08	0.02	0.003
Molybdenum	0.00001–0.005	0.001	0.0001
Nickel	0.0001–0.03	0.001	0.0005
Nitrogen	0.3–4.0	2.5	1.0
Phosphorus	0.01–0.5	0.07	0.04
Potassium	0.001–0.8	0.1	0.04
Sodium	0.02–5.0	0.05	0.01
Sulfur	0.004–4.0	0.5	0.1
Zinc	0.001–0.4	0.05	0.005

(Sources: Sayok et al. 2008; Kolli et al. 2010)

7.4 Reclamation and Management of Peat Soils

Soil reclamation usually refers to the making of a problematic soil useful by changing its physical and chemical characteristics. Here, reclamation is taken to be the conversion of an ecosystem to another ecosystem according to a definite plan for the purpose of achieving desired services and functions. Although the majority of peatlands still remain in the natural state (Joosten et al. 2012), large peatland areas were reclaimed in the past throughout the world for agriculture, horticulture, forestry, constructions and peat extraction.. In most cases the productive outputs were not sustainable and reclaimed peatlands had undergone severe degradation. Carefully reclaimed and managed peat soils can be productive under high input agriculture, horticulture and forestry but drained peat materials suffer from land reclamation treatments (Gawlik and Harkot 2000). Drainage of bogs and peats has been responsible for peatland degradation, including rapid peat oxidation, land subsidence, loss of biodiversity, lack of water storage and recharge, carbon emission, etc.

According to Ilnicki (2002), large areas of organic soils are farmed in Europe, The United States and Canada. There are 70,400 km^2 cultivated organic soils in Russia, 12,000 km^2 in Germany, 9631 km^2 in Belarus, 7620 km^2 in Poland, and 5000 km^2 in Ukraine in Europe. There are 3080 km^2 cultivated organic soils together in the United States and Canada. It has been observed in many instances that developed peat soils may be highly productive and a wide range of crops can be grown on them. O'Connor et al. (2001) reported that very high yields of maize in New Zealand were recorded in some peat soils. Common food crops grown on reclaimed peat soils are paddy, maize, cassava, sweet potato, yam, sorghum, black potato, tannia (a kind of arum), and Chinese water chestnut. Secondary crops include soybean, groundnut, yard long bean, green gram, cowpea, mungbean, velvet bean, bambara (African) groundnut, pigeon pea, winged bean, lima bean, sunflower, etc.

As peats in natural conditions give the best ecological services, they are best conserved under undisturbed natural conditions (Anon 2012). Traditional crop production is not, however, possible in peat in natural state because they are saturated with water, soft, fibric and woody, very acidic, and generally infertile. Some physical properties of peat, including high porosity and water holding capacity, good aeration when drained, and a structure that favors root penetration can, however, benefit crop productivity. But sustaining yield and preventing peat soil of reclaimed and drained peatlands from degradation offer many challenges.

7.4.1 Peatland Selection for Reclamation

Suitability of peat bogs to develop for crop production is based on some of their features, including type and depth of peat, content of wood, type of material beneath peat layer, slope and hydrology. Climate of the area has also considerable importance. Proper consideration must be given to the analysis of the nature and properties of peat well in advance of planning for draining and development of peatlands into croplands. Generally, developing fibric peat is preferred because it will break down into mesic peat within a few years due to drainage, tillage and fertilization, while mesic peat taken initially for development will decompose into the humic state within a short period. For this reason fibric peat (decomposition stage of 3 to 4 of von Post Scale) is preferred for developing peat for crop production. Von Post divided the decomposition of organic matter into 10 stages (Embrapa 2006). Stage H3 represents very slightly decomposed peat which upon squeezing releases muddy brown water, and peat of stage H4 is slightly decomposed peat which upon squeezing releases very muddy dark water; no peat passes between the fingers from any of these two peat types, plant parts are identifiable, and no amorphous material is present.

Most crops grown in reclaimed peats need a minimum depth of drainage of 1 m for healthy root growth. Greater depths of drainage may be needed for ditching to grow vegetables. Gradual break down of peat and subsidence will reduce the depth over time. So, selecting peat for reclamation should be based on average depth of water table requirement of a range of crops intended to grow there. Content of wood is also a matter of consideration for peat selection for cropping. According to Eastern Canada Soil and Water Conservation Centre (1997), 5 percent by volume of wood up to the depth to be drained may be desired; greater than 25 percent is not suitable for cultivation and ditching.

Fig. 7.5 Organic soil material on sandy substratum. (Image courtesy of John Galbraith of Virginia Tech)

Thickness of peat is another important factor for selecting peatland for crop production; thin peat tends to be wasted faster. Considerations must also be given to the materials underlying peats; sands beneath peat favor drainage and bearing capacity; silts and loams can give some problems, while clays can create severe drainage problems (Eastern Canada Soil and Water Conservation Centre 1997). Fig. 7.5 shows a peat profile where an organic layer rests on sandy mineral matter. On the other hand, Fig. 7.6 shows the entire profile composed of organic materials. After drainage peat can dry out and subside irregularly.

7.4.2 Modification of Peatland for Use

Removal of vegetation, burning the debris, installing drainage, land leveling, filling, ridging, etc. are the major operations taken for bringing peat into crop production. These operations are expressed in a phrase 'land development for peat reclamation' by many authors. Since the word 'development' has an inherent positive sense and 'reclamation' traditionally refers to minimizing soil problems, but most human activities in peat leads to some sort of its degradation in the long run, the phrase 'modification of peatland for use' is preferred here. Most land preparation operations are done manually. Traditionally, the surplus vegetation is burnt to ashes which

Fig. 7.6 A deep soil profile with organic soil material throughout more than 2 m. (Image: John Galbraith of Virginia Tech)

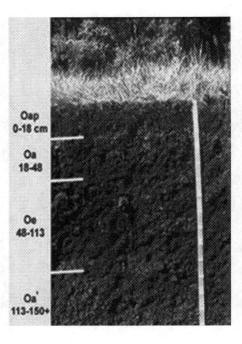

Fig. 7.7 Land cleared, drained and burned in the Tripa Peat Swamp Forest for oil palm plantation. (Image courtesy of NBC NEWS)

may improve the fertility of the peat soil temporarily. The burning, however, may also damage the surface peat layers and enhance their oxidation. Energy and cost of land preparation for peatland use vary with the type of peat, type of vegetation, depth of flooding, depth of drainage requirement, maintenance, etc. Primary drainage is achieved generally by surface ditching varying in depth, width and frequency with the drainage requirement for the desired land use (Fig. 7.7).

7.4 Reclamation and Management of Peat Soils

A perimeter ditch and several lateral ditches are dug across the area to collect water and dispose it out of the area. Ditches can be of different shapes (straight or V-shaped), width, grading, depth and frequency depending on the hydrological conditions and land use types. The closer the spacing, the more effective would be the drainage. Ditch banks need to be stabilized and this can be achieved by growing suitable grasses or herbs. Integration of surface and subsurface drainage has been done in many areas and has been found to be more effective than surface drainage alone. However, disposal of drainage water has to be done carefully so that it does not contaminate water bodies, including ponds, lakes and streams. The land is leveled by cutting and filling after draining. Often roads and culverts are constructed for some land use types and use-oriented operations.

7.4.3 Afforestation in Peatlands

Drainage and soil modification have been done in many peatland areas of the world to use the land for plantation forestry. For example, there were 200,000 ha of forest plantations mostly on blanket peat in Ireland, about the same area in Great Britain (Anderson 2010) and about 460, 000 ha of peatlands and other wetlands in Estonia (Barthelmes et al. 2015). Extensive areas of black spruce (*Picea mariana*) and tamarack (*Larix laricina*) in Alberta (Hillman 1997), about 5.7 million ha of Scots pine (*Pinus sylvestris*) and Norway spruce (*Picea abies*) (Paivanen 1991, and large areas of planted forests in the former Soviet Union and Sweden, and large areas of energy plantations and willow coppices (*Salix viminalis* and *Salix dasyclados*), alder (*Alnus* spp.) and poplar (*Populus* spp.) plantations had been developed in many countries.

The most important tree species taken for peatland afforestation are listed below.

Broadleaves Common alder (*Alnus glutinosa*), Italian alder (*Alnus cordata*), Ash (*Fraxinus excelsior*), Aspen (*Popilus tremula*), European beech (*Fagus sylvatica*), Silver birch (*Betula pendula*), Downy birch (*Betula pubescens*), Pedunculate oak (*Quercus robur*), Sessile oak (*Quercus petracea*), Poplar (*Popuius beaupre*), Norway maple (*Acer platanoides*), and Sycamore (*Ace pseudoplantanus*).

Conifers: Hybrid larch (*Larix* x *eurolepis*), Japanese larch (*Larix kaempferi*), Corsican pine (*Pinus nigra var. maritime*), Lodgepole pine (*Pinus contorta*), Macedonian pine (*Pinus peuce*), Scots pine (*Pinus sylvestris*), Norway spruce (*Picea abies*), Sitka spruce (*Picea sitchensis*), Western red cedar (*Thuja plicata*), and Irish yew (*Taxus baccata*).

Several processes start working simultaneously after draining a peatland. According to Crill et al. (2000), the net effects are not similar in all situations. Mineralization of peat occurs faster after drainage and the peat carbon storage decreases rapidly. Low temperatures in the boreal zone, however, may reduce the rate of mineralization. In boreal areas, tree stands reduces the albedo affecting radiative balance and contributing to climate warming (Lohila et al. 2010). Patterson and Anderson (2000) mentioned the conclusion of the Forestry Commission to conserve peatlands and to refrain from further forestry practices by draining peatlands.

7.4.4 Cultivation of Oil Palm in Drained Peat Soils

According to Fitzherber et al. (2008) and FAO (2009), oil palm (*Elaeis guineensis* Jacqz) is grown in about 15 M ha throughout the world. It mainly grows in tropical lowlands. According to Lester (2006), oil palm is an important source biofuel and edible oil. It is cultivated in large scales in Indonesia, Malaysia and Thailand in Southeast Asia, Nigeria in Africa, Colombia and Ecuador in South America and Papua New Guinea in Oceania (FAO 2009). Large even aged monocultures with low canopy, sparse undergrowth, and intensive use of agrochemicals are the characteristics of present day oil palm cultivation (Fitzherbert et al. 2008). Oil palm plantation has been expanded to 2.15 million hectares of peatlands throughout Malaysia and Indonesia in 2010. There are about 0.5 M ha of peatlands under oil palm cultivation in the Sarawak of Malaysia alone (Ambak and Melling 2000). The lowest projection of oil palm area by 2030 is 6 M ha; increased emission and peatland degradation may occur due to this land use change (Page et al. 2011). Widespread logging, drainage, plantation development, and expansion of fragmented landscapes are noticed in many peatlands (Silvius and Diemont 2007). The potential adverse environmental effects of growing oil palm in peatlands have raised criticism worldwide (Sheil et al. 2009).

Oil palm trees bear fruits at the age of 3 years and the yield gradually increases up to 20 years (FAO 2002). At the age of 25–30, old oil palm plantations are replaced by new plantations (Wahid et al. 2005). However, continued oil palm culture reduces the quality of soil and freshwater, and affects adversely the ecosystem services including regulation of the hydrological and biogeochemical cycles (Fitzherbert et al. 2008). Many peat swamp forests cleared for palm plantation had been lost forever; it implies the loss of carbon sequestration service through peat accumulation. Koh and Ghazoul (2010) stated that oil palm plantations are developed by replacing lowland forests; peat swamp forests represent a major proportion. The land is drained for palm plantation so that water table stays at 50–100 cm below the surface. Secondary drains of 1.5 m depth are dug perpendicular to the lines of the palm seedlings. Tertiary canals may also be needed for effective drainage. According to Jaya (2002), the major problems of developing oil palm plantation on peat include shallow water table, low bearing capacity, low fertility and inadequate root anchorage. Although mineral soils are more suitable for oil palm cultivation, oil palm is still expanding in peatland areas because two important reasons – shortage of land and high profitability of oil palm. Peat drainage and palm plantation accelerates decomposition of organic matter, increases bulk density and ash content, loss on ignition, total porosity and liquid limit, and reduces continually the content of fiber. According to Firdaus et al. (2012), gradually rising water table in mature oil palm plantations reduces the bearing capacity and restores the water content in peat soils.

Water management is a crucial factor for successful oil palm plantation in peat. The peatland must be drained but the water table must also be maintained close to the surface for successful establishment of oil palm on peat. Artificial drainage causes shrinking of peat and it is then compacted mechanically. The subsidence is regulated by manipulating the water table. When compacted properly, peat soil can

Table 7.4 Fertilizer rates for oil palm plantations

Fertilizer application (kg ha^{-1} year^{-1})[a]						
	N	P	K	Mg	B	References[b]
Immature	35–120	22–56	42–420	8.4–35	0.6–3.7	
Mature	35–245	22–98	183–581	42–105	0.6–5.6	

[a]Assuming 140 palm trees per hectare and depending on site conditions
[b]FAO (2005), Goh and Hardter (2003), Von Uexkull (2007)

attain desired capillarity, water holding capacity and anchorage; it can reduce the risk of fire and enhance growth and yield. For establishing palm, the peat is compacted by 0.5–1.0 m at each planting point, then the planting point is further compacted with a special attachment and palm seedlings are planted in the hole. This technique is known as "hole-in-hole-in-hole" planting. Oil palm has very high demands of nutrients for vegetative growth and fruit production. Large amounts of mineral fertilizers are required to support desired growth in these inherently poorly fertile peat soils, and fertilizers incur about 25 percent of the total cost of production (Caliman et al. 2007). According to Goh and Härdter (2003a, b) several factors are involved in the efficient use of fertilizer for increasing yields and enhancing profit. However, there is wide variation in nutrient requirements of oil palm, and it depends on yield target, genetic potential of the planting material used, tree spacing, palm age, soil fertility, and climate (Tarmizi and Mohd 2006). The recommended dosage of fertilizers for palm at different growth stages are shown in Table 7.4.

Rock phosphate and wood ash are added in the holes and their surrounding areas to supply N, P and K prior to planting palm seedlings. Peat soils are high in N contents but organic N is not directly available to plants. As mineralization is also low, palm plants need N fertilizers. Usually, 1.25 kg urea containing 0.6 kg N are applied annually in two installments to each plant. Phosphate fertilizers are applied annually at the rates of 300–400 g P_2O_5 per palm plant. As the peat soils are highly acidic, an annual application of 1.0–1.25 kg finely ground rock phosphate per palm plant is preferred. In the 2nd and 3rd year of establishment of palm plantation, both urea and reactive rock phosphate are added at the rate of 1.5 kg per palm plant in the following couple of years. Peat soils are very deficient in K; annually 2–4 kg K_2O per palm plant are applied as KCl in 3–4 split applications to minimize leaching losses.

Although oil palm is a very profitable crop and it provides job opportunity to millions of people in Indonesia and other countries (USDA 2007), the processes of oil palm plantation development in drained peatlands in large areas have triggered deforestation, loss of biodiversity, degradation of peat and emissions of greenhouse gas. Land clearing, road and drainage network construction, and sometimes earthworks such as terracing on undulating areas, use of agrochemicals might make the sustainability of aquatic ecosystem and hydrological functions risky (Comte et al. 2012). Mineralization of organic matter and secondary humification alters morphological, chemical, biological, and physical properties of soils (Sokołowska et al. 2011). The sorption capacity of soil is reduced and it becomes more hydrophobic in character; under such conditions, peat does not grow further, humidity decreases, and oxygen supply increases (Sokołowska and Boguta 2010). All these changes lead peat to transform into muck which contains high humic substances including humic and fulvic acids. Asing et al. (2009) reported that these substances have high sorption and cation exchange capacity. So, muck can be used in various growth media (Boguta and Sokolowska 2014).

Stichnothe and Schuchardt (2011) suggested that the environmental sustainability of oil palm cultivation may be judged on land-use change, soil quality, biodiversity and water quality impacts. Land-use conversion from forest to oil palm contributes to loss of biodiversity, increased soil erosion, nutrient loss and GHG emissions. According to Comeau et al. (2013), emissions of CO_2 from soil increased gradually along the progression of changing peat swamp forest to drained peat forest to logged peat forest to oil palm plantation. They observed that CO_2 fluxes were 28.4 Mg C-CO_2 ha^{-1} y^{-1} in the oil palm plantation, 18.5 Mg C-CO_2 ha^{-1} y^{-1} in the transitional logged drained forest and 16.0 Mg C-CO_2 ha^{-1} y^{-1} in the intact peat swamp forest.

Miettinen et al. (2012) suggested that the total area of 3.1 M ha of industrial plantations in 2010 in Malaysia and Indonesia together may be doubled in 2020; in such a situation monitoring of peat subsidence and CO_2 emissions caused by oxidation of drained peat soils is essential (Oleszczuk et al. 2008). Hooijer et al. (2012) cited that the rate of subsidence of drained peatlands ranged from 3.7 to 5.0 cm y^{-1}. The rate of subsidence varies little spatially, but it does significantly over time. At high water table, the rate of subsidence is low (van den Akker et al. 2008; Leifeld et al. 2011). The method of assessment applied for subsidence based carbon losses applied by van den Akker et al. (2008) has been used in the Dutch reporting to the United Nations Framework Convention on Climate Change (UNFCCC) and its Kyoto Protocol (van den Wyngaert et al. 2009).

7.4.5 Cropping in Naturally Drained Peat Soils

In the environmental quality context, sustainable agriculture in peat swamps in their natural state instead of draining them is desired. However, selecting economic and adaptable crop plants to swamp peat environments is an important task. Economic use of natural peatlands without disturbing their normal ecosystem functions would solve the problem of subsidence and emissions to a large extent. Sago (Sago (*Metroxylon sagus*, a smooth variety, and *Metroxylon rumphii*, the thorny variety) is a prominent crop grown in many areas of Malay-Indonesian archipelago and used for starch as food and raw material of methanol. Raffia palm (*Raphia* spp.) and papyrus (*Cyperus papyrus*) are two indigenous plant species of African peat swamps. These plants may be grown commercially if there are enough marketing facilities. Several varieties of wild rice (*Zizania aquatica* and *Zizania palustris*) are commercially grown in North America. Their cultivation could be extended to natural peat swamps and can be profitable. Some other plants including water celery (*Oenanthe javanica*) and water spinach (*Ipomoea aquatica*) may be grown commercially in peat swamps. Different species of sedges (*Carex* spp.) can be grown in natural peat swamps and are used for matting and other purposes.

7.4.6 Cropping in Artificially Drained Peat Soils

Crops grown in artificially drained peat soils are of many different types depending on the agroecological conditions, tradition, and farmer's choice as well as their experience. Soil moisture status, need of drainage, soil acidity and fertility, and climate are of the most important factors for selecting crops, soil preparation and crop management. Water management, tillage, liming and fertility management have important roles on successful crop production at the expense of the least peat soil degradation.

7.4.6.1 Crops

Large areas of drained peat soil in the temperate region are used for pastures. They need to lower the water table to about 40 cm depth; this depth is being maintained in many areas of the Netherlands for pasture for centuries to minimize the negative impacts of drainage. Fodder crops such as Napier grass can also be grown. The most adapted crop to the swamp environment is rice (*Oryza sativa*). However, growth of rice is adversely affected in oligotrophic peat soils probably because of nutrient deficiency, particularly of copper which causes empty panicles. Adapted field crops in drained peat soils include corn, sugar beet, mint, peas, grasses, and small grains. Organic soils are considered to be highly suitable for growing vegetables, including potato, onion, carrot, parsnip, lettuce, celery, cabbage, cauliflower, table beet, sweet corn, radish, yow choy (Chinese kale), choy sum (*Brassica rapa* var. *parachinensis*; a Chinese leafy vegetable of the *Brassica* genus and the Brassicaceae (cabbage) family), gai lan (Chinese broccoli), tung choy (Chinese water spinach) and spinach. Some organic soils can be developed into suitable physical medium for plant growth. However, frost can damage many susceptible crops such as sweet potato, pepper, eggplant, melon, tomato, etc. The requirement of the depth of water table depends on crops; but for most vegetables, a shallow water-table <60 cm is sufficient. Some vegetable crops including cabbage, carrots, and celery have been successfully grown in peat soils of Eastern Canada from the historical past; asparagus, beets, broccoli, cauliflower, etc. can also be grown there. Where the climate is favorable, tomatoes, peppers, cucumbers, corn, peas, beans, and egg plants can be selected. Crops such as chili, soybean and tobacco need similar management to that of vegetables and can be cultivated successfully there. Fig. 7.8 shows a good crop of cabbage in a drained peat soil.

Many horticultural crops are successfully grown on peat soils of the temperate regions; sometimes coarse textured mineral soils are mixed with peat to improve the physical condition so that plants can develop good root system. Horticultural plants can be easily planted in peat soils, even on oligotrophic peat drained to 50 cm depth in the tropical region.

Fig. 7.8 A good crop of cabbage in a peat soil of Canada. (Image courtesy of Eastern Canada Soil and Water Conservation Centre)

7.4.6.2 Tillage

Tillage operations include plowing, disking, rotovating, cultivating, harrowing, and then ridging or bedding. The objectives and methods of tillage are described in Chap. 2. The soil is prepared for seeding or planting in raised beds and ridged beds. Sometimes, the soil is rolled after tillage to ensure satisfactory soil-to-seed contact for small seeds of crops like carrots and lettuce, etc. Tillage in peat soil is done for improving hydraulic conductivity, enhancing seed germination and seedling establishment, providing adequate soil aeration, and reducing compaction, crusting and erosion. Tillage is easier in peat soils because of their inherently soft consistence and loose structure. However, there may be a high proportion of wood in some peats; removing large wood remnants manually may be needed. Deep plowing is done in some areas where mineral soil underlies peat; mixing improves the physical condition but increases the risk of rapid decomposition of organic matter and subsidence of land. Common agricultural implements designed for cultivating mineral soils are often not suitable for peat soil because of their extremely low bearing capacity. Modified equipments such as floatation tires, dual wheels, tracks, half tracks can be used.

7.4.6.3 Water Management

The water management system is one major component of the land use planning taken in hand well ahead of the desired land use in peat swamps. Water management consists of draining the entire peat swamp during land preparation, and maintaining

7.4 Reclamation and Management of Peat Soils

Table 7.5 Optimum depth of water table for different crops

Crops	Optimum water table depth, cm	Crops	Optimum water table depth, cm
Asparagus	60–90[a]	Corn	45–75[b]
Bean	45–60[b]	Lettuce	50–75[a, b]
Beet	50–75[a]	Mint	60–75[a]
Beet	60–75[b]	Onion	60–90[a]
Blueberry	30–40[b]	Onion	45–60[b]
Cabbage	75–105[a]	Parsley	30–40[b]
Cabbage	45–65[b]	Parsnip	75–105[a]
Carrot	60–90[a]	Parsnip	75–90[b]
Carrot	75–90[b]	Potato	60–90[a]
Cauliflower	50–60[b]	Potato	50–60[b]
Celery	50–75[a]	Radish	50–75[a]
Celery	40–50[b]	Radish	30–40[b]
Cereals	40–50[b]	Sod	40–50[b]
Cranberry	30–40[b]	Spinach	50–75[a]

[a]Eastern Canada Soil and Water Conservation Centre (1997)
[b]Lucas (1982)

the water table at a depth relatively safe for keeping peat decomposition at a reasonably low level during the entire period of use. Operations relating to draining the peat swamp include constructing dams, dikes, levees, canals and different types of ditches. Water management in agricultural farms on peat is aimed at keeping a balance between moisture and air in the soil for optimum yield and for controlling oxidation and decomposition of organic matter in peat. The systems used to achieve these objectives are field drains, raised beds, ridged beds and irrigation systems whichever give satisfactory results. Keeping the water table at a moderate depth may be needed to prevent peat from drying out. Peat can be irreversibly dried so that rewetting is not possible. Where prolonged dry seasons are encountered, drainage may need to be supplemented by irrigation. Table 7.5 gives the optimum depth of water table for different crops.

Characteristics of agricultural field drainage systems including their types, depth and frequency depend on crop characteristics such as depth of rooting, requirements of moisture and aeration, and on depth to which the groundwater table should be lowered, drainage facilities, etc. Most field drains for agricultural crops on peat constitute ditches of 50–80 cm depth and at 15–30 m spacing are connected at right angles to the main drains at 90–150 cm deep. Subsoil drainage systems such as tile drainage and mole drainage can be integrated with surface drains. Tile drains are vulnerable to silting up and mole channels soon close up and become ineffective. Agricultural drainage systems and their merits and demerits have been discussed in Chap. 5 on Soils with Drainage Limitations. According to Ilomets (2015), fen types are affected by the drainage for agriculture; most transitional fens have been turned into pastures after drainage.

7.4.6.4 Liming

Liming refers to the addition of carbonates, oxides, and hydroxides of usually calcium and sometimes magnesium or both to increase soil pH; these materials are called liming materials. Liming has been discussed in Chapter 11 on Acid Soils and Acid Sulfate Soils in more detail. Liming should be well judged because liming is expensive and it can create several micronutrient deficiencies. It is, therefore, customary to find a crop that is suitable and can be profitable in the current soil pH. For example, pineapple, corn, alfalfa and onions grow well in low pH (< 5.0). However, liming usually becomes essential for growing desirable crops in peat soils with low pH. The amount of lime requirement depends on current soil pH and the level to which it should be raised for the particular crop. The type of peat is also very important in this regard. Enough lime may be required to raise the pH by one unit in many peat soils that have pH values between 3.5 and 4.0 for growing vegetables and other crops. Both limestone and dolomite can be used but ground limestone is preferred because it is fast acting. As lime is relatively immobile in peat soils, it is thoroughly mixed within the depth required. Most peat soils are inherently poorly fertile; overliming there can cause deficiency of some micronutrients, produce nitrate toxicity by influencing denitrification and enhance peat decomposition.

7.4.6.5 Fertilizer Application

Only some mesotrophic and eutrophic peats can be sufficiently fertile; most peats (oligotrophic; developed from Spagnum moss) are usually very poor in fertility and deficient in N, P, K, Ca, Mg, Fe, B, Cu, Mo, Zn, and Mn. For satisfactory agricultural production in such peat soils nutrient supplementation is essential. However, the kind, amount, frequency and methods of application of fertilizers depend on the types of peat, crop and management. Fertilizers are active chemical substances; they can have several negative impacts, particularly on the quality of soil and water. They can accelerate the decomposition of the peat, enhance CO_2 emission and cause rapid peat degradation. A soil fertility management system that will ensure both economic and environmental sustainability needs to be adopted.

Other than uptake by the crops, added fertilizers have several fates, including leaching, volatilization, denitrification, immobilization, fixation, etc. Phosphates are relatively immobile in mineral soils, but according to van Beek et al. (2007), peat soils are 'hot spots' of P leaching. Litaor et al. (2006) reported that P is transported mainly as particulate P from peats containing calcareous materials. Cultivation of peat soils with the addition of P fertilizers can be a major contributor of P enrichment in surface water and groundwater (Krogstad and Bechmann 2011). Often sand is mixed with peat to reduce P leaching and to improve soil structure and heat capacity of the soil leading to better utilization of fertilizers. Split application of P in peat soil can reduce loss of P as well. Potassium leaching from peat soils poor in clay and mineral matter can create severe K deficiency. Data from different sources indicate that farmers of South East Asia use 280–560 kg N ha^{-1} for vegetables, 45–78 kg N ha^{-1} for legumes, 11–112 kg P ha^{-1} for vegetables and cereals, and amounts of K similar to N on oligotrophic peat soils. Micronutrients are applied at a rate of 5 kg ha^{-1}.

Fertilizers applied in multiple split doses can reduce loss of added nutrients, improve nutrient use efficiency and may be less risky to contaminate surrounding water bodies. However, fertilizers become more effective when applied at the time the crops need them most. Fertilizers are needed most during the maximum vegetative growth, branching, flowering (for cereals) and fruit setting (horticultural plants) stage. Farmers should ensure that in these growth stages crops must suffer from inadequate supply of nutrients. There are various methods of fertilizer application, including broadcasting, side-dressing and foliar applications. Broadcasting is done in close-growing crops such as rice, side dressing in row crops such as cabbage, in circular bands in horticultural plants and oil palm, and foliar application as low concentration solution is done where sprinkler irrigation system can be installed. Fertilizers can be added with irrigation water in drip system continuously at low rates throughout the growing season.

7.5 Peat Extraction

For almost 2000 years peat has been used as fuel for cooking and heating as alternative source of energy in many parts of the world, including those of temperate and boreal regions of Europe, such as Ireland, England, the Netherlands, Germany, Sweden, Poland, Finland and the former USSR. Peat is a cheap and easily extractable material; so, it is still a popular source of fuel in many developing countries. Large high capacity power plants have been developed using peat fuel in some areas to meet the local demand of electricity. The increasing availability and comfortable use of oil and gas as domestic and industrial fuel has decreased the need of peat fuel use. However, peat is extracted for many other purposes, such as raw materials for production of organic fertilizers, mixed organic and mineral fertilizers, as a soil conditioner, in preparing growth media for glasshouse plants, pot plants, bedding plants and vegetable plants in containers (van Schie 2000). Peat is mixed with mineral soil to improve physical properties including aggregation, porosity, and water holding capacity. Peat is used for extracting hydrocarbons, and for insulation because of its low thermal conductivity. Peat extraction was considered desirable from agricultural view point in some places where the peat is thin and underlies mineral soils good in fertility and profitable for arable use. However, peat extraction is a lucrative business (Fig. 7.9).

Peatlands are huge reservoirs of carbon. Peat extraction for fuel, horticulture, landscaping and other purposes not only causes loss of carbon but also damages the capacity of further carbon sequestration. According to Cleary et al. (2005), extraction of peat leads to a loss of 20–35 t C ha^{-1} year^{-1} in modern peat fields. Some authors (Crill et al. 2000; Waddington et al. 2002; Cleary et al. 2005) have suggested that clearing vegetation, draining the extraction site and its surroundings, the

Fig. 7.9 Peat extraction: stacking peat for drying in Somerset County. (Image courtesy of Somerset County Council)

peat collection process and storage are responsible for a great proportion of carbon loss. According to Holden et al. (2006) peat extraction exposes the subsoils which become susceptible to wind and water erosion. Peat as fuel is immediately oxidized; horticultural peat is oxidized within some years. Abandoning the sites after peat extraction prevents them from being rewetted and they become important sources of carbon emissions (Mäkiranta et al. 2007). The left over peat surface requires very long time for revegetation; the dry conditions as a result of intensive drainage cause peat decomposition and make residual materials susceptible to fires. Large carbon emissions occur as a result (Chistotin et al. 2006).

7.6 Risks Associated with Peatland Use

Artificial drainage of the wetlands where peat had developed over thousands of years is the initial operation taken for peatland reclamation for any desired use. Everywhere, artificial drainage was followed or accompanied by clearing, slashing and burning of existing vegetation, excavating, filling and ridging the land, and closing sources of water. Thus, peatland reclamation brought about changes in the whole geophysical, geochemical and biological environments of an ecosystem that reached to equilibrium through complex interactions over a very long time. From historic times, millions of hectares of peatlands had been reclaimed throughout the world, including vast peat areas in the Netherlands, Finland, Russia, Ireland UK

7.6 Risks Associated with Peatland Use

(Holden et al. 2004), Canada, the USA, Indonesia and Malaysia for pasturing, arable cropping, grazing, peat mining and forestry, oil palm and rubber plantations or horticultural production systems. Serious damages can be done by ground fires in even undisturbed organic soils, including peat and muck, drying out due to prolonged drought conditions which cause them to ignite and burn (de Groot 2012). These fires are due to smoldering combustion, and although of slow-motion, can create considerable hazards (Watts and Kobziar 2013). According to Hadden (2011), smoldering combustion is flameless and occurs by the reaction of oxygen on surface of dry peat. Watts and Kobziar (2013) observed that such fires on peat can continue for several days or even months. These fires can even be more damaging to surrounding soils and plants due to their persistence and higher heat transfer than flaming combustion (Kreye et al. 2011). Once started, smoldering combustion cannot easily be controlled because burning continues to spread gradually deeper and laterally underground over extensive areas below the surface, and leaves no sign of location and extent of smoldering (Rein 2009).

Land subsidence is almost always the consequence of peat reclamation through drainage, and peat oxidation is the main cause of land subsidence (Hooijer et al. 2011). Moreover, lowering of the water table causes increase in bulk density and compaction which is another cause of collapsing of peat surface and land subsidence. Reports indicate that the rates of land subsidence may vary from less than 1 cm year^{-1} to more than 10 cm year^{-1} worldwide. Besides subsidence, irreversible drying and the destruction of the fragile peat may ultimately occur. According to Holden et al.(2001), the collapse of macropores may generate runoff in drained peat. Thus, the loss of water, shrinkage and compaction, oxidation and loss of carbon lead to irreversible subsidence of peat. The rate of subsidence in drained peatlands under cropping depends on the types of crops and methods of cultivation. The degree of drainage, soil tillage, and fertilization are also important factors in this regard (Szajdak et al. 2002). Deverel and Leighton (2010) reported that 200–600 cm land subsidence occurred over 40–130 years in peatlands of several countries of Europe and North America and brought land surface levels close to or below sea level. Drainage removes large amounts of water from peat which causes its shrinkage. It also reduces the buoyancy of the peat layer and result in compression and increased bulk density. As mentioned by Acreman and Miller (2007), land levels were lowered by 1 cm year.$^{-1}$ in the Netherlands under normal agricultural use.

Normally, aeration as a result of drainage causes continuous and rapid oxidation and decomposition. The processes are further accelerated by agricultural operations such as tillage, mixing, fertilizing and liming, and by the alternate wetting and drying cycles. According to Jauhiainen et al. (2008), lowering of water table increases redox potential and favors microbial activity and nitrogen mineralization As a result of this, CO_2 loss by peat decomposition is enhanced and carbon emission to the atmosphere is increased (Hooijer et al. 2010). Peat may be lost or wasted by erosion and burning. Milne et al. (2006) reported average peat wastage of 1.27 cm year.$^{-1}$ and 0.19 cm year.$^{-1}$ respectively for 'thick' and 'thin' peat. Peat wastage in pasture fields in the Somerset Levels at rates of 44–79 cm a century was suggested by Brunning (2001).

Harvesting peat needs the peatland to be drained, cleared of vegetation, and leveled (Daigle and Gautreau-Daigle 2001). Peat extraction releases GHG by increasing in situ decomposition by increased oxygenation. However, emission of CH_4 is reduced (Nilsson et al. 2000; Waddington et al. 2002). According to Waddington and Warner (2001), removal of the living biomass from the peatland surface leads to the gross ecosystem production to zero. Peat extraction has altered the character of large peatland areas worldwide. Peat is used for many fuel and nonfuel purposes; *Sphagnum* peat moss is processed, packaged, and shipped to markets from Canada throughout the world mainly for horticultural use (Cleary et al. 2005). Use of tropical peatland for oil palm plantations is associated with complete replacement of the existing vegetation with palm and permanent drainage. As a result, the peat becomes a source of CO_2 emissions instead of remaining as a great carbon sink. Moreover, the inputs of carbon to the peat through biomass are stopped. Page et al. (2011) concluded that large and sustained CO_2 emissions take place from the drained peat of oil palm plantations (19–115 Mg CO_2 ha^{-1} year.$^{-1}$). In tropical regions, large areas of peat swamp forests have been drained for logging. Drained peatland areas with degraded vegetation are very susceptible to fires that cause further degradation (Hoscilo et al. (2011). According to Heil et al. (2006), catastrophic fires on peatland release large quantities of carbon into the atmosphere. Satrio et al. (2009) suggested that logging increases surface soil temperature when the peat swamp forest has been logged, thereby aerobic decomposition increases accelerating organic matter decomposition. The degree of decomposition also depends on the type of peat soil (Rezanezhad et al. 2010).

Fig. 7.10 shows a view of burning in the Tripa peat forests in Aceh, northern Sumatra, to prepare for oil palm plantations. According to Silvius et al. (2006), peat fires in South-east Asia can destroy millions of hectares of peat in one dry season. Considerable areas of peat can be lost by erosion. According to Tallis (1998) erosion of blanket peat in mires of UK occurs in three stages: the vegetation cover is disrupted and destroyed leaving the peat bare in the first stage; frost and drought reduce the cohesiveness of the exposed peat in the second stage, and removal of the friable surface layer by wind, water or oxidation in the last stage (Yeloff et al. 2006). Anshari et al. (2010) suggested that unwise land uses and land use changes adversely affect physical, chemical and biological properties of peat soils cause their degradation.

7.7 Peatland Conservation

The International Union for Conservation of Nature (IUCN) adopted the policy to allow no further loss of near natural peatlands and to restore all recoverable peatlands to a peat forming state and resilient to climate change (Bain et al. 2011). IUCN-UK Commission of Inquiry on Peatlands some strategies for rehabilitation, restoration and conservation of peatlands: (1) peatlands should be so conserved through management that it remains in a good and favorable condition and that

7.7 Peatland Conservation

Fig. 7.10 Burning in the Tripa peat forests in Aceh, northern Sumatra. (Photo AFP; published in The Sydney Morning Herald, 5 July, 2012)

healthy peatlands should be prevented from further damage; (2) partially damaged peatlands should be restored through changing land use and active management of habitat so that they are converted to a peat forming state with typical peatland vegetation and animal species; (3) severely damaged peatlands should be repaired through such major operations as replacement of woodland and revegetation on bare peat and gully blocking; and (4) loss of biodiversity should be prevented. According to SERI (2004), peatland conservation includes the processes of rehabilitation, restoration and protection.

Peatland rehabilitation The reparation of ecosystem processes, productivity and services of the former peatland; but it does not necessarily mean returning to the original state.

Peatland restoration The process of recovering degraded peatlands as near as possible to its original natural condition.

Peatland protection This involves preservation and maintenance of undisturbed peatlands so that their healthy state and ecosystem functions and services are retained in natural conditions. Protection of the healthy and natural undisturbed peatlands against degradation seems to be the best process of peatland conservation.

Restoration is a process of returning ecosystems to their original structure and composition (Lode 2001). It is always a challenging job because we often do not have enough information and records of the original landform, hydrology, species composition, ecosystem function and their interactions. In many instances where peatlands had undergone irreversible changes due to drainage and peatland mismanagement, limited or no success can be achieved through rehabilitation and restoration initiatives. Complete rewetting, which is the initial step of rehabilitation or restoration of wetland systems, is not possible in many drained peatlands because of irreversible changes in the topography and hydrology. Irregular land subsidence throughout the peatland area creates many smaller domes which preven complete rewetting large drained areas (Joosten et al. 2012). So, the highest priority is given on the protection of undisturbed mires and the prevention of further degradation of slightly degraded peatlands.

Keeping the peatlands in their natural state without unnecessary intervention so that they can perform their normal ecosystem functions properly and can retain the biological community intact is known as peatland conservation. Peatland conservation also means refraining completely from reclaiming peatlands for any purpose such as forestry, oil palm, horticulture, agriculture and harvesting. Although peatlands are always used for economic benefits, but it is not sustainable, and severe environmental problems including loss of water storage and cycling, land subsidence and enhanced CO_2 emissions. According to Joosten et al. (2012), the basic principle of peatland conservation is to "keep wet peatlands wet".

The recovery of a degraded ecosystem to recreate a naturally functioning self-sustaining system may be called ecological restoration. It aims at returning the degraded system to a protective, productive, aesthetically pleasing near natural condition but not necessarily getting back to the original state. Bonnett et al. (2009) suggested that ecological restoration or nature conservation also considers the integrity of the substrate. Restoration, in another sense, can be considered as re-creation of a peatland habitat from damaged peatlands provided that the peat is deep enough (>0.5 m), and elevating the water table to the surface and maintaining it even in the dry season are feasible. Rewetting by the raised water table is followed by recolonization of important peat-forming species such as *Sphagnum* and other peatland plant species. The process is known as revegetation.

The techniques of rehabilitation of peatlands degraded and abandoned after peat extraction include re-establishing favorable hydrological condition, propagation of *Sphagnum* moss and their colonization at the site, and maintaining water level so that regeneration of *Sphagnum* and companion species is sustained. Boudreau and Rochefort (2008) cited an example of rapid colonization of typical peatland species forming a complete moss carpet in 4–7 years depending on site conditions. If a surface crust is formed on a dried hard peat, it is broken up before introducing the moss to favor their even contact with the substrate. Sometimes, 5–10 cm of the surface may need to be milled.

Renaturalization is a process of converting degraded peatlands into a near natural wetland systems where water table is elevated, specified flora and fauna regenerate, and peat can form again. This is accomplished by rewetting and consequently changing chemical composition, hydrology and landform (Volungevicius et al. 2015). However, a few reports are there to inform the dynamics of soil organic carbon after renaturalization (Alberti et al. 2011). Secondary succession is the process of re-establishment of natural vegetation in rewetted condition (Novara et al. 2011), and it has a great significance in relation to renaturalization under altered hydrological and biogeochemical processes. The aim of renaturalization is not simply to revegetate the sites, but it is to regain the capacity of normal ecosystem functions of peatlands including carbon sequestration. Anshari et al. (2010) studied the changes in chemical and microbiological properties of degraded peat soil in the United States and Indonesia. Volungevicius et al. (2015) suggested that changes in soil organic matter in peat soil occur more intensively than in mineral soils, and observed that renaturalization caused a decrease in total organic carbon, nitrogen and phosphorus content and an increase in labile organic carbon content. According to these investigators, self renaturalization may lead to peat mineralization. However, aided renaturalization efforts in many instances have been successful. The United Nations Development Program and the Global Environmental Facility (UNDP-GEF) assisted renaturalization projects in 12 disturbed peatlands of Belarus on the overall area of 28,000 ha. The outputs of the projects are very encouraging in the climate change context. In these efforts, the hydrological regime was restored, and the peatlands that would catch fires under dry conditions every year did not catch fire for a single time even in dry months during 2010–2015. These peatlands have become popular places for fishing, collecting berries and hunting. Elevating the groundwater table to the surface and inundation reduced carbon dioxide emissions into the atmosphere by almost 500,000 tons. Rewetting, dry areas with scarcity of biodiversity flourished with abundant wetland species of flora and fauna forming productive reservoirs and mires within a short period (https://www.thegef.org/news/belarus%E2%80%99-degraded-peatlands-chance-become-mires-again). Biomass from renaturalized peatlands revegetated with reed (*Phragmites australis*), sedge (*Carex* sp.), willow (*Salix* sp.), alder (*Alnus* sp.) etc. can be good sources of energy. Reeds are also used as roofing materials and biomass also sequesters carbon. Fig. 7.11 shows a degraded peatland before and after renaturalization.

Paludiculture (Latin 'palus' = swamp) is a technique of cultivating economic plants in wet and rewetted peatlands. It facilitates peat formation and accumulation, and sustains ecosystem services associated with natural peatlands (Anon 2012). According to Wichtmann and Joosten (2007), paludiculture provides sustainable harvests and prevents peat from oxidation and decomposition. In addition, biomass can be collected also from spontaneous vegetation in natural sites and planted crops in rehabilitated sites. Joosten et al. (2012) mentioned that paludiculture can produce food, feed and fiber for local people. The biomass can also be used as raw materials for the production of biofuels and as pharmaceuticals. Joosten and Clarke (2002) noted that several edible berries (*Vaccinium, Empetrum, Rubus* and *Ribes*) and mushrooms can be produced in peatlands. Wild rice (*Zizania aquatica*) is grown in North America, bog bean (*Menyanthes trifoliata*), calamus (*Acorus calamus*) and buffalo grass (*Hierochloe odorata*) in Europe, and sago palm (*Metroxylon sagu*) in Indonesia and Malaysia (Joosten and Clarke 2002). Hunting, fishing, and aquaculture of indigenous fish species can be attractive recreational and economic activities (Wichtmann 2011).

Fig. 7.11 A renaturalized peatland of Belarus under UNDP-GEF initiative. (Source FAO; http://www.fao.org/fileadmin/user_upload/Europe/documents /Events_2010/ Woodenergy/03Chabrouskaya_en.pdf)

Study Questions
1. Classify organic soils. Explain the environmental significance of peatlands. Discuss the formation and distribution of peat soils.
2. Discuss the uses of peat. What are the risks of draining peat? Why peatlands should not be disturbed?
3. Describe soil and crop management practices for agricultural use of peat soil.
4. Comment on the objectives and techniques of peatland conservation.
5. Write notes on: (a) Peat extraction, (b) Peat burning, and (c) Engineering properties of peat soil.

References

Abdullah M, Huat BBK, Kamaruddin R, Ali AK, Duraisamy Y (2007) Design and performance of EPS footing for lightweight farm structure on peat soil. Am J Appl Sci 4:484–490

Acreman MC, Miller F (2007) Hydrological impact assessment of wetlands. Proceedings of the ISGWAS conference on groundwater sustainability, Spain, January 2006. p 225–255. http://aguas.igme.es/igme/isgwas/Ponencias%20ISGWAS/16-Acreman.pdf

Akol AK (2012) Stabilization of peat soil using lime as a stabilizer. A project dissertation submitted to the Civil Engineering Programme. Universiti Teknologi Petronas, Malaysia

Alberti G, Leronni V, Piazzi M, Petrella F, Cairata P, Peressotti A, Piussi P, Valentini R, Cristina L, La Mantia T, Novara A, Rühl J (2011) Impact of woody encroachment on soil organic carbon and nitrogen in abandoned agricultural lands along a rainfall gradient in Italy. Reg Environ Chang 11(4):917–924

Ambak K, Melling L (2000) Management practices for sustainable cultivation of crop plants on tropical peatland. Proceedings of the international symposium on tropical peatlands. Bogor,

Indonesia, 22–23 November 1999. Hokkaido University and Indonesian Institute of Sciences: Bogor, Indonesia

Anderson R (2010) Restoring afforested peat bogs: results of current research. Forestry Commission Research Note. Available at: http://www.forestry.gov.uk/pdf/FCRN006.pdf/$FILE/FCRN006.pdf

Anon (2012) Paludiculture: Sustainable productive utilization of rewetted peatlands. Institute of Botany and Landscape Ecology, University of Greifswald. www.paludikultur.de

Anshari GZ, Afifudin M, Nuriman M, Gusmayanti E, Arianie L, Susana R, Nusantara RW, Sugardjito J, Rafiastanto A (2010) Drainage and land use impacts on changes in selected peat properties and peat degradation in West Kalimantan Province, Indonesia. Biogeosciences 7:3403–3419

Asapo ES, Coles CA (2012) Characterization and comparison of saprist and fibrist Newfoundland *Sphagnum* peat soils. J Miner Mater Charact Eng 11:709–718

Asing J, Wong NC, Lau S (2009) Optimization of extraction method and characterization of humic acid derived from coals and composts. J Trop Agric Fd Sc 37(2):211–223

Bain CG, Bonn A, Stoneman R, Chapman S, Coupar A, Evans M, Gearey B, Howat M, Joosten H, Keenleyside C, Labadz J, Lindsay R, Littlewood N, Lunt P, Miller CJ, Moxey A, Orr H, Reed M, Smith P, Swales V, Thompson DBA, Thompson PS, Van de Noort R, Wilson JD, Worrall F (2011) IUCN UK Commission of Inquiry on Peatlands. IUCN UK Peatland Programme, Edinburgh

Barthelmes A, Couwenberg J, Risager M, Tegetmeyer C, Joosten H (2015) Peatlands and climate in a Ramsar context a Nordic-Baltic Perspective. Nordic Council of Ministers, Denmark

Boguta P, Sokolowska Z (2014) Statistical relationship between selected physicochemical properties of peaty-muck soils and their fraction of humic acids. Int Agrophys 28:269–278

Boguta P, Sokołowska Z, Bowanko G (2011) Influence of secondary transformation index of peat-muck soils on the content of selected metals. Acta Agrophysica 18(2):225–233

Bonnett SAF, Ross S, Linstead C, Maltby E (2009) A review of techniques for monitoring the success of peatland restoration. Natural England Commissioned Reports 086, University of Liverpool

Boudreau S, Rochefort L (2008) Plant establishment in restored peatlands: 10-years monitoring of sites restored from 1995 to 2003. In Proceedings of the 13th International Peat Congress. Available at http://www.peatsociety.org/index.php?module=shop_view_product&id=82&product_id=81 (20.1.14)

Brandyk T, Gotkiewicz J, Łachacz A (2008) Principles of rational use of peat land in agriculture (in Polish). Post Nauk Roln 1:15–26

Brandyk T, Szatylowicz J, Oleszczuk R, Gnatowski T (2003) Water-related physical attributes of organic soils. In: Parent L, Ilnicki P (eds) Organic Soils and Peat Materials for Sustainable Agriculture. CRC Press, Boca Raton

Brown PA, Gill SA, Allen SJ (2000) Metal removal from wastewater using peat. Review paper. Water Res 34:3907–3916

Brunning R (2001) Archaeology and peat wastage on the Somerset Moors. Report to the Environment Agency. Somerset County Council, Taunton

Buol SB, Southard RJ, Graham RC, McDaniel PA (2011) Soil genesis and classification, 6th edn. Wiley-Blackwell, New York

Caliman JP, Carcasses R, Perel N, Wohlfahrt J, Girardin P, Wahyu A, Pujianto DB, Verwilghen A (2007) Agri-environmental indicators for sustainable palm oil production. Palmas 28:434–445

Campos JRR, Silva AC, Fernandes JSC, Ferreira MM, Silva DV (2011) Water retention in a peatland with organic matter in different decomposition stages. R Bras Ci Solo 35:1217–1227

Charman D (2002) Peatlands and Environmental Change. John Wiley and Sons Ltd, West Sussex, England

Chistotin MV, Sirin AA, Dulov LE (2006) Seasonal dynamics of carbon dioxide and methane emission from peatland of Moscow Region drained for peat extraction and agricultural use. Agrochemistry 6:32–41. (in Russian)

Cleary J, Roulet NT, Moore TR (2005) Greenhouse gas emissions from Canadian peat extraction, 1990–2000: A Life-cycle Analysis. Ambio 34:456–461

Comeau L, Hergoualc'h K, Smith JU, Verchot L (2013) Conversion of intact peat swamp forest to oil palm plantation: Effects on soil CO_2 fluxes in Jambi, Sumatra. CIFOR Working Paper no 110. Center for International Forestry Research (CIFOR), Bogor, Indonesia

Comte I, Colin F, Whalen JK, Grünberger O, Caliman JP (2012) Agricultural Practices in Oil Palm Plantations and Their Impact on Hydrological Changes, Nutrient Fluxes and Water Quality in Indonesia: A Review. In: Sparks DL (ed) Advances in Agronomy, vol 116, pp 71–124

Couwenberg J (2011) Greenhouse gas emissions from managed peat soils: is the IPCC reporting guidance realistic? Mires and Peat 8(2):1–10

Crill P, Hargreaves K, Korhola A (2000) The role of peat in Finnish greenhouse gas balances. Ministry of Trade and Industry Finland. Studies and Reports 10/2000

Daigle J-Y, Gautreau-Daigle H (2001) Canadian peat harvesting and the environment, 2nd edn. Secretariat to the North American Wetlands Conservation Council Committee, Ottawa

de Groot WJ (2012) Peatland fires and carbon emissions. Canadian Forest Service, Great Lakes Forestry Centre, Sault Ste Marie

Deverel SJ, Leighton DA (2010) Historic, recent, and future subsidence, Sacramento-San Joaquin Delta, California, USA. San Francisco Estuary and Watershed. Science 8(2):23

Droppo IG (2000) Filtration in particle analysis. In: Meyers RA (ed) Encyclopedia of analytical chemistry. Ramtech Ltd, Tarzana, CA

Eastern Canada Soil and Water Conservation Centre (1997) Management and conservation practices for vegetable production on peat soils. University of Moncton, Moncton, Canada

Edil TB (2003) Recent advances in geotechnical characterization and construction over peats and organic soils. Proceedings of the 2nd international conferences in soft soil engineering and technology, Putrajaya (Malaysia), Embrapa EBDPA (2006) Sistema brasileiro de classificação de solos, 2nd edn, Centro Nacional de Pesquisa de Solos, Rio de Janeiro

Embrapa (2006) Brazilian system of soil classification. 2 ed. National Soil Research Center. Rio de Janeiro: Embrapa Solos, 2006

FAO (2002) Small-Scale Palm oil processing in Africa. FAO Agricultural Services Bulletin 148 ISSN 1010–1365. Rome, Italy

FAO (2005) Fertilizer Use by Crop in Indonesia, Natural Resources Management and Environment. Food and Agriculture Organization of the United Nations, Rome

FAO (2006) World reference base for soil resources 2006, A Framework for International

FAO (2009) FAOSTAT online statistical service. Food and Agriculture Organization of the United Nations, Rome, Italy. Available via URL. http://faostat.fao.org/ (Accessed August 2011)

FAO/UNESCO (1974) Soil Map of the World. Volume 1, Legend. UNESCO, Paris, p. 62

Firdaus MS, Gandaseca S, Ahmed OH, Majid NK (2012) Comparison of selected physical properties of deep peat within different ages of oil palm plantation. Int J Phys Sci 7(42):5711–5716

Fitzherbert EB, Struebig MJ, Morel A, Danielsen F, Brühl CA, Donald PF, Phalan B (2008) How will oil palm expansion affect biodiversity? Trends Ecol Evol 23(10):539–545

Forster P, Ramaswamy V, Artaxo P, Berntsen T, Betts R, Fahey DW et al (2007) Changes in atmospheric constituents and in radiative forcing. In: Solomon S, Qin D, Manning M, Chen Z, Marquis M, Averyt KB, Tignor M, Miller HL (eds) Climate Change 2007: The Physical Science Basis. Contribution of Working Group I to the Fourth Assessment Report of the Intergovernmental Panel on Climate Change. Cambridge University Press, Cambridge, UK

Fox CA, Tarnocai C (2011) Organic soils of Canada: Part 2. Upland Organic soils. Can J Soil Sci 91:823–842

Fraser LH, Keddy PA (2005) The World's Largest Wetlands: Ecology and Conservation. Cambridge University Press, Cambridge, UK

Fredriksson D (2002) SGU, Uppsala. http://www.sgu.se/geologi/jord_index.htm (cited from Kellner 2003)

Freeman C, Fenner N, Ostle NJ, Kang H, Dowrick DJ, Reynolds B, Lock MA, Sleep D, Hughes S, Hudson J (2004) Export of dissolved organic carbon from peatlands under elevated carbon dioxide levels. Nature 430:195–198

Gawlik J, Harkot W (2000) Influence of the kind of moorsh and the state of its transformation on the germination and growth of *Lolium perenne* in the pot plant experiment during spring-summer cycle. Acta Agrophysica 26:25–40

Goh KJ, Hardter R (2003) General Oil Palm Nutrition. In: Oil Palm: Management for Large and Sustainable Yields, Fairhurst, T. (eds.). Potash and Phosphate Institute, Singapore, pp. 191–230

Goh KJ, Härdter R (2003a) General oil palm nutrition. In: Fairhurst T, Hardter R (eds) Oil palm: management for large and sustainable yields. Potash & Phosphate Institute/Potash & Phosphate Institute of Canada and International Potash Institute (PPI/PPIC and IPI), Singapore

Goh KJ, Härdter R, Fairhurst T (2003b) Fertilizing for maximum return. In: Fairhurst T, Hardter R (eds) Oil Palm: Management for Large and Sustainable Yields. Potash & Phosphate Institute/ Potash & Phosphate Institute of Canada and International Potash Institute (PPI/PPIC and IPI), Singapore

Grozav A, Rogobete G (2010) Histosols and some other reference soils from the Semenic Mountains – Romania. Res J Agric Sci 42(3):149–153

Hadden RM (2011) Smoldering and self-sustaining reactions in solids: an experimental approach. PhD Dissertation. University of Edinburgh, Edinburgh

Heil A, Langmann B, Aldrian E (2006) Indonesian peat and vegetation fire emissions: study on factors influencing large-scale smoke haze pollution using a regional atmospheric chemistry model. Mitig Adapt Strateg Glob Chang 12:113–133

Helmut G (2006) Our earth's changing land: an encyclopedia of land-use and land cover change. Greenwood Publishing Group. p. 463. ISBN 9780313327841

Hillman GR (1997) Effects of engineered drainage on water tables and peat subsidence in an Alberta treed fen. In: Trettin CC, Jurgensen MF, Grigal DF, Gale MR, Jeglum JK (eds) Northern Forested Wetlands: Ecology and Management. CRC Lewis Publishers, Boca Raton, FL

Holden J, Burt TP, Cox NJ (2001) Macroporosity and infiltration in blanket peat: the implications of tension disc infiltrometer measurements. Hydrol Process 15:289–303

Holden J, Chapman P, Evans M, Hubacek K, Kay P, Warburton J (2006) Vulnerability of organic soils in England and Wales. Final technical report to DEFRA, Project SP0532

Holden J, Chapman PJ, Labadz JC (2004) Artificial drainage of peatlands: hydrological and hydrochemical process and wetland restoration. Prog Phys Geogr 28(1):95–123

Holman IP (2009) An estimate of peat reserves and loss in the East Anglian Fens Commissioned by the RSPB. Department of Natural Resources, Cranfield University, Cranfield, Bedfordshire

Hooijer A, Page S, Canadell JG, Silvius M, Kwadijk J, Wösten H, Jauhiainen J (2010) Current and future CO_2 emissions from drained peatlands in Southeast Asia. Biogeosciences 7:1505–1514

Hooijer A, Page S, Jauhiainen J, Lee WA, Lu XX, Idris A, Anshari G (2011) Subsidence and carbon loss in drained tropical peatlands: reducing uncertainty and implications for CO2 emission reduction options. Biogeosciences 8:9311–9356

Hooijer A, Page S, Jauhiainen J, Lee WA, Lu XX, Idris A, Anshari G (2012) Subsidence and carbon loss in drained tropical peatlands. Biogeosciences 9:1053–1071

Hoscilo A, Page SE, Tansey KJ, Rieley JO (2011) Effect of repeated fires on land-cover changes on peatland in southern Central Kalimantan, Indonesia, from 1973 to 2005. Int J Wildland Fire 20:578–588

Huang PM, Li Y, Sumner ME (2011) Handbook of Soil Sciences: Properties and Processes, 2nd edn. CRC Press, Boca Raton

Ilnicki P (2002) Peatlands and Peats. Wydawnictwo Akademii Rolniczej im. A Cieszkowskiego, Poznan, 606 pp. (in Polish)

Ilomets M (2015) Estonia. In: Joosten H, Tanneberger F, Moen A (eds) Mires and peatlands of Europe: Status, distribution, and nature conservation. Schweizerbart. Science Publishers, Stuttgart

IPCC (2006) In: Eggleston HS, Buendia L, Miwa K, Ngara T, Tanabe K (eds) 2006 IPCC guidelines for national greenhouse gas inventories, prepared by the National Greenhouse Gas Inventories Programme. IGES, Japan

IPCC (2007) Climate Change 2007—The Physical Science Basis, Contribution of Working Group I to the Fourth Assessment Report of the IPCC (ISBN 978 0521 88009-1 Hardback)

Islam MS, Hashim R (2009) Bearing Capacity of Stabilized Tropical Peat by Deep Mixing Method. Aust J Basic Appl Sci 3(2):682–688

Jauhiainen J, Limin S, Silvennoinen H, Vasander H (2008) Carbon dioxide and methane fluxes in drained tropical peat before and after hydrological restoration. Ecology 89:3503–3514

Jaya J (2002) Sarawak: Peat agricultural use. Retrieved from http://www.strapeat.alterra.nl/.../15%20Sarawak%20peat%20agricultural%20use.pdf

Joosten H, Clarke D (2002) Wise Use of Mires and Peatlands - Background and Principles Including a Framework for Decision-Making. International Mire Conservation Group and International Peat Society, Devon, UK

Joosten H, Tapio-Biström ML, Tol S (2012) Peatlands - guidance for climate change mitigation through conservation, rehabilitation and sustainable use, 2nd edn. Food and Agriculture Organization of the United Nations and Wetlands International Mitigation of Climate Change in Agriculture (MICCA) Programme, Rome

Kalantari B (2010) Stabilization of fibrous peat using ordinary Portland cement and additives. PhD Thesis, University of Putra, Malaysia

Kalantari B (2013) Civil Engineering Significant of Peat. Glob J Res Eng Civ Struct Eng 13(2):25–28

Kalembasa D, Pakuła K (2008) Profile differences of Fe, Al and Mn in the peat-muck soils in the upper Liwiec river valley. Acta Sci Pol Agricultura 8(2):3–8

Keddy PA (2010) Wetland Ecology: Principles and Conservation, 2nd edn. Cambridge University Press, Cambridge, UK

Kellner E (2003) Wetlands – different types, their properties and functions. Technical Report TR-04-08 www.skb.se

Koh LP, Ghazoul J (2010) Spatially explicit scenario analysis for reconciling agricultural expansion, forest protection, and carbon conservation in Indonesia. Sustainability science. Records of the National Academy of Sciences. www.pnas.org/cg/doi/10.1073/PNAS.1000530/07

Kolli K, Asi E, Apuhtin V, Kauer K, Szajdak LW (2010) Chemical properties of surface peat on forest land in Estonia. Mires and Peat 6:1–12

Kreye JK, Varner JM, Knapp EE (2011) Effects of particle fracturing and moisture content on fire behavior in masticated fuelbeds burning in a laboratory. Int J Wildland Fire 20:308–317

Krogstad T, Bechmann M (2011) Reduced P application in peat soils. www.cost869.alterra.nl/Fs/FS_P_application_peat.pdf

Langner A, Miettinen J, Siegert F (2007) Land cover change 2002–2005 in Borneo and the role of fire derived from MODIS imagery. Glob Change Biol 13:2329–2340

Lappalainen E (1996) Global Peat Resources. International Peat Society, Finland, pp 53–281

Leifeld J, Müller M, Fuhrer J (2011) Peatland subsidence and carbon loss from drained temperate fens. Soil Use Manag 27:170–177

Lester RB (2006) Plan B 2.0 rescuing a planet under stress and a civilization in trouble. (NY: WW Norton & Co.) Earth Policy Institute

Letts MG, Roulet NT, Comer NT, Skarupa MR, Verseghy DL (2000) Parameterization of Peatland Hydraulic Properties for the Canadian Land Surface Scheme. Atmosphere-Ocean 38(1):141–160

Litaor MI, Eshel G, Reichmann O, Shenker M (2006) Hydrological control of phosphorus mobility in altered wetland soils. Soil Sci Soc Am J 70:1975–1982

Lode E (2001) Natural mire hydrology in restoration of peatland functions. Doctoral thesis, Acta Universitatis Agrivuturae Sueciae, Silvestria 234, Uppsala

Lohila A, Minkkinen K, Laine J, Savolainen I, Tuovinen JP, Korhonen L, Laurila T, Tietavainen H, Laaksonen A (2010) Forestation of boreal peatlands: Impacts of changing albedo and greenhouse gas fluxes on radiative forcing. J Geophys Res 115:G04011

References

Long M, Boylan N (2012) In-Situ Testing of Peat – a Review and Update on Recent Developments. Geotech Eng J SEAGS & AGSSEA 43(4):41–55

Lucas RE (1982) Organic Soils (Histosols). Formation, distribution, physical and chemical properties and management for crop production. Michigan State University, Research Report No. 435 (Farm Science)

MacDonald GM, Beilman DW, Kremenetski KV, Sheng Y, Smith LC, Velichko AA (2006) Rapid early development of circumarctic peatlands and atmospheric CH_4 and CO_2 variations. Science 314(5797):285–288

Mäkiranta P, Hytönen J, Aro L, Maljanen M, Pihlatie M, Potila H, Shurpali NJ, Laine J, Lohila A, Martikainen PJ, Minkkinen K (2007) Greenhouse gas emissions from afforested organic soil, croplands and peat extraction peatlands. Boreal Environ Res 12:159–175

Malawska M, Ekonomiuk A, Wiłkomirski B (2006) Chemical characteristics of some peatlands in southern Poland. Mires and Peat 1(2):1–14

Miettinen J, Hooijer A, Shi C, Tollenaar D, Vernimmen R, Liew SC, Malins C, Page S (2012) Southeast Asian peatlands in 2010 with analysis of historical expansion and future projections. Glob Change Biol Bioenergy. https://doi.org/10.1111/j.1757-1707.2012.01172.x

Miettinen J, Liew SC (2010) Degradation and development of peatlands in peninsular Malaysia and in the islands of Sumatra and Borneo since 1990. Land Degrad Dev 21:285–296

Miettinen J, Shi C, Liew SC (2011) Deforestation rates in insular Southeast Asia between 2000 and 2010. Glob Change Biol 17:2261–2270

Milne R, Mobbs DC, Thomson AM, Matthews RW, Broadmeadow MSJ, Mackie E, Wilkinson M, Benham S, Harris K, Grace J, Quegan S, Coleman K Powlson DS, Whitmore AP, Sczanska-Stanton M, Smith P, Levy PE, Ostle N, Murray TD, Van Oijen M, Brown T (2006) UK emissions by sources and removals by sinks due to land use, land use change and forestry activities. Report, April 2006. Centre for Ecology and Hydrology, UK

Munro R (2004) Dealing with bearing capacity problems on low volume roads constructed on peat. The Highland Council, Transport, Environmental & Community Service, HQ, Glenurquhart Road, Inverness IV3 5NX Scotland

Murdiyarso D, Hergoualc'h K, Verchot LV (2010) Opportunities for reducing greenhouse gas emissions in tropical peatlands. Proceedings of the National Academy of Sciences, USA

Mutert E, Fairhurst TH, von Uexküll HR (1999) Agronomic Management of Oil Pals on Deep Peat. Better Crops Int 13(1):22–27

Nilsson S, Shvidenko A, Stolbovoi V, Gluck M, Jonas M, Obersteiner M (2000) Full Carbon Account for Russia. Interim Report IR-00-021. IIASA, Laxenburg, Austria

Novara A, Gristina L, La Mantia T, Rühl J (2011) Soil carbon dynamics during secondary succession in a semi-arid Mediterranean environment. Biogeosciences 8:11107–11138

Novoa-Munoz JC, Pomal P, Cortizas AM (2008) Histosols. In: Chesworth (ed) Encyclopedia of Soil Science. Springer, Dodrecht, The Netherlands

O'Connor MB, Longhurst RD, Jonston TJM, Portegys FN (2001) Fertiliser requirements for peat soils in the Waikato region. Proc N Z Grassl Assoc 63:47–51

Oleszczuk R, Regina K, Szajdak L, Höper H, Maryganowa V (2008) Impacts of agricultural utilization of peat soils on the greenhouse gas balance. In: Strack M (ed) Peatlands and Climate Change. International Peat Society, Jyväskylä, Finland

Page SE, Morrison R, Malins C, Hooijer A, Rieley JO, Jauhiainen J (2011) Review of peat surface greenhouse gas emissions from oil palm plantations in South East Asia. White Paper Number 15, Indirect Effects of Biofuel Production Series International Council on Clean Transportation, Washington, DC

Paivanen J (1991) Peatland forestry in Finland: present status and prospects. In: Jeglum JK, Overend RP (eds) Peat and peatlands diversification and innovation. Proc Symp 89. Vol 1. Peatland Forestry, The Canadian Society for Peat and Peatlands

Patterson G, Anderson R (2000) Forests and Peatland Habitats. Forestry Commission Corstorphine Road, Edinburgh. http://www.forestry.gov.uk

Prasetyo BH, Suharta N (2011) Genesis and properties of peat at Toba Highland area of North Sumatra. Indonesian J Agric Sci 12(1):1–8
Rahgozar MA, Saberian M (2016) Geotechnical properties of peat soil stabilised with shredded waste tyre chips. Mires and Peat 18(3):1–12. http://www.mires-and-peat.net/
Rein G (2009) Smoldering combustion phenomena in science and technology. Int Rev Chem Eng 1:3–18
Rezanezhad F, Quinton WL, Price JS, Elliot TR, Elrick D, Shook KR (2010) Influence of pore size and geometry on peat unsaturated hydraulic conductivity computed from 3D computed tomography image analysis. Hydrol Process 24(21):2983–2994
Richardson CJ, Huvane JK (2008) Ecological status of the Everglades: environmental and human factors that control the peatland complex on the landscape. In: Richardson CJ (ed) The Everglades Experiments: Lessons for Ecosystem Restoration. Springer, Dodrecht
Ringqvist L, Holmgren A, Oborn I (2002) Poorly humified peat as an adsorbent for metals in wastewater. Water Res 36:2394–2404
Satrio AE, Gandaseca S, Ahmed OH, Majid NMA (2009) Effect of logging operation on soil carbon storage of a tropical peat swamp forest. Am J Environ Sci 5:748–752
Sayok AK, Nik AR, Melling L, Samad AR, Efransjah E (2008) Some characteristics of peat in Loagan Bunut National Park, Sarawak, Malaysia. International Symposium, Workshop and Seminar on Tropical Peatland, Yogyakarta, INDONESIA. 27–31 August 2007 (www.geog.le.ac.uk/carbopeat/yogyaproc.html)
Schumann M, Joosten H (2008) Global Peatland Restoration Manual. Institute of Botany and Landscape Ecology, Griefswald University, Germany. http://www.imcg.net/docum/prm/prm.htm
SERI (2004) The SER International Primer on Ecological Restoration. Society for Ecological Restoration International, Tucson. www.ser.org
Sheil D, Casson A, Meijaard E, van Nordwijk M, Gaskell J, Sunderland-Groves J, Wertz K, Kanninen M (2009) The impacts and opportunities of oil palm in Southeast Asia: What do we know and what do we need to know? Occasional paper no. 51. CIFOR, Bogor
Sigua GC, Kong WJ, Coleman SW (2006) Soil profile distribution of phosphorus and other nutrients following wetland conversion to beef cattle pasture. Environ Qual 35:2374–2382
Silva AC, Horak I, Martinez-Cortizas A, Vidal Torado P, Rodrigues-Racedo J, Grazziotti PH, Silva EB, Ferreira CA (2009a) Turfeiras da Serra do Espinhaço Meridional - MG. I - Caracterização e classificação. R Bras Ci Solo 33:1385–1398
Silva AC, Horak I, Vidal Torado P, Martinez-Cortizas A, Rodrigues-Racedo J, Cmpos JRR (2009b) Turfeiras da Serra do Espinhaço Meridional - MG.II - Influência da drenagem na composição elementar e substâncias húmicas. R Bras Ci Solo 33:1399–1408
Silvius M, Diemont H (2007) Deforestation and degradation of peatlands. Peatlands Int 2:32–34
Silvius M, Kaat A, van de Bund H (2006) Peatland degradation fuels climate change. Wetlands International, Wageningen, The Netherlands
Soil Survey Staff (1999) Soil Taxonomy 2nd edn. United States Department of Agriculture. Govt Printing Office, Washington, DC
Soil Survey Staff (2010) Keys to Soil Taxonomy, 11th edn. United States Department of Agriculture and Natural Resources Conservation Service, Washington, DC
Sokołowska Z, Boguta P (2010) State of the dissolved organic matter in the presence of phosphates. In: Szajdak LW, Karabanov AK (eds) Chemical, Physical and Biological Processes Occurring in Soils. Prodruk Press, Poznań, Poland
Sokołowska Z, Szajdak L, Boguta P (2011) Effect of phosphates on dissolved organic matter release from peatmuck soils. Int Agrophys 25:173–180
Stichnothe H, Schuchardt F (2011) Life cycle assessment of two palm oil production systems. Biomass Bioenergy 35(9):3976–3984
Szajdak L, Maryganova V, Meysner T, Tychinskaja L (2002) Effect of shelterbelt on two kinds of soils on the transformation of organic matter. Environ Int 28:383–392

References

Tallis JH (1998) Growth and Degradation of British and Irish Blanket Mires. Environment Reviews 6:81–122

Tanneberger F, Wichtmann W (2011) Carbon Credits from Peatland Rewetting Climate – Biodiversity – Land Use. Schweizerbart. Science Publishers, Stuttgart

Tarmizi MA, Mohd TD (2006) Nutrient demands of Tenera oil palm planted on inland soils of Malaysia. J Oil Palm Res 18:204–209

Tarnocai C, Canadell JG, Schuur EAG, Kuhry P, Mazhitova G, Zimov S (2009) Soil organic carbon pools in the northern circumpolar permafrost region. Global Biogeochem Cycles 23:GB2023, 11PP

USDA (2007) Indonesian Palm Oil Production. United States Department of Agriculture, Washington, DC

van Beek CL, Droogers P, van Hardeveld HA, van der Eertwegh GAPH, Velthof GL, Oenema O (2007) Leaching of solutes from an intensively managed peat soil to surface water. Water Air Soil Pollut 182:291–301

van den Akker JJH, Kuikman PJ, de Vries F, Hoving I, Pleijter M, Hendriks RFA, Wolleswinkel RJ, Simões RTL, Kwakernaak C (2008) Emission of CO2 from agricultural peat soils in the Netherlands and ways to limit this emission. In: Farrell C, Feehan J (eds) Proceedings of the 13th international peat congress "After wise use –the future of Peatlands", Vol 1 Oral Presentations. International Peat Society, Jyväskylä

van den Wyngaert IJJ, Kramer H, Kuikman P, Lesschen JP (2009) Greenhouse Gas Reporting of the LULUCF Sector, Revisions and Updates Related to the Dutch NIR 2009. Alterra Report 1035-7

van Schie WL (2000) The Use of Peat in Horticulture –Figures and Developments, International Peat Society Surveys 2000. International Peat Society, Jyskä, Finland

Volungevicius J, Amaleviciute K, Liaudanskiene I, Slepetiene A, Slepetys J (2015) Chemical properties of Pachiterric Histosol as influenced by different land use. Zemdirbyste-Agriculture 102(2):123–132

von Uexkull HR (2007) Oil Palm (*Elaeis guineensis* Jacq.). East and South East Asia Program for the Potash & Phosphate Institute/International Potash Institute, Singapore

Waddington JM, Warner KD (2001) Atmospheric CO_2 sequestration in restored mined peatlands. Ecoscience 8:359–368

Waddington JM, Warner KD, Kennedy GW (2002) Cutover peatlands: a persistent source of atmospheric CO_2. Global Biogeochem Cycles 16:1–7

Wahid MB, Akmar-Abdullah SN, Henson IE (2005) Oil palm -achievements and potential. Plant Prod Sci 8:288–297

Watts AC, Kobziar LN (2013) Smoldering combustion and ground fires: ecological effects and multi-scale significance. Fire Ecology 9(1):124–132

Wichtmann W (2011) Biomass use for food and fodder. In: Tanneberger F, Wichtmann W (eds) Carbon Credits from peatland rewetting. Climate - biodiversity - land use. Schweizerbart Science publishers, Stuttgart

Wichtmann W, Joosten H (2007) Paludiculture: peat formation and renewable resources from rewetted peatlands. IMCG-Newsletter 3:24–28

Wilson D, Alm J, Laine J, Byrne KA, Farrell EP, Tuitila E-S (2008) Rewetting of cutaway peatlands: are we re-creating hot spots of methane emissions? Restor Ecol. https://doi.org/10.1111/j.1526-100X.2008.00416.x

Yeloff DE, Labadz JC, Hunt CO (2006) Causes of degradation and erosion of a blanket mire in the southern Pennines, UK. Mires and Peat 1(4):1–18

Chapter 8
Soils on Steep Slopes

Abstract Soils on steep slopes tend to be unstable due to natural and anthropogenic causes. The natural features of steep slopes that make them susceptible to failures and mass movement are gradient and shape of slope, geology, soil, vegetation and climate. Soils are more prone to mass movement in shallow and loose soils on impervious substratum, on steeper slopes, and under high intensity storms. The human actions that make soils on steep slopes unstable include development, settlement, shifting cultivation, deforestation, forest fires, soil mining, and other slope disturbances. Mass movement or landslides, which occur due to gravity, water saturation and water movement, have several forms: falls, creeps, slumps and earthflows, debris avalanches and debris flows, debris torrents and bedrock failures. These movements have on-site and off-site effects on properties, installations, house-holds, communications, crops, human lives and environmental health. Soils on steep slopes can be stabilized; or in other words, soil erosion in steeply sloping lands can be reduced by several mechanical, agronomic and agroforestry measures. Mechanical measures include drainage, contour bunds, silt fences, surface mats, grading and terracing, retention walls, slope reshaping, etc. Agronomic practices include contour cropping, contour strip cropping; and agroforestry methods are contour hedgerows, alley cropping and sloping agricultural land technology.

Keywords Slope classes · Slope failures · Mass movement · Landslides · Slope management · Gully control · Grading · Terracing · Shifting cultivation · Sloping Agricultural Land Technology · Hedgerows

8.1 Slopes and Steep Slopes

The configuration of the land is not uniform everywhere; there are hills and mountains, canyons and cliffs, mesa and valleys, plains and marshes, etc. The configuration of the land surface has an important relationship with land use, including agronomic, horticultural, forestry, aesthetic, and engineering aspects. Land surface configuration includes the slope and shape of the land. Slope is the inclination of the land surface against the horizontal plane. Slope can be simple or complex, depending on gradient, length, and aspect. Slope gradient is the magnitude of

Fig. 8.1 Slope gradient

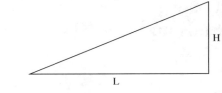

Fig. 8.2 10% slope

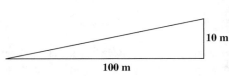

Fig. 8.3 100% slope

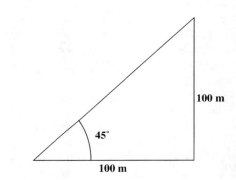

inclination of the surface of the soil from the horizontal plane. The difference in elevation between two points is expressed as a percentage of the distance between those points (Fig. 8.1).

Percentage slope gradient = $\frac{H}{L} \times 100$, where L is the horizontal distance and H is the vertical.

If the difference in elevation is 10 m over a horizontal distance of 100 m, slope gradient is 10% (Fig. 8.2). Alternatively, the slope can be expressed by the angle of inclination (degree). A slope of 45° is a slope of 100% because the difference in elevation between two points 100 m apart horizontally is 100 m on a 45° slope (Fig. 8.3).

The equivalences of the degree of slope angle and percentage gradient are shown in Table 8.1.

On simple slopes, runoff occurs from upper to lower points along the direction of inclination, and the velocity of runoff is faster on steeper slopes. On complex slopes, both direction and speed of overland flow are different in different sub-slope components. Slope complexity has an important influence on the amount and rate of runoff and on sedimentation associated with runoff. Different slope classes defined in terms of gradient and complexity by Soil Survey Division Staff (1993) are given in Table 8.2.

The slope has a shape as seen across the slope. It may be linear, convex, or concave. If the slope gradient neither increases nor decreases significantly with distance

8.1 Slopes and Steep Slopes

Table 8.1 Equivalences of the degree of slope and percentage gradient

Percentage	Angle	Angle	Percentage
0	0°00′	0°	0
5	2°52′	2°	3.5
10	5°43′	4°	7.0
15	8°32′	6°	10.5
20	11°19′	8°	14.0
25	14°02′	10°	17.6
30	16°42′	12°	21.2
35	19°17′	15°	26.8
40	21°48′	20°	36.4
50	26°34′	25°	46.6
60	30°58′	30°	57.7
70	34°59′	35°	70.0
80	38°39′	40°	83.9
90	41°59′	45°	00.0
100	45°00′	50°	19.2

Table 8.2 Slope classes

Slope classes		Slope gradient limits	
Simple slopes	Complex slopes	Lower percent	Upper percent
Nearly level	Nearly level	0	3
Gently sloping	Undulating	1	8
Strongly sloping	Rolling	4	16
Moderately steep	Hilly	10	30
Steep	Steep	20	60
Very steep	Very steep	>45	

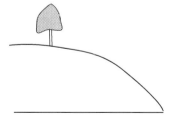

Fig. 8.4 A convex slope

down the slope, it is called a linear slope. When the slope gradient increases down the slope, it is called a convex slope (Fig. 8.4), and runoff tends to accelerate as it flows down the slope. The soil on the lower part of a convex slope is subject to greater erosion than that on the higher part. But, when the slope gradient decreases down the slope, it is known as a concave slope (Fig. 8.5). In concave slopes, the velocity of runoff decreases down the slope, and loaded sediments are deposited on the lower parts of the slope. The soil on the lower part of the slope also tends to dispose of water less quickly than the soil above it.

Fig. 8.5 A concave slope

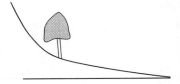

In natural landscapes, we often see assemblages of slopes of different gradients and shapes. In hilly and steep lands, slopes are more complex than nearly level and gently sloping lands. High and low hills, steep and gentle slopes, linear, convex and concave slopes can form a large unit there, which is more difficult to use, manage, stabilize and conserve.

The slope can be measured in different ways. There are simple tools such as compasses and levels (e.g., Abney level) that can be used for slope determination. The slope can be calculated from topographic maps. Surveyors and engineers determine slopes with more sophisticated equipments. Lehigh Valley Planning Commission (2008) classifies >15% slopes as steep. We can get a variety of amenities including beautiful scenes of valleys and hills covered with forests, bushes and grasses, wildlife and birds, waterfall and streams, large natural open spaces with highs and lows, and privacy from steeply sloping lands. Steep slopes can provide significant benefit to local water supplies. Stream water can be used for power generation. However, if development is poorly laid out and built, measurable consequences can happen, including significant destruction of the scenic beauty of the area by mass movement of soil and rock, runoff and erosion, slope failures, decreased water quality, flooding downstream, loss of habitats, crops and even lives, lack of safe access for emergency vehicles, and high costs for maintenance of soils, crops, and installations. Figure 8.6 shows a landscape in Vietnam with complex slopes and severely disturbed by human actions.

The land use potential of different slope gradients is shown in Table 8.3.

Depending on slopes, soils are suited for different kinds of land uses. Some soils are suitable for cultivation; they are also good for other uses such as urban, industrial, pasture, range, forest, and wildlife. Some soils are not ideal for cultivation, but they can be appropriate for pasture, range, forest, or wildlife; some soils can be used only for pasture or range and wildlife; others are only applicable for forest and wildlife. Some soils are not appropriate for uses other than only wildlife, recreation, and water-yielding uses. Grouping of soils according to their feasibilities has been done in the land capability classification by the United States Department of Agriculture. Soils on steep and very steep slopes generally belong to land capability classes IV, VI, VII and VIII. These soils are not normally suited for growing crops because of their severe susceptibility to erosion. These soils are usually shallow and stony with rock outcrops, rocky surfaces and frequent slope failures. They generally exhibit severe effects of past erosion too. In some lands of class IV and VI, range or pasture improvements can be done with seeding, liming, fertilizing, and water control with contour furrows, drainage ditches, diversions, or water spreaders. In lands of class VII and VIII, cropping, range and pasture management are not practical. All these classes of lands can, however, be used for natural and improved forestry, watershed protection, wildlife use and recreation, keeping the soils and slopes as far in their natural state as possible, and with some slope stabilization efforts (Soil Survey Division Staff 1993).

Fig. 8.6 A hilly landscape with steep and complex slopes in Vietnam bearing signs of human disturbance. (Image courtesy of Northern Mountainous Agriculture and Forestry Science Institute (NOMAFSI), Vietnam)

Table 8.3 Land use potentials of different slope gradients

Slope gradient (%)	Land use potential
0–3	Suitable for all land uses.
3–8	Suitable for medium density residential development, agriculture, industrial and institutional uses.
8–15	Suitable for moderate to low density residential development, but great care must be taken to select location for any commercial, industrial and institutional uses.
15–25	Only suitable for low density residential, limited agricultural and recreational use.
>25	Only used for open space and recreational use.

Adapted from Lehigh Valley Planning Commission (2008)

8.2 Slope Failures and Mass Movement

Slopes are naturally unstable. Overland flow of water in precipitous slopes always occurs down the slope and modifies the configuration of the land. This modification is slow in gently sloping surfaces, but abrupt in steep slopes. Such forces as gravity, wind and water, and natural or man-made disturbances can cause slope failures, mass movement, erosion, slippage or slide. The factors that influence the stability of slope include geology, vegetation, slope drainage, slope gradient and shape, soil type, human intervention to slope such as hill cutting, stream diversion, gully filling, etc. In many places, frost and high temperature separate pieces of weathered rock, and the loose material moves downhill to form piles of hillside waste, debris cones, outwash fans, and other formations.

Although all sorts of erosion including mass movement or mass wasting, gully, rill and sheet erosion can occur, mass movement is of particular concern on steep

Fig. 8.7 Vertical soil displacement in steep slopes

Fig. 8.8 Human settlements on steeply sloping hills cause slope instability, deforestation and erosion. Notice the color and turbidity of water of the Sangu river passing through this highly eroded hilly area of Bandarban, Bangladesh. (Photo courtesy of Khan Tanjid Osman)

slopes (Fig. 8.7). Mass movement of rock or soil by the force of gravity is a geological process which has shaped the Earth's surface over time and which is often accelerated by human activity. Development activities (road-building, for example) and human settlements (Fig. 8.8) on steep slopes have made them prone to mass movement. Mass movements on steep slopes are practically unstoppable once they have started and can destroy roads, railroads, buildings, and houses in their paths. Landslides often damage properties, crops, livestock, and human lives. Predominant mass movement processes are rockfalls, debris flows and landslides.

Rockfalls: Often masses of rock fall freely by the pull of gravity from an upper position of very steeply sloping hills or rocky mountains. These rock bodies are sometimes very large in size and weight, and when they fall on ground, they usually have a huge impact on the foothill. They generally bounce and move some distance and erode the places of contact. Rockfalls may damage installations on ground and sometimes cause loss of lives.

Debris flows: Debris flows refer to the movement of soil and other materials saturated with water that move as a fluid mass. They typically flow down valleys, following existing channels. The margins of debris flows are typically higher than the middle and they commonly leave piles of boulders where they end. Debris flow (or mud flow) can develop when water rapidly accumulates in the ground, such as during heavy rainfall or rapid snowmelt, changing the earth into a flowing stream of mud or slurry.

Landslides: Typical landslides are movement of large masses of rock or soil that slide along the ground surface down the slope. Landslides move as a single, and fairly coherent mass. Some landslides move very rapidly whereas others creep at very slow rates. Many mass movements which we call landslides are not actually "slides"; they are "falls" due to slope failures (there is no sliding along the slope). Large soil masses are detached from the main body and pulled down by gravity. Landslides may be quite small (several feet across), or they may be vast. Landslides and landslide-generated ground failures are among the common geo-environmental hazards in many of the hilly and mountainous terrains (Woldearegay 2013).

Landslide is defined as the downward and outward movement of slope forming materials composed of natural rock, soils, artificial fills, or combinations of these materials. It is the downslope movement of soil and rock under the influence of gravity without the primary assistance of a fluid transporting agent. Mass movement or landslides can be classified into fall, topple, slide, spread and flow of rock, soil and earth material. Swanston and Howes (1994) consider all mass movement or slope movement as landslides and classify landslides into six categories: falls, creep, slumps and earthflows, debris avalanches and debris flows, debris torrents and bedrock failures according to the depth of movement, rate of initial failure, failure mechanics, and water content of the moving material. Falls are movements that take place mainly through the air by free-fall, leaping, bounding, or rolling. They are very rapid to extremely rapid mass movements (from meters/minute to meters/second) (Fig. 8.9).

Soil creep occurs at very slow (centimeters/year) to extremely slow (millimeters/year) downslope movement of overburden. Rapid downslope movements of a mass of predominantly soil and organic debris mixed with water occur as debris flow. On the other hand, "debris torrents" are the rapid downslope movement of channelized water with high concentrations of soil, rock and organic debris. Debris avalanches are rapid, shallow landslides from steep hill slopes. Movement begins when overburden slides along bedrock or along other layers within the overburden having higher strength and

Fig. 8.9 Landslide in Ferguson, California, USA (Wikipedia)

lower permeability. Slumps and earthflows involve combined processes of earth movement (rotation of a block of overburden over a broadly concave slip surface or slump), and result in the downslope transport of the resulting mass either by a flow or a gliding displacement of a series of blocks (earthflow). Thus, the term "landslide" includes all varieties of mass movements of hill slopes including downward and outward movement of slope forming materials composed of rocks, soils, artificial fills or combination of all these materials along surfaces of separation by falling, sliding and flowing either slowly or quickly from one place to another (Fig. 8.10).

8.3 Factors Affecting Landslides

The controlling factors of erosion and slope failures on steep slopes include slope angle, soil depth, weight of soil (depth, bulk density), soil cohesion, expansive clay, internal angle of friction of soil, vegetation type and slope hydrology. Other conditions that can trigger mass movements include earthquake shaking (or other sources of vibrations), volcanic eruptions and lava flow, number and orientation of bedrock fractures, forest fires, and construction-related slope modification. Landslide activities are expected to continue in the twenty-first century for the following reasons:

8.3 Factors Affecting Landslides

Fig. 8.10 Soil creep in the Erch valley near Carnguwch church, Wales, UK. (Image courtesy of Eric Jones)

(a) increased urbanization and development in landslide-prone areas, (b) continued deforestation of landslide-prone areas, and (c) increased precipitation caused by changing climatic conditions.

8.3.1 Geology

Geology is an important factor influencing landslides. Landslides usually occur in subduction complexes, volcanic arcs, transform faults, and intraplate settings. Landslides also take place in steeply sloping surfaces having weak water saturated sediment or regolith that lies parallel to the underlying beds of sediments or rocks. Landslides often occur in places with fractures or bedding in the rocks, sediments and soils nearly parallel with the slope. Water infiltrates through fractures and provides lubrication for which the overlying soil or sediment slides along the slope downward. Fragmented materials or products of weathering are easily separated from the rock body and move down the slope more easily. The physical properties of both the materials that rest on or underlie play an important role in the occurrence of landslides. Girty (2009) suggested that if loose materials, such as regolith rest on solid fracture-free rock surface, these materials may slide down even if the degree of inclination is very small.

8.3.2 Rainfall

High intensity storms mainly trigger landslides and other slope failures (Crosta and Frattini 2003; Jakob and Wetherly 2003). Such storms produce rainfall depths above certain critical threshold values. The highest levels of erosion generally occur in storm centers with the highest rainfall densities and totals. Landslide densities tend to be higher in storms with higher total rainfalls. Prolonged wet spell is also likely to cause landslides. Generally, erosion increases as the storm intensity increases. The thickness of individual sediment pulses showed a good correlation with total rainfall for individual storms. However, there is often wide variation in landslide densities within areas of similar total rainfall. Moreover, individual hill slopes differ in their susceptibility to erosion and slope failure. In most intact slopes, there is little erosion; erosion is more severe in already eroded surfaces.

8.3.3 *Slope Gradient*

There is usually a critical slope limit for a given set of lithological, soil, hydrological and climatic conditions below which landslides do not occur. Summarizing various surveys, DeRose (1995) indicated that the limiting slope for landslide occurrence was between 18° and 24° for most areas of the hilly country of North Island. Above this limiting slope, there is an increase in the frequency of landslides reaching a maximum between 26° and 40°. There is an upper slope limit between 50° and 60°. The mean slope of landslide distribution is typically between 29° and 39°. Forested hill slopes usually suffer less from slope failures than pastures.

8.3.4 *Soil*

Soil depth (soils are often shallow on steep slopes for continuous removal by erosion), texture, structure, porosity and clay mineralogy are important edaphic factors related to slope failures. These factors influence the cohesion of soil materials subject to falls or slides. In addition, angle of internal friction, water content, pore water pressure, and gradient of the potential sliding surface are other factors that determine the stability of steep slopes. Stability in cohesive materials is controlled largely by clay mineralogy and moisture content of overburden. Under dry condition, clayey materials have high shear strength with high cohesion and angle of internal friction. Clay particles absorb water into their structures and swell. Thus, clay-rich materials (particularly expansive clays) have a high potential for accelerated deformation and ultimate failure in the presence of excess water. On the other hand, loose soil materials on impermeable substratum are also susceptible to landslides. Water infiltrates rapidly through such soils, increases the weight of the soil, and lubricates the surface of the impermeable substratum below. As a result, gliding, sliding or slipping of the soil body takes place. Such slope movement may occur even on slopes that are not very steep.

8.3.5 Natural Events

Earthquakes and volcanic activities are associated with major landslides on steep slopes throughout the world. Furthermore, ashen debris flows caused by earthquakes can also trigger mass movement of soil. Volcanoes are prone to sudden collapse, especially during wet conditions. Within minutes of a magnitude 5.1 earthquake on May 18, 1980, a huge landslide completely removed the bulge, the summit, and inner core of Mount St. Helens, and triggered a series of massive explosions (USGS 2015). Lava flow often devastates ground configuration.

8.3.6 Anthropogenic Factors

Human contribution to slope instability can be summarized as –

Deforestation: Forest canopy works as an umbrella of the soil; it incepts rain and reduces the impact of raindrops. Plant roots bind soil particles and keep the soil in place. Plant roots, stems, and litters prevent concentration of water in narrow channels, the potential gullies. A vegetation cover reduces runoff, increases infiltration, reduces sediment load in runoff and prevents deterioration of water quality. Thus, deforestation severely reduces slope stability and soil as well as water quality.

Removal of support and development: Excavating or undercutting reduces the load-bearing capacity of the slope. Development activities and human settlements on steep slopes commonly require construction of a flat site on which to put a house. This is done by the cut-and-fill technique, soil materials are removed from the uphill part of the site and placed on the downhill portion to form a level surface. The fill material may compact and settle later, and cause the cracking of foundations and walls. The extra load of a building may trigger a slope failure on unrestrained fill. Development activities on steep slope can start a cycle of erosion and flooding. Rain water that falls on forests, grass and other natural areas usually infiltrates into the soil at a relatively faster rate. Roofs, concrete, pavement and other impervious surfaces increase the amount of rainwater that runs off the land surface along steep slopes below the house and driveway. Unless appropriate measures are taken, excessive soil erosion and increased flooding can potentially occur.

Excessive load: Construction of structures or fill on a slope in excess of the tolerance limit of the slope causes failures.

Altering water courses: Concentrating the flow of runoff for water harvest, altering stream channels for power generation and other purposes, and changing the natural drainage pattern can cause slope instability.

8.4 Management of Steep Slopes

To keep development off of steep slopes is one way to protect them from erosion and mass movement. Steep slopes can be valuable community resources for recreation, forestry and wildlife habitat. Possible mechanisms for steep slope protection include creating greenways, the revegetation of deforested areas, wildlife habitat preservation, and conservation areas.

8.4.1 Mechanical Measures

Mechanical and engineering measures to protect steep slopes from erosion and mass movement include the construction of drainage ditches and improvement of natural drainage systems, contour bunds and silt fencing, grading, terracing, construction of retaining walls, etc. All these methods need physical assessment, planning, designing, budgeting and workmanship. These activities could be performed by competent construction engineers so that the risk of failures and negative impact on the environment are minimized. Some regulations and permits may also be involved for undertaking slope modification by engineering techniques.

8.4.1.1 Drainage

All drainage from development of a site in steep hillsides should be directed away from steep hillside areas and directed towards a public storm drain system as far as feasible or onto a street developed with a gutter system designed to carry surface drainage runoff. However, existing natural drainage courses on the portions of the site should not be disturbed. These natural drainage courses should be retained where feasible, but not be impacted by additional runoff from the developed portions of the site.

8.4.1.2 Contour Bunds

Contour bunds are embankments made across the slope with earth or stones to reduce runoff velocity. Contour bunds on gentle to moderately sloping lands with permeable soils intercept runoff water increase infiltration and the water content of soil and reduce erosion. They trap sediments in runoff and gradually develop into natural terraces. Contour bunds can be stabilized by close hedgerows or lines of hardy tree plantings. Gebremichael et al. (2005) reported that the mean annual soil loss of 20 Mg ha^{-1} from cultivated lands was reduced to a negligible amount with the introduction of stone bunds in the highlands of Ethiopia. According to Nyssen et al. (2000), stones lines reduce >60% of net soil losses depending on the age, soil type, slope, and climate. Figure 8.11 shows stone bunds made to save the soil in a sloping land.

8.4 Management of Steep Slopes

Fig. 8.11 Stone faced soil bund of Tigray, Ethiopia. (Image courtesy of Emni Getsu Hamed Zala)

Fig. 8.12 Silt fence

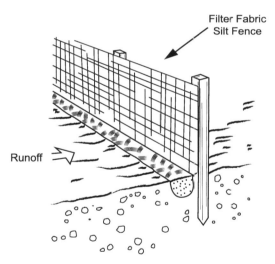

8.4.1.3 Silt Fences

Silt fences are filter barriers consisting of woven and unwoven geotextile fabric products (e.g., jute, polyethylene) anchored to vertical metal or wooden posts, which are laid out on the contour across the slope for reducing runoff velocity and filtering the sediment (Blanco and Lal 2008). Silt fences are porous filters which provide temporary erosion control but can be used before establishing a fast growing vegetation. Silt fences retain most of the total suspended solids in runoff, but are not effective in controlling a concentrated flow. Turbulent and concentrated runoff may inundate and overflow the fences. Silt fences can be integrated with contour plantations and agro-forestry. Figure 8.12 shows a silt fence.

8.4.1.4 Surface Mats

Temporary protection to soils of disturbed sites can be provided by surface mats made of permeable materials including jute, coir, paper, straw, nylon, and other synthetic materials. These mats are unrolled, spread over the soil surface and pinned with hardwood pegs. The objectives of surface mats include the protection of soil against erosion, establishment of vegetation, suppression of weeds, and stabilization of disturbed sites. Natural geotextiles (e.g., coir, jute) are preferred over synthetic materials because they are biodegradable, they do not pollute the soil or alter the solar radiation, nor do they cause any overheating of the soil surface. Stem cuttings (such as willow and cottonwood, for example) can be used as pegs or pins. A vegetation cover is established before the mats are biologically degraded. As vegetation becomes established, the biodegradable surface mats become a composite solution to erosion (Rickson 2006). Woven jute (500 g m^{-2}) and coir mat (400–900 g m^{-2}) have C-factor values <0.10 (Morgan 2005). On sloping lands, synthetic or biodegradable geotextiles are often used in conjunction with crop residues overlaid with the geotextile. Biodegradable geotextiles degrade faster than synthetic ones. Soils treated with polyacrylamide and overlaid with geotextile fabric reduce soil erosion rates to non-detectable levels as compared to soils treated with polyacrylamide alone (Blanco-Canqui et al. 2004). Mats have, however, numerous shortcomings in reducing concentrated flow erosion, especially in steep slopes (Blanco and Lal 2008).

8.4.1.5 Grading and Terracing

Slope modification is needed in many instances for different land uses including the construction of roads, tourist resorts, parks and houses. Alteration of the slope by cutting and filling of earth and creating a desired land configuration is known as grading. Usually roads, houses and other structures are built on leveled or very slightly sloping areas known as benches that have been made by reshaping the landscape. Lawns, sports grounds, livestock rearing areas, etc. can be situated in slopes. Grading is a kind of landscape architecture. The slope of the land and the size of the bench determine the amount of cut and fill in the hillside. Soil properties also determine the degree of grading; steeper grades in loose soils may not be stable. Sometimes, grading is not feasible because the natural slope of the hillside is steeper than what the grade on the cut or fill slope should be. A general rule is to retain the natural slope as far as possible on unstable soils; otherwise, grading may collapse. Figure 8.13 shows examples of desired and undesired grading techniques (Lehigh Valley Planning Commission 2008).

Terracing is a mechanical method of slope modification by cut-and-fill technique to reduce the length and degree of slope so that runoff water moves slowly along the stairs or through the safe waterways. Terracing makes undesirably sloping lands into short level to gently sloping sections. According to FAO (2000), this is also a grading technique for the reduction of soil erosion, conservation of soil and water, and

8.4 Management of Steep Slopes

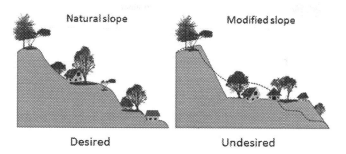

Fig. 8.13 Desired and undesired grading. (Sketch drawn following Lehigh Valley Planning Commission (2008), www.lvpc.org/pdf/SteepSlopes.pdf)

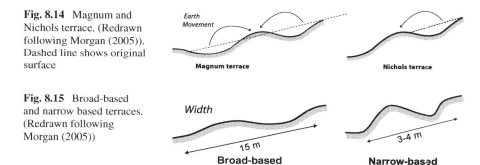

Fig. 8.14 Magnum and Nichols terrace. (Redrawn following Morgan (2005)). Dashed line shows original surface

Fig. 8.15 Broad-based and narrow based terraces. (Redrawn following Morgan (2005))

facilitation of cultivation. Moreover, this is an ancient method of slope stabilization. Some authors (Drechsler and Settele 2001; Bokhtiar et al. 2001; Kasai et al. 2001) consider it as the most widely used mechanical soil conservation practice throughout the world. According to Wheaton and Monke (2001), terracing is an effective way of conserving soil and water in sloping lands.

Types of Terraces

There are three main types of terraces: diversion terrace, retention terrace and bench terrace. As the word "diversion" indicate, the objectives of diversion terraces are to intercept runoff and to drive it away along a safe stabilized channel across the contour. There are different types of diversion terraces including Magnum terrace and Nichols terrace. The mode of construction of these two types of terraces is different; soil is taken from both sides of the embankment for making Magnum terrace while only from the upslope side only for making Nichols terrace (Fig. 8.14). Terraces can again be broad-based and narrow-based depending on the width of the embankment and channel. The width is about 15 m in broad-based terraces and only 3–4 m in narrow-based terraces (Fig. 8.15). Morgan (2005) suggested that closer spacings are feasible on steeper slopes.

According to FAO (2000), retention terraces are broad-based terraces which are constructed to retain runoff water in hillsides. They can be constructed only in permeable soils on slopes of <4.5°. Bench terraces are made of a series of alternating steps with risers in the margin of each step. On steep slopes and in loose soils (sand, loamy sand and sandy loam textures) the risers tend to collapse. So, the risers are usually stabilized with stones, concrete, debris and vegetation. If the risers are not stabilized, they can be sources of high erosion instead of controlling it by terracing. The steps or shelves are used for cultivation of different crops. Bench terraces are usually constructed in slopes <30° (about 58%) although the recommended slope for bench terracing is only 30%. Sharda et al. (2007) reported that bench terraces are being constructed even in slopes as high as 60% because of increased pressure on land. Bench terraces are of three main types: outward sloping terraces, level terraces and inward sloping terraces (ICIMOD – International Centre for Integrated Mountain Development; http://lib.icimod.org/record/27709/files/Chapter%205%20Physical%20Methods.pdf). The schematic diagrams of these three types are given in Fig. 8.16. The outward sloping terraces convert steep slopes into gentle slopes. The inward sloping terraces also have gentle slopes but drives runoff water towards the hillside instead of down the slope. Level terraces are used mainly for ponding water to cultivate crops like rice. Other crops can also be grown in level terraces on permeable soils.

The permissible degree of slope for diversion terraces is up to 7° and for retention terraces is up to 4.5°. The recommended slope for bench terraces ranges from 7° to 30°. The main factors that influence the designing of a terrace include soil and slope characteristics, rainfall characteristics, farming practices and desired land use (Sharda et al. 2007). Among soil characteristics, the depth of soil, physical properties of soil including texture, structure, bulk density, porosity and infiltration capacity and top soil distribution are very important in this regard. Important characteristics of rainfall include amount, intensity and distribution of rainfall. Generally outward sloping terraces are more suitable in low rainfall areas with permeable soils, and inward sloping terraces are constructed in areas of heavy rainfall and with relatively less permeable soils. Level terraces are preferred in areas with medium rainfall, with highly permeable soils and for growing rice in many parts of the world. The important aspects of terraces for their effective functioning include width, spacing, length and gradient. The factors that determine the width of a terrace are soil properties, slope, rainfall, and desired farming practices. As suggested by Sharda et al. (2007), the width of a terrace can be calculated from the following formula:

$$W = \frac{200 \times d}{S}$$

where W = width of the terrace in meters, d = maximum depth of the cut in meters, and S = slope of the land.

Very wide terraces are useful for the cultivation of rice while narrow terraces are adequate for raising rubber seedlings. The cost of earthwork is higher for wide terraces because it needs deep cutting and higher riser. The factors that influence the vertical

8.4 Management of Steep Slopes

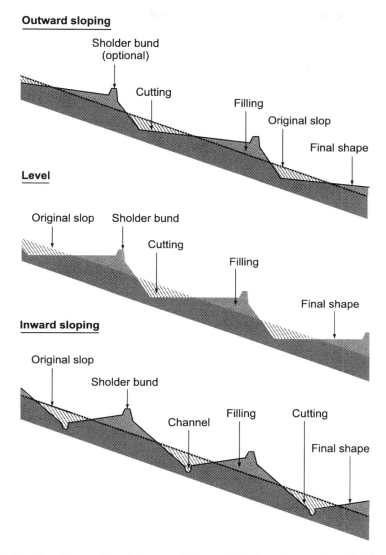

Fig. 8.16 Different types of bench terraces. (Redrawn following ICIMOD; http://lib.icimod.org/record/27709/files/Chapter%205%20Physical%20Methods.pdf)

interval between two terraces include the soil, slope, the depth of cutting and desired land use. Care should be taken during cutting of land so that the bed rock is not exposed. In addition, there must be a balance between the width and spacing of terraces. Some factors influence the length of the terrace including the shape and size of the land, the degree of dissection of the land and permeability and erodibility of the soil. Figure 8.17 shows a steep land with broad-based bench terraces used for rice cultivation.

Fig. 8.17 Rice is growing in terrace in Mu Cang Chai of Vietnam. (Source: vietnamtourism.com.vn)

It has been suggested in several reports (FAO 2000; FFTC 2004; GPA 2004) that terracing reduces runoff and soil loss by water erosion appreciably, but it has also been reported (Lasanta et al. 2001; Van Dijk and Bruijnzeel 2003) to have some disadvantages. The disadvantages of terracing include retention of too much water leading to saturation and storm runoff, higher erosion often at the foot of a terrace wall due to the steepness and the scarcity of vegetation, very high cost of primary construction and latter management of the terraces, and risks of high erosion if the slope is very steep and the soil is not stable. According to ICIMOD (1998), the major limitations of terracing are the disturbance of the natural soil, exposure of the less fertile subsoil leading to a decline in soil fertility particularly in the first several years, loss of soil during construction of terraces and, in some following years, the need of tremendous labor and investment for construction and maintenance. Moreover, terraces may not be stable in many cases and are not suitable for sandy and coarse soils and on very steep land.

8.4.1.6 Retaining Walls

Retaining walls also help to hold unstable slopes. Retaining walls are made across steep slopes to protect them against failures. Retaining walls are made in hillsides along roads and buildings, embankments, stream banks, etc. These walls provide lateral support to vertical slopes of soil which would otherwise cave, slump or slide. Retaining walls are the most common way to deal with steep slopes and can be constructed from many materials including boulders, fieldstone, concrete, treated wood, self-stacking precast concrete blocks and bricks.

8.4.2 Agronomic and Agroforestry Measures

Agronomic and agroforestry measures involve crops, cropping systems, planting methods, distribution of crop in the field, concerted approaches of growing grasses, forages, and trees together in contours, strip cropping, alley cropping and hedgerows and SALT. These are all sloping land agricultural techniques. There are some other agronomic or biological methods of erosion control and soil conservation. These methods will be discussed in relation to the management of eroded soils in lands that are not steeply sloping.

8.4.2.1 Shifting Cultivation and Alternative Farming Systems

Shifting cultivation or slash-and-burn agriculture is an ancient farming system practiced by the indigenous people in the hilly areas of Africa, South America, Oceania South Asia and Southeast Asia. Shifting cultivation is still the most widely practiced farming system in the eastern Himalayan region (Kerkhoff and Sharma 2006) including eastern hills of Bangladesh, eastern Bhutan, southwest China, northeast India, hilly Myanmar and parts of Nepal. The system is locally called, *Bukma* in Nepal, *Taungya* in Myanmar, *Kaingin* and *Lading* in the Philippines and *Jhum* in India and Bangladesh, *Chena* in Sri Lanka, *Chancar Leu* in Kampuchea and *Ray* in Lao.

In this system, a patch of forest, usually on gentle slopes or on summits of hills, is cleared, vegetation is slashed and burned, holes are dug in the soil with elementary tools, and seeds of assorted crops are sown before monsoon. The most common crops are *Oryza sativa, Sesamum indicum, Capricum frutescens, Gossypium harbaceum, Abelmoscus esculantus, Colcasia esculanta, Cucurbita pepo, Zea mays, Lablab purpureus, Ipomoea batatas, Trichosanthes anguina, Abrus precatorius, Luffa acutangula, Momordica charantia, Trichosanthes diocia, Basella alba, Pithecellobium dulce, Carica papaya, Cucurbita maxima, Benincasa hispida, Solanum tuberosum, Saccarum officinarum, Manihot esculenta, Solanum melongena, Curcuma longa, Zingiber officinale, Lageneria siceraria*, etc. (Hossain et al. 2006), depending on the respective tradition, geographical location, climate and land type. Seeds germinate and grow rainfed, and the crops are harvested after 6–8 months. The land is then left fallow. Farmers clear a new patch of forest for cultivation in the next season. In the past, shifting farmers returned to the previous land for cropping after 15–20 years. The land could recover over the fallow period, and dense secondary forests could develop through natural succession. For the growth in human population, socio-economic conditions and scarcity of suitable land, the rotation period has shrunk from 20 to about 2 years at present. This traditional system has thus become responsible for massive deforestation and soil degradation. Each year shifting cultivators were clearing about 10 million ha of forest and scrubland in tropical Asia, Africa and Latin America in the mid-1990s (Sanchez 1996). On the basis of data given in FAO and other sources, it is estimated that each year approximately 1.9 to 36×10^6 ha land of primary close forests, 3.4 to 40×10^6 ha land of secondary close forests, and 6.9 to 21.9×10^6 ha land of secondary open forests are being lost due to shifting cultivation (Ranjan and Upadahya 2006). Studies reveal that shifting cultivation adversely affects soil physical and chemical properties, reduces nutrient stocks, and accelerates soil erosion (Gafur et al. 2000) and sedimentation (Gafur et al. 2003).

Viable alternatives to shifting cultivation could be continuous cropping, SALT system, agroforestry or alley cropping, and contour strip cropping. SALT is a multi-tiered contour strip or patch cropping system involving field crops, horticultural crops, trees in different tiers separated by leguminous hedgerows. Essentially, alley cropping involves cultivation of food or field crops in between rows of trees or shrubs that are pruned during cropping to avoid shading the crops, and the prunings returned to the soil as mulch or green manure. The trees, which are usually legumes such as *Gliricidia sepium* and *Leucaena leucocephala*, but may include non-leguminous trees such as *Acioa barteri* and *Alchornia cordifolia*, are planted in rows that are 2 or 3 m apart, and food crops such as cassava, yams, cowpeas, rice and maize or sorghum are planted in the spaces between the rows of trees. Hedgerow intercropping is an intensified form of shifting cultivation, practiced by the Igbo people in south-eastern Nigeria. They developed and have been practicing this form of traditional agroforestry for several generations. The Nalaad people in the Philippines have developed a form of agroforestry which involves planting *Leucaena* hedgerows on steep slopes to control erosion. They prune the trees regularly and return the loppings to the soil as mulch for long. The systems of SALT, alley cropping and contour strip cropping have been discussed in different sections of this chapter.

8.4.2.2 Contour Farming

The contour line is an arbitrary line drawn perpendicular to the direction of the slope and plowing and planting crops, grasses, hedges and trees in the contour or across the slope is called contour farming. Contour crops reduce velocity of runoff, enhance infiltration, and trap sediments along the contour lines. Contour farming can be combined with contour bunds, ridge and furrow systems and strip cropping, all aiming at reducing runoff and erosion and the conservation of soil and water. Contour farming is most effective on slopes between 2% and 10% (slightly to gently sloping). Contour farming is not well suited to rolling topography having a high degree of slope irregularity. The effectiveness of contour farming in soil conservation depends on rainfall intensity, slope steepness, soil properties, ridge height, cover and roughness and the critical slope length. Cover, roughness and ridge height can be influenced by management. Contour lines are spaced on the basis of slope, soil, rainfall and crop type. Annual and perennial crops are planted in the ridges or furrows of the contours.

8.4.2.3 Contour Strip Cropping

Growing planned rotations of row crops, forages, small grains, or fallow in a systematic arrangement of equal width strips across a field is known as strip cropping. If the strips are placed along the contour, the system is called contour strip cropping. Contour strip cropping utilizes the variability in the character of the different crops in alternating strips in favor of the reduction of soil erosion. It provides added erosion control and plant and crop diversity because it combines contour- and strip-cropping. Strip-cropping on the contour is more effective than contouring alone for reducing soil erosion in fields with severe erosion hazard. Contour strip cropping systems can reduce soil erosion to <40% (Blanco and Lal 2008).

8.4.2.4 Alley Cropping and Contour Hedgerows

Alley cropping is an agro-forestry technique of growing trees and crops together in the same field. Tree (poplars, willows, silver maple, birches) rows are widely spaced and crops (cereals like corn, wheat, barley, oats, soybeans, potatoes, peas, beans and forages such as fescue, orchard grass, desmodium, bluegrass, ryegrass, brome, timothy, clover, alfalfa, etc.) are grown in between. Spacings of tree rows are so chosen that the mature size of the trees does not interfere for light and moisture with the crops between the rows. When light demanding crops like corn (maize) or sorghum are grown, the alleyways need to be wide enough to let in plenty of light even when the trees have matured. Non-spreading branched trees are preferred as such crops. The cropping sequence can also be planned to change as the trees grow. For instance, soybeans or corn could be grown when the trees are very small; as the tree canopy closes, forages could be harvested for hay; and finally, when the trees are fully grown and the ground is more shaded, grazing livestock or shade-tolerant crops like mushrooms or ornamental ferns could occupy the alleyways. When planted along the contours on a slopping land, alley crops provide a barrier to run off water, hold the sediments and conserve moisture.

Another agroforestry system is known as contour hedgerows. Contour hedgerows are usually double rows of close growing trees, shrubs or grass (e.g., *Vetiveria zizanioides*, *Vitex negundo*, *Leucaena leucocephala*, *Coriaria sinica*), planted along contours at 3–6 m intervals in the sloping land. Crops are grown in the space between hedgerows. It differs from alley cropping in that plants in the rows are regularly pruned to form a hedge. Usually hedgerows are pruned two to three times a year and prunings are used as green manure so that less external nutrient input is required (Isaac et al. 2003). Hedgerows reduce runoff, intercept eroded sediment from the upper slope, and conserve soil fertility. Progressive deposition of sediments along the hedgerows forms natural terraces after some years of successful maintenance of hedgerows. Contour hedgerows have proven to be an effective means for controlling soil erosion in steep slopes of many countries (Baudry et al. 2000). The system of contour hedgerows is simple, but it provides multiple benefits. Apart from its great contribution to soil conservation and soil fertility improvement, contour hedgerow intercropping technology can also diversify cropping patterns so as to ensure and increase the income of farmers. One effective way is to incorporate economic trees within hedgerows.

NRCS (2002) gives the following list of trees, shrubs and grasses suitable for planting in hedgerows:

Trees and shrubs: Alder, Arrowwood, Wild blackberry, Blackhaw, Buttonbush, Chokeberry, Black chokecherry, Coralberry, Flowering crabapple, Flowering dogwood, Red-osier dogwood, Silky dogwood, Elderberry, Hackberry, Hazelnut, Hickory, Red maple, Nannyberry ninebark, Northern red oak, White oak, Pawpaw, Persimmon, American plum, Wild raspberry, Redbud, Red cedar, Rose, Serviceberry, Spicebush, Smooth sumac, Staghorn sumac, Tea, Witch hazel, etc.
Grasses: Big bluestem, Indian grass, Switch grass, Eastern Gama grass, etc.

8.4.2.5 Sloping Agricultural Land Technology (SALT)

Sloping agricultural land technology or SALT integrates several soil conservation and agroforestry measures together in steeply sloping lands. In this technique, field crops and perennial crops are planted in bands 3–5 m wide between double rows of nitrogen fixing trees and shrubs along the contour (Fig. 8.18). It is a special kind of contour strip cropping system developed originally on a marginal site in the Philippines by the Mindanao Baptist Rural Life Center (MBRLC) in 1971. It is now practised in many parts of the world in land with slope >10% with variable success. Field crops include legumes, cereals, and vegetables, while the main perennial crops are cacao, coffee, banana, citrus, and fruit trees (MBRLC 1988). The nitrogen fixing trees and shrubs (e.g., *Lucaena leucocephala*) are thickly planted in double rows to form hedgerows. When a hedge is 1.5–2 m tall, it is cut down to about 75 cm, and the cuttings are placed in alley-ways to serve as organic fertilizers. SALT has been practiced in many countries with variable success. It may be an alternative to shifting cultivation in degraded hilly lands. Ten steps are involved in establishing a SALT farming:

Step 1. Identifying points of equal elevation with the A-frame.
Step 2. Drawing contour lines by joining the points of equal elevations.
Step 3. Plowing 1 m wide strips along the contour for planting hedgerows.
Step 4. Two furrows are laid out in each furrow, and seeds of nitrogen fixing trees and shrubs are thickly sown in the furrows. Suitable hedgerow species are *Flemingia macrophylla* (syn. *congesta*), *Desmodium rensonii*, *Calliandra calothyrsus*, *Gliricidia sepium*, *Leucaena diversifolia*, and *L. leucocephala*, etc.
Step 5. Growing crops between hedgerows as alley crops.
Step 6. Planting permanent crops such as coffee, cacao, banana, citrus; others of the same height may be planted in cleared spots of hedgerows. Permanent crops are planted in every third strip. Tall crops should be planted at the bottom of the farm while the short ones are planted at the top.

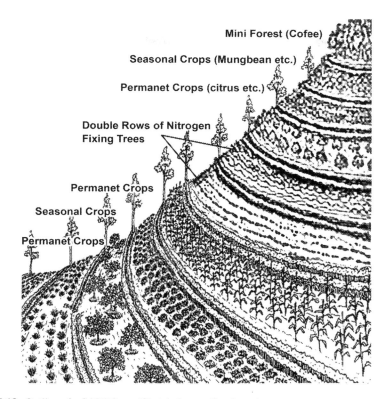

Fig. 8.18 Outline of a SALT farm. (Sketch drawn after Agnet)

Step 7. The plantation of short-term and medium-term cash crops (pineapple, ginger, sweet potato, peanut, mungbean, melon, sorghum, corn, upland rice, etc.) should be done between the strips of permanent crops.

Step 8. Trimming of hedgerows every 30–45 days by cutting to a height of 1.0–1.5 m from the ground. The prunings are piled on the soil around the crops as organic fertilizer.

Step 9. Practicing crop rotation with corn or upland rice, tubers and other crops on strips.

Step 10. Building green terraces by planting straw, stalks, twigs, branches and leaves, and also rocks and stones at the base of the rows of nitrogen-fixing trees. In the passage of time, permanent effective terraces will be formed.

8.4.2.6 Soil Bioengineering

Vegetation is often used for slope stabilization and soil and water conservation. Specialized methods for establishing vegetation on slopes have been developed, and these are called soil biotechnology or soil bioengineering systems (Wu 2005). Soil bioengineering is the use of living plant materials to perform specific engineering functions. Plants for bioengineering include grasses, shrubs and trees that may be established by conventional seeding or live planting. Mainly unrooted cuttings are taken from live plants and installed in the ground by various means and in various configurations. The plant cuttings take root and become established on the slope. Most of the systems serve the dual purpose of reducing surface erosion and reinforcing the soil. Phreatophytes (deep-rooted plants that obtain a significant portion of the water that it needs from the phreatic zone, the zone of saturation or the capillary fringe) such as willows are effective at increasing evapotranspiration. Effectiveness as soil reinforcement depends on the depth at which the cuttings can be placed and the depth to which the roots will penetrate. The following bioengineering systems are mainly used:

Live stake: Sticks are cut from rootable plant stock and tamped directly into ground. Live plants reduce erosion and remove water by evapotranspiration. Plant roots reinforce soil.

Live facine (wattling): Sticks of live plant material are bound together and placed in a trench. They are tied to the ground by live stakes. They function as mentioned earlier.

Brush mattress: Live branches are placed close together on the surface to form a mattress. They function in the same way. In addition, they provide immediate protection against erosion.

Brushlayer, branchpacking: Live branches are placed in trenches or between layers of compacted fill. They perform similar function.

Vegetated geogrid: Live branches are placed in layers between compacted soil wrapped in geogrid. The geogrid provides immediate stability. The plants reduce erosion and reinforce soil.

Soil bioengineering can be used to treat seepage zones and control erosion by stabilizing steep slopes. Artificial regeneration and revegetation can also be effective if some hardy plant species that are capable of growing in harsh conditions can be grown. Such vegetations stabilize sites, disperse runoff water and reduce erosion. Vetiver grass (*Chrysopogon zizanioides*) can be a good choice to reduce runoff, trap eroded soil materials and stabilize slopes (Fig. 8.19).

Pioneering woody species can be used in the development of soil bioengineering systems. These plants constitute the successional bridge between the herbaceous initial colonizers (seeded grasses and legumes) of a disturbed site and later seral types. Willow and cottonwood cuttings can be planted to create live fences across slopes, on stream banks and in gullies, as these species will continue to grow even if their stems are buried. Live fences can be established in swales and small drainage channels. These channels can act as sediment traps and provide clean water to downstream sites. Live gully breaks can be built with wattle fences in gullies to control the flow of water. These structures are established high in the channel where gully torrents originate, often from minor collapses of gully sidewalls. In gullies that have already torrented, live gully breaks can assist in revegetation and stabilization by providing sites where materials may be trapped and vegetation can be established. Table 8.4 gives a list of plants suitable for revegetation of disturbed sites.

8.4 Management of Steep Slopes

Fig. 8.19 Vetiver grass across the slope for its stabilization beside a highway

Table 8.4 Plants that can be used for slope stabilization

Botanical name	Common name	Height, inch	Habit
Native grass			
Bouteloua curtipendula	Sidcoat grama	12–24	Common on dry prairies; can form sod and does well on steep slopes; partial sun
Elymus canadensis	Canada wildrye	24–48	Covers an area rapidly; self seeding
Festuca ovina	Sheeps fescue	6–24	Common on sandy soils; quite drought and shade tolerant; bunch grass
Festuca rubra	Red fescue	6–24	Common lawn grass; quite drought and shade tolerant
Sporobolus cryptadrus	Sand dropseed	12–30	Dry prairie grass; grows on sand or heaver soils; drought tolerant
Schizachyrium scoparium	Little bluestem	24–48	Blue summer foliage; clump forming with arching habit; bronze and orange fall color
Panicum virgatum	Switchgrass	36–72	Native self-seeding with winter interest; good for wildlife
Herbaceous plants			
Hemerocallis spp.	Black chokeberry	12–36	Low maintenance; many flower colors' long bloom time
Hosta spp.	Hosta	4 = 48	Excellent choice for partial or complete shading
Lamium spp.	Dead nettle	8–12	White or pink flowers; leaves streaked white/silver

(continued)

Table 8.4 (continued)

Botanical name	Common name	Height, inch	Habit
Rosmarinus officinalis	Rosemary	24–48	Tough annuals; can grow on dry, poor soils, aromatic
Viola spp.	Violet	6	Beautiful flowers in spring; spreads very quickly
Shrubs			
Aronia melanocarpa	Black chokeberry	36–72	Tolerates most soils; attractive foliage, fruit and flowers
Diervilla lonicera	Bush honeysuckle	24–48	Sun or partial shade; a native tough low growing plant
Euonymus alatus	Burning bush	48–96	A favorite for highway or commercial landscapes; beautiful foliage; great fall color
Juniperus horizontalis	Creeping juniper	12–24	A great shrub for most areas; tolerates poor soil
Juniperus sabina	Savin juniper	36	Many cultivars; tolerates poor soil; sun
Rhus aromatica	Fragrant sumac	24–72	Attractive foliage; great fall color
Taxus cuspidata	Japanese yew	60–72	Great hardy plant that spreads

8.5 Formation of Gullies

A gully is a large channel created at an advanced stage of water erosion. Gullies are channels deeper than 30 cm (Fig. 8.20). They are formed by the concentration of runoff water that moves at a high velocity through a steeply sloping land. The concentration of water may occur in a previously formed rill due to heavy rainfall, and the moving water gradually cuts the soil deeper and wider unless preventive measures are taken. According to Hansen and Law (2004), gullies may occur in concentrated flow areas, such as those which can be identified on topographic maps. Gullies may originate from depressions in hill slopes and through piping of soil if it is coarse textured and loose. Often gullies can extend up to the watershed divide. Active gullies have headcuts with an abrupt drop in elevation. According to USDA (2007), the channel below the headcut is enlarged by plunging flow and erosion. Unlike rills and interrills, gully erosion seriously restricts land use. Gullies are wide and deep channels with ephemeral flows during heavy rainfall. Gullies may ultimately form streams and narrow valleys. Generally, gullies are caused (i) along roads, tracks, fence lines and firebreaks, (ii) by land clearing on unstable or steep slopes, (iii) by deep pads caused by livestock, and (iv) due to inadequate drainage control and badly designed or located drainage structures.

Morgan (2005) suggested that not all gullies develop purely by surface erosion. Gullies can form on some deforested soils by subsurface flow in natural pipes or tunnels, which is described as soil piping (Poesen 2011). It requires steep hydraulic

Fig. 8.20 Severe gully erosion on loess soils in western Iowa. (Photo courtesy of USDA-NRCS)

gradients in a soil of high infiltration capacity. Heavy storms cause the water in the pipes to break out on to the surface. Eventually, the roofs of the pipes collapse and gullying occurs. Gullies can also form where landslides have left deep, steep-sided scars which are further eroded by running water in subsequent storms. Once a gully is formed, it tends to develop further and this process is seldom inverted or halted naturally. Billi and Dramis (2003) identified two main types of gullies on the basis of morphological and hydraulic geometry characteristics: (1) discontinuous gullies developed on low gradient slopes (1–5% on an average) (2) stream gullies, formed by deep erosion processes, typically migrating upslope.

8.6 Gully Control Measures

Gullies need to be stabilized and controlled for reducing erosion and improving land use capability. For stabilization and control of gullies, the flow of water must be reduced at the source and the gullies must be refilled by some structural and vegetational measures. For example, dikes or small dams may be built at specific intervals along the gully. Structures for gully stabilization can be constructed with rock, gabions, logs, wood stakes with wire or brush, bamboo or vegetative barriers. According to Keller and Sherar (2003), a combination of physical structures along with vegetative measures is preferred for physical protection and long term support.

Fig. 8.21 A series of rock check dams used to reduce stormwater velocity

8.6.1 Rock Check Dams

Dams constructed with rock fragments or boulders across the gully to reduce flow of water, to cause sedimentation of eroded soil materials and to allow high water flow are called rock check dams. A series of rock check dams are laid across the gully at such intervals as needed depending on slope, depth, width and length of gully and the velocity of runoff (Fig. 8.21). The most important factors influencing the efficiency of rock check dam systems include the proper sizing of materials, downstream splash and plunge pool control and the frequency of structures. It is to be noted that procuring and transport of rock materials and construction of dams may involve high costs. According to USDA (2007), a control section and an energy dissipation section are needed for all rock check dams.

8.6.2 Bamboo and Rock Structures

A combined system of bamboo fences and rock dams can effectively stabilize gullies. Keller and Sherar (2003) suggested that if debris retention and gully control structures are constructed with a notched wire, runoff water may flow over the middle of the structure and the system can protect scour at the outlet of each structure. Moreover, the structures can be fixed into the firm soil banks.

8.6.3 Rock and Brush Grade Stabilization

Rock and brush grade stabilization is a technique which falls within rock dams and brush dams. It is composed of small dams built with alternating layers of rock and brush. These are short structures and are applied to small drainage areas. These dams can also intercept runoff and reduce its velocity. They can also facilitate the deposition of sediments and fill the shallow depressions.

8.6.4 Soilcrete

A material made up by mixing soil and concrete approximately at a ratio of 1:1 with a cement mixer is called soilcrete. Soilcrete is filled in cloth bags and the bags are stacked in layers across the gully. The soilcrete is hardened when moistened by flowing water or rain and form a dam effective enough to reduce the flow of water. The dams are higher at the edges and lower in the centre so that overflow of water can happen during storms. The area between two dams can be revegetated and the channel can be naturally refilled with soil material and stabilized. Such structures can be long lasting.

8.6.5 Aggregate-Filled Geotextiles

According to USDA (2007) aggregate-filled geotextiles, such as geoweb material can be used for controlling small flat gullies with low velocity of water flow. These are most effective at the channel elevation with no plunge downstream.

8.6.6 Gabions

Gabions are cages or boxes filled with rocks, concrete, sand or soils. Gabions for gully control are made with filling twisted wire baskets with rocks and stacking them one over the other across the channel (Fig. 8.22). These are relatively cheap, durable and effective in low rainfall regions. However, there are some risks involved with gabions including plunging pool development, soil piping and flow diversion around and under structures (USDA 2007).

Fig. 8.22 Gabion check dam provides sediment retention and grade control in stabilized gully channel. (Image courtesy of USDA-NRCS)

8.6.7 Land Smoothing or Reshaping

Land smoothing and reshaping are engineering works done on landscapes dissected by complex gullies. Land reshaping consists of land leveling, cutting and filling. For smoothing land surface to a desirable slope, making drainage channels to drive runoff water safely away, filling the erodible gullies by bringing soil materials from a suitable place of the landscape, and restoring large land areas to productive and aesthetic use, a combination of engineering and vegetational measures are taken. Most engineering reshaping works are done by dozers, scrappers and diggers. Fencing, ditching, trenching, subsoiling or reeping may also be needed. Integration of these engineering practices with some agronomic practices including planting trees, shrubs and herbs according to a definite landscape design is necessary. Encouraging growth of native and introduced plants by fertilizers, lime and other necessary inputs may build a rapid cover and stabilize the soil. Thus, land smoothing and reshaping are needed for long-term rehabilitation and restoration of land.

8.6.8 Vegetative Barriers

Grasses and shrubs can be grown closely across the gullies in a single row or double rows to form barriers or hedges. Vegetative barriers at requisite intervals along the gully length are effective in reducing runoff velocity, trapping sediments and allowing water to pass through slowly. The gradual deposition of sediments at the back of the hedges forms bench terraces over time. Farek and Lloyd-Reilley (2000) suggested that vegetative barriers would spread runoff and slow down its velocity so that erosion would be reduced. Vetiver grass (*Vetiveria zizanioides* (L) Nash) and switchgrass (*Panicum virgatum* L.) can form stiff, erect and permanent live barriers for controlling gullies. Narrow and moderately deep trenches are dug across a gully and vetiver clumps are transplanted closely in the moist soil. Grass barriers can be developed in concentrated flow areas to control ephemeral gully formation. Vetiver and switchgrass can also be effectively used for stabilization of waterways. De Baets et al. (2009) reported that several Mediterranean shrub and tree species are also suitable for developing vegetative barriers for the control of gully erosion.

Study Questions
1. What do you mean by steep slopes? Comment on slope gradient, slope classes and shapes. Explain why steep slopes are naturally unstable.
2. Give an account of slope failure and mass movement. Discuss the factors that affect stability of soils on steep slopes.
3. What are the types of land and soil disturbances on steep slopes?
4. Briefly describe the methods of management of soils on steep slopes.
5. What do you mean by agroforestry? Discuss the advantages of Sloping Agricultural Land Technology and Alley Cropping.

References

Baudry J, Bunce RGH, Burel F (2000) Hedgerows: an international perspective on their origin, function and management. J Environ Manag 60(1):7–22
Billi P, Dramis F (2003) Geomorphological investigation on gully erosion in the Rift Valley and the northern highlands of Ethiopia. Catena 50:353–368
Blanco H, Lal R (2008) Principles of soil conservation and management. Springer, New York
Blanco-Canqui H, Gantzer CJ, Anderson SH et al (2004) Soil berms as an alternative to steel plate borders for runoff plots. Soil Sci Soc Am J 68:1689–1694
Bokhtiar SM, Karim AJM, Khandaker Majidul Hossain T, Hossain T, Egashira K (2001) Response of radish to varying levels of irrigation water and fertilizer potassium on clay terrace soil of Bangladesh. Commun Soil Sci Plant Anal 32(17–18):2979–2991
Crosta GB, Frattini P (2003) Distributed modeling of shallow landslides triggered by intense rainfall. Nat Hazards Earth Syst Sci 3:81–93
De Baets S, Poesen J, Reubens B, Muys B, De Baerdemaeker J, Meersmans J (2009) Methodological framework to select plant species for controlling rill and gully erosion: application to a Mediterranean ecosystem. Earth Surf Process Landf 34:1374–1392
DeRose R (1995) Slope limitations to sustainable land use in hill country prone to landslide erosion. Unpublished Report, Landcare Research, Palmerston North, New Zealand
Drechsler M, Settele J (2001) Predator-prey interactions in rice ecosystems: effects of guild composition, trophic relationships, and land use changes – a model study exemplified for Philippine rice terraces. Ecol Model 137:135–159
FAO (2000) Manual on integrated soil management and conservation practices. FAO Land and Water Bulletin 8, Rome, Italy
Farek G, Lloyd-Reilley J (2000) Gully erosion control with vegetative barriers. Land and Water Magazine, March/April 2000, Ft. Dodge, IA

FFTC (2004) Soil conservation practices for slopelands. Food and Fertilizer Technology Center. [online] URL: http://www.agnet.org/library/abstract/pt2001024.html

Gafur A, Borggaard OK, Jensen JR, Petersen L (2000) Changes in soil nutrient content under shifting cultivation in the Chittagong Hill Tracts of Bangladesh. Danish J Geogr 100:37–46

Gafur A, Jensen JR, Borggaard OK, Petersen L (2003) Runoff and losses of soil and nutrients from small watersheds under shifting cultivation (Jhum) in the Chittagong Hill Tracts of Bangladesh. J Hydrol 279:293–309

Gebremichael D, Nyssen J, Poesen J, Deckers J, Haile M, Govers G, Moeyersons J (2005) Effectiveness of stone bunds in controlling soil erosion on cropland in the Tigray highlands, northern Ethiopia. Soil Use Manag 21:287–297

Girty GH (2009) Landslide. In: Perilous earth: understanding processes behind natural disasters, ver. 1.0, Department of Geological Sciences, San Diego State University

GPA (2004) Sediment mobilization, upstream erosion and agriculture. Global Programme of Action for the Protection of the Marine Environment from Land-based Activities. [online] URL: http://www.fao.org/gpa/sediments/coastero.htm

Hansen WF, Law DL (2004) Sediment from a small ephemeral gully in South Carolina. In: Proceedings, 3rd international symposium on gully erosion, University of Mississippi, Oxford MS

Hossain MM, Hossain KL, Miah MM, Hossain MA (2006) Performance of wheat cultivars as understory crop of multipurpose trees in taungya system. J Biol Sci 6:992–998

ICIMOD (1998) Bioterracing & soil conservation. Issues in Mountain Development (1998/7). ICIMOD. Kathmandu, Nepal. [online] URL: http://www.icimod.org.np/publications/imd/imd98-7.htm

Isaac L, Wood CW, Shannon DA (2003) Pruning management effects on soil carbon and nitrogen in contour hedgerow cropping with Leucaena leucocephala (Lam.) De Wit on sloping land in Haiti. Nutr Cycl Agroecosyst 65:253–263

Jakob M, Wetherly H (2003) A hydroclimatic threshold for initiation on the North Shore Mountains of Vancouver, British Columbia. Geomorphology 54:137–156

Kasai M, Marutani T, Reid LM, Trustrum NA (2001) Estimation of temporally averaged sediment delivery ratio using aggadational terraces in headwater catchments of the Waipaoa river, North Island, New Zealand. Earth Surf Process Landf 26:1–16

Keller G, Sherar J (2003) Low volume roads engineering, best management practices field guide. USDA, Forest Service, Virginia Polytechnic Institute and State University

Kerkhoff E, Sharma E (2006) Debating shifting cultivation in the eastern Himalayas. Farmers' innovations as lessons for policy. International Centre for Integrated Mountain Development (ICIMOD), Kathmandu, Nepal

Lasanta T, Arnaez J, Oserin M, Ortigosa LM (2001) Marginal lands and erosion in terraced fields in the Mediterranean mountains. Mt Res Dev 21(1):69–76. Temple PH (1972) Measurements of runoff and soil erosion at an erosion plot scale with particular reference to Tanzania. Geografiska Ann 54-A:203–220

Lehigh Valley Planning Commission (2008) Steep slopes: guide/model regulations. Pennsylvania Department of Conservation and Natural Resources, Bureau of Recreation and Conservation, USA

MBRLC (1988) A manual on how to farm your hilly land without losing your soil. Mindanao Baptist Rural Life Center, Davao del Sur

Morgan RPC (2005) Soil erosion and conservation, 3rd edn. Blackwell Publishing, Malden, MA

NRCS (2002) Conservation practice standard: hedgerow planting. Code 422, Natural Resources Conservation Service, New Jersey. efotg.sc.egov.usda.gov/references/public/NJ/NJ422.pdf

Nyssen J, Poesen J, Haile M, Moeyersons J, Deckers J (2000) Tillage erosion on slopes with soil conservation structures in the Ethiopian highlands. Soil Tillage Res 57:115–127

Poesen J (2011) Challenges in gully erosion research. Landf Anal 17:5–9

Ranjan R, Upadhaya VP (2006) Ecological problem due to shifting cultivation. Ministry of Environment and Forests, Eastern Regional Office, Bhubaneswar, India

References

Rickson RJ (2006) Controlling sediment at source: an evaluation of erosion control geotextiles. Earth Surf Process Landf 31:550–560

Sanchez PA (1996) Alternative to slash-and-burn agriculture. Agric Ecosyst Environ Special Issue 58:1–2

Sharda VN, Juyal GP, Prakash G, Joshi BP (2007) Training manual: soil conservation & watershed management, vol-II. Central Soil & Water Conservation Research & Training Institute, Dehradun, Uttaranchal, India

Soil Survey Division Staff (1993) Soil survey manual, US Department of Agriculture Handbook 18. Soil Conservation Service, Washington, DC

Swanston DN, Howes DE (1994) Slope movement processes and characteristics. In: Chatwin SC, Howes DE, Schwab JW, Swanston DN (eds) A guide for management of landslide prone terrain in the Pacific Northwest, Land Management Handbook 18. British Columbia Ministry of Forests, Victoria, BC

USDA (2007) Gullies and their control, technical supplement 14P. National Engineering Handbook Part 654 (210–VI–NEH, August 2007)

USGS (2015) 1980 Cataclysmic Eruption. US Geological Survey, US Government. [online] URL: http://volcanoes.usgs.gov/volcanoes/st_helens/st_helens_geo_hist_99.html

Van Dijk AIJM, Bruijnzeel LA (2003) Terrace erosion and sediment transport model: a new tool for soil conservation planning in bench-terraced steeplands. Environ Model Softw 18:839–850

Wheaton RZ, Monke EJ (2001) Terracing as a "best management practice" for controlling erosion and protecting water quality. Agricultural Engineering 114, Purdue University. [online] URL: http://www.agcom.purdue.edu/AgCom/Pubs/AE/AE-114.html

Woldearegay K (2013) Review of the occurrences and influencing factors of landslides in the highlands of Ethiopia: with implications for infrastructural development. Momona Ethiop J Sci 5(1):3–31

Wu TH (2005) Slope stabilization. In: RPC M, Rickson RJ (eds) Slope stabilization and erosion control: a bioengineering approach. E & FN SPON, an imprint of Chapman & Hall, London

Chapter 9
Poorly Fertile Soils

Abstract The capacity of soils to supply plant nutrients in available forms and proper balance, and in the absence of any toxicity, is known as soil fertility. All soils do not have enough capacity to provide plants with optimum nutrients required for their normal growth and development. Many soils are deficient in one or more nutrients. These soils are poorly fertile soils. Major causes of poor soil fertility include shallow depth, coarse texture, poor soil structure, high erosion, low organic matter, low activity clay, low CEC and base saturation, unfavorable chemical environment such as acidity, alkalinity, salinity, sodicity, pollution, etc. and P-fixation. Some soils are naturally poorly fertile and some soils are impoverished by soil mismanagement. Improvement and restoration of soil fertility for sustainable crop production in these soils need integrated soil and crop management efforts. The incorporation of organic residues along with chemical fertilizers, biochar amendment, green manuring, inclusion of a legume in the crop sequence, intercropping, crop rotation, cover crops, residue management and conservation tillage, liming an acidic soil, crop-livestock integration are needed in a concerted manner. No single method is enough for the management of poorly fertile soils.

Keywords Plant nutrients · Soil fertility · Nutrient depletion · Fertilizers · Organic fertilizers · Manures · Composting · Industrial fertilizers · Mixed fertilizers · Liquid fertilizers · Fertilizer application · Fertilizer losses

9.1 Soil Fertility and Plant Nutrients

Soil fertility refers to the capacity of soils to supply adequate nutrients to plants in available forms and in appropriate proportion in the absence of any sort of toxicity. Soil fertility can be natural or acquired. Natural soil fertility is the manifestation of the complex and dynamic interactions of soil physical, chemical and biological properties. It is a state of the soil system evolved by the natural processes of soil formation. Some soils are inherently poorly fertile, such as sandy soils; sands hold fewer and leach out more nutrients. In contrast, well-structured soils are more fertile; plant roots can extend to and draw nutrients from larger volumes of soil there. Soils containing high amount of easily weatherable minerals, the dominance of 2:1

type clay minerals, a pH around 6.5, high cation exchange capacity, high base saturation percentage, high organic matter content, and the absence of acidity, alkalinity, salinity, sodicity, and pollution are usually fertile. A fertile soil has a high biological activity. On the other hand, acquired soil fertility is the capacity of the soil to provide plants with adequate nutrients through human interventions. Of the seventeen chemical elements or nutrients, higher plants absorb fourteen from the soil: nitrogen (N), phosphorus (P), potassium (K), calcium (Ca), magnesium (Mg), sulfur (S), iron (Fe), manganese (Mn), copper (Cu), molybdenum (Mo), zinc (Zn), boron (B), chlorine (Cl) and nickel(Ni). These elements are present in soil in various forms (insoluble, bound, fixed, soluble and exchangeable), but plants can absorb them readily in the ionic forms (soluble and exchangeable) only. If the soil contains inadequate amount of any one or more of these nutrients, plants will suffer in growth and reproduction, and it can be said that the soil is not fertile. We can add the deficient nutrient into the soil as fertilizer and make the infertile soil fertile. If the soil is adequately fertile, plants may grow satisfactorily or not, depending on the provision of other plant requirements from the soil, including air, water and temperature, and management. If the soil is fertile but not appropriately managed, the production of crops may be low. Soil fertility is an index of nutrient availability to plants, and it is one of the factors for crop production. There are other factors that make a fertile soil productive. By appropriate management, infertile soils may be made fertile and productive. However, one should remain aware of the need of balanced status of nutrients in soil during soil fertility management; excess of some elements (Fe, Mn, Cu and Zn, particularly) may be toxic to plant roots, and excess of any one element may cause deficiency of another (for example, excess of N may cause K deficiency in some crops).

Plant tissues may contain more than 90 chemical elements; most of them are not actually needed for plant growth and development. Only about 22 elements have been reported to have specific roles on the physiology of plants. Seventeen nutrients have already been mentioned; other five elements – silicon (Si), iodine (I), vanadium (V), cobalt (Co), and sodium (Na) – have been recognized as beneficial elements to some plants (Marschner 1995). Nutrients are grouped into macronutrients and micronutrients, depending on the relative amounts of requirement. Macronutrients (C, H, O, N, P, S, K, Ca and Mg) are needed in large amounts (>1000 mg kg^{-1} shoot dry matter) and micronutrients (Fe, Mn, Cu, Mo, B, Zn, Cl and Ni) are needed in relatively small amounts (<100 mg kg^{-1} shoot dry matter) (Marschner 1995). Some chemical elements are essential for humans and farm animals, and they get these elements mainly from plant food materials. The following elements are essential for farm animals and humans:

Humans (19): C, H, O, N, P, K, Ca, Mg, S, Fe, Mn, Zn, Cu, Mo, Cl, Na, I, Co, and F.
Farm animals (18): C, H, O, N, P, K, Ca, Mg, S, Fe, Mn, Na, Zn, Cu, Mo, Cl, I, and Co.

Human and animal health may be adversely affected if these elements are not present in soil in adequate amounts. Food and Agriculture Organization of the United Nations Statistics consider that more than 99.7% of human food (calories) comes from the land (FAO 2004), while less than 0.3% comes from the marine and aquatic ecosystems. Maintaining and augmenting the world food-supply basically

9.1 Soil Fertility and Plant Nutrients

depend on the productivity and quality of all agricultural soils (Pimentel and Burgess 2013). Some examples of how soil quality affects human and animal health are given here. Soils in many Southeast Asian countries including Bangladesh are deficient in iodine (iodine is not an essential element for plants, but it is essential for humans); many people suffer from goiter there. To combat the situation, common salt is being supplemented with iodine. Ruminant animals such as cows, sheep, goats and deer can produce vitamin B_{12} if there is adequate cobalt in the diet (mainly grasses and forages although Co is not an essential element for plants) (Brunetti 2005). As the inadequacies of cobalt in soil do not affect plant production, Co may be supplemented with animal feed. Soil quality also features in any discussion of soil fertility; and the quality of crops is certainly relates to human health and nutrition.

A nutrient can be present in soil in several forms, such as dissolved in soil solution (soil water with dissolved ions, salts, acids, and bases), exchangeable cations and anions on soil colloidal surfaces, insoluble compounds, soil organic matter and soil minerals. The sum of all these forms is the total nutrient content. The average concentration of different nutrients in dry plant tissue and their available forms are given in Table 9.1 (adapted from Marschner 1995).

As pointed out earlier, only the soluble and exchangeable forms constitute the available nutrients. Available nutrients in soil are extracted with various extractants; common extractants are dilute solutions of salts, acids, alkalis or their mixtures. A small proportion of total nutrient is available to plants. For example, total P content in soils may lie between <400 and >1000 mg kg^{-1}, but available P usually remains below 25 mg kg^{-1} (only 2.5–5% of total). Plants obtain nutrients readily from the

Table 9.1 Plant nutrients and their available forms

Nutrient	% dry plant tissue	Available forms
C	45	CO_2
O	45	H_2O, O_2
H	6	H_2O
N	1.5	NH_4^+, NO_3^-, N_2^*
P	0.2	$H_2PO_4^-$, HPO_4^{2-}, PO_4^{3-}
K	0.2	K^+
Ca	0.5	Ca^{2+}
Mg	0.2	Mg^{2+}
S	0.1	SO_4^{2-}
Fe	0.01	Fe^{2+}, Fe^{3+}
Mn	0.005	Mn^{2+}
Cu	0.0006	Cu^{2+}
Mo	0.00001	MoO_4^{2-}
Zn	0.002	Zn^{2+}
B	0.002	H_3BO_3, BO_3^-, $B_4O_7^{2-}$
Cl	0.01	Cl^-
Ni	0.0001	Ni^{2+}

soil solution. The concentration of available nutrients in soil solution and the replenishing of the soil solution in the rhizosphere are of tremendous importance to plant nutrition. Wild (1996) lists concentrations of some nutrients in soil solution as 5–200 mg L^{-1} NO$_3$-N, 10–100 mg L^{-1} SO$_4$-S, 0.01–0.60 mg L^{-1} H$_2$PO$_4$-P, 10–200 mg L^{-1} Ca^{2+}, 5–100 mg L^{-1} Mg^{2+} and 1–40 mg L^{-1}K$^+$.

There are several methods of soil nutrient analysis, popularly known as soil tests. There are also methods of plant analysis and field trials for evaluating the fertility status of a soil. These methods are not described here because it is beyond the scope of the present topic. But, without some kind of analysis or observation, it is difficult to decide whether a soil is poorly fertile. Some methods are mentioned in Box 9.1.

Box 9.1 Methods for Soil Fertility Evaluation
1. Visual deficiency symptoms of plants
2. Biological tests:

    (a) Azotobacterplaque test
    (b) Aspergillusnigertest
    (c) Cunninghamellaplaque test
    (d) Carbon dioxide evolution method.

3. Chemical methods
    (a).   Qualitative and quantitative analyses:

        (i). Rapid test:soil test and plant tissue test
        (ii). Total analyses of soil and plant tissue

4. Experimental methods

    (a) Pot culture experiment under uncontrolled condition
    (b) Green houseexperiment under controlled condition
    (c) Nutrient injection and foliar sprays.

Plants need adequate supply of nutrients for satisfactory growth and yield. But, the concept of 'adequate nutrient supply' is itself arbitrary. All plants do not need the same amount of any nutrient; plants widely differ in their requirements of nutrients; even the same plant needs different amounts of nutrients at different stages of growth. Moreover, two different soils with the same nutrient level would supply different amounts of nutrients to plants if there are differences in physical and chemical conditions of the soils that affect root growth and nutrient uptake. So, a generalization of 'adequate nutrient level' is difficult. The soil testing laboratories in different parts of the world publish soil test results as low, medium and high levels for fertilizer recommendations (Table 9.2). These recommendations are crop specific, and often based on yield target, but the categories generally mean that positive crop response to added fertilizers is likely when the level of the nutrient is low or medium. On the other hand, one can get little response at high levels. The high levels indicate better soil fertility.

Table 9.2 Low, medium and high levels of different nutrients in soil

Nutrient	Levels of nutrients in soil		
	Low	Medium	High
	Percent		
Total N	<0.2	0.2–0.5	>0.5
Available nutrients	**mgkg^{-1}**		
NO_3-N	<10	10–20	>20
P (Olsen)	<8	8–16	>16
K	<150	150 – 250	>250
S (SO_4-S)	<10	10–20	>20
Ca	<400	400–1000	>1000
Mg	<24	24-60	>60
Fe	<2.5	2.5–5.0	>5.0
Mn	<0.5	0.5–1.0	>1.0
Cu	<0.25	0.25–0.50	>0.50
B	<0.5	0.5–1.0	>1.0
Zn	<0.25	0.25–0.5	>0.5

Sources: Jacobsen et al. (2002); Horneck et al. (2011); Herrera (2000)
Hill Laboratories (www.hill-laboratories.comfilefileid15530)

9.2 Poorly Fertile Soils

Soils are said to be poorly fertile if they do not have the capacity to supply one or more of the nutrients in requisite amounts in available forms and if there is any toxicity. Some soils are inherently poorly fertile; some of them have coarse texture and very low clay and organic matter contents, so that they are highly leached and cannot retain nutrients; some of them are low in easily weatherable minerals or have low reactive clays. Some soils can be shallow, so that plant roots cannot gather sufficient nutrients from enough soil volume. Major constraints to soil fertility are mentioned in Box 9.2.

Box 9.2 Major Constraints to Soil Fertility

Shallow soil: Shallow soils on bed rock, soils with root restrictive layers, soils with hard pan at a shallow depth, compacted soils.

Coarse texture: Sandy or loamy sand soils that contain low nutrients, subject to high leaching and low nutrient retention.

Poor soil structure:	Low aggregate stability, highly porous or impervious, very low or very high bulk density, unfavorable for root growth.
High erosion:	Sloping soils without vegetation covers, steep slopes, gullies, landslides.
Low organic matter:	Organic matter content less than 2%. Sole nitrogen and about half of P and S, and some micronutrients come from organic matter. Organic matter improves soil structure.
Low activity clay:	Clay mineralogical composition is very important in relation to soil fertility. Reactive clays affect almost all soil characteristics. Sesquioxide clays, kolinite, illite, montmorillonite and vermiculite clays have 0-3, 3–15, 25–40, 60–100, and 80–150 cmolc kg^{-1} CEC respectively. Soils dominated by sesquioxide and kaolinite are usually poorly fertile.
Low CEC:	Plants obtain a portion of their nutrients from exchangeable cations. Besides this, a low CEC (<10cmolc kg^{-1}) indicates that the soil has a low capacity to retain nutrients against leaching.
Low base saturation:	Exchangeable bases are available nutrients (except Na). A fertile soil should have 68% Ca^{2+} (percent of total CEC), 12% Mg^{2+}, 5% K^+, 1% Na^+, 8% H^+, and 6% other cations. Low exchangeable K^+ is a serious problem in some soils.
Low nutrient levels:	Major nutrients (N, P, K) tend to be deficient in most soils (Coyne and Thompson 2006) because of crop removal. But, often deficiencies of micronutrients such as Fe, Zn, Mn, and Cu are observed.

Unfavorable chemical environment:	Soil acidity, alkalinity, salinity, sodicity, and acid sulfate conditions are unfavorable for soil fertility.
P fixation:	P fixation is a major problem in acidic, alkaline and metalliferous soils.
Toxicity:	Organic and inorganic toxic substances may accumulate in soils under different situations. Organic toxins may accumulate in anaerobic conditions. Al-toxicity occurs mainly in strongly acidic soils of the humid tropics. A level of 1.0 cmolc kg-1 1 N KCL extractable Al is highly toxic to plant roots. Micronutrient toxicity (Fe, Mn) may occur in acidic soils and soils of mining areas.

According to Hartemink (2006), nutrient budget studies (based on inputs such as addition in biomass residues, atmospheric input, weathering, fertilizers, and outputs such as crop removal, grazing, erosion, leaching, etc.) could reveal (i) a positive balance or nutrient build-up, (ii) a negative balance or nutrient depletion, and (iii) a zero balance or a steady state of a dynamic equilibrium of nutrients. Fig. 9.1 shows the processes of nutrient inputs and outputs.

Inputs > Outputs = Nutrient build-up; Inputs < Outputs = Nutrient depletion; Inputs = Outputs = Steady state. There are some reports of nutrient build up in soils of several European countries and Russia (Giani et al. 2004) through high manure and fertilizers inputs. There are also several examples in the tropics where soil fertility has built up as a result of long-term applications of organic materials including household waste and manure (FAO 2001a; Lima et al. 2002). However, nutrient build-up is not entirely environmentally safe because nutrients may be leached from these soils into surface water and groundwater causing their pollution (Hartemink 2006).

The processes of nutrient inputs in soil include weathering of rocks and minerals, atmospheric deposition, mineralization of native organic matter, decomposition of roots, soil organisms and litter, and application of organic residues, fertilizers and lime. On the other hand, the pathways of nutrient outputs include loss by leaching and erosion, volatilization and denitrification, sorption, immobilization, burning of residues, grazing and nutrient export through the harvest of crops. If nutrient inputs exceed nutrient outputs then there is a positive nutrient balance or nutrient build up

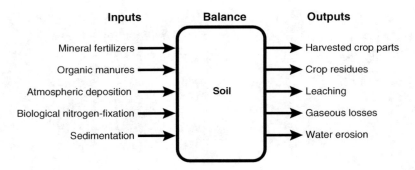

Fig. 9.1 Processes of nutrient inputs and outputs in soil

Table 9.3 Nutrient balances in some African countries

Countries	Nitrogen (kg ha^{-1} year^{-1})	Phosphorus (kg ha^{-1} year^{-1})	Potassium (kg ha^{-1} year^{-1})
Benin	−16	−2	−11
Botswana	−2	0	−2
Cameroon	−21	−2	−13
Ethiopia	−47	−7	−32
Ghana	−35	−4	−20
Kenya	−46	−1	−36
Malawi	−67	−10	−48
Mali	−11	−2	−10
Nigeria	−37	−4	−31
Rwanda	−60	−11	−61
Senegal	−16	−2	−14
Tanzania	−32	−5	−21
Zimbabwe	−27	2	−26

(Source: http://www.fao.org/docrep/006/y5066e/y5066e06.htm)

in the soil. But often nutrient outputs exceed inputs resulting in soil fertility decline. Some investigators (Drechsel et al. 2001; Dougill et al. 2002) reported widespread soil fertility decline in tropical countries, particularly Sub-Saharan Africa and the principal causes of this decline are unsustainable farming activities (Sanchez 2002; FAO 2003). The International Fertilizer Development Corporation (2003) estimated that the average losses of nitrogen, phosphorus and potassium from cultivated land might be 22, 2.5 and 15 kg ha^{-1} respectively. African farmers use very low amounts of fertilizers in comparison to other regions; for example, while only about 9 kg ha^{-1} is used in Africa, 86 kg ha^{-1} is used in Latin America, 104 kg ha^{-1} in South Asia, and 142 kg ha^{-1} in Southeast Asia (Kelly 2006). Tan et al. (2005) reported the estimated average global soil nutrient deficits in the year 2000 to be 18.7 kg N ha^{-1} year.$^{-1}$, 5.1 kg P ha^{-1} year.$^{-1}$, and 38.8 kg K ha^{-1} year.$^{-1}$ and the annual total nutrient deficits were 5.5 Tg N (1 Tg = 10^{12} g), 2.3 Tg P, and 12.2 Tg K). Nutrient balances in soils of some countries of Africa are given in Table 9.3.

9.2.1 *Processes of Nutrient Depletion*

9.2.1.1 Leaching

Nutrient leaching is the downward movement of dissolved nutrients in the soil profile with percolating water. These nutrients go below the root zone and are lost from the system at least temporarily. Leached nutrients may contribute to groundwater contamination in regions of intensive agriculture (Lehmann and Schroth 2003). Nutrient leaching from agricultural soils is a worldwide problem that has been implicated in deleterious impacts on the environment, particularly pollution of adjacent water systems (Yoo et al. 2013). Intensive leaching of nutrients results in low soil pH, especially when this occurs in degraded soils (Widowati and Utomo 2014). The transport of N and P from agricultural soils to groundwater through leaching is a potential risk to human health. Most of the N is leached as NO_3-N, which does not absorb to soil particles and is, therefore, more likely to be transported to subsurface tile drainage than in surface runoff (Owens et al. 2000; Zhao et al. 2001). The amount of nitrogen leached out from the rooting zone can be up to 80% (Lehmann et al. 2003). In general, tillage is expected to hasten decomposition of residues, resulting in more N mineralization and nitrification. Thus, more $NO3-N$ loss is expected in plowed than no-till soils (Power et al. 2001). Webb et al. (2010) estimated the vulnerability of soils to the leaching of nitrogen and phosphorus. Very low vulnerability to nitrogen leaching is confined to poorly drained soils that have significant attenuation of nitrogen through denitrification (Organic and Gley Soils). High to very high vulnerability to phosphorus leaching is confined to recent soils, stony or very stony soils, sand dunes, and shallow soils overlying rock. Extensive areas of land with moderate vulnerability consist mainly of deep loamy soils.

de Oliveira et al. (2002) estimated the amounts of leaching loss of N, K, Ca and Mg from a sandy soils cultivated with sugarcane. The mean of leached N during the experimental period of 11 months was of 4.5 kg ha^{-1}. The mean losses of K$^+$, Ca^{2+}, and Mg^{2+} were of 13, 320 and 80 kg ha^{-1} respectively. Li et al. (2008) conducted an in situ field experiment with lysimeter to study the effects of different fertilizations on the nutrient leaching loss from brown soil in the growth season of summer maize. Abundant rainfall and irrigation were the main factors affecting the leaching loss. The leaching amount was higher in the early growth period of summer maize, but decreased after then. The accumulative leaching loss of available P was only 0.148–0.235 kg ha^{-1}, while that of available K was 7.08–13.00 kg ha^{-1}. In the late growth period of summer maize, wheat stalk plus N application increased the leaching loss of soil available P and K while nitrogen application affected it slightly.

9.2.1.2 Erosion

Soil erosion is a serious global environmental problem. Soil erosion not only reduces soil fertility and crop productivity but also results in wasting of crop land. According to an estimate, the annual loss of cropland due to soil erosion may be about 10 million ha resulting in the reduction of available land for food production. The loss of cropland would aggravate food insecurity and according to the estimate of World Health Organization and the Food and Agricultural Organization, two-thirds of the world population is malnourished (Pimentel and Burgess 2013). Eswaran et al. (2002) reported an estimated annual loss of fertile soil from world agricultural systems of approximately 75 billion tons. The Central Soil Water Conservation Research and Training Institute in Dehradun, India, estimated the average annual soil loss of 16.4 t ha^{-1} and an annual total loss of 5.334 billion tonnes due to erosion in India (Anonymous 2010). Pimentel and Burgess (2013) stated that eroded soil carries away vital plant nutrients, such as nitrogen, phosphorus, potassium, and calcium. Sukristiyonubowo et al. (2002) examined nutrient losses by erosion in some catchments of Indonesia, the Philippines, Vietnam and Laos from the production of sediment yield and nutrient concentration in the sediment. They found that substantial amount of nutrients were lost in all the catchments with sediments. According to Pimentel (2006), nutrient resources are so depleted through erosion that plant growth is stunted and the overall productivity declines. Schick et al. (2000) observed that usually high losses of Ca and Mg occur due to water erosion. Since these elements are strongly adsorbed by soil colloids, they are easily transported with the sediment. Gafur et al. (2000) reported the loss of 61 kg Ca, 13 kg Mg, 13 kg K, 0.14 kg P, 0.20 kg S, 0.05 kg Cu, 6.7 kg Fe, 6.1 kg Mn and 0.065 kg Zn per hectare due to water erosion in an area of shifting cultivation in Bangladesh.

9.2.1.3 Crop Removal

Crop plants absorb nutrients from the soil for their vegetative and reproductive growth, and the amounts of nutrients taken up by plants depend on the characteristics of soil, climate and management of soil as well as crop. Among soil properties, nutrient availability, moisture content, aeration, compaction, soil temperatures, pH, salinity, etc. influence nutrient absorption the most. However, a portion of the absorbed nutrients is exported from the soil with harvested crops. Table 9.4 gives some examples of the amounts of nutrients removed by harvested crops. If the crop residues are also harvested for livestock feed or fuel, the amount of nutrients removed by a crop may be more than two times. On an average, harvested crops remove 50–100 kg ha^{-1}, 10–20 kg ha^{-1}, and 15–30 kg ha^{-1} N, P and K respectively. According to Binford (2010), the estimated three-year grain removal of P by corn was estimated to be 70–175 kg ha^{-1}.

9.3 Management of Poorly Fertile Soils

Table 9.4 Nutrient removal by harvested crops

Crops	N	P	K	S
	←	kg ha^{-1}	→	
Grain crops,				
Spring wheat	60–75	10–12	15–17	5–6
Winter wheat	60–65	11–13	14–17	7–9
Barley	80–95	15–18	20–25	7–9
Oats	60–76	12–14	15–18	5–6
Rye	60–75	11–14	15–20	5–6
Corn	95–145	18–22	22–28	7–8
Oilseed crops				
Canola	85–100	16–20	15–18	12–14
Flax	60–70	7–8	12–15	6–7
Sunflower	65–75	7–8	10–12	5–6
Pulses				
Pea	135–145	14–18	30–35	6–7
Lentil	62–74	8–10	27–35	4–5
Other crops				
Sugar beets	85–110	16–22	120–150	12–14
Potatoes	125–155	16–20	185–225	11–13
Forage crops				
Alfalfa	290–350	30–37	260–320	27–33
Clover	220–260	25–30	175–218	10–12
Barley silage	145–220	22–30	110–130	14–21
Corn silage	170–200	27–35	175–220	12–14

Converted from data compiled by the Canadian Fertilizer Institute from agronomic information obtained in Canada, 1998
http://www.cfi.ca/_documents/uploads/elibrary/d161_NU_W_01%5B1%5D.pdf [Accessed 3.1.2012]

9.2.1.4 Unbalanced Fertilizer Application

Unbalanced fertilizer application has been evidenced by long-term fertilizer consumption records of the FAOSTAT database (FAO 2001b) revealing that fertilizers are used by the farmers in an unbalanced proportion. The proportion of Nitrogen fertilizer use is increasing and the proportion of phosphorus and potassium is decreasing as evidenced from fertilizer data from 1961 to 2000. Moreover, the proportion of N fertilizer use has remained high in both developing and least developed countries (ranging between 73% and 77% and from 79% to 83% respectively), while potassium fertilizer use remained very low (between 10% and 13% and between 7% and 9% respectively). On the other hand, Tan et al. (2005) reported that a decline in average fertilizer use has taken place in developed countries (from 32.6 kg ha^{-1} year.$^{-1}$ in the year 1980 to 17.2 kg ha^{-1} year.$^{-1}$ in the year 2000), but the ratio of N, P and K fertilizers remained stable.

9.2.1.5 Residue Burning

On burning biomass residues, most nitrogen and sulfur contained in biomass are oxidized and lost into the atmosphere as volatile gases. On the other hand, most phosphorus and potassium are retained. Heard et al. (2001) described the results of a burning experiment in an uncovered container of wheat, oat and flax residues and obtained the findings mentioned above. In the experiment, some losses of P (21 percent) and K (35 percent) occurred due to the escape of smoke and ashes from the uncovered container. Loss of nitrogen and sulfur amounted to 98–100 percent and 75 percent respectively. Shifting cultivators traditionally slash the vegetation on hill slopes and burn their biomass residues to avoid competition for light, water and nutrients. The ashes add nutrients to the soil and promote crop growth. But such benefits are very temporary because the rain that follows land preparation or seeding washes the ashes and base nutrients away. Moreover, soil erosion increases when the vegetation cover is removed.

9.3 Management of Poorly Fertile Soils

Poorly fertile soils need careful management to arrest further decline in soil fertility and to improve it sustainably so that crop productivity increase gradually. To achieve these objectives an integration of soil and crop management practices is necessary. For poorly fertile soils African Soil Health Consortium (2012) adopted an integrated soil fertility management strategy which involves a set of practices that include the use of fertilizers, organic residues and improved crop varieties. All these practices must be adaptable to local conditions so that local farmers accept the technologies without hesitation. The management of inputs must also be based on some general and specific principles. The specific principles are problem oriented.

The general management principles include application of organic fertilizers, such as manures, composts, oil cakes, other non-conventional organic fertilizers and green manure. Farmers should adopt such practices as cover crops, crop rotation, conservation tillage, mulching, and soil conservation practices. Irrigation and drainage facilities should be made to avoid drought and waterlogging problems respectively. The main objective should be to build up organic matter and nutrients and restore fertility. The specific management principles include certain management options that may be necessary based on the causes of poor soil fertility. Some of such problems and management options are: (a) Unsuitable soil textures, such as sandy soils which are inherently poorly fertile and can hardly retain added nutrients; if, in contrast, the soil contains very high amount of clay, it tends to become waterlogged, and sticky as well as puddled when tilled. Soil texture is not generally changeable, so their effects may need to be modified by amendments like organic residues, biochar and other soil modifiers, irrigation, drainage and fertilizers, (b) If the soil has poor aggregation and porosity, the addition of organic residues, conservation tillage, lime and fertilizer application may be necessary, (c) If there is erosion

9.3 Management of Poorly Fertile Soils

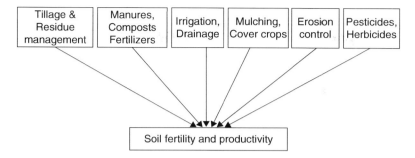

Fig. 9.2 Integrated practices for sustainable soil fertility management

hazard, engineering and agronomic soil conservation measures, cover crops, and conservation farming would be necessary to restore soil fertility, (d) The management options for soils susceptible to high leaching are addition of organic residues, low and frequent irrigation, low and frequent fertilizer application, (e) Fertility of soils containing low CEC and base saturation can be amended with organic fertilizers, biochar or bentonite, and (f) Improper soil chemical conditions, such as acidity, alkalinity, salinity, sodicity, etc. can be reclaimed with lime for acid soils, gypsum and S for alkaline and sodic soils along with tillage, irrigation and drainage.

The strategies of the Integrated Soil Fertility Management (ISFM) have increased crop productivity (Vanlauwe et al. 2010) and reduced the scale of long experienced environmental degradation in Sub-Saharan Africa (Vanlauwe et al. 2015). The benefits of ISFM mainly resulted from the combined use of organic manure and mineral fertilizers (Zingore et al. 2008), following legume-cereal rotations (Sanginga et al. 2003) and using fertilizer and manure for cereals in semiarid areas in micro-doses (Tabo et al. 2007). Once the crops are selected with some flexibility depending on resource availability and conditions of cropping, the practices shown in Fig. 9.2 are needed for sustainable soil fertility management.

9.3.1 Organic Fertilizers

Two major constraints of poorly fertile soils are their low content of organic matter and their unfavorable physical, chemical and biological conditions. Soil organic matter contains a major proportion of nutrients which are released slowly by the activity of soil microorganisms. Organic matter also retains nutrients against leaching and erosion. It favors root and microbial activity. Thus, soil organic matter is vital for soil fertility and plant growth. So, building up of soil organic matter by the addition of organic residues, organic wastes and organic fertilizers should maintain soil physical, chemical and biological fertility (Izaurralde et al. 2001). Various types of organic residues including straw, litter, corn silage, saw dust, oil cakes, bone meal, hoof meal, fish meal, slaughter house blood, kitchen wastes, etc. can be used. Tahiri and Guardia (2009) suggested that by-products of the processing of animal materials, such as skin, wool, bristle, horns, feathers, hoofs, etc. and waste by-products of livestock, meat, leather and other industries can be composted and used in agricultural production systems. Such use of these wastes may also be a way of their safe disposal to avoid environmental problems. When applied in crop field composted organic materials provide nutrients, increase soil organic matter and make soil properties favorable for crop growth. These materials contain appreciable amounts of nutrients, especially N and P (Eghball 2001; Beltran et al. 2002). Manures and organic residues of plant and animal can be composted using some particular earth worm species such as the dung worms (*Eisenia foetida*) and the red worms/red wigglers (*Lumbricus rubellus*). The earthworms eat the residues, digest them and excrete them as worm casts. When composting is complete after about 3–4 months, the heterogeneous mixture of decomposed materials and wormcasts is called vermicompost which is a rich, black natural fertilizer and soil conditioner. According to Musa et al. (2015), vermicompost can be processed further to get an extract. Vermicompost contain higher nutrient concentration than most other composts. Manures are decomposed excreta of animals (dairy cow, horse, swine, sheep, goat, rabbit, chicken). When excreta of farm animals are decomposed they form farmyard manures (FYM). Farmyard manures and poultry manures are popular organic fertilizers. Organic fertilizers have additional benefit of soil conditioning, such as improving soil structure and increasing water holding capacity (Flavel and Murphy 2006; Hassanpanah and Jafar 2012). Table 9.5 is a list of different conventional, non-conventional and potential organic fertilizers with their nutrient concentrations.

As the nutrient concentrations in organic fertilizers are usually very low, they cannot provide the whole nutrient requirement of a crop. They are also needed in large quantities (20–30 t ha^{-1}); in many situations, particularly in arid regions where biomass productivity is low, the huge amount of raw materials required may not be available. In these cases, a good option may be the combined application of a part of the nutrients as manure and a part as concentrated industrial fertilizers. Amara and Mourad (2013) have recommended a suitable combination of organic and inorganic fertilizers for sustainable agriculture. When organic fertilizers and inorganic fertilizers are applied in combination they maintain soil heath better than inorganic fertilizers alone. Such a practice also improves nutrient use efficiency and can maintain soil nutrient balance besides improving soil conditions, increasing soil organic matter and reducing fertilizer loss (Felix et al. 2012; Conacher and Conacher 1998). Liu et al. (2008) observed reduced N loss and increased N use efficiency of rice by applying combined organic and inorganic fertilizers as total basal dressing. Usman et al. (2015) recommended the combined use of organic and inorganic fertilizers to enhance and sustain maize productivity in Nigeria.

9.3 Management of Poorly Fertile Soils

Table 9.5 Nitrogen, phosphorus and potassium concentrations of different organic fertilizers

Manure	N	P	K
	←———— Percent ————→		
Animal refuse	0.3–0.4	0.04–0.09	0.08–0.25
Cattle dung, fresh	0.4–0.5	0.13–0.17	0.25–0.33
Horse dung, fresh	0.5–1.0	0.17–0.26	0.25–0.83
Poultry manure, fresh	1.0–1.8	0.60–0.77	0.66–0.75
Sewage sludge, dry	2.0–3.5	0.43–2.15	0.17–0.42
Activated Sewage sludge	4.0–7.0	0.90–1.81	0.42–0.58
Cattle urine	0.9–1.2	trace	0.42–0.83
Horse urine	1.2–1.5	trace	1.08–1.25
Sheep urine	1.5–1.7	trace	1.49–1.66
Ash, household	0.5–1.9	0.69–1.81	1.91–9.96
Ash, wood	0.1–0.2	0.34–2.54	1.25–29.88
Rural compost, dry	0.5–1.0	0.17–0.34	0.66–1.00
Urban compost, dry	0.7–2.0	0.39–1.29	0.83–1.66
Farmyard manure, dry	0.4–1.5	0.13–0.39	0.25–1.58
Filter–press cake	1.0–1.5	1.72–2.15	1.66–5.81
Rice hulls	0.3–0.5	0.09–0.22	0.25–0.42
Compost	0.30	0.09	0.33
Corn stover	0.90	0.04	0.42
Cottenseed meal	6.4	1.12	1.41
Green manure, rye	2	0.09	0.83
Green manure, oats	1.3	0.09	0.91
Hoof & horn meal	13.8	0.43	0
Bone meal, raw	7.2	9.68	0
Bone meal, steamed	2.5	11.78	0
Oyster shells	0.2	0.13	0.02
Peat moss	1.9	0.09	0.17
Sawdust	0.2	0.04	0.17
Sewage sludge	15	0.559	0.332

P and K values were calculated from P_2O_5 and K_2O by multiplying with the conversion factors 0.43 and 0.83 respectively
Source: www.extension.usu.edu/files/publications/factsheet/hg-510.pdf, www.soil.ncsu.edu/publications/Soilfacts/AG-439-18/AG-439-18.pdf

There can be three different approaches of integrated use of organic and inorganic fertilizers: (1) applying the whole recommended dose of P as organic fertilizer and adding the remaining amount of N and K as inorganic fertilizer, (2) applying 50 percent N as organic fertilizer and adding remaining amounts of N, P and K as inorganic fertilizer, and (3) applying the available amount of organic fertilizer, calculating the amounts of N, P and K applied and adding the remaining amounts as inorganic fertilizer. Examples of calculation are given in Box 9.3.

Box 9.3 Calculation of Amounts of Compost and Fertilizers Needed for Combined Application

Approach 1. Supplying the whole amount of P in compost and the remaining amount of N and K as supplementary inorganic fertilizers.

If the soil test value indicates low N, and medium P and K and if corn is the crop for which the soil test value based fertilizer recommendation is 100 kg N ha^{-1}, 60 kg P$_2$O$_5$ ha^{-1}, and 80 kg K$_2$O ha^{-1}, then the amount of compost (0.3% N, 0.2% P$_2$O$_5$ and 0.2% K$_2$O) and inorganic fertilizers according to the above approach can be calculated as shown here.

For the whole of P$_2$O$_5$ supply, the amount of compost needed = (100/0.2) × 60 = 30 t ha^{-1}. 30 t compost contains 90 kg N and 60 kg K$_2$0.

So, 10 kg N and 20 kg K$_2$O must be supplemented with inorganic fertilizers.

For 10 kg N, the amount of urea needed is (100/46) × 10 = 21.74 kg (urea contains 46 percent N).

For 20 kg K$_2$O equivalent to (20 × 0.83 =) 16.6 kg K, the amount of muriate of potash is (100/50) × 16.6 = 33, 2 kg (muriate of potash contains 50 percent K).

Therefore, the amounts of fertilizers needed are: 30 t ha^{-1} compost, 21.74 kg ha^{-1} urea and 33.2 kg K ha^{-1} muriate of potash.

Approach 2. Applying 50% N as FYM and providing the remaining N, P and K as supplementary inorganic fertilizers.

Let the fertilizer recommendation be as above and let the composition of FYM be 1% N, 0.6% P$_2$O$_5$ and 1.2% K$_2$O. 50 percent (50 kg ha^{-1}) N will be provided by 5 t FYM ha^{-1}. This 5 t FYM adds 30 kg P$_2$O$_5$ and 60 kg K$_2$O. The remaining 50 kg N, 30 kg P$_2$O$_5$ and 40 kg K$_2$O will have to be supplied as inorganic fertilizers.

50 kg N will be contained in (100/46) × 50 = 108.7 kg urea.

30 kg P$_2$O$_5$ is equivalent to 30 × 0.43 = 12.9 kg P; 12.9 kg P will be cotained in (100/20) × 12.9 = 64.5 kg TSP (TSP contains 20 percent P).

40 kg K$_2$O (40 × 0.83) = 33.2 kg K = (100/50) × 33.2 = 66.4 kg muriate of potash.

Approach 3. Applying whatever amount of compost is available. The amounts of N, P, and K likely to have been added from average composition of the compost are to be estimated. The remaining nutrients of the fertilizer recommendation for the particular soil and crop will have to be added as inorganic fertilizers.

9.3.2 Biochar Amendment

Biochar is a black charcoal like substance obtained through the pyrolysis of biomass materials. Pyrolysis is a thermo-chemical process of biomass combustion in absence of air. Almost all biomass materials, organic residues, manures and composts can be converted into biochar. Biochar is produced commercially and nowadays a cheap and affordable technology at farmers' level. Wood chips, straw, rice husk, manures, litter, and other biomass materials are heated at a temperature between 300 and 700 °C in a closed chamber with limited or no oxygen supply. The product looks like charcoal but it is different in that it contains mainly aromatic carbon compounds that are relatively resistant to biological degradation (Lehmann and Rondon 2006; Zhu et al. 2015). The properties and functions of biochar largely depend on the type of raw materials used and the conditions of pyrolysis (Nguyen et al. 2004; Bonelli et al. 2010). Biochar has recently become a very important soil amendment because it sequesters carbon (Lehmann 2007) and improves soil fertility (Steiner et al. 2007). Biochar improves physical conditions of the soil and retains water and nutrients due to their high porosity and very large surface area; it also gives a favorable habitat for soil microorganisms (Warnock et al. 2007). However, as the nature of biochar varies widely, so does plant response to biochar application (Major et al. 2009). The positive effects of biochar application to poorly fertile soils are also longer-lasting (Glaser et al. 2002; Steiner et al. 2007). Biochar is more recalcitrant than other organic amendments and has higher cation sorption capacity due to higher charge density than biogenic soil organic matter (Sombroek et al. 2003; Liang et al. 2006). Biochar amendment in soils was found to significantly increase soil pH and contents of carbon, magnesium and calcium (Soderberg 2013). It was observed that both biological nitrogen fixation and beneficial mycorrhizal relationships in common beans (*Phaseolus vulgaris*) are enhanced by biochar applications (Rondon et al. 2007). Biochar application reduces leaching loss of nutrients (Yao et al. 2012) and improves fertilizer use efficiency of highly weathered tropical soils (Yoo et al. 2013; Hardie et al. 2015).

9.3.3 Green Manuring

The plowing under a 'green crop' at a suitable stage of its growth is known as 'green manuring' because it adds organic matter and nutrients to the soil after decomposition and improves soil fertility and productivity (Boller and Hani 2004). A fast growing, low nutrient demanding legume is generally grown densely in a field for green manuring. When the plants have reached the flowering stage and the shoots are still slender and succulent they are incorporated with the soil so that they are decomposed relatively easily. The incorporation is done at least 2–4 weeks before the sowing of the next crop to ensure enough decomposition and nutritional benefit of the following crop (Sullivan 2003). Since the green manuring crop is usually a nitrogen fixing plant, one may wrongly conceive that it only contributes nitrogen and organic matter to the soil. Actually, it has multiple benefits; it reduces leaching loss of other nutrients, conserves soil moisture and reduces soil erosion by acting as a cover crop when the crop occupies the field. Green manuring can be done at the fallow period between two crops or at suitable gaps in crop rotations. Sometimes green phytomass can be composted before application. More added benefits of green manuring include increased biological activity, improved soil structure, suppressing weeds, and reducing pests and diseases (Cherr et al. 2006). Some plant species, such as chicory, lupin, red clover, used for green manuring have deep root systems that can penetrate hard soil layers at shallow depths and can decompact soils (Lofkvist et al. 2005). Green manuring has some disadvantages from farmers' perspectives too; for example, extra costs of cultivation of a non-harvestable crop and the occupation of the land for a period when short rotation vegetables could be grown. The early decomposition products of some plants, for example vetch, can have allelopathic effects of germination of seeds of the following crop (Kamo et al. 2003). (Generally, the following crops are grown for green manuring: cowpea (*Vigna unguiculata*), jack bean *(Canavalia ensiformis)*, lablab bean (*Lablab purpureus*), pigeon pea *(Cajanus cajan)*, velvet bean (*Mucuna pruriens*), white clover *(Trifolium pratense)*, red clover *(Trifolium repens)* crimson clover or Italian clover (*Trifolium incarnatum*), lucerne or alfalfa *(Medicago satvia)*, white lupin *(Lupinus albus)*, etc.

9.3.4 Industrial Fertilizers

Some chemical compounds, organic or inorganic but mostly inorganic, manufactured in industries are used as fertilizers. These fertilizers are called industrial fertilizers, chemical fertilizers, synthetic fertilizers or inorganic fertilizers (although urea, urea formaldehyde, isobutyledene diurea, etc. are organic compounds or complexes). These fertilizers are concentrated substances containing high contents of the nutrients. Almost all of them are readily soluble in water and plants can absorb immediately after application. They are highly mobile and can move to surface water reservoirs by runoff or to groundwater by leaching. These active chemical compounds can also create toxicity to plants if they are not used properly.

The Africa Soil Health consortium (2012) advised farmers to follow '4R's in fertilizer management: (i) right kind of fertilizers, (ii) at the right rate, (iii) at the right time, and (iv) following the right method of application. The readers may get an idea of these 'R's in the following descriptions. Moreover, there are appropriate government authorities including Soil Survey Department, Agricultural Extension Offices, Agronomic Divisions, etc. to advise farmers regarding all these aspects for different agroecological regions and cropping patterns.

Fertilizers have enormous contribution to the production of plant and animal food, wood, fiber and forages. Fertilizer consumption has increased over five times during the last 50–55 years (FAO 2006) and Smil (2002) estimated that nitrogen

9.3 Management of Poorly Fertile Soils

Table 9.6 Industrial fertilizers and their nutrient content

Element	Fertilizer	Formula	Content %
N	Urea	$NH_2\text{-}CO\text{-}NH_2$	46
	Ammonium nitrate	NH_4NO_3	34
	Ammonium sulfate	$(NH_4)_2SO_4$	21% N, 11% S
	Monoammonium phosphate (MAP)	$(NH_4)H_2PO_4$	10–15%N, 22% P
	Diammonium phosphate (DAP)	$(NH_4)_2HPO_4$	18%N, 20%P
	Ammonium chloride	NH_4Cl	25
	Potassium nitrate	KNO_3	13%N, 36% K
	Sodium nitrate	$NaNO_3$	16
	Urea formaldehyde	Ureaform	40
	Isobutylidinediurea	IBDU	32
	Sulfur coated urea	SCU	31–38% N, 12–22%S
	Polymer coated SCU	PCSCU	30
P	Rock phosphate	$Ca_{10}(PO_4)_6 F_x(OH)_2$	11–15%P, 25%Ca
	Phosphoric acid	H_3PO_4	23
	Ordinary super phosphate (OSP)	$Ca(H_2PO_4)_2 \cdot CaSO_4$	9% P, 12%S, 20%Ca
	Triple super phosphate (TSP)	$Ca(H_2PO_4)_2$	20% P, 13% Ca
	Monoammonium phosphate (MAP)	$(NH_4)H_2PO_4$	22% P, 10–15%% N
	Diammonium phosphate (DAP)	$(NH_4)_2HPO_4$	20%P, 18%N
	Ammonium polyphosphate	$(NH_4)_4P_2O_7$	15
K	Potassium chloride, MP	KCl	50
	Potassium sulfate	K_2SO_4	41%K, 17%S
	Potassium nitrate	KNO_3	36%K, 13%N
Fe	Ferrous sulfate	$FeSO_4 \cdot 7H_2O$	20% Fe, 19% S
Mn	Manganous oxide	MnO	48% Mn
	Manganese sulfate	$MnSO_4 \cdot nH_2O$	23–25% Mn, 17%S
Zn	Zinc sulfate	$ZnSO_4 \cdot 7H_2O$	35% Zn, 15% S
Cu	Copper sulfate	$CuSO_4 \cdot 5H_2O$	25% Cu, 13% S
Mo	Sodium molybdate	$Na_2MoO_4 \cdot 2H_2O$	40% Mo
B	Borax	$Na_2B_4O_7 \cdot 10H_2O$	11% B
Co	Cobalt sulfate	$CoSO_4 \cdot 7H_2O$	21%

fertilizers increased average per-capita food production by 40 percent in the past 50 years. There are several advantages of using industrial fertilizers: (i) definite nutrient concentrations facilitating estimation of appropriate dose from fertilizer recommendations, (ii) readily soluble, (iii) fast acting and (iv) useful in immediate correction of current nutrient deficiencies, (v) easy to store, transport, distribute, (vi) easy to apply in the field, (vii) suitable to be mixed with irrigation water, and (viii) easy to handle compared to organic fertilizers. Table 9.6 presents nutrient contents of different industrial fertilizers.

Most popular industrial fertilizers to farmers are urea for N, triple super phosphate (TSP) and ordinary super phosphate (OSP) for P and muriate of potash for K. Most nitrogen fertilizers are manufactured from ammonia (NH_3) synthesized from natural gas and air. Super phosphates are manufactured by reacting rock phosphate with sulfuric acid (OSP) and phosphoric acid (TSP). Muriate of potash is obtained by grinding natural potash ores (the chief mineral is sylvite having the formula KCl). Monoammonium phosphate (MAP) and diammonium phosphate (DAP) are also used for supplying N and P together. Such substances are called compound fertilizers.

9.3.5 Mixed Fertilizers

For the convenience of applying two or more nutrients together and balancing the nutrients for different crops, two or more fertilizers are often mixed at different ratios. These fertilizer mixtures are called mixed fertilizers. For example, a 10-10-10 mixed fertilizer is a fertilizer mixture that contains 10 percent N, 10 percent P and 10 percent K. This expression of the percentages of N, P and K in a mixed fertilizer is known as fertilizer grade and the proportion of nutrients taking 'N' as 1 is called fertilizer ratio. In a 5-10-5 mixed fertilizer the grade is 5-10-5 and the ratio is 1:2:1. Fertilizer grade and fertilizer ratio must be written clearly in the fertilizer bag.

> **Box 9.4 Some Conversion Factors***
> (Metric)Tonne ha-1 × 0.446 = (British) ton acre-1
> Ton acre-1 × 2.24 = tonne ha-1 (used as t ha-1 in this book)
> Tonne × 1.102 = ton
> Ton × 0.9072 = tonne
> Kilogram (kg) × 2.205 = Pound
> Pound × 0.454 = Kilogram (kg)
> Hectare × 2-472 = Acre
> Acre × 0.405 = hectare
> Kilogram ha-1 × 0.891 = pound acre-1
> Pound acre-1 × 1.12 = kg ha-1
> P × 2.23 = P2O5
> P2O5 × 0.43 = P
> K × 1.2 = K2O
> K2O × 0.83 = K

*The Soil Science Society of America favors a fertilizer grade based on the percentages of nitrogen, phosphorus, and potassium, but fertilizer grades based N, P_2O_5 and K_2O are also in use. However, whatever basis is used must be clearly specified by the manufacturer on the fertilizer bag. The conversion factors to get P_2O_5 from P and K_2O from K are given in Box 9.4.

Mixed fertilizers of different grades for different soils and crops can be prepared. Micronutrients can also be added to a mixed fertilizer. Soil conditioners may be used as filler materials. An example of calculating the amounts of ammonium sulfate, triple super phosphate and potassium sulfate required to prepare one metric tonne of 5-10-5 (N, P, K) grade mixed fertilizer is given in Box 9.5.

9.3 Management of Poorly Fertile Soils

> **Box 9.5 Preparation of 1 t 5-10-5 Mixed Fertilizer from Ammonium Sulfate, TSP and Potassium sulfate**
>
> To prepare 1 t of 5-10-5 (N-P-K) mixed fertilizer by using ammonium sulfate (21% N), triple superphosphate (20% P) and potassium sulfate (41% K) as sources of N, P and K, respectively the calculations are shown below.
> For 1 t or 1000 kg 5-10-5 (N-P-K) mixed fertilizer, the amounts of N, P and K are:
>
> $N = 5/100 \times 1000 = 50$ kg,
> $P = 10/100 \times 1000 = 100$ kg, and
> $K = 5/100 \times 1000 = 50$ kg
>
> For 50 kg N, amount of ammonium sulfate = $50 \times 100/21$ =   238 kg
> 200 kg P, amount of triple superphosphate = $100 \times 100/20$ =   500 kg
> 100 kg K, amount of potassium sulfate = $50 \times 100/41$ =   122 kg
>
>   Total = 860 kg
>
> For making 1000 kg 5-1-5 mixed fertilizer, we need extra 140 kg material. This extra material is known as 'filler material'. Lime or any other solid soil conditioner can be used as filler materials for mixed fertilizers.

9.3.6 Liquid and Fluid Fertilizers

Liquid fertilizers are true solutions of chemicals and can contain one or more nutrients in the same solutions. For example, anhydrous ammonia, aqueous ammonia, and solutions of urea, ammonium nitrate contain only nitrogen, while ammonium polyphosphate solutions contain N and P. Different nutrient carrier compounds can be dissolve to prepare mixed fertilizer solutions. Liquid fertilizers may be of different concentrations and can be applied in different ways. They can be used through fertigation and foliar application. Suspensions and slurries containing nutrients are called fluid fertilizers. Liquid and fluid fertilizers can also contain micronutrients and pesticides mixed in them. Liquid fertilizers are mainly used in drip and sprinkler irrigation systems with sophisticated control systems.

9.3.7 Loss of Added Fertilizers

A considerable proportion of fertilizers added to soil is lost from agricultural fields through volatilization, denitrification, erosion and leaching. Among the primary fertilizers (N, P, K), nitrogen fertilizers are lost the most. Ammonia contained in some fertilizers, such as ammonium nitrate, ammonium sulfate, ammonium phosphates, etc. or converted from some applied fertilizers, such as urea quickly produce nitrate through the action of nitrifying bacteria in arable soil. The nitrate ion is very mobile in soil and is readily leached down to the groundwater (Camberato et al. 2008). Velu and Ramanathan (2001) observed that soil conditions and fertilizer application method influenced the magnitude of N leaching from agricultural fields. Nitrate can also move to surface water reservoirs through runoff. Thus, nitrogen fertilizers along with phosphate fertilizers in agricultural fields can be important sources of water pollution (Elser et al. 2007). If the soil is poorly aerated and a deficiency of O_2 develops, some microorganisms use the oxygen of NO_3^- instead of O_2 in soil air, and convert NO_3^- to nitrogen oxide and nitrogen gas (N_2) through the microbiological process of denitrification (Bing et al. 2006). Considerable nitrogen is lost to the atmosphere in this process from wet soils. Considerable nitrogen may also be lost by volatilization of NH_3 from calcareous/sodic soils with high pH (>7.0) and in soils saturated with water. More NH_3 is volatilized if ammonical fertilizers are spread on the surface and are not incorporated well with the soil (Rochette et al. 2009). Phosphate ions are less mobile than nitrates; still some phosphate ions are leached down to the groundwater. Phosphate ions are insolubilized in low pH soil by soluble Fe and Al. This process reduced plant's availability of P. This could be overcome by liming soils. Phosphate ions are also adsorbed of colloid surfaces. A considerable amount of phosphate is carried to surface water reservoirs with sediments resulting from erosion. Such deposition of phosphate causes eutrophication of water bodies (Lewis et al. 2011). Phosphate fertilizers can thus act as non-point source of water pollution (Toor et al. 2005). Potassium ions are relatively immobile in soil. Fertilizer losses can be reduced by taking appropriate management practices.

Reducing contact of fertilizers with the soil, use of inhibitors that inhibit nitrification and denitrification, frequent and low dose fertilization, use of slow release fertilizers, controlling irrigation to reduce leaching are some strategies to minimize nutrient losses from added fertilizers. Urea can be made in small and large granules as well as large balls. Increasing granule size can reduce contact between soil and fertilizer. Large urea balls, urea-mud balls and urea-mud-neem (*Azadirachta indica*) cake balls are buried at the centre of 4 rice hills in many rice growing countries. This practice has been found to considerably inhibit nitrification and denitrification. There are some slow release nitrogen fertilizers, such as sulfur coated urea (SCU), urea formaldehyde (urea-form), isobutylidine di urea (IBDU). Nitrogen is made available slowly from these complexes so that plant roots can absorb, but leaching does not occur. However, these fertilizers are very costly and small holder farmers cannot afford them. Split application of nitrogen fertilizers for cereals (one third during soil preparation, one third during the vigorous vegetative growth and the remaining one third during flower initiation) has been found beneficial. The method of application of fertilizers also influences the loss of fertilizers and fertilizer use efficiency of crops. Application of fertilizers in narrow bands instead of broadcasting reduces leaching, volatilization and denitrification. Nitrapyrin has commonly been used as nitrification inhibitors. In an anaerobic incubation experiment, Bremer and Yeomans (1986) tested the effects of 28 nitrification inhibitors, such as nitrapyrin, etridiazole, potassium azide,

2-amino-4-chloro-6-methylpyrimidine, sulfathiazole, 4-amino-1,2,4-triazole, 2,4-diamino-6-trichloromethyl-*s*-triazine, potassium ethylxanthate, guanylthiourea, 4-nitrobenzotrichloride, 4-mesylbenzotrichloride, sodium thiocarbonate, phenylmercuric acetate, and dicyandiamide on denitrification. They observed that only potassium azide applied at rate of 10 μg g^{-1} soil retarded denitrification, and two other compounds (potassium azide and 2,4-diamino-6-trichloromethyl-*s*-triazine) applied at the rate of 50 μg g^{-1} soil inhibited denitrification. Other compound had no inhibition of denitrification.

9.3.8 Methods of Fertilizer Application

Several factors affect the selection of the right method of fertilizer application. Important of these are: types of crops, kinds of fertilizers, characteristics of soil, availability of irrigation and drainage facilities, clime and seasons, etc. Bulky organic manures are applied some weeks before sowing of seeds so that decomposition releases some available nutrients and the soil condition considerably improves. If manures are applied immediately before sowing, decomposing microorganisms may compete with the germinating seeds. However, well composted manures can be applied at any time. Again, broadcasting as basal and top dressing or with sprinkler irrigation are the options for a closely growing crop, such as mustard. In row crops, fertilizer placement in narrow bands at one or both sides of the rows may provide some protection against fertilizer loss and weed competition. In orchard plants, on the other hand, fertilizers may be placed in a ring around the base of the tree. Fertilizer solutions at low concentrations may be used as foliar application or as injection into the soil.

Thus, fertilizers may be applied as (i) basal dressing and (ii) top dressing. Basal dressing is the pre-sowing addition of fertilizers and manures. Organic fertilizers, one third of nitrogen fertilizers and the entire amount of phosphate and potash fertilizers are applied during soil preparation or along with seed in a seed driller. One third of nitrogen is applied at the most vigorous vegetative growth and the remaining one third is applied at the flowering stage, both as top dressing. Top dressing is the application of fertilizers in a standing crop. However, fertilizers are applied both as basal dressing and top dressing for perennial crops, in orchards and in forest crops. Dividing the whole fertilizer dose into several installments as for nitrogen fertilizer in field crops is known as split application.

Applying fertilizers in field, both basal and top dressing, is done in several ways depending on the types of crops and fertilizers and availability of farm facilities. When fertilizers are spread by hand or equipment over the entire field, particularly for dense and close growing crops, the method of application is called broadcasting. Broadcasting can be done during field preparation (basal dressing) and in standing crops (top dressing). In many countries, fertilizers and pesticides together or separately are broadcasted by airplanes in hilly and inaccessible terrains. Unless the crop is close growing, broadcasting fertilizers throughout the field causes considerable loss of nutrients, especially nitrogen. In row crops, such as maize or sparsely planted plants, such as in orchards, fertilizers are applied in soils close to the seeds or roots to reduce contact with the soil, to reduce loss of nutrients and to improve fertilizer use efficiency. This method is called localized placement. There are several localized placement procedures: (i) plow sole placement when fertilizers are placed in the furrow formed during plowing and covered by the next turn, (ii) deep placement when fertilizers are placed in deep reduction zone of rice fields, (iii) combined drilling placement is the placement of seed and fertilizers together in a driller; fertilizers are covered with soil and the seeds are placed over the fertilizer band, (iv) band placement can be done at (a) one side of the crop row, (b) both sides of the row and (c) around the base of fruit trees in a ring, and (v) pellet placement; it is done in paddy fields where pellets of urea, 2.5–5 cm in diameter are placed at the centers of 4 rice hills. Fertilizer solutions can be sprayed in low concentrations on the leaves of crops by sprayers or with sprinkler irrigation water; this is known as foliar application of fertilizers. Sprinkling fertilizers has the advantage of controlling the doze, frequency and time of application. Fertilizers can also be dissolved in drip irrigation water storage tank and applied in low dose to reduce loss and cost and to increase fertilizer efficiency. Sprinkler and drip systems can be automatized in sophisticated farming systems. Applying fertilizers with irrigation water is known as fertigation.

9.3.9 Cropping Systems in Relation to Soil Fertility Management

Some crops are grown in a given field simultaneously or successively within a year. The set of crops is called a cropping pattern. Some examples of cropping patterns are: corn-wheat-soybean, corn-pearl millet-potato, rice-wheat-groundnut and rice-rice-bean. There are thousands of different cropping patterns in different parts of the world depending on climatic, tradition, social values, economic situation, edaphic condition, farmer's choice, etc. Each of the cropping patterns, however, has some sort of impact on the soil. On the other hand, cropping systems are the ways the crops are arranged in field. Cropping systems are also very important in soil fertility considerations. However, the major cropping systems are (i) mono cropping where the same crop is grown successively throughout the year or over the years, (ii) multiple cropping where more than one crop are grown either simultaneously or successively, (iii) sequential cropping where one crop is sown after harvesting another, (iv) relay cropping in which a second crop is sown shortly before harvesting the previous crop, (v) mixed cropping in which more than one crop is grown simultaneously in the same field (vi) intercropping in which two or more crops are sown in different rows in the same field and (vii) crop rotation in which several crops are successively grown according to a definite plan and the pattern is repeated without major changes for a stipulated period.

9.3 Management of Poorly Fertile Soils

Crops vary in their nutrient and water requirements; some crops are shallow rooted while others have deep roots, some crops are nitrogen depleting and some others are nitrogen fixing, some crops build soil structure but there are some crops that destroy soil aggregates. Therefore, alternating crops based on such variations help improve and restore soil fertility. Alternation of crops in a regular fashion is done in crop rotations. If the soil is poorly fertile, low nutrient demanding crops that produce high biomass and leave large amounts of crop residues should be the choice.

9.3.9.1 Crop Rotation

Crop rotation is a cropping system in which a set of crops, with some flexibility of inclusion or exclusion, are repeated in the same land in a regular fashion for a definite period. A simplified example is given below:

Corn→cow pea→potato→cabbage

The selection of crops is based on the diversity in growth habit, moisture and nutrient demands, biomass production potential, depth of roots, incidence of pests and diseases, cover, etc. In any crop rotation, inclusion of a legume as a green manure crop or cover is needed whenever feasible. However, in systematic planning for crop rotation, say for 4 years, farmers are advised to divide their land into 4 plots and arrange 4 sets of crops in those plots for 1st year. In the next year, crops of the 4th plot will be planted in plot 1, of plot 2 in plot 3, and of plot 3 in plot 4. The crops will be rotated in this way for 4 years. An arbitrary example taking multiple options for each plot is given below:

Year 1

Bed 1: Underground root & stem (beetroot, parsnips, carrots, potatoes, sweet potatoes, onions, garlic)

Bed 2: Legumes (peas, beans), brassicas (broccoli, cauliflower, cabbage, Brussels sprouts)

Bed 3: Tomatoes, eggplant, peppers

Bed 4: Sweet corn, curcubits (cucumber, melons, pumpkin)

Year 2

Bed 1: Sweet corn, curcubits (cucumber, melons, pumpkin)

Bed 2: Underground root & shoot (beetroot, parsnips, carrots, potatoes, sweet potatoes, onions, garlic)

Bed 3: Legumes (peas, beans), brassicas (broccoli, cauliflower, cabbage, Brussels sprouts)

Bed 4: Tomatoes, eggplant, peppers

Year 3
Bed 1: Tomatoes, eggplant, peppers
Bed 2: Sweet corn, curcubits (cucumber, melons, pumpkin)
Bed 3: Underground root & shoot (beetroot, parsnips, carrots, potatoes, sweet potatoes, onions, garlic)
Bed 4: Legumes (peas, beans), brassicas (broccoli, cauliflower, cabbage, Brussels sprouts)
Year 4
Bed 1: Legumes (peas, beans), brassicas (broccoli, cauliflower, cabbage, Brus sels sprouts)
Bed 2: Tomatoes, eggplant, peppers
Bed 3: Sweet corn, curcubits (cucumber, melons, pumpkin)
Bed 4: Underground root & shoot (Beetroot, parsnips, carrots, potatoes, sweet potatoes, onions, garlic)

An appropriate crop rotation coupled with sound soil and crop management provides some agronomic, economic and environmental benefits including higher total crop yield, increasing soil organic matter, improving soil structure, reducing soil degradation, enhanced soil fertility, greater farm profitability, suppression of weed, reduction in the risk of diseases, better utilization of labor and resources, etc.

9.3.9.2 Inclusion of Legumes in Cropping Systems

Legumes are included in cropping systems for several reasons. It is a protein rich food crop, it improves soil fertility by fixing nitrogen, its residues can be used as fodder and fuel, it can be used as a cover crop or green manuring crop. It can be grown as a regular crop, a catch crop, an intercrop and a relay crop. Some legumes used for grains (pulses), as green manuring crops and as cover crops are listed below:

Grain Legumes:

Faba bean (*Vicia faba*), vetch (*Vicia sativa*), kidney bean (*Phaseolus vulgaris*), chick pea (*Cicer arietinum*), cowpea (*Vigna unguiculata*, syn.: *Vigna sinensis*), cluster bean (*Cyamopsis tetragonoloba*), lablab bean (*Dolichos lablab*), lentil (*Lens culinaris*), lima bean (*Phaseolus lunatus*), white lupin (*Lupinus albus*), mung bean (*Vigna radiata*, syn.: *Phaseolus aureus*), snap pea (*Pisum sativum*), groundnut (*Arachis hypogaea*), pigeon pea (*Cajanus cajan*), soybean (*Glycine max*), tepary bean (*Phaseolus acutifolius*), etc.

9.3 Management of Poorly Fertile Soils

Green Manuring Legumes:

Sweet Clover (*Melilotus officinalis*), Field Pea (*Pisum sativum* subsp. *arvense*) Lentil (*Lens culinaris*), soybean (*Glycine max*), Chickpea (*Cicer arietinum*), groundnut (*Arachis hypogaea*), pigeon pea (*Cajunus cajan*),jack bean (*Canavalia ensiformis*), tanzanian sunnhemp (*Crotalaria ochreleuca*), sunnhemp (*Crotalaria juncea*), desmodium (*Desmodium intortum Desmodium uncinatum*), horsegram (*Dolichos biflorus*), soybean (*Glycine max*), white lupine (*Lupinus albus*), alfalfa (*Medicago sativa*), sweet clover (*Melilotus alba*), velvet bean (*Mucuna pruriens*), common bean (*Phaseolus vulgaris*), berseem.clover (*Trifolium alexandrinum*), faba bean (*Vicia faba*), common vetch (*Vicia sativa*), hairy vetch (*Vicia villosa*), green gram (*Vigna radiata*), cowpea (*Vigna unguiculata*).

Legumes Used as Cover Crops:

Crimson clover (*Trifolium incarnatum*), hairy vetch (Vicia villosa), field peas (*Pisum sativum* subsp. *arvense*), subterranean clover (*Trifolium subterraneum*), red clover (*Trifolium pratense*), white clover (*Trifolium repens*), sweet clover (*Melilotus officinalis*).

Whichever the purpose of growing legumes may be, these plants contribute to the improvement of soil fertility and productivity. Many of these crops can be grown in poorly fertile soils. For example, Mapfumo et al. (2001) reported that pigeonpea was adaptable to moisture stress and low soil fertility, and in addition, it recycles nutrients efficiently. Groundnut was found to improve soil fertility even if the residues were not incorporated after harvesting the nuts; it benefitted soil by the residual effects of fixed N (Toomsan et al. 2000).

Desirable nodulation and nitrogen fixation can only assure the maximum benefit from growing legumes in poorly fertile soils. The presence of the right kind of nodulating bacteria and the efficiency of the bacteria for infecting the roots is not always guaranteed. Therefore, legume seeds often need inoculation with the appropriate bacteria (*Rhizobium* spp. and *Bradyrhizobium* spp.). Legume seeds need inoculation under three situations: (a) in soils where compatible rhizobia are absent, (b) presence of low population of compatible rhizobia, and (c) where the indigenous rhizobia are not very effective in fixing N_2. In soils, where the particular legume has not been grown earlier, the chance of getting the effective bacteria is likely to be low.

9.3.9.3 Legume Intercrops

When more than one crop is grown simultaneously in the same time in alternate rows or a similar fashion, the cropping system is called intercropping. Intercropping is a kind of crop insurance; if one crop fails due to pests and diseases, the other will reduce the economic loss (Sullivan 2003). The benefits of intercropping include resource maximization, risk minimization, soil conservation, improvement of soil fertility and higher profit. Inclusion of a legume in intercropping systems has the added benefit of providing a part of the fixed nitrogen to the companion crop. Examples of some intercropping systems involving legumes are: maize-cowpea, maize-soybean, maize-pigeonpea, maize-groundnuts, maize-beans, sorghum-cowpea, millet-groundnuts, rice-pulses, millet-cowpea, etc. The crops can be arranged in different ways, such as in alternate rows or two rows of cereals-one row legume-two rows of cereals-one row legume, etc. For intercrops both the crops may be sown or planted at the same time or one may be sown earlier than the other. For example, in pearl millet-cowpea intercrop, pearl millet is usually sown earlier. It was observed that planting four rows of cowpea to two rows of cereal was more productive, and in this system the cereal is planted first, followed by the cowpea. Many investigators (Waddington and Karigwindi 2001; Kambabe and Mkandawire 2003; Abera et al. 2005; Adeniyan et al. 2007; Waddington et al. 2007; Egbe 2010; Mucheru-Muna et al. 2010; Obadoni et al. 2010; Addo-Quaye et al. 2011; Okoth and Siameto 2011; Osman et al. 2011; Jarenyama et al. 2000; Sanginga and Woomer 2009) studied the benefits of cereal-legume intercropping.

9.3.9.4 Mycorrhiza Inoculation

Mycorrhizae are filamentous fungi that develop a symbiotic relationship with the roots of plants. This relationship is a mechanism of extending the water and nutrient absorbing apparatus of plants. From this association plant roots can draw water and nutrients from a large volume of soil. In return, the fungal partner of the association receives carbohydrates, amino acids, and vitamins essential for their growth from the host plant. So, effective mycorrhizal association is a vital factor of plant nutrition in poorly fertile soils, especially those having the deficiency of phosphorus and micronutrients. Mycorrhizal association also provides plants some resistance to drought and heat stresses, and attack by pathogens and nematodes. In many soils, however, the right fungal strains may not be available and soil conditions may not favor the development of the mycorrhizal association. In such situations, mycorrhiza inoculation has been found to increase yield of crops, even in inhospitable sites. It is a popular technology now in agriculture and forestry. For example, Beltrano et al. (2013) observed in a greenhouse experiment that arbuscular mycorrhizal fungi (*Glomus intraradices*) increased growth of pepper plants (*Capsicum annuum* L.) and helped them overcome salt stress and P deficiency. Ortega et al. (2004) observed that mycorrhiza-inoculated radiata pine (Pinus radiata D. Don) seedlings had grown in container faster than non-inoculated seedlings. In another study, they found that nursery inoculation of radiata pine with *Rhizopogon roseolus* Th. Fries and *Scleroderma citrinum* Pers. improved tree growth after transplantation in field, even at the drier site.

9.3.10 Adjustment of Soil pH

Very high (>8.5) or very low pH (<5.5) may be the cause of poor fertility in many soils. Under conditions of high acidity and high alkalinity, solubility (and hence, availability) of some of the nutrient elements is reduced and some others is increased. For example, under very acid conditions, solubility of Fe and Mn (and also Al which is not a plant nutrient) may increase to a toxic level. Under alkaline conditions, on the other hand, most micronutrients become deficient. Availability of phosphate decreases at both very acid and very alkaline conditions. So, the pH of many agricultural soils needs to be adjusted towards neutrality by proper amendments to enhance nutrient efficiency, crop productivity and economic profitability. Acidity may be reduced by lime and alkalinity may be lowered by sulfur. This aspect of soil management has been discussed in detail in Chap. 11.

9.3.11 Residue Management and Conservation Tillage

Crops produce huge amounts of residues; a part of which is usually left in the field and another part is harvested. These residues are used by farmers in many different ways, such as thatching houses, animal feed, fuel, compost, etc. The Africa Soil Health Consortium (2012) emphasized that crop residues were good sources of plant nutrients and can be used for improving soil fertility in Sub-Saharan Africa where poor soil fertility is a major "constraint to improving farm productivity and livelihood". In the White Paper on Crop Residue Removal for Biomass energy of USDA-NRCS, Andrews (2006) mentioned that crop residues have many positive roles on the quality of agricultural soils. Residues left on soil surface reduce runoff and erosion. These materials act as mulch or cover in the soil, reduce evaporation, improve infiltration, conserve soil moisture and increase drought resistance of crop plants. In cold countries, surface residues keep the soil warmer and help germination of seeds. Residues incorporated into the soil increase soil organic matter, aggregation of soil particles, water and nutrient holding capacity, release nutrients after their decomposition, and enhance biological activity in the soil. The net effect is the sustainable improvement of soil fertility.

9.3.12 Cover Crops

Some crops can be grown in the fallow period within in the cropping cycle instead of keeping the soil bare. These crops are called cover crops because they offer a vegetative cover on the soil and protect it from rainfall impacts and erosion, reduce evaporation and conserve soil moisture, recycle nutrients, improve SOM and combat weeds. Herbs and grasses can be used as cover crops. According to Hairiah

(2004), an ideal cover crop should be (i) easy to establish, with minimum inputs (ii) fast growing, (iii) adapted to variable soil types, (iv) resistant to pests and diseases, (v) able to produce enough cover to reduce erosion, (vi) capable of improving soil fertility, and (vii) with high nitrogen fixing potential. Usually legumes, such as clovers, vetch, peas, etc. are grown as cover crops. If grain legumes and cereals are grown, the residues after harvest along with other farm residues should be incorporated well with the soil to get the maximum benefit to soil fertility restoration (Styger and Fernandes 2006). Some cover crops can produce grains for human consumption and fodder for livestock. Cover cropping is a popular practice in many parts of the world including Central America, West Africa and South Africa. The practice is also known as managed fallow, and it has also been discussed in Chap. 3.

9.3.13 Integrated Crop-Livestock Farming Systems

FAO (2007) and Van Keulen and Schiere (2004) emphasized the significance of an integrated crop-livestock farming system in which both livestock and crops are produced in a coordinated manner. According to them, these components are inseparable in a unified system where both are benefitted by each other. They depend on each other; each produces some wastes which are utilized by the other. A few materials are wasted in this mutualistic system. The residues and by-products of crops are used as animal feeds, and animal wastes of the farm are composted into manure and added to the soil for improving its fertility and crop productivity. Van Keulen and Schiere (2004) reported that this kind of mixed farming system has many different forms, but it constitutes the largest category of livestock farming systems in the world. Farm animals have diverse functions, such as (i) they produce a variety of products including meat, milk, eggs, wool, hides, (ii) farm animals can be sold when they exceed the bearing capacity of the farm and if needed for the farmers to earn profit, (iii) animal power can be used for plowing, transport and for other purposes, such as milling, logging, water lifting for irrigation, etc., (iv) excreta of farm animals are used for improving nutrient cycling, and (v) they can provide fuel and biogas for farm houses. Gupta et al. (2012) suggested that this integrated system has several befits to the soil (a) reduced erosion (b) increased crop yields, (iii) higher biological activity in soil, (iv) better nutrient recycling, and (v) intensification of land use. In a review on integrated crop livestock farming system in Nigeria, Ezeaku et al. (2015) suggested that this system should be very suitable for the savannah region where livestock farming is a common practice. However, the existing cereal-livestock integrated system is a cause of gradual soil fertility depletion in that region. The authors advised to adopt cereal-legume-livestock system for improving soil fertility.

Study Questions

1. Explain soil fertility. Describe the characteristics of a fertile soil. What are the constraints to soil fertility? Why the soil organic matter level is a good indicator of soil fertility?
2. What are the available forms of the nutrients in soil? Discuss the adequate nutrient level. Delineate the principles of the management of poorly fertile soil.
3. Give a list of different organic fertilizers and discuss the benefit of using them in crop production. What are their limitations? Give an account of the process of composting.
4. How are applied fertilizers lost? How can fertilizer loss be reduced? How would you choose a method of fertilizer application?
5. Explain cropping patterns and cropping systems. Briefly discuss how cropping systems can be exploited in the improvement of soil fertility.

References

Abera T, Feyissa D, Yusuf H (2005) Effects of inorganic and organic fertilizers on grain yield of maize-climbing bean intercropping and soil fertility in western Oromiya, Ethiopia. Conference on International Agricultural Research for Development, October 11–13, 2005 Stuttgart-Hohenheim

Addo-Quaye AA, Darkwa AA, Ocloo GK (2011) Yield and productivity of component crops in a maize-soybean intercropping system as affected by time of planting and spatial arrangement. J Agric Biol Sci 6(9):50–57

Adeniyan ON, Akande SR, Balogun MO, Saka JO (2007) Evaluation of crop yield of African yam bean, maize and kenaf under intercropping systems. Am-Eurasian J Agric Environ Sci 2(1):99–102

Africa Soil Health Consortium (2012) In: Fairhurst T (ed) Handbook for integrated soil fertility management. CAB International, Nairobi

Amara DG, Mourad SM (2013) Influence of organic manure on the vegetative growth and tuber production of potato (*solanum tuberosum* L. varspunta) in a Sahara desert region. Int J Agric Crop Sci 5(22):2724–2731

Andrews SS (2006) Crop residue removal for biomass energy production: effects on soils and recommendations. White Paper, USDA-Natural Resource Conservation Service. http://www.nrcs.usda.gov/Internet/FSE_DOCUMENTS /nrcs142p2_ 053255.pdf

Anonymous (2010) India losing 5,334 million tonnes of soil annually due to erosion. The Hindu, 26 November 2010

Beltran EM, Miralles de Imperial R, Porcel MA, Delgado MM, Beringola ML, Martin, Bigeriego M (2002) Effect of sewage sludge compost application on ammonium nitrogen and nitrate-nitrogen content of an Olive Grove soils. Proceedings: 12th International Soil Conservation Organization Conference. May 26–31, Beijing, China

Beltrano J, Ruscitti M, Arango MC, Ronco M (2013) Effects of arbuscular mycorrhiza inoculation on plant growth, biological and physiological parameters and mineral nutrition in pepper grown under different salinity and p levels. Journal of soil science and plant nutrition 13:123–141

Binford GD (2010) Amounts of nutrients removed in corn grain at harvest in Delaware. 19th World Congress of Soil Science, 1–6 Aug 2010, Brisbane, Australia

Bing CAO, Fa-Yun H, Qiu-Ming X, Yin B, Gui-Xin CAI (2006) Denitrification Losses and N_2O emissions from nitrogen fertilizer applied to a vegetable field. Pedosphere 16(3):390–397

Boller E, Hani F (2004) Manures and soil amendments. Ideal book on functional biodiversity at the Farm level
Bonelli PR, Rocca D, Cerrella PA, Cukierman AL (2010) Effect of pyrolysis temperature on composition, surface properties and thermal degradation rates of Brazil nut shells. Bioresour Technol 76:15–22
Bremer JM, Yeomans JC (1986) Effects of nitrification inhibitors on denitrification of nitrate in soil. Biol Fertil Soils 2(4):173
Brunetti J (2005) Cobalt for soil and animal health. In: Wise traditions in food, farming and the healing arts. Wiston A Price Foundation, USA
Camberato J, Brad J, Nielsen RL (2008) Nitrogen loss in wet and wetter fields. Corney News Network, Purdue University. URL: http://www.kingcorn.org/
Cherr CM, Scholberg JMS, McSorley R (2006) Green manure approaches to crop production. Synth Agron J 98:302–319
Conacher J, Conacher A (1998) Organic farming and the environment, with particular reference to Australia. Biological Agriculture. Horticulture 16:145–171
Coyne MS, Thompson JA (2006) Math for soil scientists. Thomson Delmar Learning, Clifton Park, NY
de Oliveira MW, Trivelin PCO, Boaretto AE, Muraoka T, Mortatti J (2002) Leaching of nitrogen, potassium, calcium and magnesium in a sandy soil cultivated with sugarcane. J Pesq Agropec Bras 37(6):861–868
Dougill AJ, Twyman C, Thomas DS, Sporton D (2002) Soil degradation assessment in mixed farming systems of southern Africa: Use of nutrient balance studies for participatory degradation monitoring. Geogr J 168:195–210
Drechsel P, Gyiele L, Kunze D, Cofie O (2001) Population density, soil nutrient depletion, and economic growth in sub-Saharan Africa. Ecol Econ 38:251–258
Egbe OM (2010) Effects of plant density of intercropped soybean with tall sorghum on competitive ability of soybean and economic yield at Otobi, Benue State, Nigeria. J Cereals Oilseeds 1(1):1–10
Eghball B (2001) Composting manure and other organic residue. Cooperative Extension Publication (NebGuide), Institute of Agriculture and Natural Resources, University of Nebraska, Lincoln
Elser JJ, Bracken ME, Cleland EE, Gruner DS, Harpole WS, Hillebrand H, Ngai JT Seabloom EW, Shurin JB, Smith JE (2007) Global analysis of nitrogen and phosphorus limitation of primary production in freshwater, marine and terrestrial ecosystems. Ecol Lett 10:1135
Eswaran H, Lal R, Reich PF (2002) Land degradation: an overview. In: Bridges EM, Hannam ID, Oldeman LR, de Vries FWT P, Scherr SJ, Sompatpanit S (eds) Proceedings of the 2nd International Conference on Land Degradation and Desertification. Oxford University Press, New Delhi, India
Ezeaku IE, Mbah BN, Baiyeri KP, Okechukwu EC (2015) Integrated crop-livestock farming system for sustainable agricultural production in Nigeria. Afr J Agric Res 10(47):4268–4274
FAO (2001a) Lecture notes on the major soils of the world. FAO, Rome
FAO (2001b) Food and Agriculture Organization of the United Nations FAOSTAT database. (http://apps.fao.org)
FAO (2003) Assessment of soil nutrient balance: approaches and methodologies. FAO Fertilizer and Plant Nutrition Bulletin 14. FAO, Rome
FAO (2004) FAO Food Balance Sheets. FAOSTAT, Food and Agriculture Organization of the United Nations, Rome
FAO (2006) Fertilizer use by crop. FAO Fertilizer and Plant Nutrition Bulletin 17. Food and Agriculture Organization of the United Nations, Rome
FAO (2007) Tropical crop-livestock systems in conservation agriculture. The Brazilian Experience. Food and Agriculture Organization of the United Nations, Rome
Felix KN, Chris AS, Jayne M, Monicah M-M, Daniel M (2012) The potential of organic and inorganic nutrient sources in Sub- Saharan African crop farming systems. In: Whalen J (ed) Soil fertility improvement and integrated nutrient management – A global perspective. In Tech, Rijeka, pp 135–156

Flavel TC, Murphy DV (2006) Carbon and nitrogen mineralization rates after application of organic amendments to soil. J Environ Qual 35:183–193

Gafur A, Borggaard OK, Jensen JR, Petersen L (2000) Changes in soil nutrient content under shifting cultivation in the Chittagong Hill Tracts of Bangladesh. Danish J Geogr 100:37–46

Giani L, Chertov O, Gebhardt C, Kalinina O, Nadporozhskaya M, Tolkdorf-Lienemann E (2004) Plagganthrepts in northwest Russia. Genesis, properties and classification. Geoderma 121:113–122

Glaser B, Lehmann J, Zech W (2002) Ameliorating physical and chemical properties of highly weathered soils in the tropics with charcoal-a review. Biol Fertil Soils 35:219–230

Gupta V, Rai PK, Risam KS (2012) Integrated crop-livestock farming systems: a strategy for resource conservation and environmental sustainability. Indian Res J Ext Educ, Special Issue (Volume II): 49

Hairiah K (2004) Introduction to part IV: Herbaceous legume fallows. In: Cairns M (ed) Voices from the forest: farmer solutions towards improved fallow husbandry in Southeast Asia. Johns Hopkins University Press, Baltimore, MD

Hardie MA, Oliver G, Clothier BE, Bound SA, Green SA, Dugald C, Close DC (2015) Effect of biochar on nutrient leaching in a young apple orchard. J Environ Qual 44(4):1273–1282

Hartemink AE (2006) Assessing soil fertility decline in the tropics using soil chemical data. Adv Agron 89:179–223

Hassanpanah D, Jafar A (2012) Evaluation of 'Out Salt' anti-stress material effects on mini-tuber production of potato cultivars under *in-vivo* condition. J Food Agric Environ 10(1):256–259

Heard J, Cavers C, Adrian G (2001) Up in smoke—nutrient loss with straw burning. Better Crops 90(3):10–11

Herrera E (2000) Soil test interpretations. Guident A-122. http://aces.nmsu.edu/pubs/_a/a-122.pdf. Accessed 22 Dec 2011

Horneck DA, Sullivan DM, Owen JS, Hart JM (2011) Soil test interpretation guide. http://www.sanjuanislandscd.org/Soil_Survey/files/page18_3.pdf. Accessed on 3 Jan 2012

International Fertilizer Development Corporation (IFDC) (2003) Input subsidies and agricultural development: issues and options for developing and transitional economies. IFDC Paper Series No. P-29. Muscles Shoals, Alabama

Izaurralde RC, Rosenberg NJ, Lal R (2001) Mitigation of climate change by soil carbon sequestration: Issues of science, monitoring and degraded lands. Adv Agron 70:1–75

Jacobsen J, Lorbeer S, Schaff B, Jones C (2002) Variation in soil fertility test results from selected Northern Great Plains laboratories. Commun Soil Sci Plant Anal 33(3 & 4):303–319

Jarenyama P, Hesterman OB, Waddington SR, Harwood RR (2000) Relay-intercropping of sunhemp and cowpea into a smallholder maize system in Zimbabwe. Agron J 92:239–244

Kambabe VH, Mkandawire R (2003) The effect of Pigeonpea Intercropping and Inorganic Fertilizer Management on Drought and Low Nitrogen Tolerant Maize Varieties in Malawi. In: Sakala WD. Kabambe VH (eds) Maize Agronomy Research Report, 2000–2004, pp. 7–13. Record Number 20083326997

Kamo T, Hiradate S, Fujii Y (2003) First isolation of natural cyanamide as a possible allelochemical from hairy vetch Viciavillosa. J Chem Ecol 29:275–283

Kelly VA (2006) Factors affecting demand for fertilizer in Sub-Saharan Africa: agriculture and rural development discussion Paper 23. World Bank, Washington, DC

Lehmann J (2007) Nature a handful of carbon. Nature 447:143–144

Lehmann J, de Silva JP Jr, Steiner C, Nehls T, Zech W, Glaser B (2003) Nutrient availability and leaching in an archaeological Anthrosol and a Ferralsol of the Central Amazon basin: fertilizer, manure and charcoal amendments. Plant Soil 249:343–357

Lehmann J, Rondon M (2006) Bio Char soil management on highly weathered soils in the humid tropics. In: Uphoff N et al (eds) Biological approaches to sustainable soil systems. CRC Press, Florida

Lehmann J, Schroth G (2003) Nutrient Leaching. In: Schroth G, Sinlair FL (eds) Trees, crops and soil fertility. CAS International, Wallingford, UK

Lewis WM, Wurtsbaugh WA, Paerl HW (2011) Rationale for control of anthropogenic nitrogen and phosphorus in inland waters. Environ Sci Technol 45:10030–10035

Li ZX, Dong ST, Wang KJ, Liu P, Zhang JW, Wang QC, Liu CX (2008) Soil nutrient leaching patterns in maize field under different fertilizations: an in situ study [in Chinese]. Ying Yong Sheng Tai Xue Bao 19(1):65–70

Liang B, Lehmann J, Solomon D, Kinyangi J, Grossman J, O'Neill B, Skjemstad JO, Thies J, FJ L͞o, Petersen J, Neves EG (2006) Black carbon increases cation exchange capacity in soils. Soil Sci Soc Am J 70:1719–1730

Lima HN, Schaefer CER, Mello JWV, Gilkes RJ, Ker JC (2002) Pedogenesis and pre-Colombian land use of Terra Preta Anthrosols (Indian black earth) of Western Amazonia. Geoderma 110:1–17

Liu J, Xie Q, Shi Q, Li M (2008) Rice uptake and recovery of nitrogen with different methods of applying 15N-labeled chicken manure and ammonium sulfate. Plant Prod Sci 11:271–227

Lofkvist J, Whalley WR, Clark LJ (2005) A rapid screening method for good root-penetration ability: comparison of species with very different root morphology. Acta Agric Scand 55:120–124

Major J, Steinerm C, Downiem A, Lehmann J (2009) Biochar effects on nutrient leaching. In: Lehmann J, Joseph S (eds) Biochar for environmental management: science and technology. Earthscan, London, UK

Mapfumo P, Campbell BM, Mpepereki S, Mafongoya P (2001) Legumes in soil fertility management: the case of Pigeonpea in smallholder farming systems of Zimbabwe. Afr Crop Sci J 9(4):629–644

Marschner H (1995) Mineral nutrition of higher plants. Academic Press, London

Mucheru-Muna M, Pypers P, Mugendi D, Kung'u J, Mugwe J, Merckx R, Vanlauwe B (2010) Staggered maize–legume intercrop arrangement robustly increases crop yields and economic returns in the highlands of Central Kenya. Field Crop Res 115:132–139

Musa E, Sas-Paszt L, Guszek S, Ciesieska J (2015) Organic fertilizers to sustain soil fertility. In: Sinha S (ed) Fertilizer Technology I Synthesis, Chapter: 11. Studium Press LLC, USA, pp 255–278

Nguyen TH, Brown RA, Ball WP (2004) An evaluation of thermal resistance as a measure of black carbon content in diesel soot, wood char, and sediment. Org Geochem 35:217–234

Obadoni BO, Mensah JK, Emua SA (2010) Productivity of intercropping systems using *Amaranthus cruentus* L. and *Abelmoschus esculentus* (Moench) in Edo State, Nigeria. World Rural Observations 2010, 2(2). http://www.sciencepub.net/rural

Okoth SA, Siameto E (2011) Evaluation of selected soil fertility management interventions for suppression of *Fusarium* spp. in a maize and beans intercrop. Tropical and Subtropical Agroecosystems 13:73-80

Ortega U, Dunabeitia M, Menendez S, Gonzalez-Murua C, Majada J (2004) Effectiveness of mycorrhizal inoculation in the nursery on growth and water relations of Pinus radiata in different water regimes. Tree Physiology 24(1):65–73

Osman AN, Ræbild A, Christiansen JL, Bayala J (2011) Performance of cowpea (*Vigna unguiculata*) and Pearl Millet (*Pennisetum glaucum*) Intercropped under *Parkia biglobosa* in an Agroforestry System in Burkina Faso. Afr J Agric Res 6(4):882–891

Owens LB, Malone RW, Shipitalo MJ, Edwards WM, Bonta JV (2000) Lysimeter study of nitrate leaching from a corn-soybean rotation. J Environ Qual 29:467–474

Pimentel D (2006) Soil erosion: A food and environmental threat. Environ Dev Sustain 8:119–137

Pimentel D, Burgess M (2013) Soil erosion threatens food production. Agriculture 3:443–463

Power JF, Wiese R, Flowerday D (2001) Managing farming systems for nitrate control: A research review from management systems evaluation areas. J Environ Qual 30:1866–1880

Rochette P, Angers DA, Chantigny MH, MacDonald JD, Bissonnette N, Bertrand N (2009) Ammonia volatilization following surface application of urea to tilled and no-till soils: A laboratory comparison. Soil Tillage Res 103:310–315

Rondon MA, Lehmann J, Ramirez J, Hurtado M (2007) Biological nitrogen fixation by common beans (*Phaseolus vulgaris* L.) increases with biochar additions. Biol Fertil Soils 43:699–708

Sanchez PE (2002) Soil Fertility and Hunger in Africa. Science 295:2019–2020

Sanginga N, Dashiell K, Diels J, Vanlauwe B, Lyasse O, Carsky RJ, Tarawali S, Asafo-Adjei B, Menkir A, Schulz S, Singh BB, Chikoye D, Keatinge D, Rodomiro O (2003) Sustainable resource management coupled to resilient germplasm to provide new intensive cereal–grain legume–livestock systems in the dry savanna. Agric Ecosyst Environ 100:305–314

Sanginga N, Woomer PL (2009) Integrated soil fertility management in Africa: Principles, Practices and Development Process. Tropical Soil Biology and Fertility Institute of the International Centre for Tropical Agriculture, Nairobi

Schick J, Bertol I, Balbinot AA Jr, Batistela O (2000) Erosãohídricaem Canbissolo Húmicoalumínicosubmetido a diferentessistemas de preparo e cultivo do solo: II. perdas de nutrientes e carbonoorgânico. Rev Bras Ciênc Solo 24:437–447

Smil V (2002) Nitrogen and food production: proteins for human diets. Ambio 31:126–131

Soderberg C (2013) Effects of biochar amendment in soils from Kisumu, Kenya. Graduate Project, Department of Soil and Environment, Faculty of Natural Resources and Agricultural Sciences, Swedish University of Agricultural Sciences

Sombroek W, Ruivo ML, Fearnside PM, Glaser B, Lehmann J (2003) Amazonian dark earths as carbon stores and sinks. In: Lehmann J et al (eds) Amazonian Dark Earths: Origins, Properties, Management. Kluwer Acad Publ, Dordrecht

Steiner C, Teixeria WG, Lehmann J, Nehls T, deMace^do JLV, Blum WEH, Zech W (2007) Long term effects of manure, charcoal, and mineral fertilization on crop production and fertility on a highly weathered central Amazonian upland soil. Plant Soil 291:275–290

Styger E, Fernandes CM (2006) Contributions of Managed Fallows to Soil Fertility Recovery. In: Uphoff N, Ball AS, Fernandes E, Harren H, Husson O, Laing M, Palm C, Pretty J, Sanchez P, Sanginga N, Thies J (eds) Biological Approaches to Sustainable Soil Systems (Books in Soils, Plants, and the Environment). CRC Press, Boca Raton

Sukristiyonubowo, Watung RL, Vadari T, Agus F (2002) Nutrient loss and the on-site cost of soil erosion under different land use systems. Paper presented at the T" MSEC Annual Assembly. 2–7 Dec 2002. Vientiane, Lao PDR

Sullivan P (2003) Intercropping principles and pro-duction practices. Appropriate Technology Transfer for Rural Areas Publication. http://www.attra.ncat.org

Tabo R, Bationo A, Gerard B, Ndjeunga J, Marchal D, Amadou B, Annou G, Sogodogo D, Taonda JBS, Hassane O, Maimouna KD, Koala S (2007) Improving cereal productivity and farmers' income using a strategic application of fertilizers in West Africa. In: Bationo A, Waswa B, Kihara J, Kimetu J (eds) Advances in integrated soil fertility management in sub-Saharan Africa: challenges and opportunities. Kluwer Publishers, Dordrecht, The Netherlands, pp 201–208

Tahiri S, Guardia MDL (2009) Treatment and valorization of leather industry solid wastes: A review. J Am Leather Chem Assoc 104:52–67

Tan ZX, Lal R, Wiebe KD (2005) Global soil nutrient depletion and yield reduction. J Sustain Agric 26(1):123–146

Toomsan B, Cadisch G, Srichantawong M, Thongsodsaeng C, Giller KE, Limpinuntana A (2000) Biological N2-fixation and residual N benefit of pre-rice leguminous crops and green manures. Neth J Agric Sci 48:19–29

Toor GS, Condron LM, Cade-Menun BJ, Di HJ, Cameron KC (2005) Preferential phosphorus leaching from an irrigated grassland soil. Eur J Soil Sci 56(2):155–168

Usman M, Madu VU, Alkali G (2015) The combined use of organic and inorganic fertilizers for improving maize crop productivity in Nigeria. Int J Sci Res Pub 8(10):1–7

Van Keulen H, Schiere H (2004) Crop-livestock systems: old wine in new bottles? In New directions for a diverse planet. Proceedings of the 4th International Crop Science Congress, 26 September-October 2004, Brisbane, Australia

Vanlauwe B, Descheemaeker K, Giller KE, Huising J, Merckx R, Nziguheba G, Wendt J, Zingore S (2015) Integrated soil fertility management in sub-Saharan Africa: unravelling local adaptation. Soil 1:491–508

Vanlauwe B, Bationo A, Chianu J, Giller KE, Merckx R, Mokwunye U, Ohiokpehai O, Pypers P, Tabo R, Shepherd K, Smaling EMA, Woomer PL (2010) Integrated soil fertility management: operational definition and consequences for implementation and dissemination. Outlook Agric 39:17–24

Velu V, Ramanathan KM (2001) Nitrogen balance in wetland rice ecosystem as influenced by soil type. J Madras Agril 87:21–25

Waddington SR, Karigwindi J (2001) Productivity and profitability of maize + groundnut rotations compared with continuous maize on smallholder farms in Zimbabwe. Exp Agric 37:83–98

Waddington SR, Mekuria M, Siziba S, Karigwindi J (2007) Long-term yield sustainability and financial returns from grain legume-maize intercrops on a sandy soil in subhumid North Central Zimbabwe. Exp Agric 43:489–503

Warnock DD, Lehmann J, Kuyper TW, Rillig MC (2007) Mycorrhizal responses to biochar in soil—concepts and mechanisms. Plant Soil 300:9–20

Webb T, Hewitt A, Lilburne L, McLeod M (2010) Mapping of vulnerability of nitrate and phosphorus leaching, microbial bypass flow, and soil runoff potential for two areas of Canterbury. Report R10/125. Environment Canterbury Regional Council, KauniheraTaiao Ki Waitaha, 58 Kilmore Street PO Box 345 Christchurch 8140

Westover HL (1926) Farm manures. USGA Green Section 6(9):193–196

Widowati A, Utomo WH (2014) The use of biochar to reduce nitrogen and potassium leaching from soil cultivated with maize. J Degrad Min Lands Manag 2(1):211–218

Wild A (1996) Soils and the environment. Cambridge University Press, Cambridge

Yao Y, Gao B, Zhang M, Inyang M, Andrew R, Zimmerman AR (2012) Effect of biochar amendment on sorption and leaching of nitrate, ammonium, and phosphate in a sandy soil. Chemosphere 89:1467–1471

Yoo G, Kim H, Chen J, Kim Y (2013) Effects of biochar addition on nitrogen leaching and soil structure following fertilizer application to rice paddy soil. Soil Sci Soc Am J 78(3):852–860

Zhao SL, Gupta SC, Huggins DR, Moncrief JF (2001) Tillage and nutrient source effects on surface and subsurface water quality at corn planting. J Environ Qual 30:998–1008

Zhu Q, Peng X, Huang T (2015) Contrasted effects of biochar on maize growth and N use efficiency dependingon soil conditions. Int Agrophys 29:257–266

Zingore S, Delve RJ, Nyamangara J, Giller KE (2008) Multiple benefits of manure: the key to maintenance of soil fertility and restoration of depleted sandy soils on African smallholder farms. Nut Cycl Agroecosyst 80:267–282

Chapter 10
Saline and Sodic Soils

Abstract According to different estimates, there are 831 to 932 million hectares of salt affected soils – saline soils, saline-sodic soils and sodic soils. The measure of salinity is the electrical conductivity of the saturation extract (EC_e) and the measure of sodicity is the exchangeable sodium percentage (ESP) or the sodium adsorption ratio (SAR). Saline soils have $EC_e > 4$ dS m^{-1} (decisiemens per meter) at 25° C and ESP < 15 (high soluble salts and low exchangeable Na$^+$). Sodic soils have $EC_e < 4$ dS m^{-1} and ESP > 15 (low soluble salts and high exchangeable Na$^+$). Saline-sodic soils are characterized by $EC_e > 4$ dS m^{-1} and ESP > 15 (both salts and exchangeable sodium are high). These soils mainly occur in arid and semi-arid regions where precipitation to evapotranspiration ratio is low. These soils also develop in coastal regions because of the flooding of sea water, and in irrigated areas due to the rise of the groundwater table and in some impermeable soils of the humid regions due to lack of leaching. Salts adversely affect plant growth and crop yield. Crop failures are common occurrences although some plants may have some degree of tolerances. Salt tolerant crops can be grown in some low to moderately saline soils, but soil salinity may be so severe in some cases that cropping becomes impossible. Management of saline soils involves selecting salt tolerant crops, salt scraping, salt flushing, and leaching with irrigation and artificial drainage. Usually, chemical amendments are not necessary for the reclamation of saline soils, but chemical treatment is needed prior to leaching for managing sodic soils. Substances that contain soluble calcium such as gypsum and $CaCl_2$, and sulfuric acid or substances that produce sulfuric acid after application to soil such as sulfur, pyrite, ferrous sulfate, aluminium sulfate etc. are used as amendments for sodic soils. Phytoremediation of sodic soils has also been successful on some occasions.

Keywords Saline soils · Sodic soils · Saline-sodic soils · Reclamation · Salt tolerance · Salt scraping · Salt flushing · Leaching · Leaching requirement · Gypsum requirement

10.1 Characteristics of Saline and Sodic Soils

All soils contain some soluble salts. Weathering of minerals of the parent materials is the principal source of soluble salts in soils. Another major source is sea water in coastal and estuarine regions. Other sources are atmospheric deposition, capillary rise of groundwater, seepage and irrigation water. Salts are added continuously to soils, and at the same time some salts are continuously removed from the soil system by erosion and leaching where water is available. In some cases, leaching is very low (for example, in areas with low precipitation to evapo-transpiration ratio), and soluble salts accumulate in the soil. Sometimes, a layer of salts accumulate on the surface of the soil in the form of a salt crust. A variety of soluble salts are found in soils; some of these salts provide plants with nutrients, but when they accumulate in excess amounts, they become harmful to growing plants. When salts accumulate in soils above a critical limit, they are called saline soils. Plants, unless adapted to saline environment, suffer from water stress, nutrient disorders, and toxicity if there are excess soluble salts in the root zone. Some soils contain excess exchangeable sodium ions with or without excess salts; they are called sodic soils. Sodic soils are highly dispersed and impervious, extremely alkaline in reaction, and produce deficiency of some nutrients and toxicity of others.

Soil salinity can be a natural event, or it may be anthropogenic. Many soils have been made saline by faulty water management. About 23% of the world's cultivated lands are saline and 37% are sodic (Khan and Duke 2001). Nearly half of the irrigated surface is seriously affected by secondary salinity and sodicity (Flagella et al. 2002). However, the criterion for judging soil salinity is not the amount of salts present in the soil, but it is the electrical conductivity of the saturation extract of the soil (EC_e) (US Salinity Laboratory Staff: Richards 1954). The electrical conductivity is proportional to the concentration of salts in solution and it is relatively easy to determine. Saturation extract is obtained by adding water to the soil up to its maximum water holding capacity and then drawing the soil solution under suction. If the electrical conductivity of the saturation extract (EC_e) at 25 °C is >4 dS m^{-1}, the soil is saline. The criteria of sodicity are exchangeable sodium percentage (ESP) and sodium adsorption ratio (SAR).

$$ESP = \frac{\text{Exchangeable Na}^+}{CEC} \times 100, \text{and}$$

$$SAR = \frac{Na^+}{\sqrt{\frac{1}{2}\left(Ca^{2+} + Mg^{2+}\right)}}$$

A soil is said to be sodic if the ESP exceeds 15 or the SAR is higher than 13. Thus, soils may be classified on the basis of EC_e and ESP (or SAR) into four categories (Table 10.1). Richards (1954) used the unit mmhos cm^{-1} (mho is inverse of ohm; since conductivity is inverse of resistance; the unit of resistance is ohm) for

10.1 Characteristics of Saline and Sodic Soils

Table 10.1 Categories of saline and sodic soils[a]

Category of soil	EC_e, dS m^{-1}	ESP Exchangeable sodium percentage (ESP)	SAR	Common name
Saline non-sodic	> 4	< 15	< 13	Saline soil
Saline sodic	> 4	> 15	>13	Saline-sodic soil
Non-saline sodic	< 4	> 15	> 13	Sodic soil
Non-saline non-sodic	< 4	< 15	< 13	Normal soil

[a]Classical classification of saline and sodic soils according to Richards (1954).

EC_e. Later, dS m^{-1} (decisiemens per meter) is used instead of mmhos cm^{-1}. Useful conversion factors are: mmhos cm^{-1} = dS m^{-1} = mS cm^{-1} = 1000 µS cm^{-1} = 1000 µmhos cm^{-1}.

However, determination of electrical conductivity of the saturation extract (EC_e) is a tedious job for some laboratories and for routine analysis. The particular problem is associated with drawing the extract under suction. There is also possibility of variable water addition to saturate the different batches of soil samples. Some investigators, therefore, used soil:water ratios of 1:1 or 1:5 and correlated with the EC_e (electrical conductivity of the saturation extract). Data of Qadir et al. (2010) are summarized below.

Pre-leaching*   **Post-leaching****   **Combined*****
$EC_e = 1.98 \times EC_{1:1}$ $EC_e = 2.16 \times EC_{1:1}$ $EC_e = 2.06 \times EC_{1:1}$
$EC_e = 6.53 \times EC_{1:5}$ $EC_e = 2.08 \times EC_{1:5}$ $EC_e = 2.42 \times EC_{1:5}$

Number of samples, *n = 75, ** n = 105 and *** n = 180; the R^2 value ranged between 088 and 0.92

Such conversion factors were also proposed by other investigators, such as Shirokova et al. (2000) and Akramkhanov et al. (2008).

Saline soils

Saline soils contain sufficient neutral soluble salts that adversely affect the growth of most crop plants. The EC_e of 4 dS m^{-1} is still used as the standard for saline soils the world over, but the yield of most crop plants is reduced at this EC_e. Many crops exhibit yield reductions at much lower EC_e values (Munns 2005; Jamil et al. 2011), such as the EC_e of 1 dS m^{-1} (Chinnusamy et al. 2005). The Terminology Committee of the Soil Science Society of America has lowered the boundary between saline and non-saline soils to EC_e 2 dS m^{-1} in the saturation extract (Soil Survey Staff 1993). Table 10.2 presents a classification of saline soils on the basis of EC_e and plant responses.

Natural salts in saline soils include mainly neutral salts such as chlorides (Cl$^-$) and sulfates (SO$_4^{2-}$) of sodium (Na$^+$), calcium (Ca^{2+}), magnesium (Mg^{2+}) and potassium (K$^+$) although NaCl dominates. Nitrates may be present on a few occasions. Carbonates and bicarbonates are usually absent. Many saline soils contain appreciable quantities of gypsum (CaSO$_4$. 2H$_2$O) lower in the profile. The pH of saline

Table 10.2 Soil salinity classes

Soil salinity class	EC_e, dS m^{-1}	Effect on crop plants
Nonsaline	0–2	Salinity effects negligible
Slightly saline	2–4	Yield of sensitive crops may be restricted
Moderately saline	4–8	Yield of many crops are restricted
Strongly saline	8–16	Only tolerant crops yield satisfactorily
Very strongly saline	>16	Only a few tolerant crops yield satisfactorily

Source: http://www2.vernier.com/sample_labs/AWV-09-COMP-soil_salinity.pdf

soils (saturated paste) remains near neutrality but may be as high as 8.2–8.5. Saline soils have good soil structure because excess salts keep the clay particles in flocculated state. Flocculation helps particles cling together and initiate their binding into aggregates.

Saline soils generally have good physical properties; they are usually porous and permeable. Leaching these soils with low-salinity water, tend to disperse fine textured saline soils resulting in low permeability to water and air. Spotty growth of crops, burning of leaves, and white salt crusts are field indicators of soil salinity. Barren spots and stunted plants may appear in cereal or forage crops growing on saline areas.

Saline-Sodic Soils

Saline-sodic soils have electrical conductivity of the saturation extract >4 dS m^{-1} and exchangeable sodium percentage level > 15 percent. The pH values of these soils are usually less than 8.2. These soils have both excess soluble salts and excess exchangeable sodium. The dominant salts in saline–sodic soils are chlorides and sulfates of sodium, calcium and magnesium along with carbonates and bicarbonates. The physical conditions of these soils are good as long as there are high salt levels. High concentration of salts keeps the colloids flocculated and well aggregated. Leaching of salts may cause dispersion and accompanied degradation of soil structure and loss of water permeability.

Sodic soils

Sodic (non-saline) soils contain high proportion of exchangeable sodium (more than 15 percent) in colloidal surfaces which keep the particles dispersed and prevent them from binding into aggregates. So, excess exchangeable sodium has an adverse effect on physical, chemical and nutritional properties of soil. A dense layer of clay occurs at or near the surface of sodic soils. This layer, often called a clay pan, is a root restrictive layer. Many soils of ESP > 15 (or SAR > 13) have very poor soil physical properties which are unsuitable for crop growth. Contents of neutral salts are low; the salts present in sodic soils are capable of undergoing alkaline hydrolysis. Carbonates and bicarbonates of Na^+, Ca^{2+} and Mg^{2+} predominate. The electrical conductivity of saturation extracts is less than 4 dS m^{-1} at 25 °C. The pH of saturated soil paste is higher than 8.2; in some cases, it can be as high as 10.5. Dispersed and dissolved organic matter present in the soil solution of some highly sodic soils gives them a dark color for which these soils were earlier called black alkali soils.

Fig. 10.1 Schematic representation of dispersion (**a**) and flocculation (**b**)

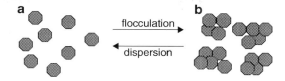

The dispersion of clay particles creates poor soil structure, and these soils are impervious and of low hydraulic conductivity. The reversible processes of dispersion and flocculation are shown in Fig. 10.1.

The total salt concentrations in sodic soils could be low or high, but the concept of sodicity is based on the high ratio of Na^+ to Ca^{2+} and Mg^{2+}. Both ESP and SAR are some forms of expression of this ratio, but the USDA Soil Salinity Laboratory criterion of ESP >15 for sodic soils is not followed universally. It has become controversial for large scale variability in sodicity levels across continents and occurrence of dispersion at variable ESP levels (Gregory and Nortcliff 2013). ESP >15% is still followed in the USA, but in Australia, an ESP of 6 is the limit. Sometimes, dispersion can even occur in ESP values <1.

The principal cause of alkaline reaction of sodic soils is the hydrolysis of either the exchangeable cations or of such salts as $CaCO_3$, $MgCO_3$, Na_2CO_3, etc. Hydrolysis of the exchangeable cations takes place according to the following reactions:

$$\text{Na-clay-Na} + H_2O \leftrightarrow \text{Na-clay-H} + Na^+ + OH^-$$

The displaced Na does not combine with, or inactivate OH^- ions, which results in an increase in the OH^- ion concentration and increased soil pH. Hydrolysis of compounds like $CaCO_3$, and $MgCO_3$ takes place according to the following reaction:

$$CaCO_3 + 2H_2O \Leftrightarrow Ca^{2+} + 2OH^- + H_2CO_3$$

Here H^+ is inactivated through combination with carbonate to form weakly ionized carbonic acid. Hydrolysis of $CaCO_3$ and of $MgCO_3$ is limited in soil due to their low solubility, and, therefore, they tend to produce a pH in soils no higher than about 8.0–8.2. Soils containing excess Na_2CO_3 have a pH greater than 8.2 and as high as 10.5. This is due to the higher solubility of Na_2CO_3. Major distinctions between saline and sodic soils are shown in Table 10.3.

10.2 Development of Salinity and Sodicity in Soils

There are natural and anthropogenic causes of soil salinity development. Natural causes include low or insufficient leaching due to aridity, parent materials of the soils, shallow groundwater table high in salts, saline seepage, capillary rise of water

Table 10.3 Major distinguishing characteristics of saline and sodic soils

Characteristics	Saline soils	Sodic soils
Chemical	(a) There are neutral salts consisting of chlorides and sulfates of sodium, calcium and magnesium. (b) the pH of saturated soil paste is less than 8.2. (c) an electrical conductivity of the saturated soil extract is >4 dS m^{-1} at 25 °C. (d) There is no well-defined relationship between pH and ESP. (e). Soils may contain significant quantities of sparingly soluble calcium compounds, e.g. gypsum.	(a) Neutral salts are generally absent. Appreciable quantities of salts capable of alkaline hydrolysis, e.g. Na_2CO_3, present. (b) the pH of the saturated soil paste is more than 8.2. (c) ESP > 15; electrical conductivity of the saturated soil extract <4 dS m^{-1} at 25 °C but may be more if appreciable quantities of Na_2CO_3 etc. are present. (d) the pH and ESP are related. (e) Sodium is the dominant soluble cation. High pH of the soils results in precipitation of soluble ca and mg such that their concentration in the soil solution is very low. (f) Gypsum is nearly always absent in such soils.
Physical	(a) for excess soluble salts the clay fraction is flocculated, and the soils have a good structure. (b) Permeability of soils to water and air is good.	(a) Excess exchangeable sodium causes dispersion of clay and poor soil structure. (b) Permeability of soils to water and air is restricted. .
Effect on plant growth	In saline soils plant growth is adversely affected: (a) Chiefly through the effect of excess salts on the osmotic pressure of soil solution resulting in reduced availability of water; (b) through toxicity of specific ions, e.g. Na, cl, B, etc.	In sodic soils plant growth is adversely affected: (a) Chiefly through the dispersive effect of excess exchangeable sodium resulting in poor physical properties; (b) through the effect of high soil pH on nutritional imbalances including a deficiency of calcium; (c) through toxicity of specific ions, e.g. Na, CO_3, Mo, etc.
Soil improvement	Improvement of saline soils essentially requires the removal of soluble salts in the root zone through leaching and drainage.	Chemical amendments followed by leaching, irrigation and drainage are needed.

due to dryness in the surface soil, coastal flooding, etc. Human induced causes are faulty irrigation system, faulty drainage system, use of saline water for irrigation, inundation with saline water for salt farming as well as shrimp farming.

It has already been mentioned that major sources of soluble salts in soils are weathering of minerals, sea water, groundwater, seepage water, irrigation water and atmospheric deposition. These salts are natural salts which mainly contain chlorides and sulfates of sodium, calcium, magnesium and potassium. If the soil is sodic, the dominant anions may be carbonates and bicarbonates and the dominant cation is sodium. There are associations of natural salt accumulations in soils with climatic regions. The most soluble salts, such as magnesium and calcium chlorides, are only

10.2 Development of Salinity and Sodicity in Soils

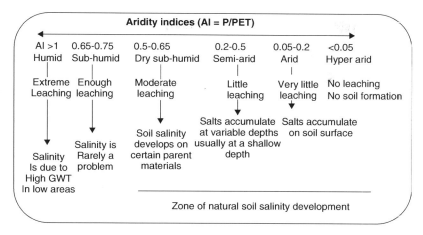

Fig. 10.2 Schematic representation of soil salinity development in arid and semi-arid lands

found under the most arid conditions; magnesium and sodium sulfates and sodium chloride are common in arid and semi-arid environments. With increasing humidity, the most soluble salts disappear and only gypsum and finally only calcium carbonate is found.

Salinity

There are two major salinization processes: natural or primary salinization and human induced or secondary salinization. Most saline soils are found in arid and semi-arid regions. As it has already been discussed, the aridity index (AI = P/PET) ranges from 0.05 to 0.65. The lower the aridity index, the lower is the availability of water for leaching soluble salts. So, at a lower aridity index, there is higher chance of accumulation of soluble salts in soil surface (Fig. 10.2).

In the arid and semiarid regions, evapotranspiration greatly exceeds precipitation, or the precipitation/evapotranspiration ratio is low. Salts released from weathering of minerals cannot be leached due to insufficient percolation. Moreover, capillary rise of water brings salts upward; water evaporates leaving the salts in the surface soil (Fig. 10.3).

Salinity also develops in soils of coastal and estuarine areas by flooding with sea water, and ponding of salt water in lagoons and depressions. In basins of humid regions, rise of the groundwater often causes soil salinity. Shallow groundwater table can contribute to the development of soil salinity in both arid and humid regions. Studies have shown that significant evaporation from soil surface can occur due to capillary rise of water from groundwater table within 1–2 m depth of soil, which contributes to the development of root zone salinity. Salinization problems can be more severe when the salinity of groundwater is high, as is usually the case in arid regions. Irrigation with salty water without enough drainage has caused salinity in many soils. In coastal regions, shrimp farmers and salt farmers pond sea water in polders in a season. These soils become highly saline in the long run.

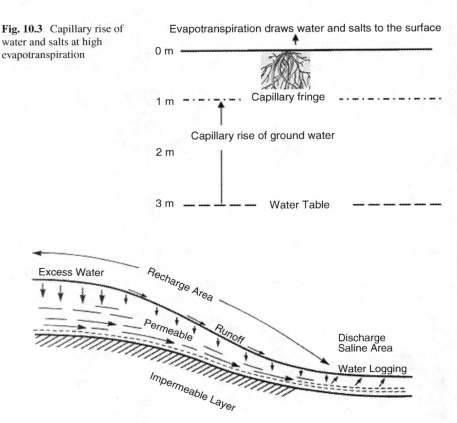

Fig. 10.3 Capillary rise of water and salts at high evapotranspiration

Fig. 10.4 Formation of saline soils through saline seeps

Saline seeps are a kind of dryland salinity which is widespread in Western Australia and in the Great Plains region of North America. It occurs extensively in Canada in the prairie provinces of Manitoba, Saskatchewan and Alberta and in the United States in the states of Montana, North and South Dakota, in South Africa, Iran, Afghanistan, Thailand and India. In this process, water infiltrates in the upper slope to a relatively impermeable layer, moves laterally downslope carrying dissolved salts and discharges at a lower point or depression. The water evaporates there leaving the salts behind. The development of saline seeps then involves two areas in the field—the recharge and the discharge areas (Fig. 10.4).

Sodicity

Sodicity is a phyco-chemical state of soil caused by the presence of high proportion (>15 percent) of exchangeable sodium ions on the surfaces of soil colloids so that they become highly dispersive and sticky and at the same time highly alkaline (pH usually 8.2 and above). Sodicity also makes the soil impermeable under wet condition, causes slacking of aggregates and can create surface crusts when dried. The predominance of Na⁺ ions in soil solution and in exchange positions can occur under certain conditions in semi-arid and arid regions, such as in some salt lakes and coastal areas. Usually, Ca^{2+} and Mg^{2+} are dominant cations in arid and semi-arid soils. When these ions are precipitated from solution as calcium and magnesium compounds, the predominance of Na⁺ ions in soil solution and exchange sites and causes sodicity. Van Breemen and Burman (2002) suggested some mechanisms of sodification of soil including (i) prolonged leaching, natural or as desalinization effort, removes salts and Ca^{2+} ions from soil solution resulting in the prevalence of Na⁺ ions, (ii) relatively greater accumulation of Na-salts that cause higher proportion of soluble and exchangeable Na⁺, (iii) accumulated Na_2SO_4 in some lake beds become reduced in wet conditions and in the presence of organic matter to Na_2CO_3 that causes sodicity, and (iv) presence of HCO_3^- ions in excess of $Ca^{2+} + Mg^{2+}$ ions in soil solutions; if Ca^{2+} are considerably lower than HCO_3^- ions, all the Ca^{2+} ions are precipitated as $CaCO_3$, and Na⁺ ions predominate in soil solution and in colloidal surfaces causing soil sodicity.

10.3 Distribution of Saline Soils

Although an earlier estimate shows that there were 932.2 million hectares of salt affected soils, including 351.5 M ha saline and 581.0 M ha sodic soils according FAO report of 2000, the total global area of salt-affected soils including saline and sodic soils was 831 million hectares (Martinez-Beltran and Manzur 2005), extending over all the continents including Africa, Asia, Australasia, and the Americas. Based on the FAO/UNESCO Soil Map of the world, the total area of saline soils is 397 million ha and that of sodic soils is 434 million ha. Most of the salt affected land lies in the arid and semiarid environment. Figs. 10.5 and 10.6 show the global distributions of saline and sodic soils.

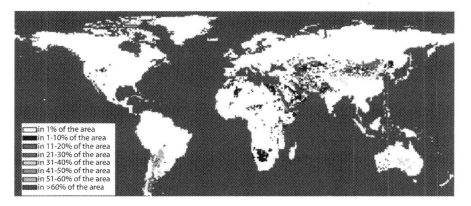

Fig. 10.5 Global distribution of saline soils. (Source FAO/UNESCO Soil Map of the World)

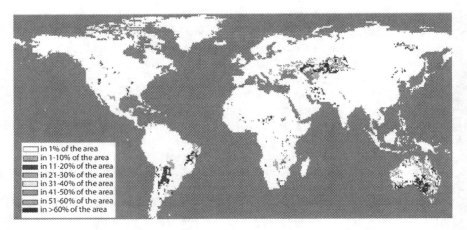

Fig. 10.6 Global distribution of sodic soils. (Source FAO/UNESCO Soil Map of the World)

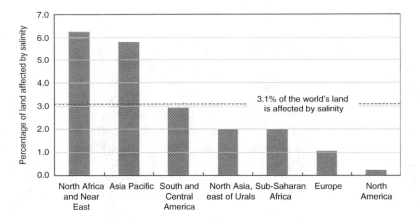

Fig. 10.7 Percentage of land area affected by salinity in different regions. (Source: FAO 2006)

The proportion of land area affected by salinity in different regions of the world is shown in Fig.10.7.

Salt affected soils occur in more than 75 countries of the world, mostly in Africa (31) and Asia (25), and some in the Americas (13). Human induced salinity has invaded over the last few decades to several major irrigation schemes throughout the world (Qadir et al. 2007), including Indo-Gangetic Basin in India (Gupta and Abrol 2000), Indus Basin in Pakistan (Aslam and Prathapar 2006), Yellow River Basin in China (Chengrui and Dregne 2001), Euphrates Basin in Syria and Iraq (Sarraf 2004), Murray-Darling Basin in Australia (Rengasamy 2006), and San Joaquin Valley in the United States (Oster and Wichelns 2003). The regional distribution of saline soils is shown in Table 10.4.

Table 10.4 Regional distribution of saline soils

Region	Area (million hectare)
Africa	69.5
Near and Middle east	53.1
Asia and Far East	19.5
Latin America	59.4
Australia	84.7
North America	16.0
Europe	20.7
Total	**322.9**

Source: FAO/UNESCO Soil Map of the World

Soil salinization is a major threat to agricultural production systems and food security in many countries. Salinity affects almost 1 billion hectare of land worldwide that is equivalent to 7% of earth's continental extent (Metternicht and Zinck 2003; Yensen 2008). Patel et al. (2011) reported that about 7 million hectares of land in India alone have saline soils. Shrivastava and Kumar (2015) noted that not only has high soil salinity affected an estimated area of 20% of total cultivated lands and 33% of global irrigated agricultural lands, but the salinized areas are increasing at an annual rate of 10 percent for various reasons including low precipitation, high evaporation, using saline water for irrigation and poor drainage. Jamil et al. (2011) reported that >50% of global arable land would be affected by soil salinity by 2050.

10.4 Field Indicators of Soil Salinity

There are some symptoms that can suggest whether salinity might be affecting a site. Some of these indicators are listed below. Readers should be aware that these indicators are not always easy to detect. One or more symptoms may develop or no symptom may be apparent at the initial stage of salinization. Moreover, symptoms usually develop in soils or plants at an advanced stage of salinity when remedial measures can be time consuming and costly. It is wiser to prevent salinity than to cure it.

- Patches of leaf-burn in the crop field.
- Patches of salt-tolerant plants developing.
- Increased soil wetness in semiarid and arid areas.
- Irregular patterns of crop growth and lack of plant vigor.
- Areas becoming bare and 'scalded'.
- White spots and streaks in the soil
- White crusting on the surface.

10.5 Effect of Soil Salinity on Plants

Soil salinity exerts its adverse effects on glycophytic plants (any plant that will only grow healthily in soils with a low content of salts) in two different ways:

(a) By reducing total soil water potential through osmotic effects of salts, and
(b) By the activity of specific ions.

The relative contribution of the two mechanisms is not always predictable because of wide variation among soils, salts and plants (Yadav et al. 2011). The initial effect of salinity, especially at low to moderate concentrations, is due to its osmotic effects. Plants cannot absorb enough water from soil if the soil water potential is very low due to the presence of high concentrations of salts in soil solution. In a way, salt stress resembles water stress. However, some plants can adapt to some degree of salinity through osmotic adjustments, so that they can maintain a potential for the influx of water (Ghoulam et al. 2002). They are salt tolerant plants.

Soil salinity has been a major concern to global agriculture throughout human history (Lobell et al. 2007). In the recent years, more saline areas are being brought under cultivation for producing food, wood, biofuel, etc. (Haque 2006). Some crops are very sensitive to even low level of soil salinity, while some others are relatively tolerant. For example, cotton, an important cash crop worldwide, is classified as one of the most salt-tolerant crops and considered a pioneer crop in reclamation of saline soils. But its growth, yield and fiber quality are negatively affected by excessive salts in the soil (Higbie et al. 2010). In general, soil salinity delays and reduces germination and emergence, decreases cotton shoot growth, and may finally lead to reduced seed cotton yield and fiber quality at moderate to high salinity levels (Khorsandi and Anagholi 2009).

Effects of salinity on the growth of crop plants have considerably been investigated worldwide (Essa 2002; FAO 2002; Munns 2005; Maghsoudi and Maghsoudi 2008; Ashraf 2009; Chaum et al. 2011; Nasser and Sholi 2012; Dong 2012; Zeinolabedin 2012; Moradi and Zavareh 2013). It is often suggested that most of the salt stresses in plant are due to the abundance of NaCl and Na_2SO_4 in soil (Amiri et al. 2010). However, there are multiple types of salts in saline soils and the composition of salts also varies greatly among locations (Tobe et al. 2002). Each salt may have a different effect on the growth of plants (Tobe et al. 2003).

Salinity greatly reduces seed germination, seedling growth and crop establishment (Azizi et al. 2011; Naseri et al. 2012; Ewase et al. 2013; Salehifar et al. 2010; Sadeghi 2010), although the effect in early vegetative growth stage is more pronounced (Botia et al. 2005; Wilson et al. 2000; del Amor et al. 2001). Reduction in length and weight of root and shoot at early growth stages in response to salinity has been reported for several crop plants (Garg and Singla 2004; Parida and Das 2005). Salinity reduces leaf number, leaf area and photosynthesis (Sadeghi 2010; Ashraf 2001; Romero-Aranda et al. 2001). High salinity reduces nodulation of legumes (Mudgal et al. 2009; Mudgal et al. 2010). Several investigators have observed adverse effects of salinity on metabolic processes related to the growth of glyco-

10.5 Effect of Soil Salinity on Plants

phytic plants (Dhanapackiam and Ilyas 2010). Salinity seriously affects leaf chlorophyll content, photosynthetic carbon metabolism and photosynthetic efficiency. Salinity drastically affects nitrogen metabolism (Mansour 2000; Santos et al. 2002) and carbon metabolism (Balibrea et al. 2003).

Specific ion effect includes ion toxicity and/or nutritional disorders caused by one or some of the ions commonly associated with salinity including the cations Na^+, Ca^{2+} and Mg^{2+}, and the anions Cl^-, SO_4^{2-} and HCO_3^-. Both Na^+ and Cl^- are toxic to plants. High concentration of NaCl acts antagonistically to the uptake of the other nutrients, such as K^+, Ca^{2+}, N and P. Increased concentration of NaCl in soil solution increases concentrations of Na^+ and Cl^- and reduces concentrations of Ca^{2+}, K^+ and Mg^{2+} in tissues of many plant species (Bayuelo-Jimenez et al. 2003). Accumulation of Na^+ in leaf tissues results in necrosis and abscission of leaves (Parvaiz and Satyawati 2008). Excess Na^+ and Cl^- interfere with the uptake of P, Fe and Zn. The high levels of Na^+ or $Na^+:K^+$ ratio can also disrupt various enzymatic processes in the cytoplasm, and the disruption in protein synthesis appears to be an important cause of damage by Na^+ (Tester and Davenport 2003). Radi et al. (2013) observed inhibition of several physiological and biochemical processes of plants due to salinity. However, the Ca^{2+} significantly affects the salinity responses of plants, particularly during the initial growth (Tobe et al. 2002, 2003).

An increase in the concentration of certain dissolved ions in soil may enhance their uptake, increase their concentrations in plant tissue and become phytotoxic above a level. This level depends on the type of plants, type of plant tissues and conditions of soils. However, ions that may create phytotoxicity in saline soil generally include Cl^-, B (as H_3BO_3 and $H_2BO_3^-$) and Al^{2+}. Specific ionic stresses may disrupt integrity and selectivity of root plasma membrane, homeostasis of essential ions and numerous metabolic activities (Zhu 2001). Many saline soils used for cereal production have low levels of plant-available Zn because Zn may be complexed with or compete with dissolved salts (CO_3^{2-}, SO_4^{2-}, Na^+) at alkaline pHs. Since crops are the principal route of most essential minerals into the human body, salinity may indirectly contribute to mineral deficiency in billions of people. The primary salinity effects give rise to numerous secondary ones such as oxidative stress, characterized by accumulation of reactive oxygen species (H_2O_2, O^{2-}, OH), potentially harmful to biomembranes, proteins, nucleic acids and enzymes (Gomez et al. 2004). Antioxidative enzymes such as superoxide dismutase, catalases, peroxidises, etc. enhance detoxication of reactive oxygen species. The relatively salt-tolerant species (e.g. pea genotypes) have increased activities of certain antioxidative enzymes (Hernandez et al. 2001).

Both Na^+ (a beneficial element) and Cl^- (a micronutrient) are potentially toxic in excessive concentrations, triggering specific disorders and causing substantial damages to crops. Under excessive Na^+ and Cl^- in rhizosphere concentration, there are competitive interactions with other nutrient ions such as K^+, NO_3^-, $H_2PO_4^-$ for binding sites and transport proteins in root cells, and latter for retranslocation, deposition and partitioning within the plant (Tester and Davenport 2003). Significantly enhanced uptake and accumulation of Na and Cl accompanied with a decrease in K concentration in the same tissues was obtained under moderate NaCl salinity

(Ondrasek et al. 2009). The high K/Na and Ca/Na ratios are essential for normal plant functioning (Chinnusamy et al. 2005); but Ondrasek et al. (2009) observed 13-fold decline in Ca/Na under salinity in radish, 113-fold in strawberry and 150-fold in muskmelon leaf tissues.

As a result of all these responses to salinity, uneven and stunted crop stands develop and crop yield is greatly reduced. According to USDA-NRCS (2011), 50 percent yield of the following fruit crops are reduced at EC_e range of 2.5–4.9 dS m^{-1}: grapefruit *(Citrus paradisi)*, orange *(Citrus sinensis)*, peach *(Prunus persica)*, apricot *(Prunus armeniaca)*, almond *(Prunus dulcis)*, plum *(Prunus domestica)*, blackberry *(Rubus sp.)*, boysenberry *(Rubus ursinus)* and strawberry *(Fragaria sp.)*. Yields of potato *(Solanum tuberosum)*, maize *(Zea mays)*, sweet potato *(Ipomoea batatas)*, pepper *(Capsicum annuum)*, lettuce *(Lactuca sativa)*, radish *(Raphanus sativus)*, onion *(Allium cepa)*, carrot *(Daucus carota)*, flax *(Linum usitatissimum)*, broad bean *(Vicia faba)*, bean *(Phaseolus vulgaris)* and turnip *(Brassica rapa)* are reduced by 50% at EC_e values between 3.6 and 5.9 dS m^{-1}.

10.6 Effects of Sodicity on Plant Growth

Sodic soils adversely affect plant growth in one or more of the following ways:

Poor physical conditions give poor tilth and adverse soil moisture-aeration relationships. Excess exchangeable sodium deteriorates the physical properties of soils. Clay particles remain dispersed (Bauder and Brock 2001), so that stable aggregates cannot develop there. For such physical conditions, sodic soils have low permeability to air and water. Dense, impermeable surface crusts may form as a result of dispersion (Falstad 2000). A compact layer of clay develops in the subsoil of some sodic soils due to migration of dispersed clay particles. This layer restricts the development of root system of plants. As a result, the growth of plants is reduced (Qadir et al. 2007; Garg and Malhotra 2008).

Secondly, sodic soils have low availability of some nutrient elements. Excess exchangeable sodium elevates the pH of soils above 8.2. Although soil pH has little direct effect on plants, the extremely alkaline pH of sodic soils lowers the availability of some essential plant nutrients, such as P, Fe, Mn and Zn, with concomitant decrease in their availability to plants. At very high soil pH of sodic soils, the concentration of the elements calcium and magnesium in the soil solution is reduced due to the formation of relatively insoluble calcium and magnesium carbonates by reaction with soluble carbonate of sodium, etc. and results in their deficiency for plant growth.

Thirdly, the accumulation of certain elements in plants at toxic levels may result in plant injury or reduced growth and even death (specific ion effects). Elements more commonly toxic in sodic soils include sodium, molybdenum and boron. Typical sodium toxicity symptoms are leaf burn, scorch and dead tissue along the outside edges of leaves. Sodium toxicity is often modified or reduced if sufficient calcium is available in the soil. For at least a few annual crops, excess sodium in soil may produce calcium deficiency.

A list of plants suffering from toxicities of sodium and boron in sodic soils is given below:

Sodium toxicity Avocado (*Persea Americana*), grapefruit (*Citrus paradisi*), orange (*Citrus sinensis*), peach (*Prunus persica*), beangreen (*Phaseolus vulgaris*), cotton (*Gossypium hirsutum*), maize (*Zea mays*), peas (*Pisum sativum*), mung (*Phaseolus aurus*), mash (*Phaseolus mungo*), lentil (*Lens culinaris*), groundnut (*Arachis hypogaea*), gram (*Cicer arietinum*) and cowpea (*Vigna sinensis*).

Boron toxicity Lemon (*Citrus limon*), blackberry (*Rubus spp.*), avocado (*Persea Americana*), grapefruit (*Citrus paradisi*), orange (*Citrus sinensis*), apricot (*Prunus armeniaca*), peach (*Prunus persica*), cherry (*Prunus avium*), plum (*Prunus domestica*), grape (*Vitis vinifera*), walnut (*Juglans regia*), pecan (*Carya illinoiensis*), cowpea (*Vigna unguiculata*), onion (*Allium cepa*), garlic (*Allium sativum*), sweet potato (*Ipomoea batatas*), wheat (*Triticum eastivum*), barley (*Hordeum vulgare*), sunflower (*Helianthus annuus*), mungbean (*Vigna radiate*), sesame (*Sesamum indicum*), lupine (*Lupinus hartwegii*), strawberry (*Fragaria spp.*), kidney bean (*Phaseolus vulgaris*), lima bean (*Phaseolus lunatus*) and groundnut (*Arachis hypogaea*).

10.7 Reclamation and Management of Saline Soils

Integrated soil, water and crop management practices are needed to manage saline soils for crop production. Many low salinity soils can be profitably used for suitable salt tolerant crops without undertaking time consuming and costly reclamation procedures. In some cases, salts may either be diluted to a tolerable limit or removed by leaching. Some soils are more profitably used for salt farming. Where there is a salt crust on the surface, decrusting may be done by mechanical scraping and with soil flushing for improving crop growth. Leaching is needed in the reclamation of most saline soils. It refers to the removal of excess salts from the soil profile with the percolating water. Leaching requires more water than normal irrigation, and the amount of extra water needed must be apprehended before leaching is done. Some soils may need decrusting before leaching. Some soils are only slightly saline; for shallow-rooted crops the salts may be driven below the root zone by temporary leaching. This technique will need less water than normal leaching. No chemical amendment is normally needed for reclamation of soils which contain excess salts only without excess exchangeable sodium.

10.7.1 Principles of the Management of Saline Soils

Management practices for saline soils may be based on the following principles:

- Selection of salt tolerant crops: A set of crops having threshold values close to the ECe of the soil under consideration should be selected for different crop seasons. It can be a viable option for sustainable management of soils with low salinity without employing significant reclamation efforts. There are some crops that can tolerate moderate to high salinity.
- Dilution of salts in the root zone: Salts affect crops via the roots (unless sprinkler irrigation is done). Salt hazard can be reduced if salt concentration in the root zone can be lowered (drip irrigation with good quality water is an efficient system of diluting salts around the roots).
- Improving soil structure with organic amendments to improve infiltration and hydraulic conductivity. Decompaction and destruction of root restrictive layers improve rooting of crop plants and hydrological conditions of the soil.
- Improving leaching of salts by irrigation and drainage. Extra water is needed to leach excess salts out of the root zone. The quality of irrigation water and draining away the salty water safely should also be important considerations.
- Reducing evaporation with mulch or cover crops. Higher evaporation concentrates soil solution and increases capillary rise of groundwater.
- Maintaining the groundwater table at a safe depth below the root zone.
- Maintaining a crop while reclamation is underway. The crop will be benefitted from the management practices and compensate for the cost of reclamation.

10.7.2 Selection of Salt Tolerant Crops

Several management practices can reduce salt levels in the soil, but it is sometimes either impossible or too costly to reduce soil salinity to the desired levels. In some cases, the only viable management option is to grow salt-tolerant crops. Salt tolerant crops are also helpful and profitable during reclamation of saline soils. Generally, tolerance of a crop to soil salinity is based on the following three criteria:

(i) The ability of the crop to survive on saline soils,
(ii) The yield of the crop on saline soils, and
(iii) Relative yield of the crop on a saline soil as compared to its yield on a nonsaline soil under similar growing conditions.

Salt tolerance on the basis of simple ability of a crop plant to survive is of limited practical significance in irrigation agriculture. Yield on saline soils is of greater agronomic importance, but the third criterion provides a better basis of comparison among diverse crops. In FAO Irrigation and Drainage Paper 29, Ayers and Westcot (1985) discussed irrigation water quality parameters related to soil salinity. They

10.7 Reclamation and Management of Saline Soils

Table 10.5 EC_e values for yield reduction potentials of different field crops

Field Crops	Yield reduction potential at EC_e, dS m^{-1}				
	10%	20%	30%	50%	100%
Barley (*Hordeum vulgare*)	8.0	10	13	18	28
Cotton (*Gossypium hirsutum*)	7.7	9.6	13	17	27
Sugarbeet (*Beta vulgaris*)	7.0	8.7	11	15	24
Sorghum (*Sorghum bicolor*)	6.8	7.4	8.4	9.9	13
Wheat (*Triticum aestivum*)	6.0	7.4	9.5	13	20
Wheat, durum (*Triticum turgidum*)	5.7	7.6	10	15	24
Soybean (*Glycine max*)	5.0	5.5	6.3	7.5	10
Cowpea (*Vigna unguiculata*)	4.9	5.7	7.0	9.1	13
Peanut (*Arachis hypogaea*)	3.2	3.5	4.1	4.9	6.6
Rice (paddy) (*Oriza sativa*)	3.0	3.8	5.1	7.2	11
Sugarcane (*Saccharum officinarum*)	1.7	3.4	5.9	10	19
Corn (maize) (*Zea mays*)	1.7	2.5	3.8	5.9	10
Flax (*Linum usitatissimum*)	1.7	2.5	3.8	5.9	10
Broadbean (*Vicia faba*)	1.5	2.6	4.2	6.8	12
Bean (*Phaseolus vulgaris*)	1.0	1.5	2.3	3.6	6.3

listed common field crops, vegetable crops and fruit crops and their yield reduction potentials at different soil salinity (EC_e) and water salinity (EC_w) levels. Based on their data Tables 10.5, 10.6, and 10.7 were constructed to show the relative tolerances of soil salinity of different field crops, vegetable crops, and fruit crops respectively. Absolute salinity tolerance, however, depends on soil types, crop growth stages, climate and cultural practices. These data can be used as guidelines for the selection of crops for soils of different salinity levels.

The maximum average EC_e value the crop can tolerate without any decline in yield is known as the threshold value. The percent loss in yield for each unit increase in EC_e above the threshold is known as slope coefficient.

The following lists of salt sensitive and salt tolerant crops can be prepared from different unpublished web sources:

Sensitive crops (threshold ECe values in parentheses) bean (1.0), carrot (1.0), strawberry (1.0), onion (1.2), almond (1.5), blackberry (1.5), plum (1.5), apricot (1.5), orange (1.7), peach (1.7), ground nut (1.8), etc.

Moderately tolerant to tolerant crops (threshold ECe values in parentheses) Red beet (4.0), harding grass (4.6), squash (4.7), cowpea (4.9), soybean (5.0), birdsfoot trefoil (5.0), perennial rye grass (5.6), durum wheat (5.7), barley forage (6.0), wheat (6.0), sorghum (6.8), sugar beet (7.0), cotton (7.7), barley (8.0), etc.

Cl$^-$ tolerant crops Perennial ryegrass, durum wheat, barley (forage), wheat, sorghum Bermuda grass, sugar beet, wheat grass, crested fairway, cotton, tall wheat grass, barley, etc.

Table 10.6 EC$_e$ values for yield reduction potentials of different vegetable crops

Vegetable Crops.	Yield reduction potential at EC$_e$, dS m^{-1}				
	10%	20%	30%	50%	100%
Beet, red *(Beta vulgaris)*	4.0	5.1	6.8	9.6	15
Squash, scallop *(Cucurbita pepo)*	3.2	3.8	4.8	6.3	9.4
Broccoli *(Brassica oleracea botrytis)*	2.8	3.9	5.5	8.2	14
Tomato *(Lycopersicon esculentum)*	2.5	3.5	5.0	7.6	13
Cucumber *(Cucumis sativus)*	2.5	3.3	4.4	6.3	10
Spinach *(Spinacia oleracea)*	2.0	3.3	5.3	8.6	15
Celery *(Apium graveolens)*	1.8	3.4	5.8	9.9	18
Cabbage *(Brassica oleracea capitata)*	1.8	2.8	4.4	7.0	12
Potato *(Solanum tuberosum)*	1.7	2.5	3.8	5.9	10
Corn, sweet (maize) *(Zea mays)*	1.7	2.5	3.8	5.9	10
Sweet potato *(Ipomoea batatas)*	1.5	2.4	3.8	6.0	11
Pepper *(Capsicum annuum)*	1.5	2.2	3.3	5.1	8.6
Lettuce *(Lactuca sativa)*	1.3	2.1	3.2	5.1	9.0
Radish *(Raphanus sativus)*	1.2	2.0	3.1	5.0	8.9
Onion *(Allium cepa)*	1.2	1.8	2.8	4.3	7.4
Carrot *(Daucus carota)*	1.0	1.7	2.8	4.6	8.1
Bean *(Phaseolus vulgaris)*	1.0	1.5	2.3	3.6	6.3
Turnip *(Brassica rapa)*	0.9	2.0	3.7	6.5	12

Table 10.7 ECe values for yield reduction potentials of different fruit crops

Fruit Crops	Yield reduction potential at EC$_e$, dS m^{-1}				
	10%	20%	30%	50%	100%
Date palm *(phoenix dactylifera)*	4.0	6.8	11	18	32
Grapefruit *(Citrus paradisi)*	1.8	2.4	3.4	4.9	8.0
Orange *(Citrus sinensis)*	1.7	2.3	3.3	4.8	8.0
Peach *(Prunus persica)*	1.7	2.2	2.9	4.1	6.5
Apricot *(Prunus armeniaca)*	1.6	2.0	2.6	3.7	5.8
Grape *(Vitus sp.)*	1.5	2.5	4.1	6.7	12
Almond *(Prunus dulcis)*	1.5	2.0	2.8	4.1	6.8
Plum, prune *(Prunus domestica)*	1.5	2.1	2.9	4.3	7.1
Blackberry *(Rubus sp.)*	1.5	2.0	2.6	3.8	6.0
Boysenberry *(Rubus ursinus)*	1.5	2.0	2.6	3.8	6.0
Strawberry *(Fragaria sp.)*	1.0	1.3	1.8	2.5	4

The following list of trees and shrubs was prepared from data of the College of Agriculture and Life Sciences, Virginia Polytechnic Institute and State University (Anon 2015):

Trees tolerant to saline soils Black walnut *(Juglans nigra)*, Eastern red cedar *(Juniperus virginiana)*, Golden raintree *(Koelreuteria paniculata)*, Southern magnolia *(Magnolia grandiflora)*, Sweetbay magnolia *(Magnolia virginiana)*, Carolina cherrylaurel (*Prunus caroliniana*), Japanese black pine (*Pinus thunbergiana*), White poplar (*Populus alba*), Red oak *Quercus rubra*), Live oak (*Quercus virginiana*), Bur oak (*Quercus macrocarpa*), Black locust (*Robinia pseudoacacia*), Japanese tree lilac (*Syringa reticulate*), Bald cypress (*Taxodium distichum*),Chaste tree (*Vitex anguscastus*), White oak (*Quercus alba*), and Pin oak (*Quercus palustris*).

Trees tolerant to salt spray Common larch (*Larix deciduas*), Sweetgum (*Liquidambar styraciflua*), Black gum (*Nyssa sylvatica*), Colorado spruce (*Picea pungens*), Austrian pine (*Pinus nigra*), Longleaf pine (*Pinus palustris*), Japanese black pine (*Pinus thunbergiana*), White poplar (*Populus alba*), Black cherry (*Prunus serotina*), Bur oak (*Quercus macrocarpa*),Willow oak (*Quercus phellos*), English oak (*Quercus robur*),Black locust (*Robinia pseudoacacia*), Weeping willow (*Salix alba*),Corkscrew willow (*Salix matsudana*), Japanese pagodatree (*Sophora japonica*), and Japanese tree lilac (*Syringa reticulate*).

Shrubs tolerant to saline soils or salt spray Red chokeberry (*Aronia arbutifolia*), Saltbush (*Baccharis halmifolia*), Littleleaf boxwood (*Buxus microphylla*), Beautyberry (*Callicarpa americana*), False cypress (*Chamaecyparis pisifera*), Summersweet (*Clethra alnifolia*), Red osier dogwood (*Cornus sericea*), Spreading cotoneaster (*Cotoneaster divaricatus*), Rockspray cotoneaster (*Cotoneaster horizontalis*), Scotch broom (*Cytisus scoparius*), Gardenia (*Gardenia jasminoides*), Rose-of-Sharon (*Hibiscus syriacus*), House hydrangea (*Hydrangea macrophylla*), St. John's wort (*Hypericum calycinum*), Chinese holly (*Ilex cornuta*), Japanese holly (*Ilex crenata*), Inkberry (*Ilex glabra*), Yaupon holly (*Ilex vomitoria*), Anise (*Illicium floridanum*), Chinese juniper (*Juniperus chinensis*), Common juniper (*Juniperus communis*), Shore juniper (*Juniperus conferta*), Creeping juniper (*Juniperus horizontalis*), Amur privet (*Ligustrum amurense*), Tatarian honeysuckle (*Lonicera tatarica*), Wax myrtle (*Myrica cerifera*), Bayberry (*Myrica pennsylvanica*), Mock orange (*Philadelphus coronaries*), Mugo pine (*Pinus mugo*), Shrubby cinquefoil (*Potentilla fruticosa*), Purple-leaf sand cherry (*Prunus x cistena*), Cherry laurel (*Prunus laurocerasus*), Beach plum (*Prunus maritime*),Pyracantha (*Pyracantha coccinea*),Indian hawthorn (*Rhapiolepis indica*),Staghorn sumac (*Rhus typhina*), Lady Banks rose (*Rosa banksiae*), Rugosa rose (*Rosa rugosa*), Scotch rose (*Rosa spinosissima*), Elderberry (*Sambucus Canadensis*), Japanese spirea (*Spiraea japonica*), Bumalda Japanese spirea (*Spiraea x bumalda*), Snowberry (*Symphoricarpos albus*), Lilac (*Syringa vulgaris*), Tamarisk (*Tamarix ramosissima*), English yew (*Taxus baccata*), Japanese yew (*Taxus cuspidate*), Highbush blueberry (*Vaccinum corymbosum*), Arrowwood (*Viburnum dentatum*), and European cranberry bush (*Viburnum opulus*).

10.7.3 Salt Scraping

Scraping is the simplest way of reclaiming a saline soil on a small scale (Fig. 10.8). In soils of arid and semiarid regions where a salt crust often develops on the soil

Fig. 10.8 Salt scraping. (Image source: Rutger2 at nl.wikipedia)

surface due to high evaporation and low leaching (Kara and Willardson 2006), this method of mechanical removal of salts is employed. In such environmental conditions, this practice is appropriate as suggested by Kang et al. (2013).

Removal of the salt layer exposes the relatively low salinity subsoil which gives a better yield. But, this process has a limited success because salts tend to accumulate again. Salts further accumulate due to lowering the ground level through salt scraping or soil desurfacing in relation to the water table. In other words, this method intensifies the problem over time. Scraping of salts improves plant growth only temporarily. Moreover, disposal of scraped salts is a problem. Scraping also involves high labor cost.

10.7.4 Salt Flushing

Salts in surface crusts in soils with low permeability are sometimes washed away by flushing with water over the surface. For flushing, the field is first ponded with water to dissolve the salts. Soil flushing is particularly feasible where flushed salts can be disposed of with irrigation or rainwater in nearby natural drainage systems such as rivers. A sufficient downward gradient is required to carry the water away. Therefore, this method is not practical in landlocked fields. Otherwise, it may pollute surrounding areas. Salt flushing is most effective initially; the efficiency decreases as salt concentrations diminish. In this method only the surface area is being somewhat remedied, the rest of the soil profile is still contaminated. This method is also of limited practical significance because the amount of salts that can be flushed away from a soil is rather small and salts keep accumulating on the surface again. Fig. 10.9 shows a field being flushed for salinity reclamation.

Fig. 10.9 A layer of salt on soil surface is being flushed. (Image courtesy of USDA.NRCS)

10.7.5 Leaching

Removal of excess salts below the root zone and through the soil profile with percolating water is known as leaching. Leaching is a natural process in well drained soils of the humid regions. In arid and semiarid regions, there is not enough rain water to leach the salts released by weathering or accumulated by capillary rise of groundwater. There leaching is aided by irrigation if enough water is available. Soil permeability is usually low and the groundwater table is very high in saline soils of the humid regions. So, artificial drainage is needed to remove excess water and salts and to lower the groundwater table. However, in areas of good natural drainage, there is no need of artificial drainage for leaching the salts. There the problem is the capillary rise of salty groundwater to the soil surface due to low precipitation and high evaporation. Leaching of salts in permeable soils is best performed by an integration of irrigation and drainage. In impermeable soils, leaching might be efficient if the permeability could be improved by organic amendments prior to the leaching effort. Moreover, several other factors are involved in the success of the leaching process. Some of these factors are: (i) electrical conductivity (ECe) of the soil, (ii) number of units of ECe to be lowered, (iii) amount of water used, (iv) salinity level of irrigation water (ECw), (v) irrigation system (vi) soil texture (vii) salt tolerance of the crop (viii) depth of roots, etc.

However, leaching is by far the most effective method of the reclamation of saline soil. Leaching should preferably be done when the soil moisture content is low and the groundwater table is deep.

Border flooding is a method of irrigation in which the field is surrounded by an earthen border to restrict the water within it. Border flooding irrigation can be used for leaching of salts from saline soils with either continuous ponding or intermittent ponding. Before flooding, the land is leveled to spread the water evenly over the field plowed to facilitate infiltration and percolation. When the continuous ponding method is used, the field is deeply flooded with a single application of water and the water is allowed to percolate. In intermittent ponding method several applications of water at suitable intervals are given in small amounts at any one time. Continuous ponding method is usually used in soils with better permeability so that percolation and leaching processes are faster. Generally, the amount of water required for such a leaching method is about one centimeter for reclaiming each centimeter of the soil profile. However, the amount of water needed for leaching may depend on soil texture. The intermittent ponding method is used in areas of low water availability. The amount of water for leaching also depends on the salinity level of the irrigation water (as discussed in the context of leaching requirement, LR). As water with a low salt-content is not always available, studies have revealed that saline groundwater can also be used for leaching, provided that the amount of water needed would be higher. There are some practices including subsoiling, subsoiling with inversion, auger hole piercing, etc. for improving permeability and breaking impermeable and compacted layers. There may be some cases where less permeable layers overlie highly permeable subsoils. Augur piercing by making about 10 cm diameter holes up to a depth of 2.5 m and then filling the holes with sand may improve rate of leaching. Use of some amendments (sand, gypsum, manures) can also be used to enhance leaching in impermeable soils.

The following factors determine the amount of water needed for salt reclamation through leaching – salt content of the soil, salinity level to achieve, depth to which reclamation is desired, type of crops grown and soil characteristics. A useful rule of thumb is that a unit depth of water will remove nearly 80 percent of salts from a unit soil depth. Thus 30 cm water passing through the soil will remove approximately 80 percent of the salts present in the upper 30 cm of soil. To leach soluble salts in irrigated soils, more water than required to meet the evapotranspiration needs of the crops must pass through the root zone. This additional irrigation water has typically been expressed as the leaching requirement (LR). Leaching requirement was originally defined as the fraction of infiltrated water that must pass through the root zone to keep soil salinity from exceeding a level that would significantly reduce crop yield under steady conditions with associated good management and uniformity of leaching.

$$LR = \frac{EC_{iw}}{EC_{dw}},$$

Where LR is the leaching requirement, EC_{iw} is electrical conductivity of irrigation water, and EC_{dw} is electrical conductivity of drainage water. Alternatively, leaching requirement can be obtained from the following equation.

$$\frac{C_s}{C_i} = \frac{\ln(\text{LR})}{\text{LR}-1},$$

10.7 Reclamation and Management of Saline Soils

where C_s is the concentration of salts in soil and C_i concentration of salts in irrigation water. This equation is independent of the pattern of plant water use, it is of general application and the authors concluded that the equation provided a better method of estimating the leaching requirements. Ayers and Westcot (1985) also presented the following equation as a guideline for calculating LR based on irrigation water salinity and crop salt tolerance.

$$LR = \frac{EC_w}{5EC_e - EC_w}$$, where EC_w is the electrical conductivity of irrigation water. The amount of water required may be calculated from the following equation:

$$\text{Amount of water to be applied} = \frac{\text{Consumptive use}}{1 - LR}$$

Consumptive use includes the amount of water transpired during plant growth plus what is evaporated from the soil surface and foliage in the crop area. It should be noted that the uniformity of water distribution over the field and the field application efficiency have not been considered in the above calculation. Several other leaching requirement models have been proposed by Corwin et al. (2007). An example of calculating the amount of water needed for leaching is given in Box 10.1.

BOX 10.1 Calculation of Leaching Requirement

Suppose, you are going to leach salts from a soil where alfalfa will be grown. The threshold ECe for alfalfa is 2.0 dS m^{-1} (salinity level for 100% yield or no yield loss) and the leaching water has an ECw of 3.1 dS m^{-1}. What is the leaching requirement? If the consumptive use of water for alfalfa in that location is 120 cm per hectare, then calculate the amount of water to be applied to meet crop needs and to leach salts.

Solution:

$$LR = \frac{3.1}{5 \times 2.0 - 3.1} = 0.45$$

Expressed in percent, it means that 45% of the irrigation water must pass through the root zone to provide sufficient leaching.

$$\text{The amount of water needed} = \frac{120}{1 - 0.45} = 218 \text{ cm ha}^{-1}$$

Answer: The LR is 0.45 and the amount of water needed is 218 cm ha^{-1}. This amount does not include distribution, conveyance, surface runoff, or other losses.

Khoshgoftarmanesh et al. (2003) conducted an experiment with five irrigation treatments to observe their effects of saline soil reclamation in Iran. The treatments were conventional irrigation practice (L0), two excess irrigation on preplanting (L1), one excess irrigation on preplanting (L2), one excess irrigation after planting (L3), and two excess irrigation after planting (L4). Changes of soil salinity were measured at two depths of 0–25 and 25–50 cm at preplant, 3–4 days after every leaching or irrigation and after harvesting. Except for treatment L0, the final ECe values of surface soil were below the EC of irrigation water. Electrical conductivity in surface soil, after plant harvesting, decreased from 67.1 dS m^{-1} before experiment) to 7.1 dS m^{-1} by L1 treatment. Jamali et al. (2012) carried out another experiment in Sindh Agriculture University for assessing the leaching effects of salts on soil. They used the following treatments: three irrigation water amounts (7.62 cm, 10.16 cm and 12.70 cm) and three interval periods (7, 14 and 21 days). The experiment was done in a Randomized Complete Block Design with four replications. The results showed that in top soil (0–30 cm) and middle soil profiles (30–60 cm), salts have been leached more than 40 percent by consuming 45.72 cm, 60 cm and 76 cm water within 7 and 14 days intervals.

10.7.6 Irrigation

Unsaturated conditions in the field during leaching can be obtained by adopting intermittent ponding or by intermittent sprinkling at rates less than the infiltration rate of the soil. Sprinkler irrigation has been found to be more effective than flooding for leaching of salts. Because of slower wetting rate under sprinkling, the zone of complete leaching at the end of irrigation extends more deeply into the profile than under flood irrigation. Drip irrigation has some advantages over flooding and sprinkler irrigation in respect to salinity control. Drip irrigation does not cause foliar accumulation of salts during irrigation. Salts are removed from the wetted areas of the rows where root density is the highest. High frequency drip irrigation applications can maintain a relatively constant soil water content and soil salinity level over time near the drip lines. Hanson and May (2004) reported that highly profitable tomato crops were obtained in saline soils through drip irrigation. However, drip irrigation system can cause salt accumulation near the periphery of the wetted areas which may be a matter of major concern if the emitter placement does not coincide well with the plant row. Salt accumulation above buried pipelines is also a concern.

Water applied in furrow irrigation moves downward and laterally from the furrows, and upward to the ridges by capillary movement. Salts are leached downward beneath the furrows, but some salts rise to the center of the ridges with water. Water is lost through evaporation and transpiration and salts accumulate. So, salts tend to concentrate in the center of the ridges in the furrow irrigation system. If seeds are placed there, they would suffer in germination and seedling emergence. Fig. 10.10 shows the distribution of salts following furrow irrigation. Seeds or seedlings must

10.7 Reclamation and Management of Saline Soils

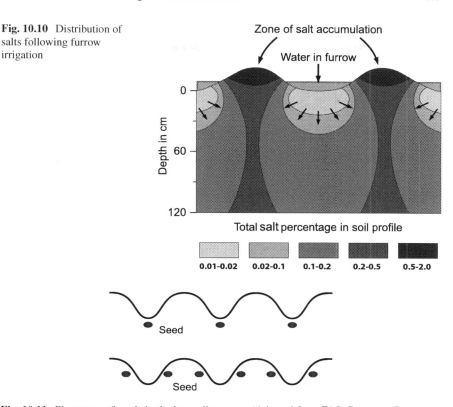

Fig. 10.10 Distribution of salts following furrow irrigation

Fig. 10.11 Placement of seeds in the low saline areas. (Adapted from FAO Corporate Document Repository; Irrigation, Water Management, Chap. 3. Furrow Irrigation. www.fao.org/docrep/S8684E/s8684e04.htm)

be placed in the low saline areas (Fig. 10.11). Certain modifications of the furrow irrigation method including planting in single/double rows or on sloping beds, are helpful in getting better stands under saline conditions. With double beds, most of the salts accumulate in the center of the bed leaving the edges relatively free of salts. Sloping beds may be slightly better on highly saline soils because seeds can be planted on the slope below the zone of salt accumulation.

Double-row bed systems require uniform wetting toward the middle of the bed. This leaves the sides and shoulders of the bed relatively free from injurious levels of salinity. Alternate furrow irrigation may be desired for single-row bed systems. This is accomplished by irrigating every other furrow and leaving alternating furrows dry. Salts are pushed across the bed from the irrigated side of the furrow to the dry side (Fig. 10.12). Care is needed to ensure enough water is applied to wet all the way across the bed to prevent build up in the planted area.

Quality of irrigation water

Natural waters contain sediments and dissolved salts. These substances have an impact on the suitability of water for irrigation. Sediments in water can clog the

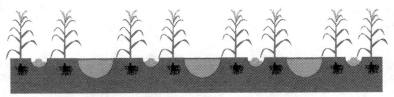

Good uniformity; salts accumulate in the center of the bed and away from plants

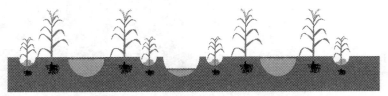

Poor uniformity; salts accumulate toward edge of bed near one row

Uniform; healthy plants with alternate furrow irrigation (salt accumulates in the dry furrow)

Fig. 10.12 Types of furrow irrigation for salinity control. (Adapted from Colorado State University Extension Fact Sheet; http://www.ext.colostate.edu/pubs/crops/00503.html)

nozzles of sprinklers and emitters of drip systems. Dissolved salts, on the other hand, can increase soil salinity and damage crops through toxicity. Salts in water are dissociated into cations and anions. The principal cations are Na^+, Ca^{2+}, Mg^{2+} and K^+ and the anions are Cl^-, SO_4^{2-}, NO_3^-, and HCO_3^-. Concentrations of Na^+ and Cl^- in water often exceed the critical levels. Therefore, both the salinity and the specific ion effects of irrigation water are important considerations in relation to salinity management. The salinity of water is expressed by the quantity of dissolved salts, or total dissolved solids (TDS), the unit being mg l^{-1}. This can be determined by evaporating a known amount of water to dryness in a glass or porcelain dish and measuring the mass gain. The higher the TDS, the higher is the salinity of water. Alternatively, the salinity of water can be expressed by its electrical conductivity (EC$_w$). Total dissolved solids can be calculated from EC$_w$ and vice-versa according to this formula:

$$TDS, mg\, l^{-1} = 640 \times ECw\left(dS\, m^{-1}\right)$$

It has been a matter of interesting debate and investigation whether saline ground water can be used for irrigation. The basic principle behind a sustainable agricultural system based on irrigation with saline waters in terms of long term crop yield is that the salt concentration in the root-zone has to be kept below a certain threshold level (Karlberg 2005). These levels are specific for each crop species. However, water of variable salinity <4 dS m^{-1} can be used for irrigation, but the leaching requirement and the amount of water needed for irrigation will also vary accordingly.

The electrical conductivity values range from <0.6 dS m^{-1} for fresh water to 1.5–3.0 dS m^{-1} for brackish water to about 45 dS m^{-1} for sea water. It has been observed that even slightly brackish water can affect the production of many salt-sensitive crops including bean and strawberry, and brackish water affects the moderately-sensitive crops corn, rice, potato and alfalfa. On the other hand, irrigation with brackish water may be feasible for salt tolerant crops, such as barley, sugar beet and cotton (Pitman and Lauchli 2002).

Often water of some degree of salinity or sodicity is used for irrigation. The use of such marginal-quality water would not only permit the horizontal expansion of irrigated agriculture, but would also reduce drainage disposal and associated environment problems (Oster and Grattan 2002). The use of saline-sodic water for irrigation without amendment tended to increase pH slightly after harvest of the crop. In Table 10.8, use of mixed saline and nonsaline water at various proportions for irrigation in different crops and their effects on yield is shown. It was adapted from Qadir and Oster (2004).

Reclaimed wastewater and drainage water are being used for irrigation increasingly. Table 10.9 presents the recommended limits for constituents in reclaimed water for irrigation.

Table 10.8 Proportion of saline water (%) that can be mixed with nonsaline irrigation water (ECw = 0.8 dS m^{-1}) to achieve a potential yield of 100 percent and 80 percent for selected crops

Crop	EC$_w$ dS m^{-1}			
	4	6	8	10
	Saline water proportion (%)			
100 percent yield				
Lettuce (*Lactusa sativa*)	2	2	1	1
Alfalfa (*Medicago sativa*)	14	9	6	5
Tomato (*Lycopersicon esculentum*)	25	15	11	9
Cotton (*Gossypium hirsutum*)	100	62	44	35
80 percent yield				
Lettuce (*Lactusa sativa*)	37	23	17	13
Alfalfa (*Medicago sativa*)	80	52	39	31
Tomato (*Lycopersicon esculentum*)	78	48	35	27
Cotton (*Gossypium hirsutum*)	100	100	100	100

Table 10.9 Recommended limits for constituents in reclaimed water for irrigation

Constituent	Long-term use (mg l^{-1})	Short-term use (mg l^{-1})
Aluminum	5.0	20
Arsenic	0.10	2.0
Beryllium	0.10	0.5
Boron	0.75	2.0
Cadmium	0.01	0.05
Chromium	0.1	1.0
Cobalt	0.05	5.0
Copper	0.2	5.0
Fluoride	1.0	15.0
Iron	5.0	20.0
Lead	5.0	10.0
Lithium	2.5	2.5
Manganese	0.2	10.0
Molybdenum	0.01	0.05
Nickel	0.2	2.0
Selenium	0.02	0.02
Vanadium	0.1	1.0
Zinc	2.0	10.0

Source: EPA 2004

10.7.7 Drainage

Irrigation water brings salts of saline soils into solution, and these salts are usually leached to the ground water or can be driven away by deep surface drainage systems. Deep ditches at suitable intervals are dug and connected to a collecting ditch in surface irrigation systems. The size of the field ditches and collecting ditches depends on the volume of water to be discharged. In this system, some land becomes unavailable for cropping. An effective subsurface drainage system must be constructed to remove the salt laden groundwater, to keep the groundwater table at a safe depth below and to prevent the groundwater table to rise at or near the root zone. Various drainage systems have been described in Chap. 5 in connection with management of poorly drained soils, and all these systems can be employed to reclaim saline soils if soil conditions, such as texture, porosity, permeability, level of soil salinity, depth of groundwater table, and irrigation facility, irrigation water availability, etc. suit them. According to Sival et al. (2009), the traditional method of salinity reclamation is to pond the field with water and to leach the salts from the soil by the sub-surface tile drainage systems. Tiles can be made of clay, concrete, plastic and other synthetic materials, such as high density polyethylene and polyvinylchloride. Presently, perforated synthetic pipes are used instead of earthen tiles which were popular in the past. The holes in the pipes are covered with thin sheets of fiber glass or spun nylon to avoid clogging and preventing inflow of soil so that the system may be long lasting. The installation of tile drainage system is, however, relatively costly and may not be very effective in poorly permeable soils. Mole drains may be suitable for subsurface drainage of clay soils, but are not usually used for soil salinity reclamation because moles can collapse in coarse textured soils.

10.8 Management of Dryland Salinity

Dryland salinity is a serious problem in Western Australia and in the Great Plains region of North America. It occurs extensively in Canada in Manitoba, Saskatchewan and Alberta and in the United States in the states of Montana, North and South Dakota. Dryland salinity is also a problem in South Africa, Iran, Afghanistan, Thailand and India. These soils range in salinity from slightly saline to severe saline conditions. Soil salinity is a major problem in areas having low precipitation to evaporation ratio. Soluble salts released by the limited chemical weathering cannot be leached due to the scarcity of water. Saline seeps often develop in places where recharge areas have light textured soils having low moisture retention and high permeability but with an underlying impermeable layer. Seepage of saline water occurs somewhere at low points in the landscape. Long-term solutions to the problem of saline seeps should include land use changes with the objective of modifying the hydrological state. In addition, site specific treatments of salt-affected land are required to restore their productivity. Management practices for saline seeps should be aimed at increasing water use in the recharge areas to decrease the excess water seepage. This can be achieved by (i) intensive cropping for increased water use in the recharge area, (ii) growing deep rooted perennials that would draw water from greater depth of the soil, (iii) drainage in the recharge area, and (iv) growing salt tolerant crops in the seep area. The management options of dryland salinity of Australia were discussed systematically by CSIRO (2000).

10.9 Reclamation and Management of Sodic Soils

Several different approaches including chemical amendments, tillage operations, crop-management practices, hydrological manipulations, and electrical currents have been used to ameliorate sodic and saline-sodic soils (Qadir et al. 2007). A number of tillage options, such as deep plowing and sub-soiling have also been used to break up the shallow, dense, sodic clay pans and or natric horizons that occur within 0.4 m of the soil's surface (Abdelgawad et al. 2004). Management of sodic soils also requires leaching through irrigation and drainage, but before leaching, excess exchangeable sodium must be replaced by some suitable cation (Ca^{2+}) so that the ESP can be lowered to the desired level along with growing suitable crops. The replacement needs amendments that contain calcium or that produce calcium ions after application to the soil. A brief account of the conventional reclamation procedures is given below.

10.9.1 Crop Selection for Sodic Soils

It has long been known from classical agronomy literature that there are wide variations in the tolerance of crops to soil sodicity: rice (*Oryza sativa*) and dhaincha (*Sesbania aculeata*) appear to be tolerant, wheat (*Triticum aestivum*) and bajra (*Pennisetum typhoideum*) are only moderately tolerant and legume crops like mash (*Phaseolus mungo*) and lentil (*Lens culinaris*) are relatively sensitive to excess exchangeable sodium. A list of plants having some degree of tolerance to excess sodium and boron (usually high content in sodic soils) is given below:

Crops tolerant to high exchangeable sodium Karnal grass (*Diplachne fusca*)**,** rhodes grass (*Chloris gayana*), para grass (*Brachiaria mutica*), Bermuda grass (*Cynodon dactylon*), rice (*Oryza sativa*), dhaincha (*Sesbania aculeata*) and sugar beet (*Beta vulgaris*).

Crops semi-tolerant to high exchangeable sodium Wheat (*Triticum aestivum*), barley (*Hordeum vulgare*), oats (*Avena sativa*), raya (*Brassica juncea*), senji (*Melilotus parviflora*), bajra (*Pennisetum typhoides*), cotton (*Gossypium hirsutum*), berseem (*Trifolium alexandrinum*), sugarcane (*Saccharum officinarum*).

Crops moderately tolerant to boron Cabbage (*Brassica oleracea capitata*), turnip (*B. rapa*), Kentucky bluegrass (*Poa pratensis*), barley (*Hordeum vulgare*), cowpea (*Vigna unguiculata*), oats (*Avena sativa*), corn (*Zea mays*), artichoke (*Cynara scolymus*), tobacco (*Nicotiana tabacum)*, mustard (*Brassica juncea*), clover, sweet (*Melilotus indica*), squash (*Cucurbita pepo*), muskmelon (*Cucumis melo*), and cauliflower (*B. oleracea botrytis*).

Crops tolerant to boron Alfalfa (*Medicago sativa*), purple vetch (*Vicia benghalensis*), parsley (*Petroselinum crispum*), red beet (*Beta vulgaris*), sugar beet (*B. vulgaris*), and tomato (*Lycopersicon lycopersicum*).

Crops very tolerant to boron (threshold values mg l^{-1} are given in parentheses) Sorghum (*Sorghum bicolor*) (7.4), cotton (*Gossypium hirsutum*) (6.0–10.0), celery (*Apium graveolens*) (9.8), and asparagus (*Asparagus officinalis*) (10.0–15.0).

A list of commercially available plant species tolerant to saline-sodic soils is prepared below on the basis of data of USDA-NRCS (2010).

Very highly tolerant Beardless wildrye, tall wheatgrass, altai wildrye, hybrid wheatgrass, slender wheatgrass and Russian wildrye.

Highly tolerant Tall fescue, western wheatgrass, fairway wheatgrass, crested wheatgrass, standard wheatgrass, Siberian wheatgrass, forage kochia, fourwing saltbush, winterfat and strawberry clover.

10.9 Reclamation and Management of Sodic Soils

Moderately tolerant Creeping foxtail, meadow brome, smooth brome, pubescent wheatgrass, intermediate wheatgrass, thick spike wheatgrass, yellow sweet clover and Cicer milkvetch.

10.9.2 Amendments for Sodic Soils

Several chemical amendments have been used most extensively for the reclamation of sodic soils. Most common amendments include (i) soluble calcium compounds such as gypsum and calcium chloride which produce Ca^{2+} ions that replace exchangeable Na^+ ions, (ii) sulfuric acid, and (iii) substances that produce sulfuric acid such as elemental sulfur, pyrite, ferrous sulfate, and aluminium sulfate. Sulfuric acid thus formed reacts with insoluble calcium carbonate commonly found in sodic soils and release soluble Ca^{2+} ions. The reactions of the various amendments and their mechanism of action are shown below.

Gypsum
Gypsum reacts with both Na2CO3, and adsorbed sodium as follows:

$$Na_2CO_3 + CaSO4 = CaCO_3 + Na_2SO_4 \,(\text{leachable})$$

$$Na - Clay - Na + CaSO_4 = Ca - Clay + Na_2SO_4 \,(\text{leachable})$$

Calcium chloride
Calcium chloride (CaCl2.2H2O) is a highly soluble salt which supplies soluble calcium directly. Its reactions in sodic soil are similar to those of gypsum.

$$Na_2CO_3 + CaCl_2 = CaCO_3 + 2NaCl\,(\text{leachable})$$

$$Na - Clay - Na + CaCl = Ca - Clay + 2NaCl\,(\text{leachable})$$

Sulfuric acid
Upon application of sulfuric acid (H2SO4) to soils containing calcium carbonate, it immediately reacts to form calcium sulfate and thus provides soluble calcium indirectly.

$$Na_2CO_3 + H_2SO_4 = CO_2 + H_2O + Na_2SO_4 \,(\text{leachable})$$

$$CaCO_3 + H_2SO_4 = CaSO_4 + H_2O + CO_2$$

$$Na - Clay - Na + CaSO_4 = Ca - Clay + Na_2SO_4 \,(\text{leachable})$$

Iron sulfate and aluminium sulfate
Both of iron sulfate (FeSO4.7H2O) and aluminium sulfate [Al2(SO4)3.18H2O] are solid granular materials soluble in water. When applied to soils, these compounds hydrolyze to form sulfuric acid.

$$FeSO_4 + 2H_2O = H_2SO_4 + Fe(OH)_2$$

$$H_2SO_4 + CaCO_3 = CaSO_4 + H_2O + CO_2$$

$$Na - Clay - Na + CaSO_4 = Ca - Clay + Na_2SO_4 \text{ (leachable)}$$

Aluminium sulfate acts similarly.

Sulfur
Sulfur (S) is not soluble in water and does not supply calcium directly for replacement of adsorbed sodium. When applied to soils, sulfur undergoes oxidation to form sulfuric acid and gives reactions as shown below:

$$2S + 3O_2 = 2SO_3$$

$$SO_3 + H_2O = H_2SO_4 \text{ (Sulfuric acid performs as above)}$$

Pyrite
Pyrite (FeS2) is another possible amendment for sodic soil reclamation. The following series of reactions occur after applying pyrite to soil:

$$\text{Step I.} 2FeS_2 + 2H_2O + 7O_2 = 2FeSO_4 + 2H_2SO_4$$

$$\text{Step II.} 4FeSO_4 + O_2 + 2H_2SO_4 = 2Fe_2(SO_4)_3 + 2H_2O$$

$$\text{Step III.} Fe_2(SO_4)_3 + FeS_2 = 3FeSO_4 + 2S$$

$$\text{Step IV.} 2S + 3O_2 + 2H_2O = 2H_2SO_4 \text{ (Sulfuric acid performs as above)}$$

The choice of an amendment depends on (a) relative effectiveness as judged from the improvement of soil properties and crop growth, (b) relative costs involved, (c) the risk of handling materials, and (d) time required for an amendment to react in the soil and effectively replace adsorbed sodium. Sulfuric acid reacts immediately with the soil calcium carbonate to release soluble calcium for exchange with sodium. Elemental sulfur must be oxidized by soil bacteria and react with water to form sulfuric acid. The formation of sizeable amounts of sulfuric acid from elemental sulfur may take several months to several years. Sulfur, sulfuric acid, iron sulfate, and aluminum sulfate do not supply calcium. They are useful for reclamation only when the soil contains lime. When the soil to be reclaimed does not contain enough lime, gypsum or materials that contain calcium are to be used. Iron and aluminium sulfates are usually too costly and have not been used for any large-scale improvement of sodic soils in the past. Because amendments like sulfur and pyrite must first be oxidized in soil to sulfuric acid, the amendments are relatively slow acting. Being the cheapest and most abundantly available, gypsum is the most widely used amendment.

There were some studies earlier to compare the effectiveness of various amendments at improving the physical and chemical properties of saline sodic and sodic soils (Amezketa et al. 2005; Hanay et al. 2004). The relative effectiveness of gypsum and sulfuric acid has received the most attention because they are widely used as reclamation amendments. Recently, crops, crop residues or industrial effluents and synthetic polymers have been included in efficiency studies (Hanay et al. 2004). Gypsum, as mentioned above, is the most commonly used amendment due to its availability and low cost (Joachim et al. 2007), but it is slow in reaction. However, its efficiency can be increased if applied at variable rates according to the gypsum requirements of the soil. According to Barros et al. (2004), application of gypsum and a mixture of gypsum and limestone were more efficient for combating soil sodicity. Agar (2011) used polyphosphogypsum (PG) as amendment to reclaim saline and sodic soils in Turkey. He added a total of 7.5 and 15.0 t PG ha^{-1} in divided doses to the field plots of 100 m^2 in three consecutive years. Results showed that the cultivation process and plant roots contributed to the improvement of soil physical and chemical properties. Application of 7.5 and 15.0 t PG ha^{-1} in divided doses respectively caused soil improvement 10 and 8 times more than gypsum treatments. Moreover, leaching 50% of the soluble salt from the soil profile required a depth of leaching water approximately 2.3 times of soil depth to be reclaimed. Divided doses of PG amendments resulted in a better reclamation process. Prapagar et al. (2012) conducted a pot experiment on saline-sodic soil in Sri Lanka with some amendments to compare their abilities of improving soil quality parameters. The treatments were: T1-control (no amendment), T2–1% cow dung (CD), T3–1% partially burnt paddy husk (PH), T4-gypsum (GYP) (100% Gypsum Requirement), T5–1% CD + GYP, T6–1% PH + GYP. The amendments T2 to T6 were applied on the soil surface and incubated at room temperature (31 ± 1 °C) for 90 days. After incubation, leaching was done for 42 days with 3 liter water per pot at 7 days intervals up to 6 cycles. After incubation and leaching of soils, onion variety Wallara-60 was grown in these pots. Results showed significant differences in bulk density, electrical conductivity (EC), pH, exchangeable Ca^{2+} and SAR among the treatments in comparison to the control. The highest reduction in SAR and EC were recorded in the treatment GYP + PH (T6). Yield of onion was also the highest in this treatment. Joachim et al. (2007) observed beneficial effect of the combined use of farm yard manure (FYM) and GYP on the reclamation of sodic soils. Equivalent quantities of some amendments for sodic soils are given in Table 10.10.

Amount of Amendments

The quantity of an amendment for sodic soil reclamation depends on the quantity of exchangeable sodium to be replaced. This will also depend on soil texture and mineralogical make up, exchangeable sodium percentage (ESP) and the crops to be

Table 10.10 Equivalent quantities of some amendments

Amendment	Relative quantity[a]
Gypsum (CaSO$_4$ 2H$_2$O)	1.00
Calcium chloride (CaCl$_2$ 2 H$_2$O)	0.85
Sulphuric acid (H$_2$SO$_4$)	0.57
Iron sulphate (FeSO$_4$.7 H$_2$O)	1.62
Aluminium sulfate (Al$_2$ (SO$_4$)$_3$.18 H$_2$O)	1.29
Sulphur (S)	0.19
Pyrite (FeS$_2$) – 30% sulphur	0.63
Calcium polysulfide (CaS$_5$) – 24% sulfur	0.77

[a]These quantities are based on 100 percent pure materials. If the material is not 100 percent pure, necessary correction must be made. Thus, if gypsum is only 80 percent pure. $1.0 \times (100 \div 80) = 1.25$, the quantity to be added will be 1.25 ton instead of 1.00 ton.

grown. The relative tolerance of a crop to exchangeable sodium and its normal rooting depth will largely determine the soil depth up to which excess adsorbed sodium must be replaced for satisfactory crop growth. Replacement of each mole of adsorbed sodium per 100 g soil will require half a mole of soluble calcium. The quantity of pure gypsum required to supply half a cmol of calcium per kg soil for the upper 15 cm soil depth will be

$$\frac{\text{molecular weight of gypsum}}{200} = \frac{172}{200} = 0.86 \text{ g kg}^{-1} \text{ soil}$$

$$= 86 \times 10^{-5} \text{ kg kg}^{-1} \text{soil}$$

$$= 86 \times 10^{-5} \times 2.24 \times 10^{6} \text{ kg ha}^{-1}$$

$$= 1926 \text{ kg or } 1.96 \text{ t ha}^{-1}$$

If it is desired to replace greater quantities of adsorbed sodium, the quantity of gypsum can be accordingly increased.

Gypsum requirement

The quantity of gypsum required for reclaiming a sodic soil is termed as the gypsum requirement (GR). The GR test is performed by mixing a small soil sample (5 g) with a relatively large volume of saturated gypsum solution and measuring the calcium lost from the solution after reaction with the soil. The decrease in calcium from the solution can be expressed on the basis of tons of CaSO$_4$.2H$_2$O per hectare 30 cm of soil. Many sodic soils contain appreciable amounts of soluble sodium carbonate in addition to excess exchangeable sodium. In such cases, the gypsum requirement test evaluates the amount of calcium required to replace the exchangeable sodium plus the amount required to neutralize all the soluble sodium carbonate in the soil. Gypsum is usually applied on the soil surface and leached, only a small fraction of the soluble carbonates reacts with applied calcium and a major fraction of the soluble carbonates are leached without reacting with applied gypsum. It does not seem necessary to add extra gypsum to neutralize soluble carbonates. Gypsum requirement can also be calculated by the following formula (Horneck et al. 2007).

$$GR = (present\ ESP - desired\ ESP) \times CEC \times 0.021$$

The factor of 0.021 assumes CEC is in meq/100 g or cmol (+charge) kg^{-1} units.

Method of application
Gypsum and other amendments except sulfuric acid can be applied by broadcasting on the surface and then incorporating within a shallow depth by disking or plowing. It is more effective than spreading gypsum only on the surface. When the problem of exchangeable sodium is only mild, gypsum applied in dissolved form was found more beneficial for the establishment of pasture in comparison to soil application treatments. Deep plowing (up to 100 cm) has been reported to be a useful practice for improving sodic soils with hardpans or dense clay subsoil layers. The success of deep plowing chiefly depends on the mixing of low-clay calcareous or gypsiferous subsoil material with high-clay B horizon material to provide a more favorable physical matrix for water movement and root penetration and to provide a source of calcium for replacement of exchangeable sodium in the profile.

Leaching a sodic soil after amendment needs more water than a saline soil because (i) chemical amendments, except sulfuric acid and calcium chloride, are less soluble than salts in soil, and (ii) sodic soils are generally less permeable than saline soils. Amelioration of sodic and saline-sodic soils with high clay content, particularly where montmorillonite is the dominant clay mineral, is technically difficult and expensive. When water is applied for leaching, the clay swells rapidly, destroying the macro-pores which provide the primary drainage pathways. The hydraulic conductivity of the fully saturated soil is usually very low and drainage cannot be provided at economic spacing (Qureshi et al. 2007). Naseri and Rycroft (2002) found that extensive swelling occurred when low salinity water (EC = 0.5 dS m^{-1}, SAR = 0.6) was used for leaching. Increasing calcium concentration in the leaching water reduced swelling during the leaching process, and controlled the dispersion.

Particle size of ground gypsum
After collecting gypsum from natural deposits it is finely ground for applying in crop fields. It is held that the finer the particles of gypsum, the more effective it is for the reclamation of sodic soils. Treatment of soil with very finely ground gypsum results in high initial hydraulic conductivity which decreased sharply with time. But treatment with 2 mm mesh gypsum helped in maintaining permeability at a higher level and for a longer period. Their results showed that higher solubility of finer particles caused them to react with free sodium carbonate, inactivating the soluble calcium due to the formation of insoluble calcium carbonate.

Organic amendments

It has been already mentioned that some organic amendments are used for reclaiming sodic and saline-sodic soils. For example, Prapagar et al. (2012) observed that the addition of organic amendments to soils is highly accepted by the farmers. Reclamation systems involving composts, green manures, crop rotation and other biological agents are termed as biodynamic systems by Ansari and Ismail (2008). In comparison to the conventional methods, biodynamically managed soils have better physical, chemical and biological properties. Often organic amendments are used in addition to gypsum (Mohamed and Abdel-Fattah 2012). Application with 10 t ha^{-1} FYM in combination with chemical amendments improved soil physical properties like bulk density, porosity, void ratio, water permeability and hydraulic conductivity and enhanced rice and wheat yields in sodic soil (Hussain et al. 2001). Other organic materials like rice straw, wheat straw, rice husk and chopped salt grass also improved these physical properties of a saline sodic soil.

Pressmud

Pressmud, also known as filter cake or filter mud, is produced in sugar mills through sedimentation of the suspended materials such as fiber, sugar, wax, ash, soil and other particles from the cane juice (Muhammad and Khattak 2009). It has been used as an ameliorant in sodic and saline-sodic soils (Barry et al. 2001). The pressmud contains considerable amount of sulfate, extracted by the method of sulfidation is called sulfidation pressmud cake. Pressmud, which contains high amount of carbonates, is extracted by the method of carbonation and is known as carbonated pressmud cake. Pressmud usually contains about 70% lime, 15–20% organic matter and 2–3% sugar (Khattak and Khan 2004). The ash comprises of oxides of Si, Ca, P, Mg and K (Partha and Sivasubramanian 2006). The organic matter is highly soluble and readily available for microbial activity and, therefore, to the soil (Rangaraj et al. 2007). Carbon dioxide produced through microbial activity may increase the solubility of lime and hence its effectiveness in reclaiming saline-sodic soils (Qadir et al. 2006). Some investigators have observed an increase in yields of various crops including maize and millet with pressmud applications (Rangaraj et al. 2007; Elsayed et al. 2008) and an improvement in soil physical, chemical and biological conditions (Barry et al. 2001).

10.9.3 Phytoremediation of Sodic Soils

The amelioration of saline, saline-sodic, sodic and calcareous sodic soils can be done by using some plants, crops or cropping systems. This is known as phytoremediation. Qadir et al. (2007) reviewed phytoremediation of sodic soils, and they pointed out that experiments with various cropping systems and on different soils for phytoremediation began in 1920s. The overall decrease in ESP under the phytoremediation treatment may even be greater than that obtainable with gypsum.

10.9 Reclamation and Management of Sodic Soils

Various cropping systems have been used. The mechanism of action of the crop plants involves the physical action of roots to modify soil aggregation, porosity and permeability, root respiration producing CO_2, enhanced proton in the root zone, production of H_2CO_3, and dissolution of native $CaCO_3$ releasing Ca^{2+} ions that replaces Na^+ ions from clay surfaces. Dissolution and precipitation kinetics of calcite are determined by the chemistry of the system. A typical reaction for the dissolution of calcite may be expressed as a function of CO_2 in the root zone:

$$CaCO_3 + CO_2 + H_2O \Leftrightarrow Ca^{2+} + 2HCO_3^-$$

The reaction is the resultant of three processes which occur concurrently: (1) conversion of CO_2 in soil solution into H_2CO_3 and its reaction with $CaCO_3$ (2) dissociation of H_2CO_3 into H^+ and HCO_3^- and the reaction of H^+ with $CaCO_3$ (3) dissolution of $CaCO_3$ resulting in Ca^{2+} and CO_3^{2-}.

$$CaCO_3 + H_2CO_3 \Leftrightarrow Ca^{2+} + 2HCO_3^-$$

$$CaCO_3 + H^+ \Leftrightarrow Ca^{2+} + HCO_3^-$$

$$CaCO_3 + H_2O \Leftrightarrow Ca^{2+} + CO_3^{2-} + H_2O$$

Amelioration takes place by all these processes if adequate leaching can be provided. Since phytoremediation ameliorates the root zone, alternating crops of differential root habit should be included in the crop sequence (Akhter et al. 2003). Again, deep-rooted crops have advantages in terms of greater depth of soil amelioration. For example, alfalfa roots can penetrate as deep as 1.2 m in the soil.

10.9.4 Management of Calcareous Saline-Sodic Soils

In calcareous (soils that effervesce upon addition of dilute HCl) saline-sodic soils, the free calcium carbonate can contribute little to the replacement of Na + ions because the calcium carbonate is insoluble in water. Instead of supplying Ca^{2+} ions by the addition of gypsum, the native calcium carbonate in these soils can be dissolved to release soluble Ca^{2+} ions that would replace exchangeable Na+. This can be achieved by (i) addition of sulfuric acid or sulfur, and (ii) biological production of carbonic acid through root and microbial respiration. The products of the reaction of $CaCO_3$ and sulfuric acid are CO_2, H_2O, SO_4^{2-}, and Ca^{2+}. The Ca^{2+} ions replace exchangeable Na^+, reduces the ESP and act as a flocculant. Any acid can dissolve soil $CaCO_3$ and release the bound Ca. Sulfuric acid is most common because it is relatively inexpensive and adds less salt to the soil. Elemental sulfur is converted to sulfuric acid by sulfur oxidizing bacteria, producing the same effect as sulfuric acid. Sulfur conversion is a biological process, however, and requires several weeks to months to take place unlike acids, which react instantly. The phytoremediation technique has been discussed in the preceding section.

10.9.5 Fertilizers for Sodic Soils

Low organic matter content, high pH and low biological activity are responsible for the general deficiency of available nitrogen in sodic soils. Considerable volatilization losses of N as ammonia may occur due to their high pH. Poor soil structure and high pH also adversely affect the transformations and availability of applied nitrogenous fertilizers, such as hydrolysis of applied urea. It is generally recommended that crops grown in sodic soils be fertilized at 25 percent excess over the rates recommended for normal soils. Sodic soils generally contain high amounts of extractable phosphorus and crops usually do not respond to applied P. Presence of sodium carbonate in these soils results in the formation of soluble sodium phosphates. However, if the soil contains significant amounts of sodium carbonate most of the soil calcium is in the calcium carbonate form and not available to the plants. Increasing soil sodicity was found to result in reduced uptake of potassium by most crops. Sodicity nearly always results in an increased uptake of sodium and decreased uptake of calcium by plants. When the exchangeable sodium levels are very high, calcium is often the first limiting nutrient. Without amendments, calcium limitation is difficult to be corrected. High pH, low organic matter content and presence of calcium carbonate strongly modify the availability of micronutrients to plants grown in sodic soils. Zinc deficiency has been widely reported for crops grown in sodic soils. Application of 10 kg $ZnSO_4 ha^{-1}$ may mitigate the deficiency of Zn in rice grown in an amended, highly sodic soil. Iron and manganese may also become deficient in sodic soils. Addition of iron and manganese salts to correct the deficiency may not be useful without changing the oxidation state by submergence.

10.10 Environmental Impact of Saline Soil Reclamation

Soil reclamation refers to the mechanical, chemical or biological eradication of the problem(s) a soil poses to a particular use. Scrapping of salt-crust, removal of excess soluble soils by leaching and removing soil acidity by liming are some examples of soil reclamation. Since soil problems are use-oriented, we can often avoid reclamation by adopting an alternative use or an innovative but sustainable management practice. Soil reclamation is usually large scale operations involving time and money for heavy machineries, chemicals, and labors. Reclamation of saline and sodic soils requires construction of bunds, ponding of water, treatment with gypsum or other soil modifiers, and installations of suitable irrigation and drainage systems. It also needs the modification of the existing environmental settings so that salinity/sodicity does not develop again. So, reclamation of saline and sodic soils may often cause drastic effects on the ecosystems (Wang et al. 2014). After an area is reclaimed and developed, intensification and diversification of cropping increase along with an increased use of agrochemicals, including fertilizers, soil modifiers and pesticides. Construction of industrial, commercial and recreational facilities

(Hadley 2009) causes significant alteration in landscape and ecosystem functionality (Lie et al. 2008). These alterations may profoundly impact physical and chemical properties of soil and water (Sun et al. 2011). Salts and nutrients leached down to the groundwater table may find their way to nearby lakes and streams and pollute them. Chemicals used to reclaim sodic soils and the metals mobilized by saline and sodic soil reclamation may degrade sediment-groundwater system (Chen and Jiao 2008). These changes are noticed over several decades following reclamation (Cui et al. 2012).

Study Questions
1. Explain the criteria classifying salt affected soils. Give an account of the distribution of salt affected soils. Write down the characteristics of saline, sodic and saline sodic soils.
2. Describe the effects of soil salinity and sodicity on plants. Mention some salt sensitive and salt tolerant crop plants with their threshold ECe values.
3. Cite the principles of saline soil management. Discuss leaching as the process of salinity reclamation.
4. What chemical substances are used as amendments of sodic soils? How can you measure the suitability of an amendment?
5. Write notes on (a) saline seeps, (b) organic amendments for sodic soil, (c) chloride and boron tolerant plants, (d) nutrient deficiency and toxicity in high pH soils, and (e) phytoremediation of sodic soils.

References

Abdelgawad A, Arslan A, Awad F, Kadouri F (2004) Deep plowing management practice for increasing yield and water use efficiency of vetch, cotton, wheat and intensified corn using saline and non-saline irrigation water. Proceedings of the 55th IEC Meeting of the International Commission on Irrigation and Drainage (ICID), September 9–10, 2004, Moscow, Russia

Agar AI (2011) Reclamation of saline and sodic soil by using divided doses of phosphogypsum in cultivated condition. Afr J Agric Res 6(18):4243–4252

Akhter J, Mahmood K, Malik KA, Ahmed S, Murray R (2003) Amelioration of a saline sodic soil through cultivation of a salt-tolerant grass *Leptochloafusca*. Environ Conserv 30:168–174

Akramkhanov A, Sommer R, Martius C, Hendrickx JMH, Vlek PLG (2008) Comparison and sensitivity of measurement techniques for spatial distribution of soil salinity. Irrig Drain Syst 22:115–126

Amezketa E, Aragüés R, Gazol R (2005) Efficiency of Sulfuric Acid, Mined Gypsum, and Two Gypsum By-Products in Soil Crusting Prevention and Sodic Soil Reclamation. Agronomy Journal 97(3):983–989

Amiri B, Assareh MH, Jafari M, Rasuoli B, Arzani B, Jafari AA (2010) Effect of salinity on growth, ion content and water status of glasswort (Salicorniaherbacea L.) Caspian J Env Sci 8(1):79–87

Anon (2015) Trees and shrubs that tolerate saline soils and salt spray drift. Publication 430–031, College of Agriculture and Life Sciences, Virginia Polytechnic Institute and State University, www.ext.vt.edu

Ansari AA, Ismail SA (2008) Biodynamic management in sodic soils. J Soil Nature 2(2):01–04

Ashraf M (2001) Relationships between growth and gas exchange characteristics in some salt-tolerant amphidiploid Brassica species in relation to their diploid parents. Environ Exp Bot 45:155–163

Ashraf M (2009) Biotechnological approach of improving plant salt tolerance using antioxidants as markers. Biotech Adv 27:84–93

Aslam M, Prathapar SA (2006). Strategies to mitigate secondary salinization in the Indus Basin of Pakistan: A Selective Review. Research Report 97. International Water Management Institute (IWMI), Colombo, Sri Lanka

Ayers RS, Westcot DW (1985) Water quality for agriculture. FAO Irrigation and Drainage Paper 29. Food and Agriculture Organization of the United Nations, Rome

Azizi M, Chehrazi M, Zahedi SM (2011) Effects of salinity stress on germination and early growth of sweet William (*Dianthus barbatus*). Asian J Agric Sci 3(6):453–458

Balibrea ME, Cuartero J, Bolarín MC, Pérez-Alfocea F (2003) Activities during fruit development of *Lycopersicon* genotypes differing in tolerance salinity. Physiol Plant 118:38–46

Barros MFC, Fontes MPF, Alvarez VVH, Ruiz HA (2004) Reclamation of salt-affected soils in Northeast Brazil with application of mined gypsum and limestone. Revista Brasileira de Engenharia Agricola e Ambiental 8(1):59–64

Barry GA, Rayment GE, Jeffery AJ, Price AM (2001) Changes in cane soil properties from application of sugar mill by-products. Proceeding Conference of the Australian Society of Sugarcane Technology. Mackay, Queensland

Bauder JW, Brock TA (2001) Irrigation water quality, soil amendment, and crop effects on sodium leaching. Arid Land Res Manag 15:101–113

Bayuelo-Jiménez JS, Debouck GD, Lynch JP (2003) Growth, gas exchange, water relations, and ion composition of *Phaseolus* species grown under saline conditions. Field Crop Res 80:207–222

Botia P, Navarro JM, Cerda A, Martinez V (2005) Yield and fruit quality of two melon cultivars irrigated with saline water at different stages of development. Eur J Agron 23:243–253

Chaum S, Pokasombat Y, Kirdmanee C (2011) Remediation of salt affected soil by gypsum and farmyard manure–importance for the production of jasmine rice. Aust J Crop Sci 5:458–465

Chen K, Jiao JJ (2008) Metal concentrations and mobility in marine sediment and groundwater in coastal reclamation areas: a case study in Shenzhen, China. Environ Pollut 151(3):576–584

Chengrui M, Dregne HE (2001) Silt and the future development of China's Yellow River. Geogr J 167:7–22

Chinnusamy V, Jagendorf A, Zhu JK (2005) Understanding and improving salt tolerance in plants. Crop Sci 45:437–448

Corwin DL, Rhoades JD, Simunek J (2007) Leaching requirement for soil salinity control: steady-state versus transient models. Agric Water Manag 9:65–180

CSIRO (2000) Management of dryland salinity: future strategic directions. Primary Industries Report Series 78, CSIRO Publishing, PO Box 1139 (150 Oxford Street), Collingwood, Australia

Cui J, Liu C, Li Z, Wang L, Chen X, Ye Z, Fang C (2012) Long-term changes in topsoil chemical properties under centuries of cultivation after reclamation of coastal wetlands in the Yangtze Estuary, China. Soil Tillage Res 123:50–60

del Amor FM, Martinez V, Cerda A (2001) Salt tolerance of tomato plants as affected by stage of plant development. Hort Sci 36:1260–1263

Dhanapackiam S, Ilyas MHM (2010) Effect of salinity on chlorophyll and carbohydrate contents of Sesbania grandiflora seedlings. Indian J Sci Technol 3(1):64–66

Dong H (2012) Technology and field management for controlling soil salinity effects on cotton. AJCS 6(2):333–341

Elsayed MT, Babiker MH, Abdelmalik ME, Mukhtar ON, Montange D (2008) Impact of filter mud application on the germination of sugarcane and small-seeded plants and on soil and sugarcane nitrogen contents. Bioresour Technol 99:4164–4168

EPA (2004) Guidelines for Water Reuse, September 2004, EPA/625/R-04/108, Environment Protection Agency, USA

References

Essa TA (2002) Effect of salinity stress on growth and nutrient composition of three soybean (Glycine max L. Merrill) Cultivars. J Agron Crop Sci 188(2):86–93

Ewase ASS, Omran S, El-Sherif S, Tawfik N (2013) Effect of salinity stress on coriander (*Coriandrum sativum*) seeds germination and plant growth. Egypt Acad J BiolSci 4(1):1–7

Falstad J (2000) Soil condition. transplant status in Burger Draw. Billings Gazette. Prepared by D.G. Steward Page. Burger Draw Comments and Recommendations. 6/06/00

FAO (2002) Global network on integrated soil management for sustainable use of salt-affected soils. FAO Land and Plant Nutrition Management Services, Rome, Italy

FAO (Food and Agricultural Organization of the United Nations) (2006) TerraSTAT database. At: <http://www.fao.org/ag/agl/agll/terrastat/>

Flagella Z, Cantore V, Giuliani MM, Tarantio E, De Caro A (2002) Crop salt tolerance physiological, yield and quality aspects. Rec Res Dev Plant Biol 2:155–186

Garg N, Singla R (2004) Growth, photosynthesis, nodule nitrogen and carbon fixation in the chickpea cultivars under salt stress. Braz J Plant Physiol 16(3):137–146

Garg VK, Malhotra S (2008) Response of *Nigella sativa* L. to fertilizers under sodic soil condition. J Med Aromat Plant Sci 30:122–125

Ghoulam C, Foursy A, Fares K (2002) Effects of salt stress on growth, inorganic ions and proline accumulation in relation to osmotic adjustment in five sugar beet cultivars. Environ Exp Bot 47:39–50

Gomez JM, Jimenz A, Olmas E, Sevilla F (2004) Location and effects of long-term NaCl stress on superoxide dismutase and ascorbate peroxidase isoenzymes of pea (*Pisum sativum* cv. Puget) chloroplasts. J Exp Bot 55:119–130

Gregory PJ, Nortcliff S (2013) Soil conditions and plant growth. Wiley Blackwell. New York

Gupta RK, Abrol IP (2000) Salinity build-up and changes in the rice-wheat system of the Indo-Gangetic Plains. Exp Agric 36:273–284

Hadley D (2009) Land use and the coastal zone. Land Use Policy 26S:S198–S203

Hanay A, Büyüksönmez F, Kızıloglu FM, Canbolat MY (2004) Reclamation of saline-sodic soils with gypsum and MSW compost. Compos Sci Utili 12:175–179

Hanson BR, May DM (2004) Effect of subsurface drip irrigation on processing tomato yield, water table depth, soil salinity and profitability. Ag Water Mgt 68:1–17

Haque SA (2006) Salinity problems and crop production in coastal regions of Bangladesh. Pak J Bot 38(5):1359–1365

Hernandez JA, Ferrer MA, Jimenez A, Barcelo AR, Sevilla F (2001) Antioxidant systems and O^{2-}/H_2O_2 production in the apoplast of pea leaves. Its relation with salt-induced necrotic lesions in minor veins. Plant Physiol 127:817–831

Higbie SM, Wang F, McD SJ, Sterling TM, Lindemann WC, Hughs E, Zhang J (2010) Physiological response to salt (NaCl) stress in selected cultivated tetraploid cottons. Int. J Agron 1:1–12

Horneck DA, Ellsworth JW, Hopkins BG, Sullivan DM, Stevens RG (2007) Managing salt-affected soils for crop production. A Pacific Northwest Extension publication Oregon State University, University of Idaho, Washington State University

Hussain N, Hassan G, Arshadullah M, Mujeeb F (2001) Evaluation of amendments for the improvement of physical properties of sodic soil. Int J Agric Biol 3:319–322

Jamali LA, Ibupoto KA, Chattah SH (2012) Effects of salt leaching on soil under different irrigation amounts and intervals. Sixteenth International Water Technology Conference, IWTC 16 2012, Istanbul, Turkey

Jamil A, Riaz S, Ashraf M, Foolad MR (2011) Gene expression profiling of plants under salt stress. Crit Rev Plant Sci 30(5):435–458

Joachim HJR, Makoi P, Ndakidemi A (2007) Reclamation of sodic soils in northern Tanzania, using locally available organic and inorganic resources. African J Biotech 6(16):1926–1931

Kang DJ, Endo A, Seo YJ (2013) Effect of soil scraping on the reclamation of tsunami-damage paddy soil. J crop Sci. Biotech 16(3):219–223

Kara T, Willardson LS (2006) Leaching requirements to prevent soil salinization. J Applied Sci 6(7):1481–1489

Karlberg L (2005) Irrigation with saline water using low-cost drip-irrigation systems in sub-Saharan Africa. PhD Thesis, KTH Architecture and the built environment, Stockholm, Sweden

Khan MA, Duke NC (2001) Halophytes- a resource for the future. Wetl Ecol Manag 6:455–456

Khattak RA, Khan RJ (2004) Evaluation, reclamation and management of saline sodic soils in Kohat division. Final Report. Deptartment of Soil and Environmental Sciences, NWFP Agriculture University Peshawar. National Drainage Programme (NDP), Govt of Pakistan

Khorsandi F, Anagholi A (2009) Reproductive compensation of cotton after salt stress relief at different growth stages. J Agron Crop Sci 195:278–283

Khoshgoftarmanesh AH, Shariatmadari H, Vakil R (2003) Reclamation of saline soils by leaching and barley production. Commun Soil Sci Plant Anal 34(19–20):2875–2883

Lie HJ, Cho CH, Lee S, Kim ES, Koo BJ, Noh JH (2008) Changes in marine environment by a large coastal development of the Saemangeum Reclamation Project in Korea. Ocean and. Polar Res 30(4):475–484

Lobell DB, Ortiz-Monsterio JI, Gurrola FC, Valenzuuela L (2007) Identification of saline soils with multiyear remote sensing of crop yields. Soil Sci Soc Am J 71:777–783

Maghsoudi AM, Maghsoudi K (2008) Salt stress effect on respiration and growth of germinated seeds of different Wheat (*Triticum aesativum* L.) cultivars. World J Agri Sci 4(3):351–358

Mansour MM (2000) Nitrogen containing compounds and adaptation of plants to salinity stress. Biol Plant 43:491–500

Martinez-Beltran J, Manzur CL (2005) Overview of salinity problems in the world and FAO strategies to address the problem. Proceedings of the International Salinity Forum, April 2005, Riverside, California

Metternicht GI, Zinck JA (2003) Remote sensing of soil salinity: potentials and constraints. Remote Sens Environ 85:1–20

Mohamed K, Abdel-Fattah (2012) Role of gypsum and compost in reclaiming saline-sodic soils. IOSR J Agric Vet Sci 1(3):30–38

Moradi P, Zavareh M (2013) Effects of salinity on germination and early seedling growth of chickpea (*Cicer arietinum* L.) cultivars. Intl J Farm Alli Sci 2(3):70–74

Mudgal V, Madaan N, Mudgal A, Mishra S (2009) Changes in growth and metabolic profile of Chickpea under salt stress. J Appl Biosci 23:1436–1446

Mudgal V, Madaan N, Mudgal A, Singh A, Kumar P (2010) Comparative study of the effects of salinity on plant growth, nodulation, and Legheamoglobin Content in Kabuli and Desi Cultivars of *Cicer arietinum* (L.). KBM. J Biol 1:1–4

Muhammad D, Khattak RA (2009) Growth and nutrient concentrations of maize in pressmud treated saline-sodic soils. Soil Environ 28(2):145–155

Munns R (2005) Genes and salt tolerance: bringing them together. Tansley rev. New Phytol 167:645–663

Naseri AA, Rycroft D (2002) Effect of swelling and overburden weight on hydraulic conductivity of a restructured saline sodic clay. Paper presented at the 17th World Congress of Soil Science, 14–21 August 2002, Bangkok, Thailand

Naseri R, Emami T, Mirzaei A, Soleymanifard A (2012) Effect of salinity (sodium chloride) on germination and seedling growth of barley (Hordeum Vulgare vulgare L.) cultivars. Int J Agric Crop Sci 4(13):911–917

Nasser JY, Sholi C (2012) Effect of salt stress on seed germination, plant growth, photosynthesis and ion accumulation of four tomato cultivars. Am J Plant Physiol 7:269–275. American

Ondrasek G, Romic D, Rengel Z, Romic M, Zovko M (2009) Cadmium accumulation by muskmelon under salt stress in contaminated organic soil. Sci Tot Enviro 407:2175–2182

Oster JD, Grattan SR (2002) Drainage water reuse. Irrig Drain Syst 16:297–310

Oster JD, Wichelns D (2003) Economic and agronomic strategies to achieve sustainable irrigation. Irrig Sci 22:107–120

Parida AK, Das AB (2005) Salt tolerance and salinity effects on plants. A review. Ecotox Environ Safety 60:324–349

Partha N, Sivasubramanian V (2006) Recovery of chemicals from pressmud- A sugar industry waste. Indian Chem Eng Section 48(3):160–163

Parvaiz A, Satyawati S (2008) Salt stress and phyto-biochemical responses of plants – a review. Plant Soil Environ 54:89–99

Patel BB, Patel Bharat B, Dave RS (2011) Studies on infiltration of saline–alkali soils of several parts of Mehsana and Patan districts of north Gujarat. J Appl Technol Environ Sanit 1(1):87–92

Pitman MG, Lauchli A (2002) Global impact of salinity and agricultural ecosystems. In: Lauchli A, Luttge U (eds) Salinity: environment – plants – molecules. Kluewer Academic Publishers, Dordrecht

Prapagar K, Indraratne SP, Premanandharajah P (2012) Effect of soil amendments on reclamation of saline-Sodic soil. Trop Agric Res 23(2):168–176

Qadir M, Noble MD, Schubert S, Thomas RJ, Arslan A (2006) Salinity induced land degradation and its sustainable management: problems and prospects. Land Degrad Dev 17:661–676

Qadir M, Oster JD (2004) Crop and irrigation management strategies for saline-sodic soils and water aimed at environmentally sustainable agriculture. Sci Total Environ 323:1–19

Qadir M, Oster JD, Schubert S, Noble AD, Sahrawat KL (2007) Phytoremediation of sodic and saline-sodic soils. Adv Agron 96:197–247

Qadir M, Wichelns D, Raschid-Sally L, McCornick PG, Drechsel P, Bahri A, Minhas PS (2010) The challenges of wastewater irrigation in developing countries. Agric Water Manag 97:561–568

Qureshi AS, Qadir M, Heydari N, Turral H, Javadi A (2007) A review of management strategies for salt-prone land and water resources in Iran.International Water Management Institute, IWMI Working Paper 125, Colombo, Sri Lanka

Radi AA, Farghaly FA, Hamada AM (2013) Physiological and biochemical responses of salt-tolerant and salt-sensitive wheat and bean cultivars to salinity. J Biol Earth Sci 3(1):B72–B88

Rangaraj T, Somasundaram E, Amanullah MM, Thirumurgan V, Ramesh S, Ravi S (2007) Effect of Agro-industrial wastes on soil properties and yield of irrigated finger millet (*Eleusine coracana* L. Gaertn) in coastal soil. Res J Agric Biol Sci 3(3):153–156

Rengasamy P (2006) World salinization with emphasis on Australia. J Exp Bot 57:1017–1023

Richards LA (1954) Diagnosis and improvements of saline and alkali soils. USDA Agriculture Handbook 60. USA

Romero-Aranda R, Soria T, Cuartero S (2001) Tomato plant-water uptake and plant-water relationships under saline growth conditions. Plant Sci 160:265–272

Sadeghi H (2010) The effects of different salinity levels on some important physiological characteristics of two wheat cultivars. 11th Iranian Crop Science Congress, Environmental Sciences research Institute, Shahid Beheshti University, Tehran, Iran

Salehifar M, Torang A, Farzanfar M, Salehifar M (2010) Comparison of salinity stress effect on germination and seedling growth in 8 lines genotypes of common bean (*Phaseolus vulgaris*). 11th Iranian Crop Science Congress, Environmental Sciences research Institute, Shahid Beheshti University, Tehra, Iran

Santos CV, IP F˜a, Pinto GC, Oliveira H, Loureiro J (2002) Nutrient responses and glutamate and proline metabolism in sunflower plants and calli under Na_2SO_4 stress. J Plant Nutr Soil Sci 165:366–372

Sarraf M (2004) Assessing the costs of environmental degradation in the Middle East and North Africa countries. Environment Strategy Notes, No. 9. Environment Department, World Bank, Washington, DC

Shirokova YI, Forkutsa I, Sharafutdinova N (2000) Use of electrical conductivity instead of soluble salts for soil salinity monitoring in Central Asia. Irrig Drain Syst 14:199–205

Shrivastava P, Kumar R (2015) Soil salinity: a serious environmental issue and plant growth promoting bacteria as one of the tools for its alleviation. Saudi J Biol Sci 22(2):123–131

Sival AA, Skaggs TH, Van Genuchten MT (2009) Reclamation of saline soils by partial ponding: simulations for different soils. Vadose Zone J 9(2):486–495

Soil Survey Staff (1993) Soil Survey Manual, Handbook 18, USDA, NRCS. US Gov Print Off, Washington, DC

Sun Y, Li X, Mander U, He Y, Jia Y, Ma Z, Guo W, Xin Z (2011) Effect of reclamation time and land use on soil properties in Changjiang River Estuary, China. Chin Geogr Sci 21(4):403–416

Tester M, Davenport R (2003) Na$^+$ tolerance and Na$^+$ transport in higher plants. Ann Bot 91:503–527

Tobe K, Li X, Omasa K (2002) Effects of sodium, magnesium and calcium salts on seed germination and radicle survival of a halophyte, Kalidium capsicum (Chenopodiaceae). Australian. J Botany 50:163–169

Tobe K, Zhang L, Omasa K (2003) Alleviatory effects of calcium on the toxicity of sodium, potassium and magnesium chlorides to seed germination in three non- halophytes. Seed Sci Res 13:47–54

USDA NRCS (2010) Plants for saline to sodic soil conditions. Technical Note: Plant Materials 9A, USDA-NRCS, Salt Lake City, Utah

USDA-NRCS (2011) Crop tolerance and yield potential of selected crops as influenced by irrigation water salinity (ECw) or soil salinity (ECe). USDA-NRCS, Washington, DC Gregory PJ, Nortcliff S (2012) Soil Conditions and Plant Growth. Blackwell-Wiley, London, UK

Van Breemen N, Burman P (2002) Soil formation, 2nd edn. Kluwer Academic Publishers, Dodrecht

Wang L, Coles N, Wu C, Wu J (2014) Effect of long-term reclamation on soil properties on a coastal plain, Southeast China. J Coast Res 30(4):661–669

Wilson C, Lesch SM, Grieve CM (2000) Growth stage modulates salinity tolerance of New Zealand spinach (*Tetragonia tetragonioides*, Pall) and Red Orach (*Atriplex hortensis* L). Annals Bot 85:501–509

Yadav S, Irfan M, Ahmad A, Hayat S (2011) Causes of salinity and plant manifestations to salt stress: a review. J Environ Biol 32:667–685

Yensen NP (2008) Halophyte uses for the twenty-first century. In: Khan MA, Weber DJ (eds) Ecophysiology of high salinity tolerant plants. Springer, Dordrecht

Zeinolabedin J (2012) The effects of salt stress on plant growth. Tech J Eng Appl Sci (TJEAS) 2(1):7–10

Zhu JK (2001) Plant salt tolerance. Trends Plant Sci 6(2):66–71

Chapter 11
Acid Soils and Acid Sulfate Soils

Abstract Soils that have pH values less than 7 are usually called acid soils or acidic soils. The less is the value of pH, the stronger is the acidity. Soils having pH values <5.5 have severe limitations to crop production. Plants may suffer from the toxicity of Al, Mn and Fe, and from the deficiency of Ca, Mg and P in acid soils. Soil acidity also hampers microbial and faunal population, growth and function. Mineralization, nitrification and nitrogen fixation are reduced due to soil acidity. Productivity of most crops is very low and non-profitable in acid soils unless ameliorated with lime and fertilizers. Acid sulfate soils are extremely acidic (pH <4.0; often below 3.0) due to the formation of sulfuric acid from pyrite (FeS_2) accumulated in coastal and brackish water environments. Pyrite undergoes redox transformations depending on hydrological conditions. Soils enriched with sulfidic materials usually have a neutral reaction as long as they are saturated with water. When they are drained, the pyrite is oxidized and the soil becomes extremely acidic. Because of their geomorphology, acid sulfate soils can be saline as well. Including 12–13 M ha acid sulfate soils, the global expanse of acid soils is 3950 M ha. Approximately 50% of the world's potentially arable lands are acidic. Selection of suitable crops, liming and other inputs including fertilizers, irrigation and drainage are necessary for sustainable use of acid soils for agriculture.

Keywords Soil reaction · Acidity · Acid sulfate soils · Phytotoxicity · Acid tolerant plants · Liming · Lime requirement

11.1 The pH Scale, Acidity and Alkalinity

The pH of a solution or suspension indicates its state of acidity or alkalinity. The term pH was taken from French 'pouvoir hydrogene' meaning 'power of hydrogen'. Sorensen (1909) defined pH as the negative (or reciprocal) logarithm of hydrogen ion activity which is equal to the hydrogen ion concentration, expressed in moles per liter, in very dilute solution.

$$pH = -\log\left[H^+\right]$$

In any dilute solution, the product of H⁺ and OH⁻ ion concentrations is 10^{-14}. Therefore, the pH may be calculated if the concentration of any one of H⁺ or OH⁻ ion is known. For example, if the concentration of OH⁻ ions in a dilute aqueous solution is 10^{-9} g l⁻¹, its pH may be calculated as follows:

$$\left[H^+\right] \times \left[OH^-\right] = 10^{-14}$$
$$\left[OH^-\right] = 10^{-9} \text{ moles l}^{-1}$$
$$\text{So,} \left[H^+\right] = 10^{-5} \text{ moles l}^{-1}$$
$$pH = -\log\left[H^+\right] = -\log 10^{-5}$$
$$= 5\log 10 = 5 \times 1 = 5.$$

The pH of the soil is 5.

Thus, the pH scale ranges from 0 to 14, with a value of 7 at neutrality. A pH value less than 7 indicates acidity, while a value greater than 7 indicates alkalinity. The pH value also indicates the relative abundance of (H⁺) and (OH⁻) ion concentrations in the system. At pH 7, H⁺ and OH⁻ ion concentrations are equal. As the H⁺ concentration increases, the OH⁻ concentration decreases. At higher H⁺ concentration, the pH is lower and the acidity is stronger. Similarly, alkalinity increases as the pH (and OH⁻ concentration) rises. Since the pH scale is logarithmic, a difference of 1 unit pH is actually a 10 fold difference. For example, a solution with pH 5 is ten times more acidic than a solution with pH 6.

11.2 Soil Reaction, Acid Soils and Acid Sulfate Soils

Soil reaction, as denoted by the pH, is the state of acidity or alkalinity in a soil. Soil is a very heterogeneous material, and its pH cannot be measured as accurately as that of a pure solution. A soil is said to have a neutral reaction if its pH lies between 6.6 and 7.3 (USDA-NRCS 1998). Soil pH generally ranges from 4 to 10. Good quality agricultural soils have pH values around 6.0–7.0. Usually, soils with pH values less than 6.5 are called acid soils. There are three categories of soil acidity: active acidity, exchangeable acidity and non-exchangeable acidity (or titratable acidity). Active acidity occurs due to H⁺ ions in soil solution on which pH is based. Exchangeable acidity is also known as potential acidity which is measured in cmole (p⁺ kg⁻¹) and consists of adsorbed H⁺ and Al^{3+} ions on clay and organic matter. Titratable acidity arises from dissociation of weak acid functional groups of soil organic matter, and de-protonated of hydroxyl-silicates. The larger the percentage of exchange sites occupied by aluminum and hydrogen, the lower is the pH, and the higher is the acidity of the soil. USDA, NRCS (1998) classify soils on the basis of pH values as:

Categories of soil acidity – alkalinity	pH range
Extremely acid	3.5–4.4
Very strongly acid	4.5–5.0
Strongly acid	5.1–5.5
Moderately acid	5.6–6.0
Slightly acid	6.1–6.5
Neutral	6.6–7.3
Slightly alkaline	7.4–7.8
Moderately alkaline	7.9–8.4
Strongly alkaline	8.5–9.0

Acid sulfate soils may, however, have pH values as low as 2. In addition to very low pH, acid sulfate soils have the predominance of free sulfuric acid produced by the oxidation of pyrites. Some soils, particularly of estuarine and costal marshes, have the accumulation of sulfidic materials which can produce sulfuric acids by oxidation on drainage. Their pH values remain almost near neutrality as long as they are saturated with water. When drained, these soils become highly acidic; they are called 'potential acid sulfate soils'. Because acid sulfate and potential acid sulfate soils usually develop under estuarine and coastal environments, some of them can be saline at the same time. For this reason, they are often included also in salt affected soils. About 67 and 79% of the world's acid top soils and subsoils respectively are highly acidic (below pH 5.0–5.2). In these soils, aluminium toxicity is a severe problem (Sumner and Noble 2003).

11.3 Global Extent of Acid Soils

Soil acidity is a major cause of declining crop production all over the world (Samac and Tesfaye 2003). However, estimates vary regarding the global extent of acid soils. There were about 3 billion hectares of acid soils in 1970s in the world (Kochian et al. 2004). After 20 years, it has been known from a survey that the extent of acid soils has increased to 4 billion hectares. Among these soils, 178 million hectares are agricultural soils (Von Uexkull and Mutert 2004). Acid soil covers 30% of the ice-free land in the world (Iqbal 2012). It is now estimated that approximately 50% of the world's potentially arable lands are acidic (Panda and Matsumoto 2007).

The regional distribution of acid soils according to the data of Sumner and Noble (2003) is given in Table 11.1. The largest areas of acid soils are in South America, North America, Asia, and Africa. In most regions, the area of acid soils far exceeds the area under cultivation, indicating that large areas of acid soils are still under natural forest or grassland vegetation.

Acid soils occur mainly in two global belts (Fig. 11.1): the northern belt, with cold, humid temperate climate, and the southern tropical belt, with warmer, humid conditions (Von Uexkull and Mutert 2004). Acid soils, in cold and temperate

Table 11.1 Regional distribution of acid soils

Regions		Area of acid soil M ha
Africa		659
Australia and New Zealand		239
Europe		391
Asia	Near East	5
	Far East	212
	Southeast and Pacific	314
	North and Central	512
America	North	662
	Central	36
	South	916
World		**3950**

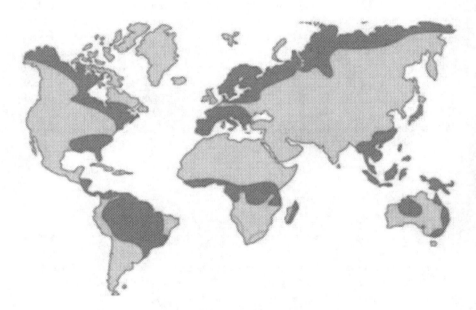

Fig. 11.1 Global distribution of acid soils (highlighted in color)

regions, are dominated by Spodosols, Alfisols, Inceptisols and Histosols and, in tropical regions, largely by Ultisols and Oxisols orders of Soil Taxonomy (Soil Survey Staff 1999). Spodosols have a subsoil accumulation of amorphous organic matter, admixed with aluminum with or without iron oxides. It is known as the spodic horizon. Spodosols are highly leached and strongly acidic. They are typically low in natural fertility. Undisturbed Spodosls usually have a surface accumulation of organic matter. Spodosols are most commonly found in cool, moist environments under coniferous forest vegetation. Spodosols are mainly soils of the temperate regions and develop on extremely basic cation–poor, unbuffered, coarse-textured

11.3 Global Extent of Acid Soils

parent materials. Large areas of Spodosols are found in northern Europe, Russia, and northeastern North America. The spodic horizon (Bh) forms due to translocation of basic cations, iron, and organic matter from the surface horizons, which, as a result, become extremely acid. Alfisols are soils developed under temperate forests of the humid midlatitudes. Eluviation is moderate and base status is fairly high (35% or higher) in these soils. Common to the humid continental and humid subtropical climates, these soils are well-developed and contain a subsurface layer of clay called an *argillic horizon*. They typically have a ochric epipedon (a light-colored, mineral soil *horizon* at the soil surface). Some Alfisols are found in the wet/dry tropical climate of Africa, South America, Australia, and Southeast Asia. Inceptisols are soils that show only the beginning of horizon differentiation. Inceptisols are widely distributed and occur across a wide range of ecological settings. They are found in all climates except aridic and under various vegetation and parent material types. They are often found on fairly steep slopes, young geomorphic surfaces, and on resistant parent materials. Histosols are soils that contain organic soil materials extending down to an impermeable layer or with an organic layer that is more than 40 cm thick and without andic properties and that usually develop from organic parent materials accumulated under wet or saturated conditions. They are mostly soils that are commonly called bogs, moors, or peats and mucks. Most Histosols occur in Canada, Scandinavia, the West Siberian Plain, Sumatra, Borneo and New Guinea. Smaller areas are found in other parts of Europe, the Russian Far East, Florida and other areas of permanent swampland. Histosols are usually moderately to strongly acidic depending on the inputs of basic cations from surrounding mineral soils. They have very high CEC values due to the highly charged nature of organic matter. The acidity is due mainly to the H^+ ions, with the Al^{3+} ions making up only a small proportion of the acidity in most Histosols. Some Histosols in fens (unforested) and swamps (forested) can have pH (in water) values in the neutral range (6–7.5) (Sumner and Noble 2003). Ultisols are highly weathered soils of humid tropics and subtropics. They are often red/yellow in color reflecting the oxidation of iron and aluminum. They have an illuvial clay layer which distinguishes them from Oxisols which do not. Ultisols are soils with poor base status (<35%). Ultisols are the dominant soils in the Southern United States, southeastern China, Southeast Asia and some other subtropical and tropical areas. Oxisols are found in warm and humid tropical and subtropical climates under broadleaf, evergreen vegetation such as rainforests. These soils are rich in low activity clays such as kaolinite and oxides of iron and aluminum or the sesquioxides. They are leached, weathered, acidic and poor in fertility. They are not well-suited for agriculture. When cleared of vegetation, the exposed surface is easily eroded. Oxisols are found almost exclusively in tropical areas, in South America and Africa (almost always on highly stable continental cratons). In the WRB system of soil classification (FAO 2006), most acid soils belong to the reference soil groups (RSGs) Acrisols, Ferralsols and Podzols. Acid soils also occur in Andosols, Arenosols, Alisols, Albeluvisols, Cambisols, Histosols, Leptosols, Plinthosols, Planosols, Fluvisols, Regosols and Umbrisols. Some Fluvisols and mine soils

(Espoli-Anthropic Regosols) contain pyrite (FeS_2), and extreme acidity develops on oxidation (Vazquez et al. 2008). Fig. 11.1 shows the global distribution of acid soils.

According to Von Uexkull and Mutert (2004), 67% of the acid soils support forests (high density of trees and close canopy) and woodlands (low density of trees and open canopy), and approximately 18% are covered by savanna (*savanna*, or savannah, is a grassland ecosystem with sparsely distributed trees and the canopy is open allowing plenty of light to the ground), prairie and steppe vegetation (grassland vegetations; prairies are grasslands with tall grasses, while steppes are grasslands with short grasses). Only 4.5% (178 m ha) of acid soils are used for arable crops. Another 33 m ha is used for perennial tropical crops.

There are about 12–13 million ha of potential and active acid sulfate soils in the world with approximately 10 million ha in the tropics (Andriesse and van Mensvoort 2006). Sulfidic materials occur in larger areas under thick covers of peat in some regions including Indonesia and Malaysia. In addition, Pleistocene, tertiary or still-older pyritic sediments, often originating from past tidal environments, occur in inland positions of Canada and the United States, Russia, Uganda, Great Britain, Denmark, Zimbabwe, Germany, and The Netherlands. Large extensions of potential and actual acid sulfate soils occur in South and Southeast Asia, in West Africa and along the north-eastern coast of South America (the Guyana's, the Orinoco, and Amazon deltas, etc.). Some acid sulfate soils are found in coastal regions of Eastern and Southern Africa, particularly in Madagascar, along the Australian coastline and in the Caribbean. In Southeast Asia, the bulk of the acid sulfate soils (approximately 5 million ha) is found in Indonesia, Thailand, and Vietnam (Andriesse and van Mensvoort 2006). In Soil Taxonomy, acid sulfate soils are classified as Sulfaquent and Sulfaquept (Soil Survey Staff 1999), and, according to WRB (FAO 2006) system, they are Thionic Gleysols and Thionic Histosols as well as Thionic Fluvisols.

11.4 Measurement of Soil pH

Soil pH can be measured in the field using dyes, paper strips and glass electrodes. In the laboratory, soil pH is measured in suspensions taking field moist or air dry soil and distilled water in the ratio 1:1 or 1:2 or 1:2.5 (20 g soil with 20 ml or 40 ml or 50 ml distilled water, respectively). The 1:2 ratio is more frequently used. Usually, dry soil has lower pH values than field moist soils. For preparing the suspension, soil is mixed with distilled water by occasional stirring for equilibration for half an hour, and the pH reading is recorded by a pH meter with glass electrode. Soil pH value increases by about 0.5 units as the soil-water ratio increases from 1:1 to 1:2. Soil pH is sometimes measured in soil suspension made with 0.01 M calcium chloride to counteract the calcium release from the soil exchange complex. Soil pH value thus obtained is generally lower than that recorded in a suspension made up with distilled water. Since the pH of distilled water is itself acidic, soil pH is sometimes measured in suspension made with neutral 1 N KCl solution. However, it is customary to mention the method used for measurement along with the pH value for a soil. Soil pH can change in different seasons of the year depending on temperature and moisture conditions. Since pH is a measure of the hydrogen ion activity [H^+], many different chemical reactions can affect it. Temperature changes the chemical activity, so most measurements of pH include a temperature correction to a standard temperature of 25° C (77° F).

11.5 Development of Soil Acidity

Acidity develops in the soil due to different natural and anthropogenic causes. Naturally acid soils are found to develop under varied conditions of climate, parent material, soil, topography and vegetation. Land use, and soil and crop management practices are the major causes of human induced soil acidification. However, the chief sources of acids or hydrogen ions in soil are: atmosphere and rainwater, mineral weathering and transformation, organic matter decomposition with the release of CO_2 and organic acids, microbial and root respiration, root secretion and release of H^+ in exchange of bases, industrial and mining wastes, and fertilizers. The processes of soil acidification include the production of organic and inorganic acids, exchange of base cations by H^+ and Al^{3+} ions and leaching of bases.

In humid areas, soils are usually acidic due to several reasons. For dissolution of atmospheric CO_2 which undergoes a reaction as shown below, rainwater becomes acidic in reaction.

$$H_2O + CO_2 = H_2CO_3$$

$$H_2CO_3 = HCO_3^- + H^+$$

$$HCO_3^- = CO_3^{2-} + H^+$$

If not loaded with dissolved or suspended bases, rainwater in equilibrium with atmospheric carbon dioxide has usually a pH around 5.5. Most soils of the humid tropics developing from neutral parent materials, such as sandstone, have a pH near this value. Similar reactions leading to the production of carbonic acid occur with water in soil, and CO_2 is produced in many different processes, including atmospheric diffusion, organic matter decomposition, and root and microbial respiration. However, these processes are operative in all soils including humid tropical soils although the rates differ with climate and vegetation.

There are also oxides of sulfur and nitrogen in the atmosphere that can produce acids in rainwater.

$$NO + \frac{1}{2}O_2 \rightarrow NO_2$$

$$3NO_2 + H_2O \rightarrow 2HNO_3 + NO$$

$$SO_2 + O_2 \rightarrow SO_3 + H_2O \rightarrow H_2SO_4$$

About one-fourth of the acidity of rain is accounted for by nitric acid (HNO_3). Most of the remaining 75% of acidity in rainwater is accounted for by the presence of sulfuric acid (H_2SO_4). In some soils of the industrialized regions, acid rain is a cause of soil acidification. Acid rain is actually polluted rainwater. In some areas of the United States, the pH of rainwater can be 3.0 or lower, approximately 1000 times more acidic than normal rainwater. In 1982, the pH of a fog on the West Coast of the United States was measured at 1.8 (Casiday and Frey 2013). Such an extreme acidity in rainwater is caused by the presence of acid forming gases emitted from industries and the combustion of fossil fuels.

In humid areas, the soils are intensively weathered and leached. Weathering decomposes the minerals and leaching removes the suspended and soluble products of weathering. The finer fraction of soil is dominated by low activity 1:1 type of clay like kaolinite and oxides of iron and aluminium or the sesquioxides. These soils have low CEC and BSP, and hence low buffering capacity which increases their tendency to be acidified. These characteristics are common in Oxisols and Ultisols. Aluminium ions (Al^{3+}) are adsorbed on exchange complexes which can be hydrolyzed to produce H^+ ions. Hydrogen ion replaces adsorbed bases and makes the soil more acid (pH below 5.5).

$$Al^{3+} + 6H_2O \leftrightarrow Al(H_2O)_6^{3-} \leftrightarrow Al(OH)(H_2O)_5^{2+} + H^+$$

In high rainfall areas (where precipitation exceeds evapo-transpiration and there is substantial downward movement of water), the basic cations such as Ca^{2+} and Mg^{2+} are leached out of the soil leading to its acidification.

Some other inorganic acids than carbonic acid can be formed in soils and become responsible for soil acidification. These acids may include nitric, hydrochloric and sulfuric acids. One of the most significant inorganic acidification reactions in soil is sulfur oxidation. In some areas such as mine spoil and mangrove reclamation areas, sulfur content is naturally high and acidification is a serious problem. In estuarine areas and mangrove swamps, a significant amount of pyrite (FeS_2) is present in the soil. Under submerged or saturated conditions these soils have pH values near neutrality, but if reclaimed or drained, the pyrite is oxidized producing sulfuric acid which gives the soil an extremely low pH (<3.0). The following reaction readily occurs, producing 2 hydrogen ions for every sulfur ion oxidized. Unless a plentiful supply of a liming material is available, the soil pH can be driven to a very low value. These soils are known as acid sulfate soils and classified in the great group Sulfaquent and Sulfaquept (Soil Survey Staff 1999).

$$2FeS_2 + 7O_2 + 6H_2O = 4SO_4^{2-} + 8H^+ + 2Fe(OH)_2$$

11.5 Development of Soil Acidity

Although a small amount of organic acids enter the soil from atmospheric deposition and canopy throughfall, most organic acids arise from root exudation, lysis, and by the release from soil microorganisms (Ryan et al. 2001; Jones et al. 2003). Kumari et al. (2008) observed the production of citric, oxalic, formic and maleic acids during decomposition of rice straw in soil. These acids may not be stable under aerobic conditions. Their contribution to soil acidity has not been studied, but citric acid and oxalic acid were found to contribute significantly to the solubilization of P from tricalcium phosphate and rock phosphate.

Many forms of organic matter can also be acidifying, depending on the plant from which the organic matter is derived. Some plants contain significant quantities of organic acids or produce organic acids during decomposition of their residues. Residues of some plants are low in bases; growing such plants for a long period may turn a site acidic. Thus, vegetation can be an important factor influencing soil reaction in the long run. They contribute to acidity or alkalinity by their removal of cations and anions from the soil, their exudation and secretion, and their residues and decomposition products. Leguminous plants are particularly acidifying because they take up more cations in comparison to anions than non-leguminous plants. Legumes take up little nitrate from the soil because most of their nitrogen needs are satisfied by microbial nitrogen fixation within the plant structure. Tang and Rengel (2003) have given a list of the acidifying potential of a number of crop plants including legumes and cereals. Some tree species are acid forming due to low base contents in their litter.

Fertilizers can be a major source of hydrogen ions to soil. Nitrogen and phosphorus fertilizers are particularly important in this respect. Most nitrogen fertilizers are based on ammonium (NH_4^+), and the eventual conversion of ammonium to nitrate (NO_3^-) is accompanied by release of hydrogen ions.

$$NH_4^+ + 2O_2 = NO_3^- + 2H^+ + H_2O$$

That is, two hydrogen ions are produced for every ammonium ion converted to nitrate. A popular nitrogen fertilizer is urea [CO (NH$_2$)$_2$] which, in soil with pH less than 6.3, is decomposed as follows:

$$CO(NH_2)_2 + 2H^+ + 2H_2O = 2NH_4^+ + H_2CO_3$$

That is, two hydrogen ions are consumed for each urea molecule decomposed. This tends to increase pH in the surrounding soil, but the ammonium ion is then converted to nitrate as indicated above, with four hydrogen ions being released by the two ammonium ions. Thus, while there may be a short term increase in pH when urea is applied, the overall reaction is acidifying. Each of anhydrous ammonia, urea and ammonium nitrate produces an average of 1.8 lbs. of calcium carbonate neutralizable acidity for each pound of nitrogen applied and nitrified in the soil. Ammonium sulfate, which contains two ammonium ions, releases an average of 5.4 lbs. of calcium carbonate neutralizable acidity per pound of ammonium nitrogen applied and nitrified. Rates of acidification can be as high as 40 kmol H^+ ha^{-1} $year^{-1}$ in production systems receiving high rates of ammoniacal N fertilizers (Sumner and Noble 2003).

Ammonium nitrogen, whatever may be the sources – mineralization, biological fixation or fertilizers, readily undergoes nitrification under aerobic conditions in the soil. The nitrate ions that are formed may be absorbed by plant roots and leached beyond plant roots. The fact that for each negatively charged nitrate ion that is absorbed, one negatively charged hydroxyl ion is excreted by the root maintains the electrical balance, and there is no soil acidification. But, if nitrate leaches, equivalent amount of positively charged cations (Ca^{2+}, Mg^{2+}, K^+) are also leached to maintain electrical balance. This produces soil acidity.

Triple super phosphate or monocalcium phosphate [$Ca(H_2PO_4)_2$] is a popular phosphate fertilizer. It reacts in soil with water to form dicalcium phosphate ($CaHPO_4$) and phosphoric acid (H_3PO_4). Phosphoric acid gives off hydrogen ions as shown in the following reactions:

$$Ca(H_2PO_4)_2 + H_2O = CaHPO_4 + H_3PO_4$$

$$H_3PO_4 = H^+ + H_2PO_4^- = 2H^+ + HPO_4^{2-} = 3H^+ + PO_4^{3-}$$

Phosphorous fertilizer is usually placed in bands around plant rows. Because of the tendency for H_2PO_4 to give up some of its hydrogen ions, very low pH values can occur in the band. This acidity then gradually diffuses into the soil surrounding the band.

Acids produced in several processes in soil may react with calcium carbonate in soils rich in lime, or those developing from calcareous parent materials, such as limestone or dolostone.

$$CaCO_3 + H_2CO_3 \leftrightarrow Ca^{2+} + 2HCO_3^-$$

Carbonic acid also accelerates the dissolution of primary silicate minerals such as pyroxene, and releases Ca^{2+} and Mg^{2+} (Boettinger 2005) which may contribute to the resistance of the soil to acidification and lead to calcification.

$$CaMgSi_2O_6 + 4H_2CO_3 + 2H_2O \leftrightarrow Ca^{2+} + Mg^{2+} + 4HCO_3^- + 2H_4SiO_4$$

Lime can neutralize other organic and inorganic acids which are produced in soils through various chemical and biochemical processes. This phenomenon resists the acidification of soils, particularly in arid environments and in calcareous soils.

$$CaCO_3 + heat \rightarrow CaO + CO_2 \uparrow (gas)$$

$$CaCO_3.MgCO_3 + heat \rightarrow CaO.MgO + 2CO_2 \uparrow (gas)$$

$$CaO + H_2O + H_2SO_4 \rightarrow CaSO_4 + 2H_2O$$

Table 11.2 Processes contributing to the development of soil acidity

Sources of acidity	Reactions
Carbon dioxide	$H_2O + CO_2 = H_2CO_3 = H^+ + HCO_3^- = 2H^+ + CO_3^-$
Ammonium	$NH_4^+ + 2O_2 + NO_3^- + 2H^+ + H2O$
Phosphate fertilizer	$Ca(H_2PO4)_2 + H_2O = CaHPO_4 + H_3PO_4$
	$H_3PO_4 = H^+ + H_2PO_4^- = 2H^+ + HPO_4^{2-} = 3H^+ + PO_4^{3-}$
Pyrite	$2FeS_2 + 6H_2O + 7O_2 = 4SO_4^{2-} + 8H^+ + 2Fe(OH)^2$
Organic matter	Decomposition and production of organic acids
Nutrient uptake by roots	Release of H^+ ions in exchange of bases
Acid rain	Reaction in atmosphere of water

Adapted from Harter (2007)

Properties of parent materials (types, composition, texture, degree of weathering) also influence soil pH. There are acidic parent materials (granites, rhyolites, diorites) containing greater proportion of quartz, feldspar and sesquioxide minerals. Usually these soils are less fertile because of low base status. Basic materials usually have low content of soluble Al and Fe and high representation of Ca and Mg. These soils are neutral to slightly alkaline in reaction and have high buffer capacity.

For detailed information on factors and mechanisms of soil acidification, interested readers can consult *Handbook of Soil Acidity* by Rengel (2003). The processes of development of soil acidity are shown in Table 11.2.

11.6 Buffering Capacity of Soils

Buffering capacity refers to the ability of a solution to resist changes in its pH on addition of a small amount of acid or alkali. Soils have considerable buffering capacity, i.e. their pH values tend to remain unchanged upon addition of a little amount of acid or alkali. Buffering capacity of the soil can be demonstrated by making a soil water suspension and adding dilute acid slowly from a burette in it under a pH electrode. The pH reading will not change upon addition of some dilute acid solution for some time; it will very slowly go down after a while due to its buffering capacity, and when the buffering capacity is diminished, the pH reading will sharply fall. Buffering capacity in soil occurs due to the presence of some weak acids and weak bases in soil solution, soluble acidic anions such as NO_3^-, HCO_3^-, SO_4^{2-}, basic cations such as Ca^{2+} and Mg^{2+} and exchangeable cations and anions and active groups on humus. Soils with high clay and organic matter content (i.e. higher CEC) have high buffering capacity. Calcareous soils often have high buffering capacities because free $CaCO_3$ effectively neutralizes acid. Buffering is an important soil property; soils with a high buffering capacity need a great deal of liming or acidifying effort to alter pH. Having a high buffering capacity is good if the soil has a desirable pH; but it can be a problem if the soil needs pH modification.

11.7 Effects of Acidity on Soil Processes

Soil pH determines the chemical environment of soil. Some important physical and most chemical and biological processes in the soil are affected by soil acidity. These processes are related to nutrient availability and nutrient uptake by plants.

11.7.1 Solubility and Availability of Chemical Elements

All plant nutrients except C, H, and O are absorbed from the soil by higher plants. Since they are chemical elements, their solubility and availability is influenced by the chemical environment in the soil. Such important elements include Fe, Mn and other micronutrients, and Ca, Mg, P and other macronutrients. Solubility of some other elements which have no known essential function in plants such as aluminium also affect plant performance by their corrosive effects on roots.

11.7.1.1 Aluminium

Aluminium has an atomic number of 13, an atomic mass of 27, one oxidation state (+3) and one naturally occurring isotope (^{27}Al). Aluminium is a lithophile element and is the most abundant metal in the lithosphere. Aluminium is present in soils as soluble and insoluble oxides and hydroxides, in aluminium containing minerals including aluminosilicates, as organic complexes, and as exchangeable cations on colloidal surfaces. Important minerals of Al in the soil include sillimanite Al_2SiO_5, corundum Al_2O_3 and kaolinite $Al_2Si_2O_5(OH)_4$. It is a major constituent in many rock forming minerals, such as feldspar, mica, amphibole, pyroxene and garnet. Secondary clay minerals, including kaolinite and smectite, and Al-hydroxides, including gibbsite, nordstrandite and bayerite, are formed by the weathering of aluminium containing primary minerals. These substances may control the equilibrium concentration of Al in soil solutions. Aluminium is present in various forms and undergoes complex chemical reactions in soil solution depending on its pH. Soil pH is the single most important factor controlling the amount of Al^{3+} in the soil solution. Aluminium solubility increases as the pH begins to drop below 6.0. In most soils, the increase in aluminium in solution may seriously affect crops at soil pH below 5.5. Soluble Al rises abruptly in soils as pH drops below 5.0. The amount of dissolved aluminum is about 1000 times greater at pH 4.5 than at 5.5.

Aluminium has high ionic charge and a small ionic radius. So, it has higher reactivity than other metals in soil solution. Hydrolysis of Al occurs at higher rates in soils with pH >4.0; most of the total Al is hydrolyzed at pH near 5.0. Enhanced Al hydrolysis decreases the charge density of the Al molecule and leading to polymerization of Al units (Menzies 2003). At higher soil solution pH than 4.0, mononuclear species $AlOH^{2+}$, $Al(OH)_2^+$, $Al(OH)_3$, and $Al(OH)_4^+$ of Al and soluble Al-complexes

11.7 Effects of Acidity on Soil Processes

with inorganic ligands, such as sulfate and fluoride, AlF_2^+, AlF_3^+, $Al(SO)_4^+$ predominate. Many organic compounds of Al are also formed. Complexation of Al occurs mainly with organic functional groups including COOH, phenolic-, enolic-, and aliphatic-OH groups. These transformations of Al in soil are mainly pH dependent. The ionic strength, kind and amount of competing cations and the presence of organic ligands are also involved in the formation of these complexes.

Exchangeable Al, as determined by the amount extracted with an unbuffered neutral salt solution such as 1 M KCl, is the major reserve of labile Al in the soil. It enters the soil solution through exchanges with other cations. The proportion of CEC satisfied with Al^{3+} is known as aluminium saturation which is strongly pH dependent, decreasing with increasing pH to low levels at pH 5.5. The degree of Al saturation has been found to be a more successful predictor of Al toxicity than the amount of exchangeable Al^{3+}. However, exchangeable Al^{3+} and its hydrolyzed-polymerized forms $\left(Al(OH)_x (H_2O)_{6-x}^{(3-x)+}\right)$ produce the acidity in most soils as they hydrolyze further toward $Al(OH)_3$. The Al^{3+} ion is strongly adsorbed on clay surfaces, and it is very slowly exchangeable. Only $Al(H_2O)_6^{3+}$ is considered truly exchangeable, and it is present in soils in appreciable amounts only at pH <5.5 (Bohn et al. 2001).

11.7.1.2 Iron

Iron is an important element in the lithosphere regulating oxidation and reduction processes in soil and transfer of electron as well as energy in plants. There is about 5% iron in the lithosphere. There are different forms of iron is soil: total Fe, DTPA-extractable Fe, soil solution plus exchangeable Fe, Fe adsorbed onto inorganic sites and oxide surfaces, and iron bound by organic sites (Sharma et al. 2008). Iron oxides and oxyhydroxides are of widespread occurrence in soils and rocks. Their dissolution in soil solutions releases Fe^{2+} and Fe^{3+} ions that can be absorbed by plant roots and that may interact with other substances modifying their availability to plants.

Iron can occur in either the divalent (Fe^{2+}) or trivalent (Fe^{3+}) states in soil. Iron occurs predominantly as Fe^{3+} oxides in soils. The divalent state (or ferrous state) can be oxidized to the trivalent state (or ferric state), where it may form oxide or hydroxide precipitates. Free iron minerals that occur in soil and form pedogenically include Hematite (αFe_2O_3), Maghemite (γFe_2O_3), Magnetite (Fe_3O_4), Ferrihydrite ($Fe_2O_3 \times n\,H_2O$), Goethite ($\alpha FeOOH$), Lepidocrocite (($\gamma FeOOH$), Ilmenite ($FeTiO_3$), Pyrite (FeS_2), Ferrous sulfide (FeS), Jarosite ($KFe_3(SO_4)_2(OH)_6$) (Kabata-Pendias and Pendias 1992). In the highly weathered soils of arid, semiarid and tropical areas, the most common iron mineral is goethite, followed by hematite and maghematite. These minerals are usually inherited from parent materials and are stable in an oxidizing environment. Magnetite is a magnetic iron oxide that usually occurs as a sand-sized mineral and is mostly inherited from parent material. Ferrihydrite is a common, but unstable, soil mineral and is easily transformed to hematite in warm regions and to goethite in humid temperature zones. Goethite is responsible for the brownish to yellowish color of many soils; hematite has a red color. In submerged soils containing sulfur and plenty of organic matter to supply energy for microbial reduction as in acid sulfate soils, the dominant iron containing minerals are pyrite, ferrous sulfide and jarosite. Iron oxides occur in soils as discrete particles and as coatings on silicate clay particle surfaces.

If the soil is aerated and when the soil reaction is alkaline, a little iron is found in soil solution. According to Hersman et al. (2001), the solubility products of Fe(III) (hydr)oxides range from 10^{-39} to 10^{-44}, which limit equilibrium concentration of Fe(III) in aqueous solution to estimated 10^{-17} M if complexing ligands are not present. Soluble Fe^{3+} decreases abruptly with the increase in pH, and it is not available above pH 4. The solubility of Fe^{2+} also decreases with the increase in pH but much less abruptly than Fe^{3+}. The E_h (redox potential which is a function of soil moisture status) also affects the solubility of Fe^{2+}. Therefore, Fe^{2+} is potentially the most available form of soluble inorganic Fe under reduced conditions above pH 4. Organic matter, clay minerals, and hydrous oxides of iron and manganese also affect the solubility and distribution of Fe. Humic substances strongly adsorb or form complexes with iron at pH > 3.

Some chemolithotrophic bacteria (for example, *Acidithiobacillus ferooxidans*, can oxidize dissolved Fe^{2+} in low pH systems (pH <3). They oxidize inorganic compounds, like Fe^{2+} to generate energy and use CO_2 as a source of carbon. *A. ferrooxidans* and some other bacteria are important in sulfur and iron oxidation in acid sulfate soils. *A. ferrooxidans* can increase the rate of iron oxidation as much as five orders of magnitude relative to strictly abiotic rates. Fe^{2+} is converted to Fe^{3+} at low pH, but at pH <3, Fe^{2+} oxidation is a slow process unless it is catalyzed by iron oxidizing bacteria like *A. ferrooxidans* and *Leptospirillum ferrooxidans*. Fe^{3+} oxidizes pyrite and results in the generation of even greater acidity than when oxygen is the primary oxidant (Bingham and Gagliano 2006). Microbial transformations of iron include enzymatic oxidations and reductions, and the formation of chelates and complexes with proteins, amino acids, other organic acids, etc.

11.7.1.3 Manganese

Manganese is a micronutrient and is present in the soil in soluble and insoluble forms and as minerals. The common Mn containing minerals include pyrolusite (MnO_2), rhodochrosite ($MnCO_3$) and manganite ($MnO(OH)$) and some oxides. It is present in some other minerals, such as garnet, olivine, pyroxene, amphibole and calcite as an accessory element. McLennan and Taylor (1999) reported an average upper crustal abundance of 600 mg Mn kg^{-1} and a bulk continental crust average of 1400 mg Mn kg^{-1} Manganese can exist in several oxidation states including +2, +3, +4, +6 and +7. Manganese (II) is the most predominant oxidation state of Mn in soils. The Mn (III) and Mn (IV) states are less abundant. However, Mn (III) may be an intermediate product in soil redox reactions but the persistence of Mn (III) in soil

11.7 Effects of Acidity on Soil Processes

solution is low. The ligands that form complexes with Mn(III) may be decomposed and the solubility of Mn (III) and Mn (IV) oxides is low.

Manganese solubility in soil is influenced by pH, redox potential (E_h) and organic amendments. Potter et al. (2004) suggested that Mn toxicity to plants may occur in soils with high Mn reserves. The amounts of readily reducible Mn in soil, pH and availability of electrons govern the concentrations of manganese in soil solution. The reaction is shown below:

$$MnO_2 + 4H^+ + 2e^- \rightleftharpoons Mn^{2+} + 2H_2O$$

Generally MnO_2 is more stable Mn oxide in soils than Mn_2O_3 and Mn_3O_4 The reduction of Mn (III) and Mn (IV) oxides produce soluble Mn (II). Hue et al. (2001) stated that some soils may contain considerably high and phytotoxic levels of Mn at pH < 6.0.

In soil solution, hydrated Mn^{2+} is the dominant inorganic Mn species. Ion pairs with SO_4^{2-}, HCO_3^-, and Cl^- are formed, but only sulfate is the most dominant in acid soils. Mn(II) can form complexes with a range of organic ligands as well.

11.7.1.4 Phosphorus

Phosphorus is a very critical element in soil system both in respect of its reactions in the soil and its role in plant nutrition. In soils, there are two main fractions of phosphorus – organic and inorganic phosphorus. Rock phosphate is the primary source of inorganic phosphorus in soil that contains the apatite minerals having the general formula $Ca_{10}(PO_4)_6(OH,F,Cl)_2$. Among the three apatite minerals, Ca-fluorapatite [$Ca_3(PO_4)_3F$] is the most common, and Ca-chlorapatite [$Ca_3(PO_4)_3Cl$] and Ca-hydroxyapatite [$Ca_3(PO_4)_3(OH)$] are rare. Important inorganic P compounds include aluminium, iron, manganese and calcium phosphates. The sum of the amounts of all forms of phosphorus in soil is known as total phosphorus, a very little proportion of which is present in soil solution (0.03–0.3 mg P l^{-1}) in fertile arable soils (Sims 2000). Under natural conditions, the weathering and dissolution of rocks and relatively insoluble P-containing minerals is a slow process (Batjes 2011). In acid soils, various forms of iron, aluminium and manganese oxides strongly bind P, while in calcareous soils, P is mainly found in the form of Ca-compounds (Ryan and Rashid 2006).

Phosphorus has a unique tendency of being immobilized in soil. Almost all phosphorus that is released from mineral weathering, mineralization of organic phosphorus and applied as fertilizers is converted to insoluble forms (and so unavailable or difficultly available to plants) within a short time. Therefore, phosphorus is present in soil largely in insoluble forms. Some phosphate ions can be released to soil solution from these insoluble forms depending on soil pH. Solubility of phosphorus is the maximum at pH 6.5 (Tan 2010). There can be three phosphate anions in soil solution depending also on the pH – $H_2PO_4^-$, HPO_4^{2-} and PO_4^{3-}. The primary orthophosphate ($H_2PO_4^-$) predominates in acid soils, and the other forms increase in abundance as the pH rises from 5.5. HPO_4^{2-} and PO_4^{3-} ions dominate in alkaline soils. All the phosphate anions $H_2PO_4^-$, HPO_4^{2-} and PO_4^{3-} can exist together in soil solution, and at pH 7.2, the proportion of $H_2PO_4^-$, HPO_4^{2-} is almost equal. These two phosphate ions are available forms of phosphorus to plants, $H_2PO_4^-$ being preferable to HPO_4^{2-}.

Phosphate anions are very reactive in soil solution; they react with soluble constituents to form insoluble compounds and are bound to soil particle surfaces so strongly that plant roots cannot readily absorb them. There are two terms for the explanation of this phenomenon – phosphate retention and phosphate fixation. Phosphate retention and phosphate fixation are often used interchangeably. Retention refers to the part of adsorbed phosphate that can be extracted with dilute acids, and fixation refers to the part which cannot be extracted with dilute acids. Retained phosphate is relatively available, while fixed P is unavailable to plants. Both P retention and P fixation are included in another term 'phosphate sorption'. In acid soils, phosphate ions react with soluble Al, Fe and Mn to produce their insoluble phosphates and hydroxy-polyphosphates. Phosphate ions can be adsorbed on compounds of these elements as well. Above pH 5.5, phosphate ions can react with Ca^{2+} to form insoluble calcium phosphate. In alkaline and calcareous soils, most P is fixed with calcium. Phosphate ions can be adsorbed and fixed on surfaces of clay colloids, particularly 1:1 type of clay, against exposed -OH groups.

Thus, phosphate is an immobile element in soil, and usually do not leach to the groundwater to any significant extent. Most of the applied phosphorus remains in the plow layer. Heavy application of P fertilizers in crop fields to meet P deficiency may lead to P build-up in the soil which, on erosion, can contribute to the eutrophication of surface waters.

11.7.1.5 Sulfur

Elemental sulfur and sulfide compounds are predominant in some soils, including acid sulfate soils, soils of mining areas, organic soils, and in soils in the immediate vicinity of phosphate fertilizer granules. Sulfur oxidation and the production of sulfuric acid leading to the creation of soil acidity as great as pH 1.5 to 2 is a serious problem in drained acid sulfate soils around estuarine and coastal areas of the tropics and subtropics. When these soils are flooded again for a considerable period, sulfate is reduced again, and the soil reaction reaches neutrality. Acid sulfate soils contain reduced sulfur compounds, particularly pyrite (FeS_2) accumulating from sea water and organic residues. Prior to drainage, they have a neutral reaction as unripe sulfidic clays, but they become extremely acidic when drained (raw acid sulfate soils) due to the (microbial) oxidation of reduced sulfur compounds to sulfuric acid. After the complete oxidation of the sulfur compounds and dissipation of the sulfuric acid, which often contaminates adjacent water bodies, sulfate soils remain strongly acid (ripe acid sulfate soils). Both raw and ripe varieties contain high levels of exchangeable Al^{3+} originating from acid weathering of minerals and high basic cation status derived from their riverine origin. The overall reaction can be shown as (CSIRO 2003):

$$2\,FeS_2 + 9\,O_2 + 4\,H_2O \rightarrow 8\,H^+ + 4\,SO_4^{2-} + 2\,Fe(OH)_3\,(solid)$$

Other products of sulfur transformation in acid sulfate soils include elemental S (yellow solid), H_2S (smelly gas), FeS (gray or black solid), Fe_2O_3 (hematite; red solid), FeO.OH (goethite; a brown mineral), schwertmannite (a brown mineral) and H-Clay.

Oxidation of metal bound sulfides in mining spoils is generally a slow process and may need several years to complete oxidation. These soils suffer long from acidity. Sulfur oxidation in mining spoils of arid and semi-arid regions is a further slow process because of the scarcity of water, but plants suffer little from acidity in these soils because of the presence of bases (Bohn et al. 2001)

11.7.1.6 Pant Micronutrients

Plant micronutrients include iron (Fe), manganese (Mn), zinc (Zn), copper (Cu), molybdenum (Mo), boron (B), zinc (Zn), chloride (Cl), and nickel (Ni). Maximum availability of iron and manganese occurs at pH ranges of 4.0–6.5 and 5.0–6.5 respectively. Boron, copper and zinc become more available at pH between 5.0 and 7.0. Molybdenum availability is reduced by soil acidity; it may even become deficient in strongly acidic soils and its maximum availability is found at pH 7.0–8.5. In some acid soils, high levels of soluble iron, aluminum and manganese may be toxic to plants. In acid soils, liming may improve plant growth by reducing the solubility of these elements while increasing the availability of molybdenum.

11.7.2 Microbial Processes

Organic matter decomposition and mineral transformations are microbiological processes. Since the population, community dynamics and functions of microorganisms are affected by soil pH, these processes are also influenced by soil acidity.

11.7.2.1 Nitrogen Mineralization and Nitrification

Organic nitrogen is bound in soils in proteins and other complex substances that are not readily available sources of nitrogen for terrestrial ecosystems. These compounds need to be converted to chemically simpler forms by soil microorganisms. The process is known as nitrogen mineralization, and more specifically, ammonification (nitrogen mineralization and ammonification are often used synonymously), because ammonia (in alkaline soils) and ammonium ions (in acid soils) are produced through a series of microbial metabolic activities. Research findings have shown that soil microorganisms responsible for ammonification and further conversion of ammonia into nitrate (this process is known as nitrification) are sensitive to soil acidity. Nitrogen transformation in acid soils is significantly different from neutral and alkaline soils (Kresovic et al. 2010). It is generally believed that mineralization, nitrification and denitrification are all affected by soil pH for its effect on the composition, growth and activity of microorganisms responsible for these processes.

However, Zhang et al. (2013) found significantly higher gross rates of N mineralization in acidic forest soils (pH <5) of southern China than neutral to alkaline Northern forest soils probably due to higher organic matter and total nitrogen contents in those soils. The rates of autotrophic nitrification and NH_3 volatilization were significantly lower in acidic soils in their study. Southern acidic soils had a much higher capacity for retaining inorganic N than northern soils, as indicated by their significantly lower autotrophic nitrification rates and significantly higher rates of NO_3^- immobilization. Zhao et al. (2007) stated that both ammonium oxidation and nitrification rates were exponentially correlated with soil pH (which revealed the importance of soil pH to the nitrification process.

De Boer and Kowalchuk (2001) stated that chemoautotrophic nitrification is the function of a group of real bacteria including *Nitrosomonas, Nitrosococcus, Nitrosospira, Nitrosobus, Nitrosovibrio* and *Nitrobacter*. These bacteria are sensitive to pH of soil and water, the minimum threshold value being 4.5. As the nitrification is significantly reduced in acid soils there may be a consequent accumulation of nitrite. In other words, the rate of oxidation of NH_4-N is higher than the oxidation of NO_2-N in the soil. Shen et al. (2003) observed that a high level of ammonium nitrogen in soil can also inhibit *Nitrobacter* activity leading to nitrite accumulation in the soil and can create nitrite toxicity to plants in acid soils. According to Jakovljevic et al. (2005), a toxic level of aluminum (more than 50 mg kg^{-1}) in acid soil can significantly reduce nitrification activity. Therefore, nitrification can be enhanced by liming acidic soils. However, Kresovic et al. (2010) suggested that an increase of pH in the soil increases biological denitrification as well. It has already been mentioned above that nitrification was inhibited at low pH, and raising pH of cultivated soil by liming increased nitrification. Prosser (2011) observed that nitrifying bacteria required an optimum pH of 7.5–8.0. However, De Boer and Kowalchuk (2001) suggested that there are examples of high nitrification rates in very acidic forest soils, the physiological basis of which has not yet been understood.

11.7.2.2 Nitrogen Fixation

Soil pH has significant effect on biological nitrogen fixation, both symbiotic and non-symbiotic. Soil acidity affects the survival and persistence of nodule bacteria in soil and reduces nodulation. Nitrogen fixation by the symbiotic process depends on the occurrence, survival and efficiency of *Rhizobium* strains in soils (Adamovich and Klasens 2001). For example, *Sinorhizobium meliloti* and *Rhizobium galegae* are highly sensitive to acid pH and soluble Al when the critical soil pH is 4.8–5.0 (Lapinskas 2004). Legumes and *Rhizobium* can form an efficient symbiosis and fix enough nitrogen when soil pH is no less than 5.6–6.1. Soil acidification was found to inhibit the root-hair infection process and nodulation (Ambrazaitienė 2003). Suryantini (2014) reported that pH less than 5 inhibits nodulation and N fixation in soybean. Groundnut is usually nodulated by slow-growing *Bradyrhizobium* spp. (Were et al. 2012) although it has been found recently that fast growing rhizobia also nodulate (Taurian et al. 2006). Soil acidity and associated nutrient deficiencies and heavy metal toxicities are amongst stressful edaphic factors that severely limit the growth, survival and metabolic function of rhizobia (Mohammadi et al. 2012). Some rhizobia withstand pH below 4.5–5.0, but acidity inhibits host legume growth, root colonization and nodulation. The legume *Leucaena leucocephala* is poorly adapted in acid soils. Kisinyo et al. (2005) suggested that soil acidity and P deficiency are the major causes of poor *Leucaena leucocephala* establishment in tropical soils. They observed in a greenhouse experiment that lime and P fertilizers increased nodulation and N contents in *L. leucocephala*. Lime significantly increased soil pH. High acidity is associated with toxicities of aluminum, iron and manganese and deficiencies of phosphorus and molybdenum. In legumes, P deficiency interferes with nodulation, N_2 fixation, growth and grain yield. Lapinskas (2008) reported that *Rhizobium leguminosarum* bv. *trifolii* was widely distributed in slightly acid soils with pH_{KCl} 5.6–6.0. The average content of rhizobia was 540.0×10^3 cfu g^{-1} of the soil. There were fewer *Rhizobium leguminosarum* bv. *viciae* and significantly fewer *Sinorhizobium meliloti* and *Rhizobium galegae*. *Rhizobium* significantly declined in soils with pH_{KCl} between 4.1and 5.0. Most of the nitrogen was fixed at soil pH_{KCl} 6.1–7.0. In this case, *Rhizobium galegae* accumulated 196 to 289 kg N ha^{-1}. Liming soil had a positive effect on nitrogenase activity in red clover.

Acidity affects some species of algae, including the nitrogen fixing blue-green algae. Blue-green algae are rare in soils with pH values below 4.4. Nayak and Prasanna (2007) reported that cyanobacteria were more in number at high pH in rice fields. Among free living bacteria, all members of the genus *Azotobacter* fix nitrogen, and *Azotobacter* are able to develop on media with a pH range of 4.5–8.5. The optimum pH for growth and nitrogen fixation by *Azotobacter* is near or slightly above neutrality (pH 7.0 to 8.5). Barnes et al. (2007) suggested that populations of *Azotobacter* sp. in soil exceed several thousand cells per gram of neutral or alkaline soils, but these bacteria are generally absent or occur in very low numbers in very acid soils (pH < 5.0).

11.8 Effect of Soil Acidity on Plants

Some plants, such as magnolias, camellias, azaleas and rhododendrons can grow well in acid soils. But, most plants, especially crop plants, suffer from soil acidity. Many soils are naturally very acid and infertile to great depths in the profile.

Cultivation of such soils without inputs (lime and fertilizers) results in very low yields (Sumner and Noble 2003). The unfavorable conditions of plants in acid soils arise from nutritional disorders, toxicities, deficiencies, or unavailability of some essential nutrients. Acid soils, besides having a lower pH, are low in bases like calcium, magnesium, and potassium; low in phosphorus; and high in iron, manganese and aluminum. According to Kochian et al. (2004), the primary limitations to plant growth on acid soils are toxic levels of aluminum and manganese as well as suboptimal levels of phosphorous. Thus, plant growth is restricted by acidity in surface soil and subsoil for one or more of the following: toxicity of H, Al, Fe or Mn, deficiency of Ca, Mg, P and Mo and reduced mineralization, nitrification, nodulation, and mycorrhizal infection although Chen et al. (2012) considers Al toxicity and phosphorus deficiency as the main constraints for crop production in acid soils. They also regarded that P addition might be capable of alleviating Al toxicity in plants. However, increasing pH of strongly acid soils by liming may remove Al toxicity and improve P availability.

11.8.1 Phytotoxicity in Acid Soil

11.8.1.1 Hydrogen

It has been observed in solution culture experiments that H^+ can limit plant growth (Menzies 2003). As noted by Zu et al. (2012), high concentration of H^+ in the soil solution can inhibit root growth (Polomski and Kuhn 2002), disrupt the functions of the plasma membrane (Vitorello et al. 2005) or increase Al^{3+} toxic levels (Ma 2007). Crop production in acid soils may be limited by one or more of the following: (i) reduced uptake of Ca, Mg, and P; (ii) toxicity of Al^{3+} and Mn^{2+}; and (iii) damage of roots due to corrosive effects of H^+ (Menzies 2003).

11.8.1.2 Aluminium

Nearly 95 years ago, Hartwell and Pember (1918) first postulated that soluble aluminium is a major inhibitor of plant growth in acid soils. The negative effects of the high levels of soluble Al on plant growth have been widely reported (Matsumoto 2002; Langer et al. 2009). It is actually the most common and most severe limitation to plant growth in acid soils. Aluminum toxicity generally occurs in soils of pH below 5.0 and rarely occurs above it. The solubility of Al-containing minerals increases exponentially below pH 5.0; thus, the probability of Al toxicity to plants becomes higher as pH decreases. Excess aluminium in the growing medium causes a reduction in both root and shoot growth (Menzies 2003). Aluminium can inhibit root growth at the organ, tissue, and cellular levels at micromolar concentrations (Ciamporova 2002). When plants are exposed to high Al levels, elongation of the main axis of the root is inhibited, and roots become thickened, stubby, brown, brittle, and occasionally necrotic. As a result, the plants become stunted and can show deficiency symptoms of several nutrients including phosphorus which is already in short supply in acid soils. However, there are no identifiable and specific symptoms of Al toxicity suitable for use as a diagnostic indicator (Menzies 2003).

Zheng (2010) suggested that aluminium in acid soils will be solubilized into ionic forms, especially when the soil pH falls below 5. These ionic forms of Al have been shown to be very toxic to plants, initially causing inhibition of root elongation by destroying the cell structure of the root apex, and thereby affecting water and nutrient uptake by roots causing serious hindrance to plant growth and development.

11.8.1.3 Manganese

As an essential micronutrient, manganese takes part in both structure and function of the plant. It is a constituent of photosynthetic proteins and enzymes. It contributes to photosynthesis by participating in the water-splitting system of photosystem II (PSII) and provides necessary electrons (Buchanan et al. 2000). But, excess Mn damages the photosynthetic apparatus. Therefore, Mn acts both as an essential and a toxic element depending on its concentration in soil solution (Kochian et al. 2004; Ducic and Polle 2005). Mn toxicity usually occurs in acid soils because of its higher solubility with decreasing pH. In high pH soils, Mn (III) and Mn (IV) forms predominate but these forms are not available (absorbable) to plants (Rengel 2000). From an experiment on wheat (*Triticum aestivum* cv. Arina), Page and Feller (2005) reported that Mn is rapidly translocated to shoots from roots and generally accumulates in shoots of plants. Page et al. (2006) also observed similar tendency of Mn translocation in young white lupine plants (*Lupinus albus*). Excess accumulation of Mn in plant tissues can alter several physiological processes including enzyme activity, and absorption, translocation and utilization of other nutrients, such as Ca, Mg, Fe and P (Ducic and Polle 2005; Lei et al. 2007).

There are differences in tolerance to Mn among plant species and even within varieties of the same species. For example, Mn toxicity was observed when leaf Mn concentration exceeded 150 mg kg^{-1} in bean (*Phaseolus vulgaris*), 650 mg kg^{-1} in clover (*Trifolium subterraneum*), 1000 mg kg^{-1} in watermelon (*Citrullus lanatus*), and 5000 mg kg^{-1} in lowland rice (*Oryza sativa*) (Hue and Mai 2002). Hue et al. (2001) reported that leaf Ca/Mn ratio is a better predictor of Mn toxicity than leaf Mn concentration alone.

11.8.1.4 Iron

Iron toxicity to plants occurs due to excess availability of iron either at high soil acidity or under prolonged soil reduction by waterlogging or both. Generally, it is found in acid sulfate soils (Sulfaquepts, Sulfaquents) and strongly acid Oxisols,

Ultisols, and Histosols. In acid sulfate soils, iron toxicity is an important growth limiting factor. Iron toxicity symptom in rice includes the appearance of small brown spots on the lower leaves starting from the tips. Later, the whole leaf turns brown, purple, yellow or orange. There is, however, varietal difference in the sensitivity or tolerance of plants to excess iron. Selection of tolerant varieties of crops should be a better option than soil amelioration from economic and environmental points of view.

11.8.2 Deficiency of Nutrients

11.8.2.1 Calcium and Magnesium

In highly weathered acid soils of the humid tropics (Oxisols, Ultisols), acid weathering of rocks and minerals releases Ca, Mg, K, Fe, Al, etc. in soluble forms, but Fe and Al mostly get precipitated as oxides and hydroxides and are retained in the soil. On the other hand, plenty of bases are leached out of the soil resulting in the deficiency of Ca and Mg. Liming acid soil increases the pH of the soil and at the same time can remove calcium deficiency, if there is any. Calcium deficiency occurs rarely. Acid rain has been found to deplete soil calcium. Acid deposition provides (1) hydrogen ions, which displace cations adsorbed to soil surfaces, and (2) sulfate and nitrate ions, which tend to keep base cations dissolved in soil water that eventually leaches out and drains into streams and lakes.

11.8.2.2 Phosphorus

Phosphorus availability in soil depends on soil organic matter, pH and active Al, Fe and Ca. It is most available to crop plants in soils of slightly acid to neutral in reaction (Ch'ng et al. 2014). Phosphorus deficiency to plants occurs usually in soils in most acid soils due to fixation of inorganic P by Al and Fe (Adnan et al. 2003). Vance et al. (2003) reported that in spite of being high in total P many soils are inherently poor in available phosphorus. According to Smith (2001), organic P comprises a large proportion of total P and organic phosphates are not currently available to plants. Vance (2001) reported that P deficiency limits crop productivity on >40% of the world's arable lands. According to Kochian et al. (2004), iron and aluminium oxides and hydroxides bind phosphates through chemical precipitation or physical adsorption. This process considerably reduces the availability of applied phosphate fertilizers to growing plants (Syers et al. 2008). Therefore, only a little amount of P remains in soil solution for plant's uptake at any given time. So, Balemi and Negisho (2012) stated that phosphorus is one of the most inaccessible nutrients in the soil. Tilman et al. (2002) suggested that applying chemical fertilizer alone cannot desirably increase crop production in P-limiting soils.

11.8.2.3 Molybdenum

Molybdenum availability is low in acid soils. Iron and aluminium hydroxides strongly hold molybdate ions so that availability of molybdenum is reduced there. Plants take up molybdenum as the molybdate (MoO_4^-) ions. High levels of sulfate (SO_4^{2-}) in acid sulfate soils can suppress molybdate uptake by ion antagonism (Schulte 2004). Occasionally, the harmful effect of soil acidity on leguminous plants seems to be caused by Mo deficiency rather than by Al toxicity. Mo is required by the *Rhizobium* in the N_2 fixation process of legumes. The main symptoms of molybdenum deficiency in non-legumes are stunting and failure of leaves to develop a healthy dark green color. The leaves of affected plants show a pale green or yellowish green color between the veins and along the edges. In advanced stages, the leaf tissue at the margins of the leaves dies. The older leaves are the more severely affected. In cauliflowers, the yellowing of the tissue on the outer leaves is followed by the death of the edges of the small heart leaves.

11.9   Effects of Soil Acidity on Soil Fauna

The structure and function of living soil communities remain in equilibrium with the chemical conditions of the soil. Soil acidity is the most important factor that influences the balance between groups of living organisms. Soil fauna cannot effectively cope with large changes in soil pH. The population and activity of worms and termites usually decrease as the soil becomes more acidic. Abundance of earthworms decreases in acidic soil conditions. Generally, deep burrowing and soil eating worm species cannot tolerate low soil pH.

11.10   Management of Acid Soils

Soil acidy is one of the major soil management problems (Li et al. 2001) for two reasons: (i) high levels of soil acidity reduce farm profitability by increasing costs of production and reducing yields of crops and pastures (Scott et al. 2000b) and (ii) soil acidy affects nutrient balance in the soil by immobilizing phosphorus and increasing aluminium and manganese toxicity, consequently facilitating soil degradation and reducing plant production (Trapnell and Malcolm 2004; Upjohn et al. 2005). However, management of acid soils involves (i) the choice of tolerant crops, (ii) reducing acidity and Al/Mn toxicity by liming, (iii) correcting P deficiency with fertilizers and lime, and (iv) enhancing cation exchange and buffering by the use of manures. The most popular and common option of acid soil management is liming; but it is costly, and often risky in environmental health context. If some crops that can cope with the present acidity level of the soil are found, liming can be avoided. But, when soil pH falls below 5.5, liming becomes necessary to correct Al/Mn toxicity and P deficiency. Profitable crop production is not possible without liming in many strongly acid soils. Some crop plants, such as wetland rice varieties, can alleviate Mn toxicity by oxidizing soils in their rhizosphere. Therefore, selection of suitable crops and integration of appropriate liming procedure remains to be the key elements in the management of acid soils. Liming is beneficial because it increases pH, reduces Al/Mn toxicity, increases Ca and Mg supply, reduces P deficiency, enhances microbial activity, improves nitrogen fixation, etc. Acid soils can be very productive if lime and nutrients are applied at proper time and in appropriate quantity, and adequate irrigation and drainage are provided.

11.10.1 Selection of Crops for Acid Soils

Some crops such as tea, rubber and oil palm are relatively resistant to aluminium toxicity (Spaargaren 2008). Other Al-tolerant plants are whistler, diamond bird, wheat, rye-grass, tall fescue, subterranean clover, chickory, narrow leaf lupins, oats, triticale, cereal rye, cocksfoot, paspalum, yellow and slender serradella, consol love grass, etc. These plants can grow satisfactorily in acid soils without liming. However, durham wheat, barley, lentils, chickpeas, lucerne, strawberry, berseem, buffel grass, canola, red clover, balansa clover, white clover, and tall wheatgrass are very sensitive to high levels of aluminium found in acid soils. Some plants are tolerant to high levels of Mn (700 to >1000 mg kg^{-1}); these plants include sub-clover, cotton, cowpea, soybean, wheat, barley, rice, sugar cane, tobacco, sunflower, most pasture grasses, oats, triticale, and cereal rye. Mn sensitive plants include leucerne, pigeon pea, barrel and bar medics, white clover, strawberry, clover, chickpea, canola, etc. These plants should not be planted in acid soils with Mn levels >700 mg kg^{-1} because their growth is stunted there (Upjohn et al. 2005). A list of acid tolerant plants of different pH preferences is given in Table 11.3.

11.10.2 Liming

Neutralizing soil acidity and increasing soil pH by incorporating materials containing carbonates, oxides, hydroxides and similar compounds or complexes of Ca or Mg is known as liming. These materials are called agricultural lime. Liming materials are relatively insoluble and, therefore, are incorporated much before the next cropping season to favor their dissolution and release of sufficient Ca^{2+} or Mg^{2+} ions that would neutralize the acidity of the soil. The Ca^{2+} or Mg^{2+} ions replace exchangeable H+ ions so that the pH of the soil increases. The successful neutralization of soil acidity by liming depends on (i) the rate of lime, (ii) the purity of lime, (iii) the fineness of lime particles, and (iv) the degree of mixing with the soil (Anon 2006; Rodríguez et al. 2009). Lime requirement of soil depends on the number of units of soil pH to be raised (desired pH – current soil pH) along with some

11.10 Management of Acid Soils

Table 11.3 List of some acid tolerant plants

Suitable pH	Plants
	Vegetable crops
4.5–5.5	Radish, sweet potato
5.5–6.5	Endive, parsley, pepper, rhubarb, soybean
	Fruit crops
4–5.5	Blueberry, cranberry, raspberry
5–6.5	Apple, grape, strawberry
	Woody plants
4.0–5.0	Spruce, black, azalea
4.5–6.0	White birch, heather, rhododendron, balsam fir, hemlock, pine, Jack
5.0–6.5	Beech, oak, pine, tamarack
	House plants
4.5–5.5	Achmines, Adiantum, African violet, aloe, Amarylis, Aphelandra, Aurucaria, Norfolk pine, azalea, begonia, caladium, Calathea, Crossandra, cyclamen, dieffenbachia, Epiphyllum, gardenia, hydrangea, impatiens, Maranta, Peperomia, Pilea, Polypodium, primula, Rechsteineria, Saxifraga, Scindapsus, Streptocarpus, Syngonium, Zygocactus
5.5–6.5	Anthurium, bromeliad, Cattleya, Columnea, cymbidium, cypripedium, daffodil, gladiolus, hyacinth, iris, narcissus, Phalaenopsis, Platycerium, Thipsalidopsis, tulip, Vanda

Source: http://www.coopext.colostate.edu/TRA/PLANTS/acidlove.shtml

other factors, such as buffering capacity (resistance to change in pH), cation exchange capacity (CEC) and soil texture. Besides reducing soil acidity, liming has some additional benefits including (i) improvement of physical conditions, (ii) reduction in toxicity of some elements such as Al, Fe and Mn by precipitating them as their hydroxides, (iii) decreased solubility and toxicity of some heavy metals, such as cadmium copper, nickel, and zinc, (iv) increased availability of some nutrients, such as Ca, Mg and P, (v) enhanced nitrogen fixation by legumes, etc. (Truog 2004; Prochnow 2008).

The ultimate aim of liming is to improve crop productivity by reducing soil acidity. Scott et al. (2000a) observed that liming acid soils under cropping and pasture rotations increased crop yield and persistence of pastures. Mullen et al. (2006) and Brennan and Li (2006) reported improvement of yields of acid tolerant wheat, barley and canola by liming of acid soils.

11.10.2.1 Liming Materials

Carbonates such limestone or calcite ($CaCO_3$) and dolomite ($MgCO_3.CaCO_3$) are cheaper and easier to handle than other liming materials. These materials are, therefore, most commonly used for liming agricultural soils. Both limestone and dolomite are secondary rocks or minerals obtained from ores and ground to fine particles to enhance their reactivity in soil. Although dolomite is relatively costlier and slower in action, it has the advantage of addition of Mg along with Ca. In the United States, more than 90 percent of liming materials are calcite and dolomite. Sometimes ground marl and oyster shells, which contain considerable amounts of Ca and Mg, are used as liming materials, especially in coastal areas.

Oxides of Ca and Mg (CaO or MgO) are also used as liming materials. They include burned lime, unslaked lime, and quicklime. Crushed calcite limestone or dolomite limestone are baked in a furnace to produce these materials by driving off carbon dioxide (CO_2). Although these materials have high reactivity in soil and are the most efficient in raising soil pH rapidly, they are difficult to handle because of their caustic nature. Their cost is also higher than carbonate materials. One ton of calcium oxide has the neutralizing power of 1.8 tons of calcite. When oxides of Ca react with water they form hydroxides. Calcium hydroxides (hydrated lime or slaked lime) are also used for liming. They are also quick in acting but difficult to handle. They are more expensive than limestone or dolomite as well. Some miscellaneous materials including by-products of industrial processes can be used as liming materials. Basic slag from blast furnaces and lime sludge from sugar processing plants are two such examples. However, their contaminants are a matter of concern.

Lime can be applied as slurries known as fluid lime. Fluid lime is prepared by suspending crushed carbonates or oxides of Ca and/or Mg in water. Sometimes a small amount of clay is added to keep the materials in a state of suspension. Usually, lime particles in suspension are smaller than 100 mesh, so that a high quality lime is applied more uniformly and can act fast.

11.10.2.2 Quality of Lime

The chemical composition or purity and the degree of fineness of lime are the criteria of the quality of lime. One quality parameter of lime is the calcium carbonate equivalence (CCE). The CCE of a liming material is determined by comparing with the CCE of pure calcium carbonate which is assigned theoretically a value of 100 (molecular weight of $CaCO_3$ is 100 too). The CCE values of some common liming materials are given in Table 11.4. Atomic weights of constituting elements and degree of impurity affect the CCE. The CCE is taken to compare the acid neutralizing capacity of different liming materials.

Table 11.4 Calcium carbonate equivalence of common liming materials

Liming material	Composition	CCE
Calcitic limestone	$CaCO_3$	100
Dolomitic limestone	$CaMg(CO_3)_2$	109
Calcium oxide	CaO	179
Calcium hydroxide	$Ca(OH)_2$	136
Slag	$CaSiO_3$	80

Source: http://learningstore.uwex.edu/assets/pdfs/a3671.pdf

11.10 Management of Acid Soils

However, the degree of fineness or particle size primarily determines the neutralizing power of a liming material because finer materials react more rapidly than coarser materials. Usually, agricultural lime contains particles of different sizes. As a general rule, particles that are retained by a 10-mesh screen is 0 percent reactive, those passing through 10-mesh and remain on 60-mesh are 50 percent reactive, and particles passing through 60 mesh is 100 percent reactive. Now, let us calculate the fineness factor of a liming material of which 10 percent particles remain on 10-mesh, 30 percent particles pass through 10 mesh but remain on 60-mesh and 60 percent particles pass through 60-mesh.

$$\text{The fineness factor}(FF) = \left[(0.10 \times 0.0) + (0.30 \times 0.50) + (0.60 \times 1.0)\right] = 0.75$$

We can further calculate the effective calcium carbonate factor (ECC) by multiplying the FF with CCE (USDA, NRCS 1999). If the CCE is 0.90 (90 percent calcium carbonate equivalent), the ECC = [0.75 (fineness) × 0.90 (CCE)] = 0.675. Thus, if 2.5 t ha^{-1} of lime is applied, the effective amount of calcium carbonate will be (2.5 × 0.675 = 1.69 t ha^{-1}.

11.10.2.3 Mechanism of lime Action

Liming materials are sparingly soluble or insoluble substances. They slowly undergo dissolution in the soil and release CA^{2+} (or Mg^{2+}) ions which replace the H^+ ions in the exchange sites on soil colloids. The dissolution of limestone in the soil can be shown as:

$$CaCO_3 + H_2O + CO_2 = Ca^{2+} + 2HCO_3^-$$

The Ca^{2+} ions replace aluminium (Al^{3+}) and hydronium (H_3O^+) ions from colloid surfaces. These ions and those aluminium and hydronium ions already present in soil solution of acid soils are neutralized by the $H_2CO_3^-$ ions produced from lime.

$$HCO_3^- + H_3O^+ = CO_2 + 2H_2O$$

The replacement of H_3O^+ ions by Ca^{2+} can be shown as:

| Colloid | H3O$^+$
H3O$^+$ | + Ca^{2+} | ⟷ | Colloid | Ca^{2+} + H3O$^+$ |

11.10.2.4 Lime Requirement

The amount of lime required to raise soil pH to the desired level depends on many factors in addition to the number of units of pH to increase. For example, a famer has two different fields having the same soil pH, say 5.5, but one soil is sandy loam and the other is clay. If he likes to increase pH of both the soils to 6.5, he needs different amounts of lime. The clay soil needs more lime because it has higher cation exchange capacity and higher number of H^+ and Al^{3+} ions to replace. It has been mentioned earlier that the amount of lime to be added depends on clay content, cation exchange capacity, base saturation percentage and buffering capacity. Therefore, the amount of lime needed cannot be estimated from soil pH alone (current pH, desired pH and number of units of pH change) and lime requirement of each soil requires separate determination. If existing pH, desired pH, clay content, CEC, exchangeable Al^{3+} and H^+ and percent base saturation are known, computer programs can estimate lime requirement at present.

Two different methods are generally used for the determination of lime requirements. These methods are known as the SMP (Shoemaker-McLean-Pratt) and Adams-Evans buffer methods. The SMP method is applied to soils with significant reserves of exchangeable Al, and the Adams-Evans buffer has been designed for soils that are coarse-textured, with low cation exchange capacities and organic matter contents. However, the Mehlich lime buffer has been developed for Ultisols, Histosols, Alfisols, and Inceptisols. The SMP buffer was modified by Sikora (2006), and all these buffers can be employed satisfactorily for the estimation of lime requirement of soils. The amount of lime required to increase soil pH by one unit varies from 2 to 5 t ha^{-1}.

11.10.2.5 Applying Lime

Lime is generally applied after harvesting a crop and well ahead of planting the next crop. Lime is evenly spread over the soil surface and incorporated well with the soil within the root zone. Simple spreading of lime on surface does not help because it can increase pH of 1–2 cm surface soil due to low solubility of lime. Usually in soils requiring high amount of lime, one half of the lime is applied before tillage and the other half after tillage. The soil is tilled well to mix the lime with the soil for quick action. In soils that require very high amount of lime (very strongly acid soil), one half of lime may be applied in the first year and the other half in the next year.

11.11 Management of Acid Sulfate Soils

Exposed acid sulfate soils are often characterized by the presence of spotted yellow coloration of jarosite [$KFe_3(SO_4)_2(OH)_6$]. Rain and flood water can flush sulfuric acid formed by oxidation of pyrite into nearby waterways, killing fish, other aquatic organisms and vegetation. Strong acidity and Al^{3+} toxicity are the problems associated with the use of acid sulfate soil for crop production. Some shrimp farms have been established in acid sulfate soils of Asia including Bangladesh and the Philippines. Unsatisfactory growth of phytoplankton, shrimp kills, damage of the gills and obnoxious odor are problems in shrimp farming in acid sulfate soils.

Potential acid sulfate soils can be profitably used for paddy rice cultivation if fresh irrigation water can be provided.

11.11.1 Reclamation of Acid Sulfate Soils

For reclamation of acid sulfate soils, oxidation of pyrite and removal of acids by irrigation and drainage have been employed with varying success. Aeration, flooding and disposal of acid water should be done with great caution so that reclamation work does not lead to the degradation of environment. Acids in acid sulfate soils may be neutralized by liming well ahead of utilization for cropping. Thoroughly mixing the appropriate amount and type of lime into disturbed acid sulfate soils will neutralize any acid produced. Hydrated lime is often more appropriate for treating acid waters due to its higher solubility. Once the acid water has been treated to pH 6.5–8.5 and metals have been reduced to appropriate levels, it can usually be safely released from the site at a controlled rate to prevent significant changes to the quality of offsite waters. Quicklime, sodium bicarbonate dolomite, and some industry by-products such as basic slag may also be used.

Hydraulic separation is suitable for coarse-textured soils containing iron sulfides. Sluicing or hydrocycloning, is used to hydraulically separate the sulfides from the coarser textured materials. This could be an effective form of management when the sediments contain less than 10–20 percent clay and silt, and have low organic matter content. The separated sulfidic material extracted via the process requires special management involving either neutralization or strategic reburial. Ex-situ soil oxidation, neutralization and washing have been tried in some instances. This method is very costly and farmers find little interest.

Under undisturbed and waterlogged conditions acid sulfate soils are almost harmless. They can support luxuriant mangroves. These soils may better be kept in their natural state. Deeply seated sulfidic horizons must not be exposed for shrimp pond construction.

11.11.2 Aquaculture in Acid Sulfate Soils

Excavation of soil for fish or shrimp pond construction in coastal regions often exposes sulfidic materials to the surface. On exposure, the pyrite is oxidized to produce sulfuric acid and make the soil acidic. They may have pH values of 5–7 when wet, but when dried, pH may fall to 2 or 3. Acid-sulfate soils should not be used for aquaculture ponds if other alternatives are available (Boyd et al. 2002). According to an estimate, at least 60 percent of the fishponds in the Philippines are affected by acid sulfate conditions. For intensive shrimp culture in ponds on acid sulfate soils a planned sequence of operations are needed. These include filling, draining, drying the ponds, liming the bottom soil and tilling to mix the lime with the soil. Tilling of bottom soil after spreading about 1 t ha^{-1} agricultural lime may be done with a tooth harrow and a draft animal or a small power tiller. Long narrow paddies are constructed on the top of the dikes and seawater is pumped into them. Regular monitoring of the pH of the soil and water is needed.

11.12 Risks of Overliming

Liming of most tropical soils may be viewed both as calcium fertilization and acid soil reclamation, and the target pH should probably not exceed about 6.0 (Harter 2007). At such a pH level, the aluminum and manganese concentrations in soil solution remain above the toxic level for plants. However, overliming can reduce toxicity of these elements too, but it may further increase pH and can create molybdenum toxicity. Overliming can also cause deficiency of some micronutrients such as copper, zinc, boron, and manganese. This is because of reduced solubility of these nutrients at higher pH levels. Harter (2007) suggests that the most important problems of overliming tropical soils is physical rather than chemical. Overliming can cause a destabilization of soil structure resulting in reduced permeability and the lack of adequate drainage. Since sesquioxide stabilized aggregates are typically formed in the humid tropical soils, the reduced permeability due to aggregate destabilization, can result is wet soils and complete change in the ecosystem.

Study Questions
1. What are the various categories of acid soils? How and why do some soils become acidic? Which soil orders are generally acidic?
2. What are the effects of soil acidity on soil processes related to the growth and nutrition of plants?
3. What do you mean by actual and potential acid sulfate soils? How do they form? What are their problems?
4. Give examples of some acid tolerant crops. Discuss liming materials and the quality of lime. Explain the benefits of liming.
5. What problems are associated with aquaculture in acid sulfate soils? Describe the techniques of management of acid sulfate soils.

References

Adamovich A, Klasens V (2001) Symbiotically fixed nitrogen in forage legume–grass mixture. Grassl Sci Eur 6:12

Adnan A, Mavinic DS, Koch FA (2003) Pilot-scale study of phosphorus recovery through struvite crystallization-examining to process feasibility. J Environ Eng Sci 2(5):315–324

Ambrazaitiene D (2003) Activity of symbiotic nitrogen fixation in the *Dystric albeluvisols* differing in acidity and fertilization. Agric Scientific Articles 83(3):173–186

References

Andriesse W, van Mensvoort MEF (2006) Acid Sulfate soils: distribution and extent. In: Lal R (ed) Encyclopedia of soil science, 2nd edn. CRC Press, Columbus, Ohio

Anonymous (2006) Lime quality-does it matter? Potash and Phosphate Institute, Agri-briefs, agronomic news items, Norcross

Balemi T, Negisho K (2012) Management of soil phosphorus and plant adaptation mechanisms to phosphorus stress for sustainable crop production: a review. J Soil Sci Plant Nutr 12(3):547–562

Barnes RJ, Baxter SJ, Lark RM (2007) Spatial covariation of *Azotobacter* abundance and soil properties: a case study using the wavelet transform. Soil Biol Biochem 39(1):295–310

Batjes NH (2011) Overview of soil phosphorus data from a large international soil database. Report 2011/11, Plant Research International (PRI), Wageningen UR, and ISRIC – World Soil Information, Wageningen

Bingham J, Gagliano W (2006) Acid mine drainage. In: Lal R (ed) Encyclopedia of soil science. CRC Press, Boca Raton

Boettinger JL (2005) Calcification. In: Lal R (ed) Encyclopedia of soil science, volume 1, 2nd edn. CRC Press, Boca Raton

Bohn HL, McNeal BL, O'Connor GA (2001) Soil chemistry, 3rd edn. John Wiley and Sons, Inc. New York

Boyd CE, Wood CW, Thunjai T (2002) Aquaculture pond bottom soil quality management. Aquaculture collaborative research support program. Oregon State University, Corvallis, Oregon

Brennan J, Li G (2006) MASTER – economic analysis, Primefacts, profitable and sustainable primary industries. NSW Department of Primary Industries, Orange, NSW

Buchanan B, Grusen W, Jones R (2000) Biochemistry and molecular biology of plants. American Society of Plant Physiologists, Rockville

Casiday R, Frey R (2013) Acid rain: inorganic reactions experiment. worldacidrainsecrets.blogspot.com/2013_01_01_archive.html

Ch'ng HY, Ahmed OH, Ab. Majid NM (2014) Improving phosphorus availability in an acid soil using organic amendments produced from agroindustrial wastes. Sci World J 2014, Article ID 506356, 6 pages. https://doi.org/10.1155/2014/506356

Chen RF, Zhang FL, Zhang QM, Sun QB, Dong XY, Shen RF (2012) Aluminium-phosphorus interactions in plants growing on acid soils: does phosphorus always alleviate aluminium toxicity? J Sci Food Agric 92(5):995–1000

Ciamporova M (2002) Morphological and structure responces of plant roots to aluminium at organ, tissue, and cellular levels. Biol Plant 45:161–171

CSIRO (2003) Acid Sulfate Soil Technical Manual 1.2, CSIRO Land & Water, Australia

De Boer W, Kowalchuk GA (2001) Nitrification in acid soils: micro-organisms and mechanisms. Soil Boil Biochem 33:853–866

Ducic T, Polle A (2005) Transport and detoxification of manganese and copper in plants. Braz J Plant Physiol 17:103–112

FAO (2006) World reference base for soil resources 2006, A framework for international classification, correlation and communication. FAO–UNESCO–ISRIC. FAO, Rome

Harter D (2007) Acid soils of the tropics. ECHO Technical Note 48, ECHO, 17391 Durrance Road, North Fort Myers, Florida 33917, USA

Hartwell BL, Pember FR (1918) The presence of aluminum as a reason for the difference in the effect of so-called acid soils on barley and rye. Soil Sci 6:259–277

Hersman LE, Forsythe JH, Ticknor LO, Maurice PA (2001) Growth of *Pseudomonas endocina* on Fe(III) (hydr)oxides. Appl Environ Microbiol 67(10):4448–4453

Hue NV, Mai Y (2002) Manganese toxicity in watermelon as affected by lime and compost amended to a Hawaiian acid Oxisol. Hort Sci 37:656–661

Hue NV, Vega S, Silva JA (2001) Manganese toxicity in a Hawaiian Oxisol affected by soil pH and organic amendments. Soil Sci Soc Am J 65:153–160

Iqbal MT (2012) Acid tolerance mechanisms in soil grown plants. Malays J Soil Sci 16:1–21

Jakovljevic M, Kresovic M, Blagojevic S, Antic-Mladenovic S (2005) Some negative chemical properties of acid soils. J Serb Chem Soc 70:765–770

Jones DL, Dennis PG, Qwen AG, van Hees PAW (2003) Organic acid behavior in soils – misconceptions and knowledge gaps. Plant Soil 248:31–41

Kabata-Pendias A, Pendias H (1992) Trace elements in soils and plants, 2nd edn. CRC Press, Boca Raton, FL

Kisinyo PO, Othieno CO, Okalebo JR, Kipsat MJ, Serem AK, Obiero DO (2005) Effects of lime and phoshorus application on early growth of *Leucaena* in acid soils. Afr Crop Sci Conf Proc 7:1233–1236

Kochian LV, Hoekenga OA, Pineros MA (2004) How do crop plants tolerate acid soils? Mechanisms of aluminium tolerance and phosphorus efficiency. Annu Rev Plant Biol 55:459–493

Kresovic M, Jakovljevic M, Blagojevic S, Zarkovic B (2010) Nitrogen transformation in acid soils subjected to pH value changes. Arch Biol Sci 62(1):129–136

Kumari A, Kapoor KK, Kundu BS, Mehta RK (2008) Identification of organic acids produced during rice straw decomposition and their role in rock phosphate solubilization. Plant Soil Environ 54(2):72–77

Langer H, Cea M, Curaqueo G, Borie F (2009) Influence of aluminum on the growth and organic acid exudation in alfalfa cultivars grown in nutrient solution. J Plant Nut 32:618–628

Lapinskas E (2004) Liucernų inokuliavimo efektyvumas, priklausomai nuo gumbelinių bakterijų štamų adaptacijos dirvožemio rūgštingumui. Vagos LŽŪU mokslo darbai 63(16):20–25

Lapinskas E (2008) Biological nitrogen fixation in acid soils of Lithuania. žemės ūkio mokslai 15(3):67–72

Lei Y, Korpelainen H, Li C (2007) Physiological and biochemical responses to high Mn concentrations in two contrasting *Populus cathayana* populations. Chemosphere 68:686–694

Li GD, Helyar KR, Conyers MK, Cullis BR, Cregan PD, Fisher RP, Castleman LJC, Poile GJ, Evans CM, Braysher B (2001) Crop responses to lime in long-term pasture-crop rotations in a high rainfall area in south-eastern Australia. Aust J Agric Res 52:329–341

Ma JF (2007) Syndrom of Aluminum toxicity and diversity of Aluminum resistance in higher plants. Int Rev Cytol 264:225–252

Matsumoto H (2002) Plants under aluminum stress: toxicity and tolerance. In: Waisel Y, Eshel A, Kafkafi U (eds) Plant roots: the hidden half. Marcel Dekker, Inc, New York

McLennan SM, Taylor SR (1999) Earth's continental crust. In: Marshall CP, Fairbridge RW (eds) Encyclopedia of geochemistry. Kluwer, Dodrecht

Menzies NW (2003) Toxic elements in acid soils: chemistry and measurement. In: Rengel Z (ed) Handbook of soil acidity. Marcel Dekker, Inc, Madison Avenue, New York

Mohammadi K, Sohrabi Y, Heidari G, Khalesro S, Majidi M (2012) Effective factors on biological nitrogen fixation. Afr J Agric Res 7(12):1782–1788

Mullen CL, Scott BJ, Evans CM, Conyers MK (2006) Effects of soil acidity and liming on lucerne and following crops in central-western new South Wales. Aust J Exp Agric 46:1291–1300

Nayak S, Prasanna R (2007) Soil pH and its role in cyanobacterial abundance and diversity in rice field soils. Appl Ecol Environ Res 5:103–113

Page V, Feller U (2005) Selective transport of zinc, manganese, nickel, cobalt and cadmium in the root system and transfer to the leaves in young wheat plants. Ann Bot 96:425–434

Page V, Weisskopf L, Feller U (2006) Heavy metals in white lupin: uptake, root-to-shoot transfer and redistribution within the plant. New Phytol 171:329–341

Panda SK, Matsumoto H (2007) Molecular physiology of aluminium toxicity and tolerance in plants. Bot Rev 73:326

Polomski J, Kuhn N (2002) Root research methods. In: Waisel Y, Eshel A, Kafkafi U, Dekker M (eds) Plant roots: the hidden half. Marcel Dekker, New York

Potter G, Bajita-Locke J, Hue N, Strand S (2004) Manganese solubility and phytotoxicity affected by soil moisture, oxygen levels, and green manure additions. Comm Soil Sci Plant Anal 35:99–116

Prochnow LI (2008) Optimizing nutrient use in low fertility soils of the tropics in times of high fertilizer prices (Brazil). Better Crops 3:19–21

Prosser JI (2011) Soil nitrifiers and nitrification. In: Ward BB, Arp DJ, Klotz MG (eds) Nitrification. American Society for Microbiology press, Washington, DC

Rengel Z (2000) Uptake and transport of manganese in plants. In: Sigel A, Sigel H (eds) Metal ions in biological systems. Marcel Dekker, New York

Rengel Z (ed) (2003) Handbook of soil acidity. Marcel Dekker, Inc. Madison Avenue, New York

Rodríguez JAC, Hanafi MM, Omar SRS, Rafii YM (2009) Chemical characteristics of representative high aluminium saturation soil as affected by addition of soil amendments in a closed incubation system. Malays J Soil Sci 13:13–28

Ryan J, Rashid A (2006) Phosphorus. In: Lal R (ed) Encyclopedia of soil science, vol 2. Taylor and Francis, New York

Ryan PR, Delhaize E, Jones DL (2001) Function and mechanism of organic anion exudation from plant roots. Annu Rev Plant Physiol Plant Mol Biol 52:527–560

Samac DA, Tesfaye M (2003) Plant improvement for tolerance to aluminium in acid soils- a review. Plant Cell Tissue Organ Cult 75:189–207

Schulte EE (2004) Soil and applied molybdenum. http://corn.agronomy.wisc.edu/Management/pdfs/a3555.pdf

Scott JF, Lodge GM, McCormick LH (2000a) Economics of increasing the persistence of sown pastures: costs, stocking rate and cash flow. Aust J Exp Agric 40:313–323

Scott BJ, Ridley AM, Conyers MK (2000b) Management of soil acidity in long-term pastures of south-eastern Australia: a review. Aust J Exp Agric 40:1173–1198

Sharma BD, Chahal DS, Singh PK, Raj-Kumar (2008) Forms of iron and their associations with soil properties in four soil taxonomic orders of arid and semi-arid soils of Punjab, India. Commun Soil Sci Plant Anal 39:2550–2567

Shen QR, Ran W, Cao ZH (2003) Mechanisms of nitrite accumulation occurring in soil nitrification. Chemosphere 50:747–753

Sikora FJ (2006) A buffer that mimics the SMP buffer for determining lime requirement of soil. Soil Sci Soc Am J 70:474–486

Sims T (2000) Soil fertility evaluation. In: Sumner ME (ed) Hand book of soil science. CRC Press, Boca Raton

Smith FW (2001) Plant response to nutritional stresses. In: Hawkesford MJ, Buchner P (eds) Molecular analysis of plant adaptation to the environment. Kluwer Academic Publishers, The Netherlands

Soil Survey Staff (1999) Soil taxonomy: a basic system of soil classification for making and interpreting soil surveys. In: Agriculture Handbook No. 436, U.S, 2nd edn. Government Printing Office, Washington, DC

Sorensen SPL (1909) Enzymstudien. II: Mitteilung. Über die Messung und die Bedeutung der Wasserstoffionenkonzentration bei enzymatischen Prozessen. Biochemische Zeitschrift (in German) 21:131–304

Spaargaren O (2008) Alisols. In: Chesworth W (ed) Encyclopedia of soil science. Springer, Dordrecht

Sumner ME, Noble AD (2003) Soil acidification: the world story. In: Rengel Z (ed) Handbook of soil acidity. Marcel Dekker, Inc, Madison Avenue, New York

Suryantini (2014) Effect of lime, organic and inorganic fertilizer on nodulation and yield of soybean (*Glycine max*) varieties in Ultisol soils. J Exp Biol Agric Sci 2(1):78–81

Syers JK, Johnston AE, Curtin D (2008) Efficiency of soil and fertilizer phosphorus: reconciling changing concepts of soil phosphorus behaviour with agronomic information. FAO Fertilizer and Plant Nutrition Bulletin 18,108, FAO, Rome

Tan KH (2010) Principles of soil chemistry, 4th edn. CRC Press, Boca Raton

Tang C, Rengel Z (2003) Role of plant cation/anion uptake ratio in soil acidification. In: Rengel Z (ed) Handbook of soil acidity. Marcel Dekker, Inc, Madison Avenue, New York

Taurian T, Ibañez F, Fabra A, Aguilar OM (2006) Genetic diversity of rhizobia nodulating *Arachis hypogaea* L. in central Argentinean soils. Plant Soil 282:41–52

Tilman D, Cassman KG, Matson PA, Naylor R, Polasky S (2002) Agricultural sustainability and intensive production practices. Nature 418:671–677

Trapnell L, Malcolm B (2004) Net benefits from liming acid soils, Proceedings of Contributed Papers, 2004 AFBM Network Conference

Truog E (2004) Acidic soils. In: Gardiner DT, Miller RW (eds) Soils in our environment, 10th edn. Pearson Education, Inc, Upper Saddle River

Upjohn B, Fenton G, Conyers M (2005) Soil acidity and liming, Agfact AC 19, 3rd edn. Orange, NSW, NSW Department of Primary Industries

USDA-NRCS (1998) Soil quality indicators: pH. National Soil Survey Center in cooperation with the Soil Quality Institute, NRCS, USDA, and the National Soil Tilth Laboratory, Agricultural Research Service, USDA)

USDA NRCS (1999) Liming to improve soil quality in acid soils. Soil Quality – Agronomy Technical Note No 8. Washington, DC. http://www.nrcs.usda.gov/Internet/FSE_DOCUMENTS/nrcs142p2_053252.pdf

Vance CP (2001) Symbiotic nitrogen fixation and phosphorus acquisition: plant nutrition in a world of declining renewable resources. Plant Physiol 127:390–397

Vance CP, Uhde-Stone C, Allan D (2003) Phosphorus acquisition and use: critical adaptation by plants for securing non-renewable resources. New Phytol 15:423–447

Vazquez FM, Arbestain MC, Chesworth W (2008) Acid soils. In: Chesworth W (ed) Encyclopedia of soil science. Springer, Dordrecht

Vitorello VA, Capaldi FR, Stefanuto VA (2005) Recent advances in Aluminum toxicity and Esistance in higher plants. Braz J Plant Physiol 17:129–143

Von Uexkull HR, Mutert E (2004) Global extent, development and economic impact of acid soils. Plant Soil 171:1–15

Were BA, Otinga AN, Okalebo JR, Onyango DA, Ogega JK, Othieno C, Pypers P, Okello DK, Merckx R (2012) Improving biological nitrogen fixation (BNF) by groundnuts (*Arachis hypogeal* L.) grown in acid soils amended with calcitic and dolomitic limestones. Research Application Summary, Third RUFORUM Biennial Meeting 24–28 September 2012, Entebbe, Uganda

Zhang JB, Cai ZC, Zhu TB, Yang WY, Müller C (2013) Mechanisms for the retention of inorganic N in acidic forest soils of southern China. Scientific reports 3, article number: 2342. https://doi.org/10.1038/srep02342

Zhao W, Cai ZC, Xu ZH (2007) Does ammonium-based N addition influence nitrification and acidification in humid subtropical soils of China? Plant Soil 297:213–221

Zheng SJ (2010) Crop production on acidic soils: overcoming aluminium toxicity and phosphorus deficiency. Ann Bot 106(1):183–184

Zu C, Wu HS, Tan LH, Yu H, Yang JF, Li ZG et al (2012) Analysis of correlation between soil pH and nutrient concentrations across Hainan black pepper advantage region. Chin J Trop Crops 33:1174–1179

Chapter 12
Polluted Soils

Abstract Soils can be polluted with several organic and inorganic pollutants. Organic pollutatnts include hazardous persistent organic compounds such as polycyclic aromatic hydrocarbons (PAHs), polychlorinated biphenyls (PCBs), polychlorinated naphthalines (PCNs), polychlorinated dibenzodioxins (PCDDs), polychlorinated dibenzofurans (PCDFs), and other persistent organic substances. Inorganic pollutants mainly include heavy metals such as lead (Pb), cadmium (Cd), mercury (Hg), zinc (Zn), copper (Cu) and nickel (Ni), as well as the metalloid arsenic (As), and radioactive substances or radionuclides. The sources of soil pollutants are mainly anthopogenic. There are also point and diffuse sources; the point sources include municipal wastes, industrial wastes, medical wastes, agricultural wastes, composts and sludges, agrochemicals, domestic wastes and nuclear wastes. Organic and inorganic soil pollutants can be toxic to soil organisms, plants and animals. Some of the soil pollutants enter into the food chain, and can adversely affect human health. Moreover, soil pollutants can be transferred to surrounding air and water through volatilization, runoff, dust storms and leaching. In these ways, the quality of air and water, both surface and groundwater, can be degraded. There are several historic events where soil pollution had devastating effects on native population. Thus, remediation of polluted soils has become a dire necessity in many areas of the world. Meanwhile, some useful methods have been developed for the prevention of soil pollution, including waste management and waste disposal, and remediation of organic pollutants, heavy metals and radioactive pollutants.

Keywords Soil pollution · Wastes · Agrochemicals · Pollutants · Persistent organic pollutants · Heavy metals · Waste management · Remediation

12.1 Soil Pollution

Pollution refers to the presence, of a substance or energy, native or introduced, in the environment above a threshold level that is poisonous or harmful to organisms. Marcel van der Perk (2006) pointed out that although 'pollution' and 'contamination' are often used synonymously, but contamination actually means the introduction of a chemical substance which was not originally present in the system. Thus, soil pollution can be defined as the presence of one or more chemical substances in concentrations that can cause harm to organisms of the soil, and those that depend on soil. The concentration of the substance is usually elevated by human activity (Kabata-Pendias and Pendias (2001). Disposal of wastes, particularly wastes containing hazardous chemicals, in the ground and indiscriminate use of agrochemicals, including fertilizers, pesticides and veterinary pharmaceuticals, mainly causes soil pollution. Traffic and vehicle exhausts, leakage of underground septic tanks into the soil; mining activity and use of polluted irrigation water are the additional causes of soil pollution. Atmospheric deposition and acid rain can pollute soil of some industrialized countries.

12.2 Sources of Soil Pollutants

Most pollutants come to soils from point sources. These point sources include industrial discharges, inappropriate waste disposal systems, or accidental spills of hazardous substances during the transportation or handling, etc. Valentın et al. (2013) suggested that diffuse contamination in soil occurs from some agricultural and forestry practices, transportation and improper waste and wastewater management. Atmospheric deposition of particulates and low-volatile compounds are also diffuse sources of soil pollution. According to Alloway (2012), major sources of soil pollutants are industrial, municipal, agricultural, domestic and nuclear wastes.

12.2.1 *Municipal Wastes*

Municipal wastes include household wastes, market wastes, hospital wastes, livestock and poultry wastes, slaughterhouse wastes, etc. Some of these wastes are biodegradable such as food and kitchen wastes - meat trimmings or vegetable peelings, yard or green wastes, paper, etc. These materials undergo rapid biological decomposition in the soil; so they are not soil pollutants. There are some recyclable materials such as glass, plastics, metals and aluminum cans. These materials, if separated from other wastes, can be recycled in industries; but in most developing countries waste management systems are not so well developed and these materials are disposed of indiscriminately with other wastes. They can degrade soil physically and chemically. Some of these materials including metallic substances and metal scraps can cause serious soil pollution. Municipal wastes also include some inert wastes (construction and demolition wastes) and composite wastes (clothing and plastics), which do not usually cause soil polution but degrade soil physical properties and limit soil use. Municipal wastes contain some hazardous materials (medicines, paint, batteries, electric bulbs, containers of fertilizers and pesticides and e-wastes such as damaged computers and their accessories, cell phones, etc. A large proportion of municipal wastes include non-biodegradable materials; for example

12.2 Sources of Soil Pollutants

polyethylene and plastic sheets, bags and bottles, which hampers water movement and natural drainage, tillage and planting operations. Municipal solid wastes can be divided also into durable goods and non-durable goods. Examples of durable goods are tires and furniture, while those of nondurable goods are newspapers, paper containers and packaging cartons, yard waste, food, etc.

12.2.2 Sewage Sludge

Sludge is a "mixture of water and solids separated from various types of water as a result of natural or artificial processes", and sewage sludge is the "sludge from urban waste water treatment plants" ((EU 2000). Here, 'urban waste water' refers to domestic waste water containing solid and liquid wastes of human metabolism and household activities and can often be mixed with industrial waste water. Urban sewerage systems transport domestic sewage, industrial effluents and storm-water runoff from urban areas are transported through the urban sewage system. Sewage sludge contains organic substancess, inorganic solutes and particulate materials. The solid materials are separated from water by different treatment methods, and when the water becomes environmentally safe it is discharged into streams or lakes. The solid residue is often dumped in open places or in landfills, and incinerated, or composted for use as organic fertilizer in crop fields. Large amounts of organic matter in composted sludge can improve physical, chemical, and biological properties of soil and can enhance its productivity. Use of sewage sludge as fertilizer can also reduce some demand of industrial nitrogen and phosphorus fertilizers and increase crops yield. According to McGrath et al. (2000), application of sewage sludge in crop fields has the risk of contaminating soils with some heavy metals and can create toxicity to plants.

Several toxic and hazardous contaminants such as metals, pathogens, and organic pollutants may be present in sewage sludge (Werle and Dudziak 2014). Harrison et al. (2006) carried out a survey of literature on contaminants of sewage sludge and found 516 organic compounds of 15 different classes. There were pesticides and their residues, PAHs (polycyclic aromatic hydrocarbons) and PCBs (polychlorinated biphenyls) which may pose great risk. According to McGrath et al. (2000), the average concentrations of adsorbable organohalogens (AOX) and nonylphenole (NP) and the ranges of concentrations of some other major organic contaminants in sewage sludge are:

AOX (adsorbable organohalogens)	75,890 mg kg^{-1} dm
NP (nonylphenole)	46 mg kg^{-1} dm
LAS (linear alkylbenzene sulfonates)	< 1 to 424 mg kg^{-1} dm
DEHP (Di(2-ethylhexyl) phthalate)	4 and 170 mg kg^{-1} dm
PAH (polycyclic aromatic hydrocarbons)	1–10 mg kg^{-1} dm
PCB (polychlorinated biphenyl)	0.105 mg kg^{-1} dm

On the other hand, Paulsrud et al. (2000) observed insignificant content of PCDD (Polychlorinated dibenzodioxins) and PCDF (polychlorinated dibenzofurans).

The concentrations of heavy metals in sewage sludge vary within a wide range. For example, some investigators (Szymański et al. 2011; Li et al. 2012) mentioned the following ranges of some heavy metals.

Cadmium	1–3410 mg kg^{-1}
Chromium	10–990,000 mg kg^{-1}
Copper	80–2300 mg kg^{-1}
Nickel	2–179 mg kg^{-1}
Lead	13–465 mg kg^{-1}
Zinc	101–49,000 mg kg^{-1}.

12.2.3 Composts

Composts may contain some potential soil pollutants including heavy metals and organic compounds. These contaminants were inherited from feedstock. Hogarh et al. (2008) investigated heavy metal concentrations in several compost samples and observed concentrations of Ni, Zn, Cu and Cd within Australian permissible standards, but lead (Pb) concentration was significantly higher than the safe limit. According to Australian standard the safe limit of Pb in compost is 150 mg kg^{-1}. Analyzing 183 livestock feeds and 85 animal manure samples from commercial farms in England and Wales for their zinc and copper contents, Nicholson et al. (1999) observed the following ranges of concentrations:

Sample	Concentrations mg kg^{-1} dm	
	Zinc	Copper
Pig feeds	150–2920	18–217
Poultry feeds	28–4030	5–234
Pig manures	c.500	c.360
Poultry manures	c.400	c.80
Cattle manures	c.180	c.50

Benisek et al. (2015) mentioned that the information about organic pollutants in composts is scarce. Some organic pollutants are non-persistent; they could be degraded during composting. According to Umlauf et al. (2011) persistent organic pollutants such as PCBs, PCDDs and PCDFs could accumulate in soil if contaminated compost is continually applied in a crop field. Brandli et al. (2007) studied some POPs such as PCBs, PAHs, PCDD and PCDF in composts inherited from organochlorine insecticides, constituents of personal care products and industrial chemicals in composts. Safe limits for the content of PCDDs, PCDFs, PCBs, PAHs and others POPs in composts have been established in many European countries (Saveyn and Eder 2014). Yard waste and composts can be contaminated by the

12.2 Sources of Soil Pollutants

overuse and persistence of some insecticides. The composting process cannot degrade some organochlorine pesticides and some herbicides. Bezdicek et al. (2001) reported the presence of clopyralid and picloram in composts. Residues of 2,4-D (2,4-dichlorophenoxy acetic acid), dicamba (2-methoxy-3,6-dichlorobenzoic acid) and MCPP (2-(4-chloro-2-methylphenoxy propionic acid) were also detected in composts.

12.2.4 Medical Wastes

Medical wastes include huge quantity of wastes produced in the healthcare facilities such as hospitals, clinics, diagnostic centres, etc. Almost 75 percent of these wastes are similar to municipal wastes; these are biodegradable and seem to be innocent in the context of soil pollution. The remaining 25 percent constitue hazardous substances such as sharp instruments (knives, scissors, blades), human tissues (blood, body parts), chemicals (disinfectants, diagnostic reagents, solvents, salines), pharmaceuticals (medicines, antibiotics, hormones), infectious wastes (cultures of infectious agents, secretions, excreta), heavy metals. radioactive wastes (liquids from radiotherapy, contaminated glassware, packages, or absorbent paper, excreta from patients treated or tested with radioactive substances) and miscellaneous wases (batteries, photographic developers and fixers, X-ray plates, pressurized containers, gas cylinders, spilled or unused medicines, expired drugs leftover cytotoxic drugs, equipment, etc.). According to International Committee of the Red Cross 2011) 10–25 percent medical wastes pose the risk of soil pollution. Some substances in hospital wastes are highly hazardous and may contain genotoxic wastes consisting of mutagenic, teratogenic or carcinogenic substances.

12.2.5 Veterinary Pharmaceuticals

Various types of pharmaceuticals such as hormones, vaccines, antibiotics and metals are used in high intensity livestock farming for improving meat production, for prevention and treatment of infectious diseases and for controlling parasites (Tolls 2001). In the EU, Antibiotics and anthelmintics are the most important groups of veterinary pharmaceuticals used in the European Union. According to Kay and Boxsall (2000), the total usage of antibiotics may be around 5000 t; 3500 t is used for therapeutic purposes. About 1500 t is applied with feed for promoting growth of the farm animals (Alder et al. 2000). The use of antibiotics in livestock is even higher in the United States; in 1985 the estimated amount was about 8300 t. Animals receive veterinary pharmaceuticals mixed with feed, through injection or external application. Some of these pharmaceuticals can be excreted unaltered or as transformation products; often intermediate transformation products are more hazardous for the environment. Kay and Boxsall (2000) mentioned that a considerable portion of these pharmaceuticals and their intermediaries can reach the soil through urine, feces or manure. Hamscher et al. (2000) observed the presence of about 10 μg tetracycline kg^{-1} in liquid manure-treated agricultural fields. Kolpin et al. (2000) detected trimethoprim and sulfamethoxazole in 30% of the water samples suspected to be contaminated with antibiotics in the United States. Osterman et al. (2014) observed the following average concentrations of antibiotics in agricultural soils: sulfamethazine– 110 μg kg^{-1}, chlortetracycline– 111 μg kg^{-1} and enrofloxacin– 62 μg kg^{-1}.

Chen et al. (2012) investigated the occurrence of several antibiotics in manure samples collected from four swine farms of eastern China. The maximum concentrations of tetracycline, oxytetracycline, chlortetracycline, doxycycline and sulfadiazine in the manure samples were 98.2×10^3, 354.0×10^3, 139.4×10^3, 37.2×10^3, and 7.1×10^3 μg kg^{-1} respectively. Tetracyclines are the most heavily used antibiotics in livestock farming. IWW (2014) found in literature the presence of about 713 veterinary pharmaceuticals, including 142 transformation products in the environment. Sixteen pharmaceutical substances were found thoughout the world; these included the pain relievers such diclofenac, ibuprofen, paracetamol, naproxen, aspirin, ofloxacin,; antibiotics such as sulphamethazole, ciprofloxacin, trimethoprim, norfloxacin; sex hormones such as estrone, 17β-estradiol, 17α-ethinyl estradiol and estriol; lipid lowering drug such as clofibric acid and anti-epileptic drug such as carbamazepine. The accumulation of veterinary pharceuticals and their transformation products in soil may be harmful to human health because crop plants can take up considerable amounts of such toxic drugs as antibiotics, anti-parasitics, antifungals and others. Kemper (2008) noted the possibility of multidrug resistence in *Salmonella* spp. and *Escherichia coli*. Belgium, Ireland, the Netherlands, Switzerland, Denmark, Germany and UK are the countries that face the highest risk of contamination from veterinary pharmaceuticals (Torre et al. 2012).

12.2.6 Industrial Wastes

Industrial wastes include a variety of materials such as synthetic chemicals, acids, salts, solvents, oil, metals, stones, concrete, paints, wood chips, etc. All these wastes are not hazardous; some the constituents are, however, ignitable, corrosive, reactive and some materials in the industrial wastes are very hazardous. Chemical industries are sources of acids and bases, spent solvents, reactive waste, wastewater containing organic constituents, etc. Printing industries releases heavy metals, waste inks, solvents, ink sludges containing heavy metals, etc. Petroleum refining industries can add wastewater containing benzene and other hydrocarbon, sludge from refining process. Metal industries are the principal sources of heavy metals.

12.2 Sources of Soil Pollutants

12.2.7 Agrochemicals

Agrochemicals are a group of chemical compounds or their mixtures used to produce and protect crops. They include fertilizers (mainly industrial or synthetic fertilizers), pesticides (including insecticides, herbicides, and fungicides), soil conditioners (mainly particle aggregating and soil stabilizing organic polymers), liming and soil acidifying materials (limestone, dolomite, gypsum, sulfur), hormones and other growth promoting chemicals. Fertilizers, manures, pesticides and their residues are often responsible for soil pollution. Phosphate fertilizers are usually contaminated with arsenic, cadmium, manganese, uranium, vanadium and zinc to variable extents depending on their source rocks and their processing. Poultry manure and pig manure may contaminate soil with zinc, copper and arsenic.

Pesticides are being used in agriculture at increasing rates world wide to protect plants from harmful organisms (plant, microorganism, insects, etc.). These substances undergo several changes in the soil including dissolution, adsorption and elution with the soil solution and colloidal phases. Thus, they move within soil, water, atmosphere systems and their mobility depends on their solubility, adsorbability, and volatility.

12.2.7.1 Insecticides

A large number of organophosphorus, organochlorine, carbamate compounds are used as agricultural insecticides.

Organophosphorus Compounds

Organophosphorus pesticides include fumigants, contact poisons, and systemic insecticides, which act on the central nervous system of the target organisms. Some of the frequently used organophosphate pesticides include tetraethyl pyrophosphate (TEPP), sarin, malathion, dibrom, chlorpyrifos, temephos, diazinon and terbufos. Modern organophosphate pesticides are toxic to target organisms, but they are less toxic to mammals. Fig. 12.1 shows the chemical structures of TEPP and sarin.

Organochlorines

Probably the first synthetic organochlorine insecticide was DDT (dichloro-diphenyltrichloroethane), which was developed during 1940s and had been widely used to combat malaria, typhus, and the other insect-borne human diseases. It became very

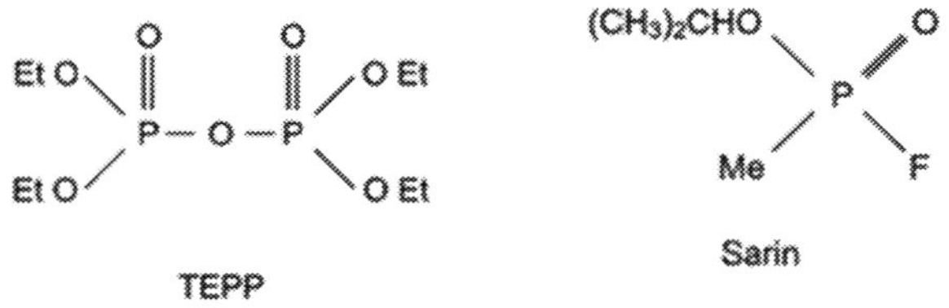

Fig. 12.1 Structure of TEPP and sarin

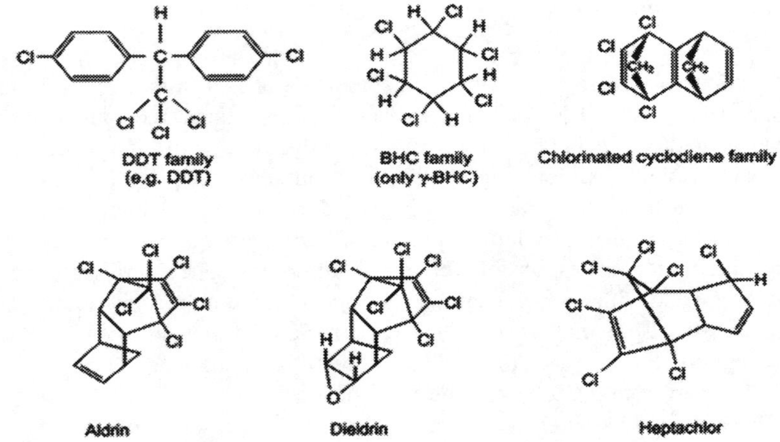

Fig. 12.2 Structures of some organochlorine pesticides

popular because of its success in eradicating malaria from many different countries. Later scientific investigations revealed that DDT and its residues are very persistent, they accumulate in fatty tissues of animals including human, they also kill non-target and beneficial organisms as well, and they can move long distances through the air. Now, production and use of DDT are restricted. Organochlorines are cheap and effective. Other members of this group include BHC (Benzene hexachloride) family and the cyclodiene family. The main compound with insecticidal property of the BHC family is lindane. Aldrin, dieldrin and heptachlor belong to the cyclodiene family. These organochlorines are persistent and poisonous chemicals. Fig. 12.2 shows structures of some organochlorine insecticides.

Carbamates

Carbamates are derivatives of carbamic acid (NH_2COOH) and include urethanes, aldicarb, carbaryl, propoxur, oxamyl and terbucarb. They are applied on crops or on soil as systemic insecticides. Carbamates have variable persistence ranging from few hours to several months. They can also be toxic to non-target organisms. Structures of some carbamates are given in Fig. 12.3.

12.2.7.2 Herbicides

Herbicides are chemicals that kill weeds. Herbicides are often considered essential amendments, particularly in no till and minimum tillage systems in which weed infestation can cause serious reductions of crop growth. There are several organic and inorganic herbicides. Important organochlorine herbicides are 2,4-D (2,4-dichlorophenoxyacetic acid), 2,4,5-T (2,4,5-trichlorophenoxyacetic acid), and MPCA (2-methyl-4, 6-dichlorophenoxy-acetic acid) ((Fig. 12.4). The US Army used a mixture of 2,4-D and 2,4,5-T in equal proportion as a defoliator of forests

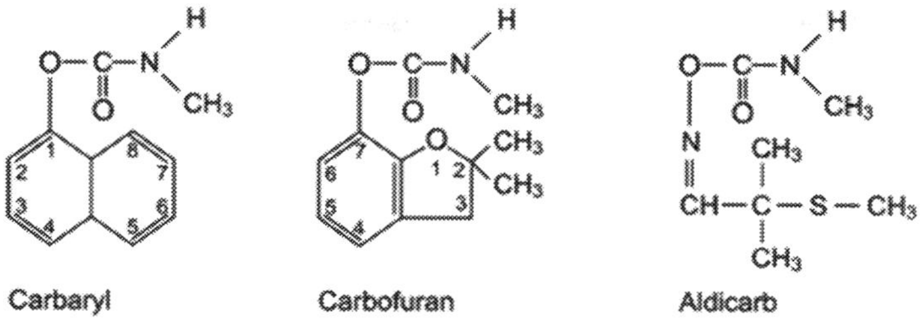

12.2 Sources of Soil Pollutants

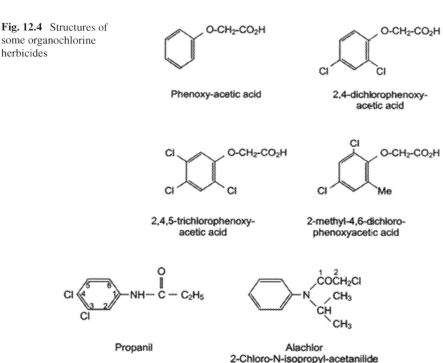

Fig. 12.3 Structures of carbamate pesticides

Fig. 12.4 Structures of some organochlorine herbicides

Fig. 12.5 Structures of some herbicides derived from aniline

during their war agaist Vietnam. They sprayed this mixture in the code name of 'Agent Orange' in millions of hectares of forest to defoliate tress so that they can locate enemies hiding under forest cover. Its effects on human health had been long lasting; Vietnamese people still suffer from the damages caused by this substance. There are also derivatives of aniline among Organochlorine herbicides also include aniline derivatives such as propanil and alachlor (Fig. 12.5).

Glyphosates are organophosphorus herbicides and are used widely in agriculture for controlling weeds. Glycine is modified to glyphosate. Fig. 12.6 shows the structure of a glyphosate herbicide. Some urea derivatives are also used as herbicides (Fig. 12.7).

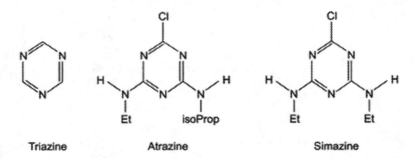

Fig. 12.6 Structures of glycine and glyphosates

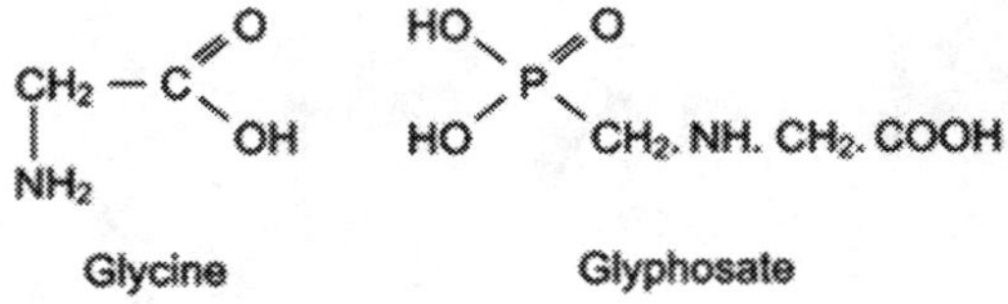

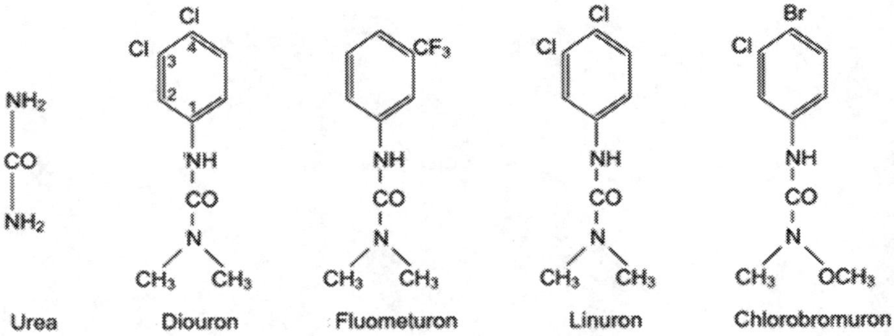

Fig. 12.7 Structures of urea and some of its derivatives used as herbicides

Fig. 12.8 Structures of some triazine derivatives

Atrazine is a derivative of triazine (Fig. 12.8) and is used as a herbicide. It is of a great concern because it can contaminate groundwater. It is persistent for some weeks in soils; and it can cause cancer and other health disorders in human (West 2014). Atrazine can kill many plant species other than weeds, and it can cause immunosuppression, hermaphroditism and sex reversal in male frogs. It is banned in Europe, but it is still used in many other countries to control weeds in several crops.

12.2.7.3 Fungicides

Many natural or synthesized, inorganic and organic chemical substances are used for controlling plant diseases and preventing damage of seeds and grains by pathogenic fungi. These substances, mainly chemical compounds, are known as

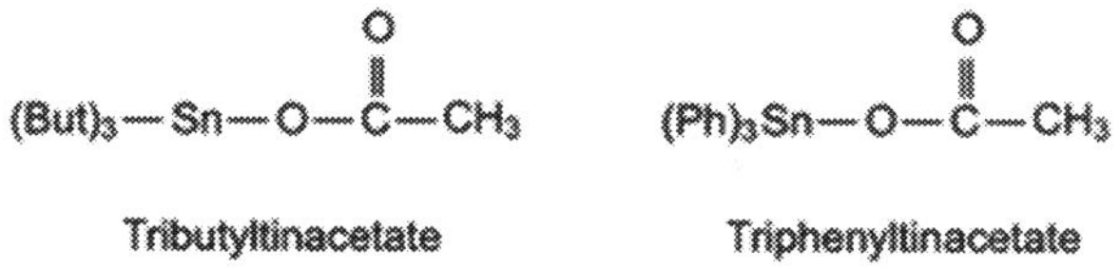

12.2 Sources of Soil Pollutants

Fig. 12.9 Two organotin fungicides

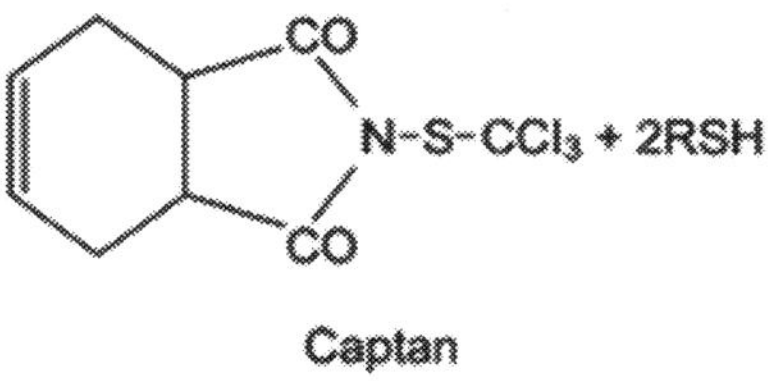

$(But)_3—Sn—O—\overset{\overset{O}{\|}}{C}—CH_3$ $(Ph)_3Sn—O—\overset{\overset{O}{\|}}{C}—CH_3$

Tributyltinacetate   Triphenyltinacetate

Fig. 12.10 Structure of Phthalimide

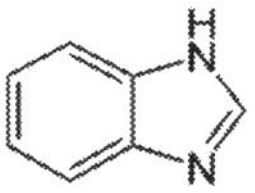

N−S−CCl₃ + 2RSH

Captan

Fig. 12.11 Structure of benzimidazole

fungicides and can be used also as fumigants. There are several categories of fungicides such as inorganic and organic chemical compounds of heavy metals including copper (Bordeaux mixture) and tin (tributyltinacetate or triphenyltinacetate; Fig. 12.9), derivatives of phthalic acid (Captan; Fig. 12.10), and benzimidazoles (Fig. 12.11).

In the above account the chemical nature and structures of some agrochemicals, particularly the pesticides are described. Most of these pesticides have no specific targets; they can be considered as broad spectrum pesticides. Many of these pesticides can retain their chemical integrity and functional characteristics for a long time after application to soils and crops. Some of them are soluble and mobile; they can be transported to long distances. According to Navarro et al. (2007), many organic pesticides can persist for a long period in soils, surface waters, groundwaters, aquifers, and aquatic sediments and pollute them. Some of the transformation products in soil and water are more persistent and hazardous than the original compounds. These persistent organic pollutants (POPs) include many organochlorine compounds and their residues. These organochlorine pesticides and their residues can build up in soil and water and can cause harm to organisms of both soil and water, and eventually to humans. Some persistent organochlorine pesticides of concern for environmental pollution and human health include beta-hexachlorocyclohexane, Gamma-hexachloro-cyclohexane (lindane), Hexachlorobenzene, Dieldrin, Mirex, Heptachlor epoxide, Oxychlordane, trans-Nonachlor, o,p,-DDT, p,p,-DDT, p,p,-DDE. Production of some of the organochlorine pesticides has been restricted in the United States and European countries.

Usually, desirable herbicides should kill the weeds after they are applied in field and be degraded after the season of application. Still, some herbicides persist and cause soil pollution. The length of stability of the pesticides in soil varies widely from one moth to more than 1 year. For example, the herbicides 2,4-D, glufosinate, glyphosate, MPCA may persist for about 1 month; Acetochlor, alachlor, bentazon, butylate, DCPA, dimethanamid, EPTC, flumetsulan, foramsulfuron, halosulfuron, lactofen, linuron, mesotrione, metolachlor, metribuzin, naptalam, siduron for 1–3 months; Atrazine, benefin, bensulide, bromoxynil, clomazone, diuron, ethalfluralin, homesafen, hexazinone, imazaquin, imazethapyr, isoxaflutole, oryzalin, pendimethalin, primisulfuron, prodiamine, pronamide, prosulfuron, simazine, sulfentrazone, terbacil, topramezone, trifluralin for 3–12 months; and Bromacil, chlorsulfuron, imazapyr, picloram, prometon, sulfometuron, tebuthiuron for >12 months.

Some specific persistent herbicides in several chemical groups are mentioned here: **S-triazines** – atrazine (AAtrex, Atrazine), hexazinone (Velpar), prometon (Pramitol), simazine (Princep); **Dinitroanilines** – benefin (Balan), oryzalin (Surflan), pendimethalin (Pendimax, Prowl), prodiamine (Barricade), trifluralin (Treflan, Tri-4, Trilin); **Phenylureas** – diuron (Karmex, Direx); **Uracils** – bromacil (Hyvar-X), terbacil (Sinbar); **Imidazolinones** – imazapyr (Arsenal), imazaquin (Scepter), imazethapyr (Pursuit); **Sulfonylureas** – chlorimuron (Classic), chlorsulfuron (Telar), nicosulfuron (Accent), primisulfuron (Beacon), prosulfuron (Peak), sulfometuron (Oust), etc.

The persistence of an organic compound and its derivatives in soil depends on many factors, including the chemical nature of the compounds such as chemical composition, structure and bonding; physical and chemical characteristics of soil such as type of clay, pH and CEC; sorption of the compounds on soil colloids that restricts the accessibility of microorganisms to the compounds; presence of appropriate biological agents of degradation, etc.

12.2.8 Mining Wastes

Huge amounts of wastes are generated in mining processes including extraction, beneficiation, and processing of minerals. There are two major types solid mining wastes – waste rock and tailings. Mining wastes can be liquid and fluid too. These materials can be divided into several groups of hazardous mining wastes, such as (i) acid mine drainage containing acids and metals generated during processing of ores, mainly sulfide ores; (ii) tailings that contain dangerous chemical substances; (iii) wastes having hazardous materials generated during processing of metalliferous ores; (iv) wastes that have dangerous chemical substances generated during processing of non-metalliferous mining materials; (iv) drilling muds and wastes mixed with oil or containing dangerous substances. Soil contamination due to mining activities occurs from the mines themselves, for example by acid mine drainage, and from wastes of ore processing and petroleum refining industries.

12.2.9 Atmospheric Deposition

Dust particles, ashes, smoke, sea salt, water droplets, gases and some organic materials can form aerosols which can be concentrated in areas of industry, volcanoes, dry and windy areas. Aerosols can be transported to long distances and some of the suspended materials are deposited on lands from the air and the process is called atmospheric deposition. There are three processes of atmospheric deposition – wet deposition, dry deposition and cloud deposition. Wet deposition of all components of aerosols can take place with precipitation. Rainwater dissolves the gases like oxides of sulfur and nitrogen, chlorides, fluorides, ammonia etc. and fall on land and vegetation. Airborne mass of particulate SO_4^{2-}, NH_4^+, H^+, and Pb residing with sub-micrometer aerosol can be deposited by wet deposition. Sulfur dioxide (SO2) and nitrogen oxides (NO, NO2) react in the atmosphere with water to form sulfuric and nitric acids respectively and fall on ground with rainwater as acid rain. Acid rain occurs mostly in the northern hemisphere where industrial activity releases huge quantity of gases that form acids with water. Dry deposition is the direct fall of particles and gas to vegetation, soils, and surface water. The particle-associated substances usually contain K, Ca, Mg, Al and Si and can be deposited by gravitational sedimentation of larger particles. Luo (2009) reported atmospheric deposition of As, Cr, Hg, Ni and Pb to soils. The atmospheric concentrations of Zn, Cu, Pb, and Cd in the city of Amman, Jordan were 344, 170, 291, and 3.8 ng m^{-3}, respectively and the levels of dustfall deposition of these elements were 505, 94, 74 and 3.1 μg g^{-1} respectively (Momani et al. 2000). Atmospheric deposition of heavy metals, PAHs and PCBs can also occur. Atmospheric radioactive fallout of various radionuclides, such as radioiodine (^{131}I), radiocaesium (^{134}Cs, ^{137}Cs) and radiostrontium (^{90}Sr) occurred due nuclear bomb tests in the open air during 1950 and 1960. According to Marcel van den Park (2013) radiological accidents such as Chernobyl accident also caused radioactive fallout and soil pollution.

12.2.10 Traffic

Soils along major highways can be polluted by traffic due to emission of particulate matters and gaseous pollutants. The pollutants can be retained or transformed in the soil or subsequently modified by soil microorganisms (Wesp et al. 2000). Metals in high concentration are often found in roadside soils. Lead has been found to be present at elevated levels in street dust and roadside soils. Other metals including Cu, Fe, Zn, and Cd can be added to roadside soils as a result of mechanical abrasion and wear of many alloys, wires and tires.

Traffic can also be a source of some toxic and persistent organic pollutants. Some of these contaminants such as PCBs, PCDDs and PCDFs are hazardous. Tsai et al. (2014) mentioned that chronic exposure to PCDDs and PCDFs can cause human health risks including cancer mortality. Polychlorinated dibenzo-para-dioxins

(PCDDs) and polychlorinated dibenzofurans (PCDFs) are potentially highly toxic and can cause enzyme induction, reproductive and developmental toxicity, immunotoxicity, adverse endocrine effects, chloracne, and tumor promotion (King-Heiden et al. 2012). Human fertility has been suggested to be adversely affected by exposure to pollution from traffic (de Rosa et al. 2003).

12.2.11 Leaded Petrol and Vehicle Exhausts

Leaded petrol and vehicle exhausts are important sources of Pb and some other metals. Lead poisoning from vehicle exhausts has internationally been recognized. Lead is a neurotoxin, and according to Koller et al. 2004, it has some risk to brain and kidney damage, hearing impairment and diminished cognitive development in children even at lead levels previously considered safe. For boosting octane levels, lead was added to gasoline in the early 1920s. Although addition of lead to gasoline is no longer required due to discovery of a new technology and lead as a gasoline additive has been banned, leaded gasoline use is still allowed for use by aircraft, race cars, and farm equipment. Amanda (2010) reported that approximately 75% of the lead used in leaded gasoline enters the atmosphere as a fine lead dust emitted from the exhaust pipe. Mohammed (2009) reported high concentrations of Pb, Cd, Cu and Zn in the dust specimens in air of Kirkuk city due to heavy traffic, burning of petroleum trash and vehicle exhausts. Onianwa (2002) reported that Nigeria still runs gasoline of Pb concentrations of 0.66 g per litre and it is estimated that about 2800 metric tones of vehicular gaseous Pb emission is deposited to urban areas in Nigeria annually. When fossil fuels are combusted in vehicle engines, they also release hazardous polycyclic aromatic hydrocarbons (PAHs).

12.3 Nature of Soil Pollutants

12.3.1 Organic Pollutants

Many organic pollutants that find their way into the soil with the pesticides were mentioned in section 10.2.6. Here, some organic pollutants from other sources are discussed. For the clarification of the relationships among several organic pollutants, some of them are mentioned again. Several organic pollutants are found in the soil including chlorinated solvents including polychloroethylene (PCE, CCl_2CCl_2), trichloroethylene (TCE, C_2HCl_3), dichloroethylene (DCE, $CHClCHCl$), and vinyl chloride or chloroethylene (VC, CH_2CHCl), monoaromatic hydrocarbons, chlorinated aromatic compounds, polynuclear aromatic hydrocarbons (PAHs) and polychlorinated biphenyls (PCBs). The monoaromatic hydrocarbons include members of BTEX (benzene, toluene, ethylbenzene, and xylenes). The examples of chlorinated aromatic compounds are hexachlorobenzene and pentachlorophenol. Many organic pollutants can be very resistant to chemical or biological degradation; they are called persistent organic pollutants (POPs) (Pedro et al. 2006). The POPs may belong to three categories: (i) pesticides, e.g. DDT, (ii) industrial chemicals such as polychlorinated biphenyls (PCBs) and polycyclic aromatic hydrocarbons (PAHs), and (iii) unintentional by-products, such as dioxins and furans. The POPs exhibit the following common properties: (i) high toxicity, (ii) persistence in soil for years, even decades, (iii) volatility and long distance transport through air and water, and (iv) tendency of accumulation in fatty tissues. The PAHs consist of several hundreds of individual compounds. The PCBs comprise 209 organochlorides with 2 to 10 chlorine atoms attached to biphenyl, which is a molecule composed of two benzene rings. The PCBs are produced by the partial or complete chlorination of the biphenyl molecules. Although PCBs were first synthesized in 1864, their commercial production began in the United States in 1929 for use in the electrical industries. Bentum et al. (2012) noted that the PCBs were first identified as environmental contaminants in 1966. The production of PCB was banned in 2001 by the Stockholm Convention on Persistent Organic Pollutants (POPs). An important group of soil contaminants consists of polychlorinated dibenzo-p-dioxins (PCDDs) and polychlorinated dibenzofurans (PCDFs). The general chemical formulae of PCDDs and PCDFs are $C_{12}H_8$-nO_2Cl_n and $C_{12}H_8$-$nOCl_n$ respectively, where n represents the number of chlorine atoms in the molecule (Environment Agency 2007). According to HPA (2008), 17 of the 210 PCDDs and PCDFs are highly toxic. The most toxic PCDD is 2,3,7,8-tetrachloro-p-dibenzodioxin (2,3,7,8-TCDD). Chlorinated naphthalines (CNs) and polychlorinated naphthalines (PCNs) are generally mixtures of several congeners, and are persistent, bioaccumulative, and extremely toxic to organisms. The NPEs are used in a wide variety of industrial applications and consumer products. They are converted to more resistant and more toxic NP in soil. The BTEX members (Benzene, Toluene, Ethylbenzene, and the Xylenes) are natural-natural compounds, and are the most hazardous components of gasoline. Phthalate esters (PAEs) are also found to contaminate soils from sources such as plastics, automotive, clothing, cosmetics, lubricants and pesticides. The PAEs are persistent and can enter the food chain and endanger human health. Structures of some POPs are given in Figs. 12.12, 12.13, 12.14, 12.15, 12.16, 12.17, and 12.18.

Even if POPs may be present in low concentrations in soil, chronic exposure to them may be of a particular concern (Valentın et al. 2013). Sikka and Wang (2008) suggested that some POPs are endocrine disrupters and harm the reproductive system of organisms. Other effects in mammals associated with organochlorine compounds such as PCBs, hexachlorobenzene (HCB), TCDD, toxaphene and DDT include immunotoxicity, dermal effects, carcinogenicity, vitamin and thyroid deficiencies and mass mortalities by infectious diseases (Vos et al. 2000). Although the effects of chronic exposure of humans to low levels of POPs are difficult to predict, exposure of children to PCBs and PCDD/Fs may be linked to an elevated risk for infectious diseases. According to Domingo et al. (2002), emissions of PCDDs and PCDFs by municipal solid waste incinerators cause concern to the populations living in the vicinity of these facilities.

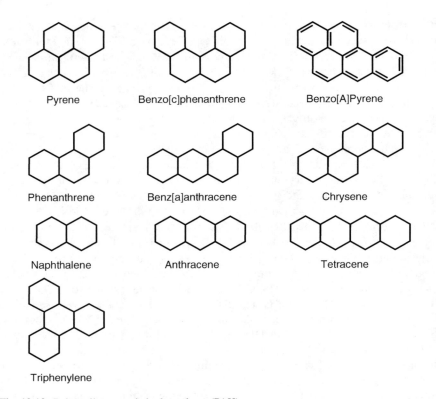

Fig. 12.12 Polycyclic aromatic hydrocarbons (PAH)

Fig. 12.13 Polychlorinated biphenyl (PCB)

Fig. 12.14 Dioxins and Furans

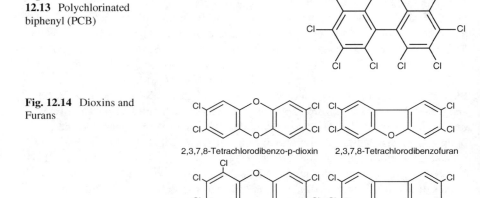

2,3,7,8-Tetrachlorodibenzo-p-dioxin   2,3,7,8-Tetrachlorodibenzofuran

1,2,3,7,8-Pentachlorodibenzo-p-dioxin   2,3,4,7,8-Pentachlorodibenzofuran

12.3 Nature of Soil Pollutants

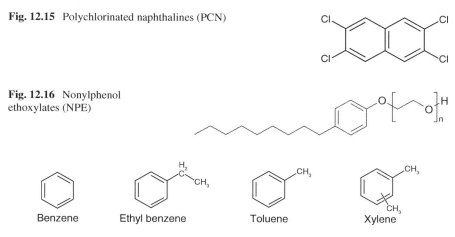

Fig. 12.15 Polychlorinated naphthalines (PCN)

Fig. 12.16 Nonylphenol ethoxylates (NPE)

Fig. 12.17 Structures of BTEX

Fig. 12.18 Structures of Phthalate ester and Bisphenol A

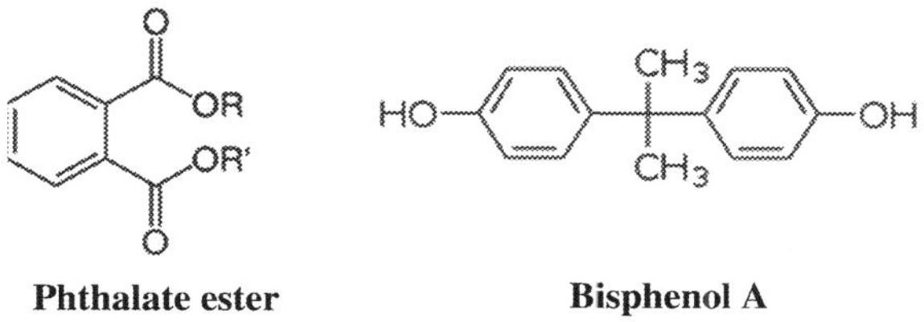

Organic pollutants undergo several transformation reactions including biological, photochemical, and chemical degradation, volatilization, and leaching in soil, and adsorption onto soil minerals and organic matter (Semple et al. 2003). Soil organisms absorb and accumulate some of these into their tissues; this tendency of chemicals to concentrate in living tissues is expressed as a bio-concentration factor (BCF) which is defined as the ratio of the chemical concentration in biota to that in its environment at steady state. Bioconcentration is the process of increasing concentration of chemical in an organism compared to that in soil. Pollutants move along the food chain through dietary uptake and the pollutants are further concentrated. Transfer of a pollutant from one trophic level to another leading to increased concentration is defined as biomagnification. Doick et al. (2005) mentioned that bioaccumulation of pollutants depends on a number of factors, including biological diversity and abundance, soil characteristics and solubility, polarity, hydrophobicity, and molecular structure of the pollutant. The most important pathway for humans to bioaccumulating organic compounds is represented by cattle through dairy products and beef. Lipophilic and persistent organic chemicals accumulate in the milk and beef fat. The pollutants are bioaccumulated in grasses and biomagnified in meat and milk of cattles. So, there is bioconcentration-accumulation - biomagnification along the grass-cattle-human food web. Biomagnification occurs as pollutants move from a lower trophic level to a higher trophic level.

12.3.2 Heavy Metal Pollutants

Generally, metals (or metalloids) that have a density greater than 5 g cm^{-3} and an atomic mass higher than that of calcium are called heavy metals (or metalloids). Some heavy metals are environmentally very important for their effectts on biota. Some of them are essential nutrients (for example, Zn, Cu and Ni for plants; Zn, Cu, Se and Cr for animals) and some others have no known essential role, structural or physiological, on organism (for example, Pb, Cd, Hg, Sn, Ag, etc. and the metalloid As). The essential heavy metals are known as micronutrients because they are needed in very small amounts. All the heavy metals, essential or non-essential, can exert toxic effects on plants and animals if their concentrations are high in the soil and exceed critical levels. On chronic exposure, manganese, mercury, lead, and arsenic can affect the central nervous system; mercury, lead, cadmium, and copper can damage the kidneys or liver, and nickel, cadmium, copper, chromium can cause disorders of skin, bones, or teeth (Zevenhoven and Kilpinen 2001).

Heavy metals enter into the soil system through anthropogenic or geogenic pathways. The geogenic process refers to the release of bioavailable and chemically reactive metals through the weathering of rocks and minerals. Some soils are naturally metalliferous soils that include mainly serpentine soils and calamine soils. Serpentine soils develop from ferro-magnesian ultramafic rocks, which are significantly enriched in Ni, Cr, and Co. Several ultramafic rock types (peridotites; dunite, wehrlite, harzburgite, lherzolite) composed of ferromagnesian silicates can produce serpentine soils under suitatble conditions of weathering and soil formation. Secondary alteration by hydration of these ultramafic rocks within the Earth's crust produces serpentinite, the primary source of serpentine soil. Serpentine actually refers to a group of minerals, including antigorite, chrysotile, and lizardite, etc. Serpentine soils have elevated levels of heavy metals such as nickel, cobalt, and chromium (Alexander et al. 2007). A few plants have evolved the capacity to survive in the serpentine soils (Burrell et al. 2012). These plants are adapted to such conditions; the genetic and molecular mechanisms underlying this adaptation have not, however, been exactly known (Brady et al. 2005). The calamine soils are formed from calcareous loess parent materials and are enriched in Pb, Zn and Cd, and sometimes carry high concentrations of As and/or Cu. There are also some other mettaliferrous soils derived from argillites and dolomites containing high concentrations of Cu and Co sulfides and some oxides, carbonates and silicates, and soils derived from various Se-rich rocks.

Anthropogenically heavy metals and metalloids are introduced in to the soils through emissions and wastages from industries, mining operations, disposal of metal wastes and metal scraps, paints, fertilizers, manures, sewage sludge, pesticides, wastewater irrigation, residues of fossil fuels, spillage of petrochemicals, and through atmospheric deposition (Khan et al. 2008). Heavy metals are not degraded biologically or chemically (Kirpichtchikova et al. 2006), and they persist in soils for a long time (Adriano 2003). The major anthropogenic sources and their heavy metal (metalloid) contaminants include chemical wastes (As, Pb, Cr, Cd, Ba, Zn, Mn, Ni), metal finishing/plating (Cr, Pb, Ni, Zn, Cu, Cd, Fe, As), pharmaceutical wastes (Pb, Cr, Cd, Hg, As, Cu), mining and smelting (Pb, As, Cr, Cd, Cu, Zn, Fe, Ag), battery recycling (Pb, Cd, Ni, Cu, Zn), paint (Pb, Cr, Cd, Hg), nuclear processing (Ra, Th, U), etc.

The order of abundance of the common heavy metals in soil is usually Pb, Cr, As, Zn, Cd, Cu, and Hg. According to Buekers (2007), these metals may exist in soils in different chemical forms (speciation) that differ in their bioavailability, mobility and toxicity. The soil environment also influences the behavior of the metals such as bioavailability and mobility. Some important soil properties in these regards are texture, porosity, infiltration, permeability, organic matter content, cation exchange capacity, pH, redox potential, and temperature. The reactions that the metals undergo and that influence metal distribution in soil include (i) precipitation and dissolution, (ii) ion exchange, adsorption, and desorption, (iii) aqueous complexation, and (iv) biological immobilization and mobilization. Thus, the main forms of heavy metals in soil are soluble, exchangeable, lattice-bound, precipitates and insoluble complexes with other metal oxides and carbonates (Aydinalp and Marinova 2003). The soluble and exchangeable fractions are bioavailable and cause toxicity to organisms. The mobility of metals, however, depend both on metal and soil characteristics. For example, Cd, Ni, and Zn are very mobile in acid soils; Cr is moderately mobile; and Cu and Pb are practically immobile. On the other hand, in neutral to alkaline soils Cr is highly mobile, Cd and Zn are moderately mobile and Ni is immobile.

Plant roots absorb heavy metals and distribute them to various plant parts though different mechanisms. Roots can retain a portion of the absorbed heavy metals; a portion may be precipitated and accumulated in vacuoles or translocated to shoots. The heavy metals which are essential micronutrients perform important and specific physiological functions in living organisms. Other metals do not have any positive role in biological systems; rather, they can interfere with normal physiological functions and cause toxicities. After uptake of metals, some plants can prevent their toxicities by metal compartmentalization and binding to intracellular ligands. Cho et al. (2003) explained that (i) metals can be bound to extracellular exudates and cell wall constituents; (ii) metals can move from cytoplasm to vacuoles; (iii) metals can form complexes inside the cell by organic acids, amino acids, phytochelatins, and metallothioneins; (iv) plant metabolism can allow adequate functioning of metabolic pathways and rapid repair of damaged cell structures. However, there are some plants that can tolerate very high concentrations of heavy metals in the growth media. Some plants, known as metal hyperaccumulators, can accumulate huge amount of metals/metalloids in their shoots. There are several transporter proteins including MT-protein which translocate metals from roots to shoots. Plants like *Arabidopsis halleri*, *Silene paradoxa* and *Silene vulgaris* contains high levels of MT. On the other hand, the Ni-accumulating *Noccaea caerulescens* has high ABC transporter proteins responsible for metal sequestration (Visioli et al. 2012). A plant is, however, known as a metal hyperaccumulator (i) if it can accumulate >100 mg Cd kg^{-1} or >1000 mg Pb kg^{-1}, or >10,000 mg Zn kg^{-1} in its shoot dry matter; (ii) the metal concentrations in shoots of a hyperaccumulator plant must be higher than that in roots; (iii) the ratio of metal concentrations in shoots:roots must be >1; and (iv) despite this high concentration of metals in tissues the plant is able to grow and reproduce normally and without any signs of toxicity (Yanqun et al. 2005). Such potential of some plants has evolved a phytotechnology, known as phytoextraction, for remediation of sites contaminated with As, Cd, Cu, Ni, and Pb, and others (Anjum et al.2011). This aspect has been discussed in relation to phytoremediation of polluted soils in Section 12.5.2.3.

The chemistry of some important heavy metals and metalloids in soil and their effects are briefly discussed below:

Arsenic
Arsenic is metalloid that has an atomic number of 33, atomic mass of 75 and density of 5.72 g cm^{-3}. It is naturally present in all types of rocks in the range of 1–5 mg kg^{-1}. There are more than 200 minerals that contain As. There are arsenides, arsenates, arsenites, sulfides, oxides, and elemental As. The dominant minerals include arsenopyrite (AsFeS), realgar (AsS) and orpiment (As$_2$S$_3$). The mean concentration of As in the earth's crust is 5 mg kg^{-1} (Matschullat 2000). Arsenic exists in different oxidation states – III, 0, III, V. In the aerobic environment the dominant oxidation state is As(V) and there it exists in various protonation states including H$_3$AsO$_4$, H$_2$AsO$_4^-$, HAsO$_4^{2-}$, and AsO$_4^{3-}$. In the reducing conditions the dominant oxidation state is As(III) and the principal protonation forms are H$_3$AsO$_3$, H$_2$AsO$_3^-$, and HAsO$_3^{2-}$, AsO$_3^{3-}$. Mobility of As in soil increases as pH increases. Arsenic takes part in reactions with inorganic and organic substances in the soil. Arsenates can be adsorbed on iron oxyhydroxides and As can form chelates with organic acids such as methylarsinic acid (CH$_3$)AsO$_2$H$_2$ and dimethylarsinic acid (CH$_3$)$_2$AsO$_2$H. Arsenic is one of the most toxic elements encountered in the environment (Dermatas et al. 2004). Concentrations of As in non-contaminated soils should be <10 mg kg^{-1} (Fitz and Wenzel 2002). But phytotoxicity is likely to be observed at 20 mg As kg^{-1} soil (Huq et al. 2003). Crop plants like maize, ryegrass, rape and sunflower grown in arsenic contaminated soils have been shown to accumulate As in the roots (Gulz 2002). Still, the accumulation of As in aboveground biomass may exceed the tolerance limit for food (0.2 mg As kg^{-1}) and fodder crops (4 mg As kg^{-1}) in the Swiss standard (Gulz 2002). However, As exerts the most damaging effect through groundwater contamination. In south-east Asian countries like Bangladesh, China, India and Nepal groundwater has severely been contaminated with As. The USEPA and WHO standard is 0.01 mg l^{-1} of As for drinking water; in Bangladesh most shallow tube well water exceeds the national standards of 0.050 mg l^{-1}. The cause of this contamination is geochemical; the sedimentary rocks there contain high As. Drinking of As contaminated groundwater causes large scale occurrences of arsenicosis in parts of Bangladesh and India (Alam et al. 2002; Chakraborti et al. 2002).

Cadmium
Cadmium is a metal that has an atomic number of 48, atomic mass of 112.4, and density 8.65 g cm^{-3}. It is a malleable and soft metal that occurs ubiquitously in rocks and soils (Alloway 2012). The common minerals of cadmium are greenockite (CdS), octavite (CdSe), and monteponite (CdO) (Kabata-Pendias and Mukherjee

12.3 Nature of Soil Pollutants

2007). Cadmium has exclusively the oxidation state of +2. Cadmium reacts with oxygen, sulfur and many common anions such as chloride, nitrate and carbonate to form salts. The average abundance of Cd in the earth's crust is 0.1–0.2 mg kg^{-1}. Phosphatic rocks may contain in excess of 200 mg Cd kg^{-1} (Garret 2002). The average content in soils ranges from 0.2 to 0.3 mg kg^{-1}. Cadmium concentration in soil solution is relatively low ranging from 0.2 to 6.0 µg l^{-1} (Kabata-Pendias and Pendias 2000). The Cd content in soil depends largely on the parent material. Cadmium can also be added to soils through atmospheric deposition, fertilizers, pesticides, and irrigation water. Generally, soils derived from sedimentary rocks contain the largest quantities of Cd. The chemical behavior of Cd largely depends on pH of soil. Cadmium solubility increases as soil pH decreases, and very little adsorption of Cd by soil colloids, hydrous oxides, and organic matter takes place. Cadmium is adsorbed by the soil solids or is precipitated in soils with pH values >6. Cadmium forms soluble complexes with inorganic and organic ligands, in particular with chloride ions. If there is cadmium available in the soil, it is readily absorbed by plants and translocated to the above-ground parts. The uptake of Cd decreases as the soil pH increases. On the other hand, soil salinity and lime induced zinc deficiency enhance the uptake of Cd by plants (Smolders 2001). Cadmium is absorbed and accumulated by food chain crops and it is regarded as one of the most toxic trace elements in the environment.

Lead

Lead has an atomic number of 82, atomic mass of 207.2, density of 11.4 g cm^{-3}. Most common minerals of lead are Galena (PbS),cerussite (PbCO$_3$) and andanglesite (PbSO$_4$). Galena is the principal ore mineral, usually found in association with sphalerite (ZnS). Lead is found in the earth's crust in the range of 10 to 30 mg kg^{-1}. According to Kabata-Pendias and Pendias (2001), lead concentration in surface soils ranges from 10 to 67 mg kg^{-1} with the mean Pb concentration worldwide of 32 mg kg^{-1}. Soils nearer to industrial sites can be contaminated with Pb. Contamination of lead can also occur from leaded fuels, old lead plumbing pipes, etc. Traunfeld and Clement (2001) noted that lead is highly immobile; without remedial action, high lead levels in soil will never return to normal. Lead can have Pb (II) and Pb (IV) states. Lead (II) compounds are predominantly ionic (e.g., Pb^{2+} SO$_4^{2-}$), whereas Pb (IV) pounds tend to be covalent (e.g., tetraethyl lead, Pb(C$_2$H$_5$)$_4$).

Lead exists in several forms in soil solution including free metal ion, Pb^{2+}, and soluble organic and inorganic complexes. The pH of the soil influences greatly the forms of Pb in solution. At pH values greater than 7 the dominant form is PbCO$_3$. Other species include PbOH$^+$, Pb(OH)$_2$, Pb(OH)$_3$, PbCl$_3$, PbCl$^+$, PbNO$_3^+$ and Pb(CO$_3$)$_2$ but their concentrations are very low. Lead exists as Pb^{2+} in solution at pH 4 (Markus and McBratney 2001). Plant roots absorb a small proportion of lead present in soil solution. Also, a little amount of lead usually enters the food chain. Only a few plants can lake up and accumulate extraordinarily high lead levels in their shoots. The availability of lead for plant uptake includes speciation of lead, soil pH, cation-exchange capacity, root characteristics and mycorrhizal association. The Casparian strips in roots act as barriers of lead transport to shoots; so most lead that is taken up by plants accumulates in roots. Excessive lead accumulation in tissues of non-tolerant plants impairs physiological, and biochemical functions. Pourrut et al. (2011) pointed out that excess lead caused phytotoxicity by changing cell membrane permeability, by reacting with active groups of different enzymes involved in plant metabolism and by reacting with the phosphate groups of ADP or ATP, and by replacing essential ions. However, vegetable and fruit crops including corn, beans, squash, tomatoes, strawberries, and apples seldom exceed safe level of lead.

Mercury

Mercury is a liquid metal at standard temperature and pressure. It has an atomic number of 80, atomic mass of 200.6 and density of 13.6 g cm^{-3}. The most important source of mercury is the naturally occurring mineral, cinnabar (HgS). The concentration of Hg in the earth's crust is very low with the range of 0.02 to 0.06 mg kg^{-1} (Kabata-Pendias and Mukherjee 2007). The Environment Agency (2007) mentioned that the concentrations of total mercury in soils of the United Kingdom ranged from 0.07 to 1.22 mg kg^{-1} with a mean value of 0.13 mg kg^{-1}; urban soils contained higher total mercury in the range of 0.07 to 1.53 mg kg^{-1}, with a mean of 0.35 mg kg^{-1}.

Mercury exists in mercuric (Hg^{2+}), mercurous (Hg_2^{2+}), elemental (Hg^0) and alkylated forms (methyl/ethyl mercury). Mercurous and mercuric mercury are more stable under oxidizing conditions. Organic or inorganic Hg may be reduced to elemental Hg under reducing conditions, and may then be converted to alkylated forms which are soluble in water and volatile in air and are the most toxic forms of mercury. Kabata-Pendias and Mukherjee (2007) stated that about 1–3 percent of total mercury in surface soil is in the methylated form. Dimethyl mercury is a highly toxic and volatile compound; monomethylated mercury compounds are also volatile and, due to their relatively high mobility compared with inorganic forms, they are the most crucial mercury species for environmental pollution. Monomethylated mercury compounds are most likely to be found in soil as a result of natural microbial transformation of inorganic mercury. Mercury accumulates in roots after being absorbed by plants indicating that the roots function as a barrier to mercury translocation. Mercury concentration in aboveground parts of plants depends largely on a foliar uptake of Hg^0 volatilized from the soil. Uptake of mercury has been found to be plant specific in bryophytes, lichens, wetland plants, woody plants, and crop plants. Inhalation of volatile mercury may cause mercury poisoning in humans.

Chromium

Chromium has an atomic number 24, atomic mass 52, density 7.19 g cm^{-3} and does not occur naturally in the elemental form. The mineral chromite ($FeCr_2O_4$) is most commonly the ore forming mineral of chromium and is mined. Chromium is released from mining operations and by disposal of Cr containing wastes. Usually Chromium (VI) is found in contaminated soils and water with the major species chromate (CrO_4^{2-}) and dichromate ($Cr_2O_7^{2-}$). These anions can be precipitated easily some metal cations including Ba^{2+}, Pb^{2+}, and Ag^+. Chromium (III) can be a dominant form at low pH (<4). The Cr^{3+} ion can form complexes with NH_3, OH^-, Cl^-, F^-, CN^-, SO_4^{2-}, and soluble organic ligands. Chromium (VI) is the more toxic form of chromium and is also more mobile.

Nickel

Nickel has an atomic number 28, atomic weight 58.69 and density 8.908 g cm^{-3}. The forms of Ni that can be present in soil depend on soil reaction and redox conditions. For example, in acidic soils nickelous ion Ni (II) predominates while in neutral to slightly alkaline solutions, it precipitates as nickelous hydroxide $Ni(OH)_2$. On the other hand, stable nickelo-nickelic oxide, Ni_3O_4 is formed in alkaline and very oxidizing conditions. Nickel is an essential element for plants, but a very small amount is needed. According to Wuana and Okieimen (2011), Ni can be very harmful if the concentration far exceeds the maximum tolerable limit. Nickel uptake by plants depends on the acidity of soil or solution, the concentration of Ni^{2+}, the presence of other metals and organic matter composition (Chen et al. 2009). The typical background levels of some heavy metals for non-contaminated soils are arsenic 3–12 mg kg^{-1}, cadmium 0.1–1.0 mg kg^{-1}, copper 1–50 mg kg^{-1}, lead 10–70 mg kg^{-1}, nickel 0.5–50 mg kg^{-1} and zinc 9–125 mg kg^{-1}. The unsafe levels of these metals in soils for growing vegetables are >50, >10, >200, >500, > 00, >200 mg kg^{-1} respectively. (http://www.aleastern.com/forms/Heavy%20Metal%20Interpretation.pdf.). The threshold levels of As, Cd, Cr, Co, Pb and mercury are 0.07, 1.70, 1,00,000, 660, 80, 18, and 1600 mg kg^{-1} respectively in soils in human health concerns. (California Office of Human Health Assessment, https://oehha.ca.gov/risk/chhsltable.html).

12.3.3 *Radionuclides*

Igwe et al. (2005) stated that the elements that have an atomic number greater than bismuth - 83 is unstable and radioactive; these elements are called radionuclides. Marcel van der Perk (2006) suggested that radionuclides disintegrate or change spontaneously with a loss of energy in the form of ionizing radiation. There are more than sixty radionuclides in nature; they can be of three categories such as primordial (initial radionuclides existing since the earth was formed and which have not completely decayed, e.g. ^{238}U, ^{235}U and ^{232}Th), cosmogenic (radionuclides produced by natural processes by interaction of cosmic radiation with the atomic nuclei of the atmosphere or with extraterrestrial matter coming down to earth as meteors or meteorites; e.g. americium-241, cesium-137, Fe-60, Cobalt-60, etc.) and anthropogenic (nuclear explosions, nuclear accidents, routine releases from nuclear reactors, releases due to nuclear weapon production programs or nuclear fuel reprocessing installations, and nuclear medicine) (Knox et al. 2000).

12.3.4 Toxicity of Heavy Metals

When living organisms are exposed to high levels of heavy metals, they usually suffer from physiological disorders and develop different kinds of toxicity symptoms. The exposure can take place through direct contact, uptake, ingestion and food intake. Soil organisms and plant roots have the immediate contact of heavy metals in soil solution and exchangeable metal ions on soil colloids. Plants absorb these metals by the roots; they suffer themselves and make animals that eat these plants to suffer more from toxicity of metals along the food chain. The heavy metals are bioaccumulated and biomagnified along the food web. According to Petruzzelli et al. (2010), ingestion of the soil may result in significant exposure to toxic substances.

12.3.4.1 Toxicity of Heavy Metals/Metalloids to Soil Organisms

Soil microorganisms need a very small amount of cobalt, chromium, nickel, iron, manganese and zinc; they are essential for their normal growth and physiological as well as biochemical functions including enzymatic reactions and osmotic regulation (Bruins et al. 2000; Hussein et al. 2005). Whether essential or non-essential, heavy metals are toxic to soil microorganisms at high levels (Pawlowska and Charvat 2004). High concentrations of heavy metals adversely affect microbial community structure and dynamics, diversity and functions (Filip 2002; Kelly et al. 2003). Excess heavy metals disrupt cellular and genetic structures (Bruins et al. 2000). Lanno et al. (2004) pointed out the routes of exposure of earthworms to elevated levels of heavy metals in the soil. These are direct dermal contact and ingestion of pore water, polluted food and soil particles. Hobbelen et al. (2006) reported bioaccumulation of Cd, Cu and Zn by the earthworms *Lumbricus rubellus* and *Aporrectodea caliginosa*. According to Hirano and Tamae (2011), earthworm's ability to bioaccumulate heavy metals could be used in the bio-monitoring of soil pollution (heavy metals also adversely affect arthropods, mollusks and nematode. Several investigators (Li et al. 2006; Georgieva et al. 2002; Liang et al. 2006; Navas et al. 2010)) have found that excess heavy metals such as Cu, Zn and Cd adversely affect structure, diversity and maturity of nematode community in polluted soils. Faunal composition, diversity and structure are also affected by elevated levels of heavy metals in soil.

12.3.4.2 Toxicity of Heavy Metals/Metalloids to Plants

Symptoms of phytotoxicity of heavy metals include chlorosis, growth reduction, yield depression, reduced nutrient uptake, and disorders in plant metabolism. Heavy metals can also inhibit nitrogen fixation in leguminous plants (Dan et al. 2008). Phytotoxic symptoms of heavy metal include chlorotic leaves due to impairment of

12.3 Nature of Soil Pollutants

chlorophyll synthesis (Fodor et al. 2002), alteration of the ratio of chlorophyll a:chlorophyll b (Viehweger and Geipel 2010), reduced photosynthetic activity (Küpper et al. 2007), dwarfism and disruption of root ultrastructure (Barcelo et al. 2004) and severe oxidative damage to biomolecules like lipids, proteins and nucleic acids (Mudgal et al. 2010a). Several investigators (Fodor et al. 2002; Viehweger and Geipel 2010; Küpper et al. 2007; Barcelo et al. 2004) observed detrimental effects of the common heavy metals on plants including (i) inhibition of seed germination of mung bean by arsenic; (ii) reduction in photophosphorylation, ATP synthesis, mitochondrial NADH oxidation, and electron-transport system affecting seed germination and seedling growth by cadmium and lead; (iii) severe wilting and chlorosis of plants by chromium; and (iv) retardation of photosynthesis, protein sysnthesis, impairment of cell division by mercury, and (v) leaf necrosis due to nickel toxicity.

12.3.4.3 Toxicity of Heavy Metals to Humans

Humans are exposed to the heavy metals in soil through: (a) inhalation (for gaseous and particulate matters suspended in air); (b) dermal contact; and (c) ingestion by contaminated food). Children often swallow contaminated soil materials. However, food is a principal source of essential and toxic heavy metals for humans. The dietary contribution for toxic metal intake has been extensively studied (Santos et al. 2004). Metals that enter nasopharyngeal, tracheobronchial, or pulmonary compartments may be transported to the gastrointestinal tract (Mudgal et al. 2010b). The heavy metal ions in human body interfere and often impair synthesis and reactions of proteins, enzymes, hormones, etc. (Schoof 2003). Heavy metals form complexes with amino acids (glutathione (GSH), cysteine, histidine) and proteins (metallothioneins, transferrin, ferritin, lactoferrin, hemosiderin, ceruloplasmin, melanotransferrin). Toxic effects of metals may be chronic or acute (teratogenic, mutagenic and carcinogenic). Yu (2005) reported that heavy metal toxicity may lead to impairment of the functions of the central nervous system and oxidative metabolism, injury to the reproductive system, or alteration of DNA leading to carcinogenesis. Heavy metals can also disrupt or destroy cellular structure and function. Yu (2005) mentioned the specific toxic effects of As, Cd, Pb. Hg and Ni. Arsenic in drinking water causes arsenicosis which is a very common disease in south east Asian countries including India and Bangladesh. Arsenic may cause cancers of the bladder, kidney, skin, liver, lung, colon and lymph. Cadmium is known widely for the deadly disease 'Itai-itai' that was caused in Japanese people in 1945 due to consumption of rice grown with Cd-contaminated irrigation water. The other toxic effects of Cd include renal tubular dysfunction, high blood pressure, lung damage, and lung cancer. One of the most widely known toxic effects manifested by Cd poisoning is nephrotoxicity. The central nervous system, kidneys and lungs are damaged by Pb. It may also cause anemia, nausea, anorexia, and abdominal cramps, muscle aches and joint pain, difficulty in breathing, asthma, bronchitis, and pneumonia, damage to the fetus, miscarriage. Toxicity of Hg in human is exhibited by paraesthesis, ataxia, dysarthria, and deafness, salivation, loss of appetite, anemia,

Table 12.1 Toxicity of metals on human health

Metal	Effects
Lead	Hypertension and chronic kidney disease
Cadmium	Human carcinogen
Aluminium	Liver dysfunction, asthmatic conditions
Copper	Brain and liver damage
Zinc	Hemolytic anaemia
Iron	Hemochromatosis, Conjunctivitis
Chromium	Cr VI is carcinogenic
Mercury	Kidney disease, kidney failure
Arsenic	Brain damage, lung cancer

Adapted from Levine et al. (2006)

gingivitis, excessive irritation of tissues, nutritional disturbances, and renal damage. Severe Hg poisoning from contaminated fish occurred in Minamata of Japan and from contaminated wheat in Iraq. Methyl mercury fungicide was sprayed on wheat. More than 6000 children and adults were poisoned, with about 500 deaths. Nickel can cause vomiting and headaches, and it can affect the fetus after crossing the placental barrier. Inhalation of Ni compounds may cause lung, sinonasal and laryngeal carcinomas. Bradl (2005) noted toxic effects of Cr, Cu, and Se on human. Chromium impairs growth, alters immune system, and decreases reproductive functions. Both Cr(III) and Cr(VI) are potent human carcinogens. Perforations and ulcerations of the septum, bronchitis, decreased pulmonary function, and pneumonia. Toxicity of Cu is rare, intake of high amount of Cu for a long time may be toxic to children. Selenium causes nausea, vomiting, and diarrhoea. Selenosis occurs due to chronic oral exposure to high concentrations of Se. Major signs of selenosis are hair loss, nail brittleness, and neurological abnormalities. According to Duruibe et al.(2007), zinc is considered to be relatively non-toxic. However, excess amount can impair growth and reproduction. Table 12.1 shows themmajor toxic effects of heavy metals on human health.

Since the principal route of human exposure to heavy metals is contaminated food materials, and a major proportion of our food is of plant origin, the contents of heavy metals in plant foods are of particular concern. Crops grown in metal contaminated soils can absorb and accumulate considerable quantities of heavy metals in their edible parts. Fu et al. (2008) observed Cd contents in more than 30 percent of rice samples collected from different parts of China above the national maximum allowable concentration. A study revealed that the concentrations of Cd, Zn, Pb and Cr in roots, stems and leaves of the vegetable plants green amaranth (*Amaranthus viridis*) and waterleaf (*Talinum triangulare*) grown on poultry dumpsite of Nigeria were 0.62–2.74, 50.67–102.98, 2.27–7.21, 0.64–4.45 mg kg^{-1} respectively; some of these values were above the safe levels (Adefila et al. 2010). Bagdatlioglu et al. (2010) observed that in vegetables such as parsley, onion, lettuce, garlic, peppermint, spinach, broad bean, chard, purslane, and fruits such as tomato, cherry, grape and strawberry grown in Manisa region of Turkey contained high levels of heavy metals. The concentrations of Fe, Cu, Zn, Pb and Cd ranged from 0.56 to 329.7, 0.01 to 5.67, 0.26 to 30.68, 0.001 to 0.97 and 0 to 0.06 mg kg^{-1} respectively. Davarynejad et al. (2010) reported contamination of several fruits, including orange, mango, almond, lemon, sweet orange, grapefruits, papaya, muskmelon, apple, quince, grape, strawberry, banana, pineapple, carambola, longan, date palm and apricot with heavy metals. According to Graffham (2006), the safe level of Cd for vegetables (except leafy vegetables) and fruits is 0.05 mg kg^{-1}, for leafy vegetables, fresh herbs and mushrooms is 0.20 mg kg^{-1} and for potato it is 0.10 mg kg^{-1}. CAC (2003) reported safe level of Pb in cereals and legumes as 0.20 mg kg^{-1} and Graffham (2006) mentioned safe levels of Pb for vegetables, excluding brassica, leafy vegetables, fresh herbs and mushroom as 0.10 mg kg^{-1}, for brassica and leafy vegetables as 0.30 mg kg^{-1}. USDA (2003) set 50 mg Zn kg^{-1} in grains and 100 mg Zn kg^{-1} in beans as safe levels. On the other hand, permissible limits in spices according to WHO for Cu, Ni, Zn, Fe, Pb, and Hg are 50, 50, 100, 300, 100, and 10 mg kg^{-1} respectively (Nkansah and Amoako 2010).

12.4 Prevention of Soil Pollution

'Pollution prevention is the reduction or elimination of wastes and pollutants at their source.' For prevention of soil pollution at the fist place the following principles may be followed.

Principles
- Excavation, grading, and paving activities should be scheduled for dry weather periods.
- Cleaning up leaks, drips and other spills immediately.
- Reducing waste generation and recycling all wastes if possible.
- Managing maximum wastes at the place of waste generation and disposing wastes in a safe and appropriate manner.
- Wastes that cannot be recycled should be dumped in a protected landfill or disposed of as hazardous waste.
- Covering the dump sites with suitable covers so that scavengers do not scatter wastes and rainwater does not wash pollutants into soils.
- Preventing pollution at the source, if possible.
- Applying fertilizers judiciously and using low-persistence pesticides.
- Assessing contaminant loads of composted sludge or treated wastewater before applying in crop fields.
- Monitoring the build-up of pollutants in soils periodically.
- Making people aware of the consequences of generating unnecessary wastes and unhealthy disposal of wastes.

12.4.1 Waste Management

One major step of prevention of soil pollution is the waste management which involves waste treatment in industries, collection of wastes from generation sites or temporary collection sites, screening and sorting of recycleable and non-recycleable wastes, disposal of non-recycleable and hazardous wastes and recycling of wastes. In developed countries there are waste treatment plants in large industries, but, in small industries and in industries of under-developed countries, there are very few treatment plants. Municipal wastes are often dumped in open places because it is the easiest, cheapest waste disposal system but it is most risky. The organic wastes are fermented and spread obnoxious smell to the surroundings; scavengers spread waste materials to distant places and pollute air, water and soil. Untreated wastes contain more hazardous contaminants. Sorting divides recycleable wastes into metals, glasses, papers and other organic wastes before they are sent to recycling facilities. Organic wastes can be composted for use in agricultural fields. Hazardous wastes and materials that cannot be recycled should be disposed of in a proper way. Municicipal solid wastes are disposed of in landfills or incinerated. Combustion reduces waste to ash by about 75% by weight for disposal in a landfill. Most of municipal solid wastes can be composted and used in crop fields but the heavy metal content of composted wastes and sludges need to be monitored.

12.4.1.1 Landfill

A landfill unit is a discrete area of land or an excavation that receives wastes. Landfills should be so built that wastes cannot harm the natural environment through pollution. The objectives of constructing landfills include prevention of pollutants from entering the soil and polluting ground water. Generally, protected chambers are built for disposal of municipal solid wastes, household hazardous wastes, municipal sludge, municipal waste combustion ash, infectious wastes, industrial wastes, and mining wastes. However, some landfill units are dangerous places; there are continuous release of volatile gases such as methane and carbon dioxide produced by anaerobic microorganisms, and leakage in the form of leachates. Prevention of leakage is done by the use of clay liners and synthetic liners like plastics.

12.4.1.2 Composting Municipal Wastes

Reducing the amounts of organic residues and stabilizing them as brown to black friable materials rich in humic substances through microbial digestion and decomposition is known as composting and the product is called compost. Compoosts contain large amounts of organic matter and considerable amounts of plant nutrients. They are, therefore, good fertilizers and soil conditioners. Almost all types of organic residues such as municipal wastes, animal excreta, crop residues, food

12.4 Prevention of Soil Pollution

wastes, poultry wastes, etc. can be composted. Two major benefits of composting are: (i) large reduction in volume and mass of the wastes, and (ii) converting materials that encourage growth of insects, and flies and spread obnoxious odor into an odorless and valuable soil amendment. The process recycles the wastes efficiently and improves soil fertility and productivity. Some investigators (Pokhrel and Viraraghavan 2005; Montemurro et al. 2005) suggested that composting is a good of disposal of huge amounts of municipal solid waste (MSW) and can be used successfully for conditioning soil and supplying plant nutrients. However, agricultural fields receiving MSW compost should be regularly monitored because municipal composts can contain some heavy metals.

12.4.1.3 Composting Sewage Sludge

Huge sewage sludge is produced in urban and industrial processes. It is rich in organic matter, and nutrients. Therefore, sewage sludge is usually dewatered and composted for application in agriculturl fields as organic fertilizer and soil conditioner. It is an important way of disposal and recycling of huge amounts of sludge, and it can improve soil fertility and productivity. The composting process largely destroys the agents of diseases and pests. However, sludge can contain some heavy metals and persistent residues of pesticides. The metals Zn, Cu, Pb, Ni and Cd are of primary concern because excessive amounts of these metals may reduce plant yields or impair the quality of food. Repeated applications of compost from sewage sludge on crop fields can build up heavy metals to a significant level and can decline yield and crop quality. Appropriate rates of sewage sludge can improve physical properties and fertility of soil and growth of plants by the re-use of nutrients. However, sewage sludge amendment slightly increases Cu content of soils and Zn content of plants. Saruhan et al. (2010) suggested that sewage sludge with caution because sewage sludge contains high concentrations of potentially toxic elements such as Zn, Ni, Cd, and Cu that may create problems in an agricultural soil (Madyiwa et al. 2002). According to Siuta (1999) and Moreno et al. (1999), the acceptable limits of Cd, Cr, Cu, Hg Ni, Pb and Zn in composted sludge for application to agricultural fields are 10, 500, 800, 5, 100, 500 and 2500 mg kg^{-1} respectively.

12.4.1.4 Incineration

Generally, almost all municipal solid wastes (MSW) are burned in the incineration plants in industrialized countries of Europe as well as in Japan, the USA and Canada. It is a process that usually involves the combustion of raw or residual MSW. Typically, incineration is done with full supply of oxygen at temperatures higher than 850 °C and the organic waste is converted into carbon dioxide and water. Any non-combustible materials such as metals, glass, etc. that can be present in the solid waste remain as a solid, known as bottom ash. Incineration with energy recovery is a well-established technique for municipal waste treatment. Incineration plants can be equipped with grate firing systems, as in pyrolysis, gasification or fluidized bed plants or their combinations. The incineration process emits several gases including CO_2, N_2O, NOx (oxides of nitrogen), NH_3 and organic C. The incineration of 1 Mg of municipal waste in incinerators is associated with the release of about 0.7 to 1.2 Mg of carbon dioxide. Bottom ash may contain some heavy metals and dioxins (Friends of the Earth 2002). Bottom ash from incinerators can be buried in landfill. It is used for engineering purposes in the Netherlands, Denmark, France and Germany.

12.5 Remediation of Polluted Soils

Remediation refers to the improvement of a contaminated site through prevention, minimization, or mitigatation of damage to human health or the environment. Remediation involves the development and application of a planned approach that removes, destroys, contains or otherwise reduces the availability of contaminants to receptors of concern. Remediation uses physical, chemical, and biological clean-up methods to remove or contain a toxic spill or hazardous materials from a contaminated site. In general, cleanup falls into one of the following three categories: (i) **Removal:** Harmful chemicals are removed from contaminated soil, (ii) **Treatment:** Contaminats are chemically treated to change them into less harmful substances, and (iii) **Containment:** Harmful chemicals are left in the ground, but steps are taken to prevent them from moving into clean soil and to prevent people from coming into contact with them.

Thus, there are *in situ* (on site) and *ex situ* (off site) techniques of remediation for soil pollution. Remediation at the contamination site without soil excavation, soil removal, treatment and refilling is known as *in situ* remediation. In this method, the chance of causing further environmental harm is minimized. These methods are relatively expensive and slow (Ward et al. 2003) or limited by the production of secondary waste streams that require subsequent disposal or treatment. In *ex situ* remediation the soil is excavated, soil materials covering the depth of contamination are collected and brought to the treatment shed/site, decontaminated by various physical and chemical methods and refilled in its original place.

12.5.1 *Remediation of Organic Pollutants*

12.5.1.1 Physical and Chemical Methods

Many advanced soil remediation techniques for organic pollutants have been commercialized and adopted in industrialized and developed countries. These include Gas Phase Chemical Reduction (GPCR), Mechan-ochemical dehalogentation (MCD), and Thermal Desorption. Some promising techniques such as Base

Catalyzed Decomposition (BCD) and sonic technology are still either at the laboratory stage or at the pilot study. However, due to the financial constraints, many advanced technologies are unlikely to be adopted by the developing countries (Li 2007). Here, the conventional methods of remediation of organic pollutants such as soil washing, thermal desorption, chemical oxidation and decomposition, vitrification, etc. are discussed.

Soil washing
Soil washing is an *ex situ* remediation technique. Soil materials are dumped into the treatment shed and separated into different size classes, e.g. sand, silt and clay. Since pollutants principally remain adsorbed on clay surfaces, this fraction is washed with water and other solvent mixtures, including dichloromethane, ethanol, methanol and toluene (Rababah and Matsuzawa 2002). Additionally, surfactants, which are amphiphilic molecules, are used to facilitate the washing (Wick et al. 2011). Surfectants can be anionic, cationic, nonionic or zwitterionic depending on the charge on the polar head (Mulligan et al. 2001). Surfactants can improve desorption, apparent aqueous mobility and bioavailability of hydrophobic organic compounds such as PAHs (Mata-Sandoval et al. 2002). According to Ahn et al. (2008), TWEEN 40, TWEEN 80, Brij 30, DOWFAX 8390 and STEO 330 are some surfectants that have been effective for removal of PAH from soil. Vegetable oil has been found to be effective as a non-toxic, biodegradable and cost-effective alternative to these conventional solvents and surfactants (Gong et al. 2006).

Adsorption on activated carbon
Activated carbon is one of the best adsorbents for many organic chemicals because of its hydrophobicity, high specific surface and microporous structure (Vasilyeva et al. 2006). It is widely used for cleaning up drinking water and contaminated wastewater. Activated carbon has also been recommended for reducing the phytotoxicity of many herbicide residues and other chemicals in agricultural soils. Activated carbon is applied for the removal of volatile organics such as petroleum hydrocarbons, chlorinated solvents, polychlorinated biphenyls, pentachlorophenol, etc. Activated carbon amendment has become a promising option for the *in situ* remediation of soils for organic pollutants (Bes and Mench 2008). It reduces the bioavailability of organic contaminants due to its strong sorption properties (Bucheli and Gustafsson 2000). It is cost-effective and environmental friendly. It does not release new amounts of pollutants (Hilber and Buchel 2010). Smol et al. (2014) determined six PAHs including (benzo(b) fluoranthene, benzo(k)fluoranthene, benzo(a)pyrene, dibenzo(a,h)anthracene, indeno(1,2,3,c,d)pyrene and benzo(g,h,i) perylene) from environmental samples. These PAHs are carcinogenic according to USEPA (2000). The investigators prepared model solutions and treated them with three types of sorbents: quartz sand, mineral sorbent and activated carbon. After sorption processes the concentration of hydrocarbons in solution decreased, and the maximum (96.9 percent) decrease was caused by activated carbon. Vasilyeva et al. (2006) observed that activated carbon helped overcome toxicity during bioremediation of 3,4-dichloroaniline (DCA), 2,4,6-trinitrotoluene (TNT), and polychlorinated biphenyls (PCB) by transferring them to a less toxic soil fraction. It resulted in

accelerated biodegradation of DCA and promoted strong binding through accelerated microbial reduction of its nitro groups and catalytic chemical oxidation of the methyl group and polymerization of TNT. Alternative carbon materials like biochar (Yu et al. 2009), coke breeze (Millward et al. 2005), fly ash (Burgess et al. 2009), etc. have also been tried.

Oxidation with hydrogen peroxide

According to Goi et al. (2009), chemical oxidation is a promising and innovative process for degrading an extensive variety of hazardous organic compounds at waste disposal and spill sites. The hydrogen peroxide oxidation technique has been employed successfully for the remediation of organic pollutants including chlorophenols, polycyclic aromatic hydrocarbons, diesel and transformer oil (Goi et al. 2009). It can be employed *in situ* or *ex situ* but the *in situ* technique is advantageous because it is fast and safe, and the products of reaction are usually harmless (e.g. H_2O, CO_2, O_2, halide ions). It is also cheap due to the scope of generating hydrogen peroxide electrochemically on site. The oxidation process of hydrogen peroxide is catalyzed naturally by iron oxide minerals (hematite Fe_2O_3, goethite FeOOH, magnetite Fe_3O_4 and ferrihydrite) present in the soil. In some instances, hydrogen peroxide along with Fe (II) is used. This is known as Fenton's reagent which can be prepared with different concentrations (3 to 35%) of hydrogen peroxide along with ferrous iron (Fe II) as a catalyst to oxidize organic chemicals (Flotron et al. 2005). Peroxide (H_2O_2) decomposes into highly reactive nonspecific hydroxyl radicals with the help of ferrous iron. The pH requirement for the reaction is 3–5. If the soil has a high enough iron oxide minerals hydrogen peroxide alone can effectively oxidize the organic pollutants in soil (Watts et al. 2002). Low soil permeability, incomplete site delineation and soil alkalinity may limit the applicability of the hydrogen peroxide oxidation technique (Goi et al. 2009).

Ozone treatment

Ozone treatment is an effective soil remediation technique for organic pollutants. The major benefits of ozone remediation include (i) *in-situ* destruction of pollutants, (ii) relatively rapid remediation, (iii) contaminants are destroyed instead of being transferred from one phase to another, (iv) no hazardous by-products are produced, and (v) micro-bubbles act to extract pollutants from both groundwater and soil pores. According to Pierpoint et al. (2003), the introduction of ozone into the soil is effective for remediation of soil contaminated with pesticides. Generally, ozone is transported into the soil through columns packed with various materials, including sand and aquifer materials (Choi et al. 2000). Ozone can degrade hydrocarbons, and polyaromatic hydrocarbons. Column studies by Pierpoint et al. (2003) revealed that ozone could rapidly degrade aniline and trifluralin in soil. Choi et al. (2001) also mentioned that ozone treatment for *in situ* remediation of organic pollutants such as PAHs can be done with the injection of gaseous or aqueous ozone into the soil. Ozone transforms organic pollutants into more soluble and more biodegradable oxygenated intermediates. Ozone also oxidizes organic matter and can release some of the sequestered PAHs making them more available to biodegradation (O'Mahony et al. 2006). The presence of metal oxides can catalyze the formation of hydroxyl radicals which are more aggressive oxidants than ozone. The efficiency of ozone treatment can be increased if integrated with other techniques such as extraction before ozonation, or ozonation followed by biodegradation. Disadvantages of ozonation include the production of intermediates which can be more toxic than the parent compound. Ozonation can also destroy the indigenous microbial degraders (Wick et al. 2011).

Permeable reactive barrier technology
A Permeable Reactive Barrier or PRB is constructed with iron metal or zero-valent iron (Fe^0) in soil across the predominant direction of movement of groundwater. While passing through this barrier organic molecules react with Fe^0 and are dehalogenated or detoxified (Sharma and Reddy 2004). Thus, it is a very effective method of dehalogenation or detoxification of organic contaminants in groundwater. The Fe^0 has been effective against chlorinated ethylenes, halomethanes, nitroaromatic compounds, pentachlorophenol, chlorinated pesticides such as DDT, polychlorinated biphenyls, atrazine, and other organic compounds containing reducible functional groups or bonds. The most common contaminants treated with Fe^0 include trichloroethylene, carbon tetrachloride, dichloroethylene (cis and trans), tetrachloroethene, and vinyl chloride (Muegge 2008).

The capacity of zero-vaent iron particles to degrade and detoxify organic pollutants can be greatly increased if their particle size can be made into the nanoscale. These iron particles are known as nano iron particles or NIP. Nanoscale zerovalent iron particles (NIP) can reduce the contaminants very rapidly because of their small size and large specific surface area compared to conventional granular zero-valent iron (Raychoudhury and Scheytt 2013). Nanoscale zero-valent iron particles (NIP) have become superior to PRB and iron filings; contaminants need not pass through the PRB. Nano zero-valent iron particles can be injected directly into the contaminated soil zones. The extremely small size and the enhanced reactivity due to high surface area to volume ratio make NIP as excellent choice for *in situ* subsurface remediation. Reddy (2010) has developed inexpensive and environmentally-benign lactate-modified NIP that are stable and capable of transporting in soils and groundwater and dehalogenating organic pollutants such as pentachlorophenol and dinitrotoluene.

Electrokinetic remediation
Electrokinetic (EK) remediation method has especially been developed for the removal of contaminants in soil, sediments and sludge. It can, however, be applied to other solid porous materials as well. The method involves the application of electric current of low intensity to the porous matrix to be decontaminated. The general principle is to insert two electrodes – the anode and the cathod in the contaminated matrix and pass electricity to induce the mobilization and transportation of contaminants through the porous matrix towards the electrodes, where they are collected, pumped out and treated (Cameselle et al. 2013). The electrokinetic remediation technology can be applied for both heavy metals and organic pollutants (Lu et al. 2005). The mechanisms of organic contaminant transport under induced electriacal potential include (i) electroosmosis which can be defined as the bulk movement of

pore fluid through the electrical double layer in clayey soils, generally occurring from anode to cathode, (ii) electromigration – transport of polar organic compounds or complexes within the pore fluid towards oppositely charged electrodes; (3) electrophoresis – transport of charged colloids, micelles, bacterial cells, etc. within the pore fluid towards oppositely charged electrodes, and (4) diffusion – transport of chemicals due to concentration gradients. Electroosmosis is the major transport process for non-polar organic compounds, while electromigration is the dominant transport process for ionic compounds. Saichek and Reddy (2005) cited examples of electrokinetically enhanced remediation of soils contaminated by hydrophobic organic compounds. Yap et al. (2011) mentioned that electrokinetic technique can be integrated with the Fenton treatments specifically for soils contaminated with polycyclic aromatic hydrocarbons. There are certain limitations of the electrokinetic remediation of organic contaminants of soil. For example, the more dangerous and persistent organic contaminants are not soluble in water. So, electromigration of such substances are not possible and their electrokinetic remediation is not satisfactory. In some cases, the *in situ* remediation of organic pollutatnts can be very slow or unsatisfactory in soils of low permeability. In that case, enhancement of EK technology is achieved by several means including combination with Fenton technique, EK combined with surfactants / co-solvents technique, EK combined with bioremediation method and EK combined with ultrasonic remediation method (Huang et al. 2012).

12.5.1.2 Bioremediation of Organic Pollutants

The elimination, attenuation, and transformation of pollutants into less harmful substances by biological processes is known as bioremediation (Shukla et al. 2010). In the actual sense, any process that uses microorganisms, fungi, green plants or their enzymes to return the natural environment altered by contaminants to its original condition is bioremediation (Kensa 2011). Bioremediation can be employed for the decontamination and detoxification of soil, water, and sediments. Bioremediation technologies can also be *ex situ* and *in situ* (Boopathy (2000). The most common bioremediation process of the environment is biodegradation which has been historically used for degradation of organic wastes and organic compounds. Recently bioremediation has become an effective, affordable, socially acceptable and widely applicable remediation technology for organic contaminants in soil, sediment, sludge and water. Biodegradation is based on the capabilities of biological agents such as microorganisms (yeast, fugi, bacteria) and plants for removing pollutants (Vidali 2001; Megharaj et al. 2011). There are two major mechanisms of bioremediation: (i) Microbial remediation and (ii) Phytoremediation. According to Perelo (2010), microbial remediation involves biostimulation (activation of viable native microorganisms) and bioaugmentation (artificial inoculation of viable microorganisms). On the other hand, phytoremediation is based on the dergadative and extractive capabilities of plant roots and other tissues. These new techniques promise to be of lower impact and more cost efficient than traditional management strategies. This

12.5 Remediation of Polluted Soils

technology offers an attractive alternative to other conventional remediation processes because of its relatively low cost and environmentally-friendly method. Some investigators distinguish between phytoremediation and bioremediation. To them, bioremediation includes the processes of eliminating pollutants only by the activity of microorganisms. According to Kang (2014), bioremediation and phytoremediation, are promising technologies that utilize natural resources to eliminate toxic organic contaminants.

Soils and sediments are usually contaminated with complex mixtures of organic compounds. For examples, Greenberg et al. (2005) cited an example of soil contamination with creosote which is a source of PAHs and contains more than 100 aromatic compounds. According to them, it could be very difficult to use a single technique to completely remove all the components of such complex mixtures. In such situations, through an understanding of the nature of contaminating substances, it should be wise to select multiple remediation processes strategically to remove contaminants rapidly and completely. Greenberg (2006) developed a multi-process remediation system composed of volatilization, photooxidation, microbial degradation and phytoremediation processes to eliminate soil pollution with complex mixtures of recalcitrant contaminants rapidly and completely. The techniques applied included land farming (aeration and light exposure), microbial remediation (introduction of contaminant degrading bacteria) and phytoremediation (degradation, extraction, volatilization and assimilation with plants). The overall remediation process was found to be greatly improved in this multi-process system. Land farming was chosen because it is a fast and effective method for removal of volatile chemicals such as naphthalene, acenaphthene, and acenaphthylene (Greenberg et al. 2005).

Microbial remediation
A wide variety of microorganisms are known to degrade persistent organic compounds including polyaromatic hydrocarbons (PAHs), polychlorinated biphenyles (PCBs) and pesticides and their residues in soil. These organisms can be thermophiles, alkaliphiles, halophiles, etc. (Seo et al. 2009). Microorganisms can interact chemically and physically with soil organic contaminants and result in structural changes that reduce their toxicity or degrade them completely (Wiren-Lehr et al. 2002; Diez 2010). Bacteria (Singh 2006), fungi (Gadd 2001) and actinomycetes (Diez 2010) are effective microbial transformers and pesticide degraders. Fungi are biotransformers of pesticides and other xenobiotics by changing their structures so that they are converted to less harmful intermediaries that undergo further degradation in soil by bacteria into nontoxic molecules (Gianfreda and Rao 2004). Microorganisms transform contaminant organic molecules through reactions related to their metabolic processes. They attack the pollutants enzymatically and convert them to harmless products. Microbial remediation is accomplished by (1) complete oxidation of organic contaminants, i.e. mineralization, (2) biotransformation of organic chemicals into less toxic simpler metabolites, or (3) reduction of highly electrophilic halo- and nitro- groups and production of less toxic compounds. Contaminant degrading microorganisms can naturally exist in the contaminated

sites but in many places they do not have enough capacity to detoxify persistent organic compounds. In that case indigenous microorganisms can be stimulated by supplying moisture, nutrients, hormones, etc. The phenomenon is known as biostimulation. Alternatively, effective strains of organisms can be introduced into the contaminated site. This is known as bioaugmentation. Suitable conditions of pH, temperature, moisture, and at adequate supply of vitamins and nutrients, microorganisms can biodegrade and biotransform complex hazardous organic chemicals into simpler and harmless ones (Sinha et al. 2009). *In situ* bioremediation may also involve bioventing (bioventing is the process of supplying oxygen to contaminated soil with the objective of stimulating microbial degradation of contaminants) and biosparging (injection of air under pressure below the water table to increase groundwater oxygen concentrations and enhance the rate of biological degradation of contaminants by naturally occurring bacteria).

At present more than 100 genera of microbes are used for controlling organic pollution of soil. Glazer and Nikaido (2007) mentioned that these organisms belong to at least 11 different prokaryotic divisions. According to Vidali (2001), thegroups of contaminants that can be eliminated by bioremediation include chlorinated solvents (Trichloroethylene, Perchloroethylene), PCBs (4-Chlorobiphenyl, 4,4-Dichlorobiphenyl), Pentachlorophenol, BTEX (benzene, tolune, Ethylbenzene, xylene), PAHs (naphthalene, antracene, Fluorene, pyrene, nenzopyrene), and pesticides (atrazine, carbaryl, carbofuran, coumphos, diazinon, glycophosphate, parathion, prpham, 2,4,D).

Bacteria

A large group of bacteria can effectively degrade persistent organic compounds aerobically and anaerobically, although the anaerobic process proceeds relatively slowly. Anaerobic bacteria can be used for bioremediation of polychlorinated biphenyls (PCBs) in river sediments, dechlorination of the solvent trichloroethylene (TCE), and chloroform. Some examples of aerobic bacteria that have been found to degrade pesticides and both aliphatic and polyaromatic hydrocarbons are *Pseudomonas*, *Alcaligenes*, *Sphingomonas*, *Rhodococcus*, and *Mycobacterium*. *Pseudomonas oleovorans* is reported to degrade tetrahydrofuran (THF). *P. oleovorans* DT4 can also utilize and biotransform THF and BTEX compounds (Zhou et al. 2011). Generally, the PCBs that contain less than six chlorines can be degraded aerobically (Ellis et al. 2003) while biphenyls which have higher number of chlorines in the molecule, for example in Arochlor 1260, are resistant to microbial degradation under aerobic conditions, but may be dechlorinated under anaerobic conditions (Master et al. 2002). There are some aerobic bacteria that utilize methane for carbon and energy. They can degrade a broad range organic compounds include ing the chlorinated aliphatics trichloroethylene and 1,2-dichloroethane through the enzyme methane mono-oxygenase.These bacteria are called methylotrophs. A list of bacteria and the pollutants they degrade effectively and rapidly are given in Table 12.2.

12.5 Remediation of Polluted Soils

Table 12.2 A list of bacteria demonstrated to degrade different organic pollutants of soil

Bacterial species/strains	Degrading contaminants	References
Achromobacter sp. NCW	Carbazole	Guo et al. (2008)
Alcaligenes denitrificans	Fluoranthene	Weissenfels et al. (1990)
Acidovorax delafieldii P4–1	Phenanthrene	Balashova et al. (1999)
Bacillus cereus P21	Pyrene	Kazunga and Aitken (2000)
Burkholderia cepacia BU-3	Naphthalene, phenanthrene, pyrene	Kim et al. 2006
Burkholderia xenovorans LB400	Benzothiopene, benzonate	Denef et al. (2005)
Chryseobacterium sp. NCY	Carbazole	Guo et al. (2008)
Cycloclasticus sp. P1	Pyrene	Wang et al. (2008)
Janibacter sp. YY-1	Dibenzofuran, fluorine, dibenzithiopene, phenanthrene, anthracene	Yamazoe et al. (2004)
Marinobacter NCE312	Naphthalene	Hedlund et al. (2001)
Mycobacterium sp.	Pyrene, benzo(a)pyrene, fluoranthene	Lee et al. (2007)
Mycobacterium sp. RJGII-135	Pyrene, benzo(a)pyrene, benz[a]anthracene	Schneider et al. (1996)
Mycobacterium sp. PYR-1, LB501T	Fluoranthene, pyrene, phenanthrene, anthracene	Ramirez et al. (2001), van Herwijnen et al. (2003)
Mycobacterium flavescens	Pyrene, fluoranthene,	Dean-Ross et al. 2002; Dean-Ross and Cerniglia (1996)
Mycobacterium vanbaalenii PYR-1	Phenanthrene, pyrene, dimethylbenz[a]anthracene	Kim and Freeman (2005), Chávez et al. (2004)
Mycobacterium sp. KMS	Pyrene	Miller et al. (2004)
Pasteurella sp. IFA	Fluoranthene	Sepic and Leskovsek (1999)
Polaromonas naphthalenivorans CJ2	Naphthalene	Pumphrey and Madsen (2007)
Pseudomonas sp. BT1d	3-hydroxy-2-formylbenzothiophene	Bressler and Fedorak (2001)
Pseudomonas sp. B4	Biphenyl, chlorobiphenyl	Chávez et al. (2004)
Pseudomonas sp. HH69	Dibenzofuran	Fortnagel et al. 1990
Pseudomonas sp. CA10	Carbazole, chlorinated dibenzo-*p*-dioxin	Habe et al. (2001)
Pseudomonas sp. NCIB 9816-4	Fluorine, dibenzofuran, dibenzothiophene	Resnick and Gibson (1996)
Pseudomonas paucimobilis	Phenanthrene	Weissenfels et al. (1990)
Pseudomonas vesicularis OUS82	Fluorine	Weissenfels et al. (1990)

(continued)

Table 12.2 (continued)

Bacterial species/strains	Degrading contaminants	References
Pseudomonas putida CSV86	Naphthalene, Methyl naphthalene, phenanthrene	Mahajan et al. (1994)
Pseudomonas fluorescens BS3760	Phenanthrene, coronene, benz[a]anthracene	Balashova et al. (1999)
Pseudomonas saccharophilia	Pyrene	Kazunga and Aitken (2000)
Pseudomonas aeruginosa	Phenanthrene	Romero et al. (1998)
Ralstonia sp. SBUG 290 U2	Dibenzofuran, naphthalene	Becher et al. (2000)
Rhodanobacter sp. BPC-1	Benzo[a]pyrene	Kanaly et al. (2002)
Rhodococcus sp.	Pyrene, fluoranthene	Dean-Ross et al. 2002
Rhodococcus erythropolis D-1	Dibenzothiophene	Matsubara et al. 2001
Staphylococcus sp. PN/Y	Phenanthrene	Mallick et al. (2007)
Stenotrophomonas maltophilia VUN 10,003	Pyrene, fluoranthene, benzo[a]pyrene, benz[a]anthracene, dibenz[a,h]anthracene, corene	Juhasz et al. (2000), Juhasz et al. (2002)
Sphingomonas yanoikuyae R1	Pyrene	Kazunga and Aitken (2000)
Sphingomonas yanoikuyae JAR02	Benzo[a]pyrene	Rentz et al. (2008)
Sphingomonas sp. LB126	Fluorine, phenanthrene, fluoranthene, anthracene	Pinyakong et al. (2000)
Sphingomonas sp.	Dibenzofuran, dibenzothiophene, carbazole	Gai et al. (2007)
Sphingomonas paucimobilis EPA505	Fluoranthene, naphthalene, anthracene, phenanthracene	Mueller et al. (1990)
Sphingomonas wittichii RW1	Chlorinated dibenzo-*p*-dioxin	Nam et al. (2006)
Terrabacter sp. DBF63	Dibenzofuran, chlorinated dibenzothophene, chlorinated dibenzo-*p*-dioxin, fluorene	Habe et al. (2004), Habe et al. (2001), Habe et al. (2002)
Xanthamonas sp.	Pyrene, benzo[a]pyrene, carbazole	Grosser et al. (1991)

Fungi

A large number of fungi have been reported to possess the capacity of degrading and detoxifying organic pollutants of soil and water. For example, Ellegaard-Jensen (2012) cited a large number of fungi that can act on a single substrate diuron, a persistent herbicide, and can transform it into less toxic form or can degrade it. The fungi are members of the ascomycetes - *Botrytis cinerea, Beauveria bassiana, Beauveria bassina, Aspergillus niger, Alternaria* sp., *Phoma cf. Eupyrena,* the basidimycetes fungi - *Rhizoctonia solani, Bjerkandera adusta, Phanerochaete chrysosporium, Coriolus versicolor,* Basidiomycete strain Gr177, and the zygomycetes fungi - *Cunninghamella echinulata* Thaxter, *Cunninghamella elegans, Mortirella*

12.5 Remediation of Polluted Soils

isabellina, Cunninghamella elegans, Mortierella isabellina, Mortierella sp., *Mucor* sp., *Cunninghamella elegans, Mortierella* sp., and its several strains. Kjøller and Struwe (2002) suggested that saprotrophic fungi produce a variety of extracellular enzymes which help fungi degrade organic pollutants. However, there are some limitations of biodegradation by white rot fungi because of their requirement of high temperatures. Therefore, they do not have enough capabilities compared to indigenous organisms in the environment (Baldrian 2008; Gao et al. 2010). According to Marco-Urrea and Reddy (2012), the extracellular lignin modifying enzymes system consisting of lignin peroxidase, manganese peroxidase, and laccase is responsible for degradation of lignin and other organic compounds by fungi. Since these enzymes are relatively non-specific, white rot fungi possess a unique ability to degrade a variety of organic pollutants such as dioxins, polychlorinated biphenyls, petroleum hydrocarbons, trinitrotoluene, herbicides and pesticides (Pointing 2001; Reddy and Mathew 2001). The white-rot fungi, *Phanerochaete chrysosporium* and *Trametes* versicolor are widely used for lignin biodegradation because of their fast growth, easy handling and good ligninolytic properties (Mougin et al. 2002). The extracellular peroxidase enzyme systems produced by white-rot fungi are responsible for the degradation of a variety of aromatic xenobitics, including chlorophenols, pesticides and dyes (Tortella et al. 2005; Rubilar et al. 2007). Tortella et al. (2008) observed that the Chilean native white-rot fungus *Anthracophyllum discolor* Sp4 possesses a high ligninolytic activity and has a great potential for xenobiotic degradation. Filamentous fungi of the genera Fusarium, Penicillium, Trichoderma, Aspergillus, Neosartorya, Pseudallescheria, Cladosporium, Pestalotiopsis, Phoma and Paecillomyces can synthesize extracellular oxidative enzymes related to lignin degrading enzyme systems of crude oil contaminants (Naranjo et al., 2007). D'Annibale et al. (2006) suggested that external nutritional supply like saw dust, fertilizers or plant debris is always required to accelerate fungul activity.

Enzymes

Enzymes are active biological agents that possess a great potentiality to transform and detoxify organic pollutants. Enzyme mediated transformations of organic compounds can occur at a detectable rate. They can be employed in detoxification treatments and restoration of polluted environments (Rao et al. 2010). The use of enzymatic proteins may overcome most disadvantages related to the use of microorganisms in remediation of organic pollutants (Nicell 2001; Gianfreda and Bollag 2002). Enzymes can have either narrow or broad specificity. Thus, they can be applied to sites contaminated with a large range of different compounds in mixture. They can bring extensive transformations of structural and toxicological properties of contaminants, and they can even cause complete mineralization of POPs. Enzymes may have advantages over microbial remediation as well. They can be used under extreme conditions limiting microbial activity. The most important classes of enzymes for remediation of polluted soils are: hydrolases, dehalogenases, transferases and oxidoreductases mainly produced by bacteria, fungi, plants and microbe-plant associations (Rao et al. 2010). Transformation of different xenobiotic substances by some of these enzymes has been tested under laboratory conditions (Whiteley and Lee 2006).

Phytoremediation of organic pollutants

Phytoremediation refers to the use of living plants, which through their metabolic activities, remove, transform, degrade, or immobilize toxic substances present in polluted soils, sediments, ground water, and wastewater in wetlands. According to Trapp and Karlson (2001), phytoremediation is a novel technique to clean up polluted soils using plants. Theoretically, phytoremediation methods are relatively cheap, are accepted by the public both because soils are not disturbed and toxic by-products are not needed to dispose of, and are environment-friendly (Chen et al. 2013b). Phytoremediation could be used to treat different types of contaminants including petroleum hydrocarbons, chlorinated solvents, pesticides, explosives, heavy metals and radionuclides in soil and water (Truu et al. 2015). Here in this section phytoremediation of organic pollutants is discussed. According to Susarla et al. (2002) phytoremediation can be applied to degrade, assimilate, metabolize, or detoxify hydrocarbons, pesticides, and chlorinated solvents, and the in situ method is applicable to low to medium level of contamination of soils, sediments and shallow groundwater. The chemical compounds taken up into plants may be metabolised, accumulated, or volatilized into air. So, there are several phytoremediation methods including phytoextraction (the uptake of contaminants in plant roots and their concentration in harvestable tissues), rhizofiltration removal of toxic substances through uptake by mass of roots of aquatic plants), rhizodegradation (biodegradation of pollutatns by metabolites in exudates of roots) and phytodegradation (biodegradation of pollutants by plant enzymes within plant tissue), and phytovolatilization (the uptake of contaminants by plants and their subsequent release into the atmosphere in a volatile form). Susarla et al. (2002) concluded that phytoremediation has the potential for providing the most cost-effective and resource-conservative approach for sites contaminated with hazardous chemicals, but phytoremediation prescriptions must be site-specific. A large number of plant species have been found to be promising candidates for the phytoremediation of organic pollutants (Gerhardt et al. 2009). Since the methods are very slow, and for unfavorable soil conditions in many instances, there had been a few examples of successful applications of phytoremediation (Trapp and Karlson (2001). However, Campos et al. (2008) cited examples of successful phytoremediation in many cases. Moreover, several techniques including soil modification and use of biotechnology and genetic engineering are being undertaken for the enhancement of phytoremediation. The following plants are commonly used for phytoremediation of organic pollutants: *Hamamelis virginiana, Ulmus pumila, Robinia pseudoacacia, Populus deltoids, Galega orientalis Quercus* spp. (Barnswell 2005), *Salix viminalis, Leucaena leucocephala* (Trapp and Karlson 2001) *Ipomoea batatas Festuca arundinacea Arabidopsis thaliana* (Doty et al. 2003) *Zea mays* (Zand et al. 2010) *Populus* spp. *Solanum tuberosum, Cucurbita pepo* (Campos et al. 2008) *Morus rubra* (Inui et al. 2001) *Oryza sativa* (Kawahigashi et al. 2007) *Glycine max* (Njoku et al. 2009) *Nicotiana tabaccum Salix* spp. (Zand et al. 2010).

12.5 Remediation of Polluted Soils

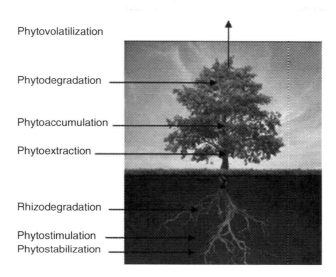

Fig. 12.19 Mechanisms of phytoremediation of organic pollutants in soil

Mechanisms of phytoremediation

Phytoremediation usually involves the steps (i) uptake, (ii) translocation, (iii) transformation, (iv) compartmentalization and often mineralization within the plant tissue (Schnoor et al. 1995). The uptake, distribution and transformation of organic compounds by plants are affected by several factors including physical and chemical nature of the compound (solubility, molecular weight), plant characteristics (root system, enzymes) and environmental conditions (temperature, pH, organic matter, and soil moisture content) (Suresh and Ravishankar 2004). It has already been mentioned that the chief mechanisms of phytoremediation include (i) phytodegradation (ii) phytoextraction (iii) rhizodegradation (iv) phytovolatilization and (v) rhizofiltration. The mechanisms of phytoremediation of organic pollutants in soil are shown in Fig. 12.19.

The phytoremediation methods applicable for different organic pollutants are given in Table 12.3.

Phytodegradation

Soil organic contaminants are absorbed by plants and enzymes in the plant tissue transform and degrade them. Polychlorinated biphenyls (PCBs) have been metabolized by sterile plant tissues. Phenols have been degraded by plants such as horseradish, potato (*Solanum tuberosum*), and white radish (*Raphanus sativus*) that contains peroxidase (Roper et al. 1996). After reaching the plant rhizosphere contaminants migrate into the root. The pollutant enters into the xylem stream and crosses the suberised casparian strips in the root endodermis. Transformation, metabolism and detoxification of the contaminant occur there (Schroder et al. 2002; Chen et al. 2013b). The P-450 enzymes catalyse phase-I transformation reactions

Table 12.3 Mechanisms of phytoremediation and their degradable organic pollutants of soil and groundwater

Mechanism	Contaminants to be degraded
Phytoextraction/ phytoaccumulation	BTEX (benzene, ethyl benzene, tolune, xylenes), pentachlorophenol, shot-chained aliphatic compounds
Phytodegradation	Nitrobenzene, nitroethane, nitromethane, nitrotolune, picric acid, RDX, TNT, atrazine, chlorinated solvent (chloroform, carbon tetrachloride, hexachloroetane, tetrachloroethane, dichloro ethane, vinyl chloride, trichloroethanol, dichloroethanol, DDT, dichloroethene, methyl bromide, tribroethene, chlorine and phosphate based pesticides, chlorinated biphenols, other phenols, nitriles.
Rhizoregradation	Phenol, chlorinated solvents, polycyclic aromatic hydrocarbons, BTEX (benzene, ethyl benzene, tolune, xylenes), petroleum hydrocarbons, atrazine, alachlor, PCB, tetrachloroethane, trichloroethane.
Phytovolatilization	Chlorinated solvents (tetrachloroethane, trichloroethene, tetrachloromethane),
Rhizofiltration	Various organic compounds.

Adapted from Susarla et al. (2002)

like hydroxylation, sulfoxidation and N and O-dealkylation and Glutathione-S-transferases are responsible for conjugation reactions leading to detoxification.

Rhizodegradation

Rhizodegradation is the degradation of contaminants in the root zone, either due to microbial activity or by roots, or by both. Rhizodegradation is a complex mechanism of phytoremediation which involves complex interactions of roots, root exudates, rhizosphere soil and microbes. Plant roots release some enzymes, nutrients, hormores and metabolites with the exudates. The enzymes may directly act on some pollutants, and transform and mineralize them. Moreover, the exudates, enzymes and metalolites secreted by the plants and the aeration of the rhizosphere soil may stimulate the growth and activity of rhizosphere microbial communities (phytostimulation) which degrade the organic pollutants at an accelerated rate. Plants also protect rhizosphere microorganisms against abiotic stresses (Kuiper et al. 2004). For such rhizosphere effect, enlarged number, diversity and metabolic activity of microbes develop resulting in the degradation of contaminants more efficientlyl (Kent and Triplett 2002). Root exudates contain organic acids (lactate, acetate, oxalate, succinate, fumarate, malate, and citrate), sugars and amino acids, and also some secondary metabolites (isoprenoids, alkaloids, and flavonoids). Root exudates can act as energy sources by microorganisms. Exudates can induce the expression of specific catabolic genes in microorganisms necessary for the degradation of the contaminant (Singer et al. 2003). For instance, plant secondary metabolite salicylate has been linked to the microbial degradation of PAHs (naphthalene, fluoranthene, pyrene, chrysene) and PCB (Master and Mohn 2001), while terpenes can induce the microbial degradation of toluene, phenol, and TCE (Kim et al. 2002). The weed plant *Senecio glaucus* was found growing in sand polluted with up to 10% petroleum in Kuwait. Roots of these plants were associated with huge number of

12.5 Remediation of Polluted Soils

Table 12.4 List of some plants and organic contaminants they can degrade by the roots

Plants	Pollutants
Prairie grasses	PAHs
Alfalfa	Pyrene, anthracene, phenanthrene
Sugar beet	PCBs
Senecio glaucus	Oil
Barley	2,4-D
Wheat	2,4-D
Grasses	Naphthalene
Oat, lupin, rape, pepper, radish, pine	Pyrene

Adapted from Nwoko (2010)

oil-degrading bacteria (*Arthrobacter*), which took up and detoxified alkanes and aromatic hydrocarbons (Trapp and Karlson 2001). Rhizodegradation are frequently used for the remediation of organic contaminations, among them petroleum, PAH, BTEX, TNT, chlorinated solvents and pesticides (USEPA 2000). Bramley-Alves et al. (2014) observed that subantarctic native tussock grass, *Poa foliosa* (Hock. f.), was very tolerant to Special Antarctic Blend (SAB) diesel fuel. This grass has the capacity to stimulate the hydrocarbon degrading bacteria in the rhizosphere. Latter, they used this grass for successful remediation of soils of the world heritage Macquarie Island accidentally contaminated with SAB spillage (Table 12.4).

Phytoextraction

Phytoextraction is the removal of a contaminant from the soil, ground water or surface water by live plants. Phytoaccumulation occurs when the contaminant is taken up by plant roots and is accumulated in the aboveground parts. The plants used for phytoextraction are tolerant to the contaminant, can grow satisfactorily in the contaminated soil and do not show any symptom of toxicity even at very high concentrations of the contaminant in plant tissue. The technique of phytoextraction is preferably used for heavy metals. However, some plants can take up organic contaminants, and the uptake of organic pollutants depends on properties and bioavailability of organics, size and shape of the root system, and evapotranspiration rate (Schwitzguébel et al. 2002). Phytoextraction depends on the absorption, translocation, and metabolism of organic pollutants in plants. Some organic compounds are able to enter into plant cells by penetrating cell membrane easily. *Meaicago osativa* and *Tagetes patula* are potential candidates for the phytoremediation of soils contaminated with PAEs and PAHs (Fu et al. 2012). A number of enzymes are involved in the transformation and sequestration of organic pollutants in plants. For example, cytochrome P450 enzymes participate in the oxidative process for emulsifying highly hydrophobic pollutants (Page and Schwitzguébel 2009). Glutathione-S-transferases (GSTs) catalyze the conjugation between toxic organic pollutants and sulfhydryl (-SH) group of glutathione (GSH). GST-pollutants conjugates can be further transported and sequestered from cytosol to vacuoles in plant cells (Cummins et al. 2011).

Phytovolatilization
Phytovolatilization is the process of diffusion of substances from leaves into the air. Some organic compounds and some metals/metalloids (Se, Hg) are absorbed and transported to the leaf from which they volatilize into the atmosphere. In the way to the leaves organic compounds may undergo some kind of biochemical transformation and the products of transformation and residual contaminants enter into the leaves. The process was shown to remediate m-xylene, chlorobenzene (Baeder-Bederski et al. 1999), trichloroethene and other volatile compounds (Burken and Schnoor 1998), but also for organically bound mercury (USEPA 2000). Various plants can volatilize such organic soil pollutants as carbon tetrachloride (CCl_4), ethylene dibromide (EDB), trichloroethylene (TCE), methyl tert-butyl ether (MTBE), 1,2-dichloroethane (DCA), etc. Since these compounds are photo-reactive, they are readily degraded in the atmosphere.

Rhizofiltration
Rhizofiltration is usually used to remediate surface water, groundwater, and wastewater contaminated with low concentration of contaminants including metals/metalloids and organic compounds. Plant roots absorb these contaminants from the surrounding water and accumulate in roots or translocate to shoots or both. For rhizofiltration, plants are grown hydroponically, and when the root system has satisfactorily developed, they are transferred to the polluted water. Another option is to grow the plants capable of rhizofiltration in the polluted water from the beginning. The roots or whole plants are harvested when they become saturated with the contaminants for disposal by landfill after inceneration. Both terrestrial and aquatic plants can be used for *in situ* or *ex situ* rhizofiltration (Etim 2012). Although Mahendran (2014) suggested that low concentration of hydrophobic organic chemicals can be removed by rhizofiltration, this mechanism of phytoremediation is primarily suitable for heavy metals/metalloids and other inorganic substances.

12.5.2 Remediation of Heavy Metal Pollutants

Several critical reviews of the possible sources, chemistry, hazards and best available remedial strategies for a number of heavy metals and metalloids (lead, chromium, arsenic, zinc, cadmium, copper, mercury and nickel) have been done (Stegmann et al. 2001; Bradl and Xenidis 2005 and Wang et al. 2010; Wuana and Okieimen 2011). Contamination of soils by heavy metals is one of the most important environmental issues throughout the world and the clean up of these soils is a difficult task (Oustan et al. 2011). Remediation methods of soils contaminated with heavy metals can be of three different classes or their combinations – physical, chemical and phytoremediation (Zhou and Song), based basically on two fundamental principles: (i) complete removal of contaminants from polluted sites and (ii) transformation of the pollutants into less toxic or harmless forms (Jiang et al. 2011). Some methods can be applied *in situ* (the soil is not removed from its place) and

12.5 Remediation of Polluted Soils

some others are *ex situ* (soil is excavated up to the depth of contamination, carried to the treatment shed, separated into fractions, the most contaminated fraction is treated, and placed again in its original place). The major chemical methods include immobilization, precipitation, and ion exchange. However, most physical and chemical methods including solidification, stabilization, electrokinetics, encapsulation, vitrification, and soil washing as well as flushing are expensive and cannot easily make the soil suitable for plant growth (Marques et al. 2009).

12.5.2.1 Physical Methods

The principal physical methods of soil remediation are soil washing, flushing, encapsulation, vitrification, and electrokinensis. The methods are briefly discussed below.

Soil washing

Soil washing is used often as an *ex situ* method of remediating metal contaminated soils due to some advantages including: (i) contaminants can be removed completely, (ii) it is usually sustainable, (iv) it may be cost-effective and (v) it may produce recyclable material or energy (GOC 2003). Soil washing systems offer the greatest promise for application to soils contaminated with organic and inorganic contaminants, but the main target contaminant group is heavy metals (Arwidsson et al. 2010). Washing solution may be water, surfactants, chelating or complexing agents, reducing agents and acid/alkaline solutions (Singh et al. 2014). However, heavy metals (and their compounds) are sparingly soluble and remain mainly in the adsorbed condition. Therefore, washing the soils with water alone cannot satisfactorily extract heavy metals or metalloids adsorbed on particle surfaces or precipitated into soils because of their relative immobility. The extractability of metals may be enhanced by adding some chemical agents with washing solutions (Dikinya and Areda 2010). Several chemical agents including surfactants, cosolvents, cyclodextrins, chelating agents and organic acids are used for soil washing (Zvinowanda et al. 2009). Chelating agents can increase the solubility of metals in water, replace them from the soil particles and concentrate the contaminants into a smaller volume (Gitipour et al. 2011). The principal chelating agents used for metal extraction and soil washing generally include ethylene-diaminetetraacetic acid, nitrilotriacetic acid, diethylenetriaminepentaacetic acid and citric acid (Bilgin and Tulun (2015). Moon et al. (2016) used several soil washing solutions such as hydrochloric acid (HCl), nitric acid (HNO_3), sulfuric acid (H_2SO_4), tartaric acid ($C_4H_6O_6$) and ethylenediaminetetraacetic acid ($C_{10}H_{16}N_2O_8$, EDTA) for bench-scale soil washing experiments with some heavy metals (Pb, Cu, Zn). The concentrations of the washing solutions ranged from 0.1 to 3 M with a liquid-to-solid ratio of 10. The soil washing results showed that hydrochloric acid (HCl) was the best washing solution at 3 M for heavy metal removal. Sun et al. (2014) carried out soil washing experiments with citric acid as washing reagent for the remediation of soils highly contaminated with heavy metals, and then activated carbon was used in absorption processing for leaching solution. Lead was reduced by 36 percent, Cu by 47.74 percent, and Cd by 61.88 percent. The factors that affect efficiency of soil washing include contact time, pH, concentration of extract and agitation speed. The efficiency of a washing agent, for example EDTA, depends mainly upon the type of soil, contamination type, contamination period and metals present in soil (Karthika et al. 2016).

Soil flushing

Soil flushing is an *in situ* method of decontamination similar in principle to soil washing in that heavy metal contaminants are removed by water, solvents, surfactants or chelating agents. However, water or solvents are injected into the soil to dilute, solubilize, mobilize, or release sorbed metals which leach to the ground water. The metals move with groundwater to the extraction well through which contaminated water is extracted and treated to decontaminate. The number, location, and depth of the injection and extraction wells depend on several factors including hydraulic conductivity of soil, metal mobility, metal speciation, metal sorption, efficiency of solvents and chelating agents, and engineering considerations. Extracted ground water could necessitate further treatment in order to meet discharge standards before it is either recycled or released to a publicly owned wastewater treatment works or stream. Purification of the flushing solution and separation of surfactants from the recovered flushing fluid is a key factor in the costing of soil flushing. There are several limitations in soil flushing including (i) low permeability soils restricts metal movements, (ii) surfactants can adhere to soil and reduce porosity, (iii) the contaminants can spread beyond the capture zone, and (iv) treatment of contaminated water may be costly.

Encapsulation

Encapsulation refers to the construction of an impermeable barrier across the underground polluted zone so that contaminants (gas, liquid or metals) are contained and do not spread to uncontaminated areas. Several construction methods can be adopted. Examples are cut-off slurry walls using cement-bentonite-water slurries, thin walls, sheet pile walls, bored-pile cut-off walls, jet grouting curtains, injection walls, frozen barriers, etc. Encapsulation is one of the easiest ways to prevent spread of metal pollution in soils and to dispose of hazardous wastes. An easy way of encapsulation is to fill the contaminated soils and wastes in three quarters of a leak-proof container. Materials such as cement, plastic foam or bentonite clay is poured into the container until completely filled. The contents are allowed to harden, and when they are hardened, the container is sealed and may be landfilled, stored or buried. Chemical or pharmaceutical wastes together with sharps and other hazardous materials may also be encapsulated.

Stabilization/Solidification

Stabilization of pollutants is the process of converting contaminated materials into more chemically stable and less mobile constituents so that their toxicity and further spreading are reduced. On the other hand, solidification involves the addition of reagents to contaminated soil or sludge to form solid products. Among the various treatment techniques of stabilization/solidification, treatment of effluent containing hazardous wastes with Portland cement-based solidifying/stabilizing agents is an effective option for remediating soil pollution with metals, other inorganic and organic substances (Goyal and Chauhan 2015). By the process of stabilization/solidification polluted materials are transformed into forms that become insoluble and intoxicated (Batchelor 2006). Stabilizing/solidifying can significantly reduce solubility and mobility of pollutants by their sorption and encapsulation as well as their destruction (Caldwell et al. 1999). According to the United States Protection Agency, stabilization/solidification processes are utilized across the world and are the best available technology neutralizing a large number of hazardous mineral waste, soils, slurries and sludges contaminated with toxic metals (Shi and Fernandez-Jimenez 2006). In the processes of stabilization/sodification polluted materials and wastes are mixed with appropriate solidifying/stabilizing mixtures based on such agents as cement, fly ash, blast furnace slag, calcium carbonate, Fe/Mn oxides, charcoal, zeolite, water glass, mortar sand, hydrated lime, and organic stabilizers such as bitumen, composts, and manures, or a combination of organic-inorganic amendments, etc. Effective binding mixture is prepared depending on kind of wastes or pollutants to be neutralized. After mixing a cement-based formulation, the contaminated soil is turned into almost concrete which reduces the flexibility of the treated material for re-use (Pensaert et al. 2008).

Vitrification

The process of vitrification refers to the *in situ* transformation of a substance into a non-crystalline amorphous solid or glass. Vitrification of contaminated soil is generally done by heating the soil to more than 1200–2000 °C in the contaminated zone by passing electric current through electrodes until it liquidizes. The liquefied material is rapidly cooled to produce a vitrified solid. In this process the contaminated soils are converted into chemically inert and stable glass by a thermal treatment process. Large electrodes are inserted into soils containing significant levels of silicates. The electrodes are usually placed in 3–5 m^2. Generally, electrodes are inserted into the soil in two different ways. The electrodes can be inserted through pre-constructed holes covering the contaminated soil volume. In the other way, graphite is placed on the surface soil and connects the electrodes through which high electric current is passed. The soil gradually melts, and the electrodes sink further into the ground causing deeper soil to melt (Fig. 12.20).

After the electric current is turned off, the melted soil cools and vitrifies making a solid block of glass-like material. Larger areas are treated by fusing together multiple individual vitrification zones (Wuana and Okieimen 2011). As the soil is vitrified, the original volume of soil shrinks and causes subsidence of the soil surface. The sunken area is filled with clean soil (USEPA 2001).

Electrokinesis

The same principles as described under Section 12.5.1.1 also apply here for *in situ* and *ex situ* remediation of soil polluted with heavy metals through electrokinesis. Electrokinesis involves the application of a low level DC current or voltage gradient

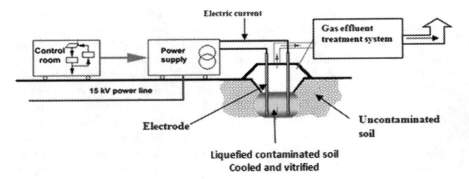

Fig. 12.20 Shematic diagram of vitrification

across electrodes placed in the contaminated areas of saturated soils. The contaminants move either towards the cathode or anode depending on the kind of charge. There are several advantages of the electrokinetic remediation: (i) the method is simple and requires simple equipment; (ii) it is safe; people in the vicinity are not exposed to contaminants; (iii) it can be used for remediation of different media including soil, sediment, sludge and groundwater; (iv) it can be used for a wide range of contaminants such as metals, radionuclide and polar organic compound, and (v) it is cost effective (Reddy 2013). The electrokinetic remediation is most suitable for clayey soils because of conduction of electric current. Moreover, experimental results revealed a better cleaning efficiency of electrokinesis in saturated and less dense soils (Greičiūtė and Vasarevičius (2007). Metals are not very mobile in soil. Moreover, metal ions have a strong tendency to be sorbed on soil colloids. For enhancing efficiency of electrokinetic remediation of heavy metals, several conditioning agents such as HNO_3, HCl, H_2SO_4, citric acid, oxalic acid, ascorbic acid, EDTA, etc. are used (Iannelli et al. 2015).

12.5.2.2 Chemical Remediation

Heavy metals are present in soil in various forms including soluble ions, sorbed on colloidal surfaces, chemical precipitates, insoluble compounds and minerals. When soluble and bioavailable forms exceed the threshold levels, toxicity occurs. Although physical remediation processes for metals such as washing/flushing, encapsulation, etc. use a number of enhancing chemical agents, there are some purely chemical processses of cleaning up of heavy metals from contaminated soils. These processes include immobilization, precipitation, chelation, ion exchange, etc. Ion exchange is applied maily to remove metals from water or wastewater by passing through resin columns. Chelation has been described in connection with soil washing/flushing as well.

Immobilization

Immobilization as a chemical process refers to the conversion of a soluble, biologically available or absorbable and mobile element, ion or compound into an insoluble and immobile forms so that plants cannot absorb it or other biota cannot be exposed to it. Immobilization could be used as an *in situ* remediation method involving cheap materials such as lime or dolomite. Immobilization may reduce environmental risk (McGowen et al. 2001) because the products are of low solubility minerals and precipitates and the process may have a long-term effect (Basta and McGowen 2004). Many immobilization studies were done using chemical amendments dolomite (Trakal et al. 2011), limestone and fly ash (Yun and Yu 2015), di-ammonium phosphate, vermicompost and zeolite (Abbaspour and Golcin 2011). Other popular amendments are clay, cement, minerals, phosphates, organic compost, and microbes (Finzgar et al. 2006). The application of liming materials increases soil pH and decreases solubility and bioavailability of metals. Correia (2014) considered the immobilization process as chemical passivation which involves both organic and inorganic amendments to provide a shielding effect over heavy metals, constricting their mobility and chemical interaction with other compounds in the soil.

Precipitation

Chemical precipitation of heavy metals can be expressed by the equation $M^{2+} + 2(OH)^- = M(OH)_2$ (Barakat 2011). It is often used for removing metals from industrial effluents. Metals can be precipitated by the addition of coagulants such as alum, lime, iron salts and other organic polymers. Chemical precipitation with lime is the most frequently used process for treating effluents. Precipitation with metals in wastewater with lime has several advantages including its simplicity, inexpensive equipment requirement, and convenient and safe operations (Wuana and Okieimen 2011).

12.5.2.3 Bioremediation

Biological methods of removing metals and reducing their toxicity are called bioremediation. Bioremediation involves natural processes that encourage the growth of natural vegetation, favor the establishment or reestablishment of plants on polluted soils and in most situations do not interfere with economic utilization of the polluted soil. Bioremediation is also economic (Chibuike and Obiora 2014), although it may take a long time for complete elimination of metal toxicity. Heavy metals are not degradable by biotic activity but their bioavailability, and hence toxicity, can be reduced by biological transformation. By the change of oxidation state, some heavy metals may be precipitated and some others may be volatilized (Garbisu and Alkorta (2003). As bioremediation of heavy metals can be accomplished by using microorganisms and plants, the process is divided into microbial remediation and phytoremediation.

Microbial remediation

Some microorganisms can be used for the remediation of heavy metals. Often native microorganisms are not highly capable; in that case extraneous microorganisms are used (Prescott et al. 2002). Depending on the chemical nature of the metal pollutants, microbial species or strains are selected (Dubey 2004). According to Watanabe et al. (2001), several types of pollutants are encountered in a contaminated site, and diverse types of microorganisms may be required for its remediation. Microorganisms can reduce toxicity of metals by (i) changing their valence states; (ii) precipitating by their by extracellular chemical substances; (iii) volatilization; and (iv) reducing by extracellular enzymes (Garbisu and Alkorta (2003). For example, more toxic selenate and selenite are reduced to the much less toxic elemental selenium (Garbisu et al. 1997). Some bacteria, algae and fungi can convert Se into dimethylselenide or trimethylarsine and can remove them through volatilization (White et al. 1997). Sulfate reducing bacteria convert sulfate to hydrogen sulfide and then to insoluble metal sulfides such as zinc sulfide and cadmium sulfide (Iwamoto and Nasu 2001). Some bacteria such as *Bacillus subtilis, Pseudomonas putida*, and *Enterobacter cloacae* can reduce highly toxic Cr (VI) to the less toxic Cr (III) (Garbisu et al. 1998).

Phytoremediation

Although most plants suffer in growth from high metal concentrations in the growing media, a few groups of plants have the unique ability to tolerate high concentrations of heavy metals in the soil. Tolerance to heavy metals in plants is exhibited by their ability to survive without any noticeable growth reduction in a soil contaminated with elevated concentrations of heavy metals. The concentrations of metals have become so high in some places, naturally or anthropogenically, that most plants fail to grow or the growth is severely stunted (Shah and Nongkynrih (2007). According to Baker (1981), the tolerance of those plants to these metals arises through three different mechanisms: (i) metal entry from soil into roots or metal transport from roots to shoots are restricted (exclusion), and maintenance of a low concentration in shoot over a wide range of concentrations of metals in soil; (ii) metal concentrations in the shoot maintain a linear relationship with metal concentration in soil solution (inclusion); and (iii) accumulation of metals in plant tissues without any symptom of toxicity at whatever concentration of metal is there in the soil solution. The plants that can tolerate high levels of toxic heavy metals are known as metallophytes. There are two types of metallophytes: obligate metallophytes and facultative metallophytes. Obligate metallophytes can only survive in the presence of high concentration of heavy metals, and facultative metallophytes can tolerate such conditions but are not confined to them. Metallophytes are commonly specialised flora found on spoil heaps of mines. Some plants can sequester large amounts of heavy metals in their biomass. Such unique metabolic capabilities of plants have evolved a cheap and environment-friendly socially acceptable technology of remediation of pollutants – phytoremediation. Plants can clean up many kinds of pollutants including heavy metals, pesticides, explosives, petroleum-hydrocarbons and oil (Ferreiro et al. 2014; Nichols et al. 2014), but the removal of

12.5 Remediation of Polluted Soils

heavy metals/metalloids and reducing their toxicity will be discussed in this section. The chief mechanisms of phytoremediation of metals are phytostabilization, phytovolatilization, and phytoextraction.

Phytostabilization

Phytostabilization is the *in situ* metal inactivation by means of revegetation either with or without non-toxic metal-immobilizing soil amendments Stabilization of metals often refers to the reduction in mobility and conversion of soluble forms of heavy metals to insoluble forms within the root zone through root-mediated precipitation or sorption. For example, under natural conditions lead is precipitated as sulfate at the plant roots. Plant roots can change soil pH by the production of CO_2 by root respiration and microbial decomposition of root exudates. Plant root exudates may contain a number of organic acids that increase mobilizatization and phytoavailability of metals in soil. This may cause increased (i) uptake and accumulation of metals in tolerant/hyperaccumulator plants; (ii) toxicity in sensitive plants; and (iii) leaching and contamination of groundwater. The kinds of metals and their chemistry, soil conditions, and the biodiversity of existing or the introduced vegetation would actually determine the fate.

Phytovolatilization

Some plants take up some heavy metals/metalloids (for example, As, Hg and Se) from the soil, transport them to the aerial parts and diffuse them into the atmosphere through the stomata; the process is known as phytovolatilization (Masayuki et al. 2007). After being released to the air, mercury and selenium are transported away with the wind. In humid areas, Hg may be redeposited in the soil with rain water. Phytovolatilization efforts should be carried out in places far from human habitation. So far, no plant has been identified to have Hg accumulating ability (Raskin and Ensley 2000). Phytovolailization remains to be the only mechanism to remediate Hg from contaminated soil. Some plants have been genetically modified to incorporate Hg phytovolatilization ability. Examples of such transgenic plants are *Nicotiana tabacum, Arabidopsis thaliana*, and *Liriodendron tulipifera* which can volatilize Hg from polluted soil (Meagher et al. 2000). These plants are genetically modified with the gene for mercuric reductase, or merA. Some plants have also been modified with organomercurial lyase (merB) gene for the detoxification of methyl-Hg (USEPA 2000). Although genetic modification with merA and merB gene is not acceptable to USEPA, plants altered with merB are more acceptable because the gene prevents the introduction of methyl-Hg into the food chain. Plants that can effectively be used for phytovolatilization of Se from contaminated soils are *Brassica juncea* and *Brassica napus* (Bañuelos et al. 1997).

Phytoextraction

There are some plants that can absorb, transport and accumulate very high concentrations of metals in their aboveground parts; more than one and up to four orders of magnitude in other adjacent plants (Reeves and Baker 2000). A number of plants of different families have the unique ability to grow on metalliferous soils and to accumulate extraordinarily high amounts of heavy metals in the aerial organs, far in excess of the levels found in the majority of species, but without suffering from phytotoxic effects (Rascio and Navari-Izzo 2011). These plants are known as metal hyperaccumulator plants (Reeves and Baker 2000; Reeves 2006). Jaffré et al. (1976) reported for the first time high concentration of Ni in latex (25.74 percent on a dry weight basis) of New Caledonia tree species *Sebertia acuminate* but Brooks et al. (1977) are given the credit of coining the term 'hyperaccumulation'. This ability of uptake and accumulation of very high concentration of metal/metalloids has been observed in approximately 500 (Kramer 2010) to 700 (Xi et al. 2010) plant species. Using hyperaccumulator plants for cleaning up of metals from contaminated soils is known as phytoextraction (Marques et al. 2009) or phytoaccumulation. Some metals can be locked up for a long time in the wood of metal accumulating plants. This can be called metal sequestration (Osman and Kashem (2016). Metal hyperaccumulator plants can extract and remove toxic metals but do not destroy structure and fertility of the soil. Plants that accumulate concentrations in aboveground parts of >10,000 mg kg^{-1} Mn or Zn; > 1000 mg kg^{-1} As, Co, Cr, Cu, Ni, Pb, Se or Ti; and >100 mg kg^{-1} Cd without any symptom of toxicity are hyperaccumulator plants of the respective metals. A list of metal hyperaccummulator plants is given in Table 12.5.

Some hyperaccumulator plants can accumulate very high amounts of heavy metals. Siegel (2002) cited an example of *Thalspi calaminare* and *Phyllanthus serpentinus* which accumulated 39,600 mg Zn kg^{-1} and 38,100 mg Ni kg^{-1} respectively in their leaves. In addition, hyperaccumulator plants have some other characteristics, such as (1) the metal concentrations in shoots must be greater than that in roots (transfer factor is >1); plants have the ability of absorbing and transporting metals for storage in their aboveground parts (Wei et al. 2002); (2) the metal concentrations in aboveground plant parts must be 100–500 times higher than those of the same plant species from non-polluted environments (Yanqun et al. 2005); and (3) the concentrations of heavy metals in shoots are greater than that in soils indicating higher metal bioaccumulation ability (McGrath and Zhao 2003). Phytoexraction may often be a slow process and require very long time for soil remediation. Two major factors are responsible for this low pace of phytoextraction – (i) Although some hyperaccumulators such as *Brassica napus, Brassica juncea*, and *Brassica rapa* grows fast and with high biomass potentials (Ebbs and Kochian 1997), most hyperaccumulator plants are generally slow growing and produce small biomass; this reduces the efficiency of the remediation process (Van Ginneken et al. 2007), and (ii) metals have very low mobility in soil and the concentrations of metals in soil solution around roots are low. The limited bioavailability of various metallic ions restricts their uptake/accumulation by plants. Thus, enhancement of the phytoextraction process may be needed in many cases. The plant itself can enhance metal bioavailability. For example, plants can extrude H+ via ATPases, which replace cations from colloidal surfaces making metal cations more bioavailable (Taiz and Zeiger 2002).

12.5 Remediation of Polluted Soils

Table 12.5 A list of some metal hyperaccumulator plants

Metal / metalloid	Plant species	Family	Reference
Arsenic (As)	Pteris vittata	Pteridaceae	Danh et al. (2014)
	Callitriche lusitanica	Plantaginaceae	Favas et al. (2012)
Cadmium Cd)	Arabidopsis halleri	Brassicaceae	Bert et al. (2002)
	Beta vulgaris	Amaranthaceae	Chen et al. (2013a)
	Bidens pilosa	Asteraceae	Sun et al. (2009)
	Myriophyllum heterophyllum	Haloragaceae	Sivaci et al. (2008)
	Pfaffia glomerata	Amaranthaceae	Gomes et al. (2013)
	Phytolacca americana	Phytolaccaceae	Zhao et al. (2011)
	Potamogeton crispus	Potamogetonaceae	Sivaci et al. (2008)
	Rorippa globosa	Brassicaceae	Sun et al. (2010)
	Sedum alfredii	Crassulaceae	Tian et al. (2011)
	Sedum alfredii	Crassulaceae	Tian et al. (2011)
	Thlaspi caerulescens	Brassicaceae	Banasova and Horak (2008)
	Thlaspi praecox	Brassicaceae	Vogel-Mikuš et al. (2006)
	Berkheya coddii	Asteraceae	Keeling et al. (2003)
Cobalt (Co)	Crotalaria cobalticola	Fabaceae	Oven et al. (2002)
	Haumaniastrum robertii	Lamiaceae	Brooks (1998)
	Brassica juncea	Brassicaceae	Diwan et al. (2008)
Chromium (Cr)	Gynura pseudochina	Asteraceae	Mongkhonsin et al. (2011)
	Leersia hexandra	Poaceae	Zhang et al. (2007)
	Phragmites australis	Poaceae	Calheiros et al. (2008)
	Salsolakaki	Amaranthaceae	Gardea-Torresday et al. (2005)
Copper (Cu)	Commelina communis	Commelinaceae	Wang and Zhong (2011)
	Crassula helmsii	Crassulaceae	Kupper et al. (2009)
	Silene paradoxa	Caryophyllaceae	Mengoni et al. (2003)
	Silene vulgaris	Caryophyllaceae	Van Hoof et al. (2001)
	Sorghum sudanens	Poaceae	Wei et al. (2008)
Nickel (Ni)	Alyssum lesbiacum	Brassicaceae	Singer et al. (2007)
	Alyssum murale	Brassicaceae	Broadhurst et al. (2004)
	Berkheya coddii	Asteraceae	Moradi et al. (2010)
	Bornmuellera kiyakii	Brassicaceae	Reeves et al. (2009)
	Hybanthus floribundus	Violaceae	Bidwell et al. (2004)
	Senecio coronatus	Asteraceae	Boyd et al. (2002)
	Stackhousia tryonii	Celastraceae	Bhatia et al. (2004)
	Streptanthus polygaloides	Brassicaceae	Jhee et al. (2005)
	Thlaspi geosingense	Brassicaceae	Persans et al. (2001)
	Celosia cristata	Amaranthaceae	Cui et al. (2013)
Lead (Pb)	Helianthus annuus	Asteraceae	Walliwalagedara et al. (2010)

(continued)

Table 12.5 (continued)

Metal / metalloid	Plant species	Family	Reference
	Hemidesmus indicus	Apocynaceae	Chandrasekhar et al. (2005)
	Plantago orbiguyana	Plantaginaceae	Bech et al. (2012)
	Sesbania drummondii	Fabaceae	Sahi et al. (2002)
	Thlaspi praecox	Brassicaceae	Vogel-Mikuš et al. (2006)
	Astragalus bisulcatus	Fabaceae	Freeman et al. (2006)
Selenium (Se)	Brassica juncea	Brassicaceae	Orser et al. (1999)
	Brassica oleracea	Brassicaceae	Tamaoki et al. (2008)
	Iberis intermedia	Brassicaceae	Leblanc et al. (1999)
	Stanleya pinnata	Brassicaceae	Freeman et al. (2006)
	Arabidopsis halleri	Brassicaceae	Deinlein et al. (2012)
Zinc (Zn)	Arabis gemmifera	Brassicaceae	Kubota and Takenaka (2003)
	Arabis paniculata	Brassicaceae	Zeng et al. (2011)
	Picris divaricata	Asteraceae	Du et al. (2011)
	Potentilla griffithii	Rosaceae	Hu et al. (2009)
	Sedum alfredii	Crassulaceae	Lu et al. (2013)
	Thlaspi caerulescens	Brassicaceae	Banasova and Horak (2008)
	Thlaspi praecox	Brassicaceae	Vogel-Mikuš et al. (2006)

Sometimes chelators are used to mobilize metals. For example, EDTA (ethylene diamine tetra acetic acid) assists in mobilization of Pb, Cd, Cr, Cu, Ni, and Zn) and their subsequent accumulation in *Brassica juncea* (Indian mustard) and *Helianthus anuus* (sunflower) (Turgut et al. 2004). The ability of other metal chelators such as CDTA (Glycine, N,N'-1,2-cyclohexanediylbis[N-(carboxymethyl)-, monohydrate), DTPA (Diethylenetriaminepentaacetic acid), EGTA ((ethylene glycol-bis(β-aminoethyl ether)-N,N,N',N'-tetraacetic acid), EDDHA (Ethylenediamine-N,N'-bis((2-hydroxyphenyl)acetic acid), and NTA (Nitrilotriacetic acid) to enhance metal accumulation has also been tested in various plant species (Lombi et al. 2001b). The risks associated with using certain chelators are excess mobility and and leaching and contamination of groundwater. The phytoextraction efficiency of plants may be enhanced by promoting plant growth (Zhang et al. 2007) with the help of metal tolerant plant growth promoting rhizobacteria (PGPR) inhabiting in the rhizosphere soil (Ma et al. 2011). Hansda et al. (2014) also suggested that association of PGPR and hyperaccumulator plant roots can improve plant growth through secretion of various regulatory chemicals and facilitate sequestration of toxic heavy metals. The genetic engineering of plant-associated bacteria can improve metal extraction. Genetic manipulation of plant hormone level may also improve hyperaccumulator plant biomass (Eapen and D'Souza 2005).

Genetic engineering for phytoremediation of heavy metals
Genetic engineering can be a useful tool to overcome a variety of limitations of phytoremediation. This involves transfer of genes form an organism that cannot be

12.5 Remediation of Polluted Soils

crossed by conventional breeding methods (Berken et al. 2002). In this method, a foreign piece of DNA is stably inserted into the genome of a cell to produce a, mature transgenic plant. This piece of DNA can be taken from any organism ranging from bacteria to mammals or other plants. According to Eapen and D'Souza (2005), the possible areas of genetic manipulation could be – (1) The transfer of human metallotionein gene in tobacco which resulted in enhanced Cd tolerance and pea metallotionein gene to *Arabidopsis thaliana* resulted in increased Cu accumulation; (2) The synthesis of phytochelatin in *Brassica juncea* overexpressing different enzymes involved in extract more Cd, Cr, Cu, Pb, and Zn than wild plants; (3) The overexpression of citrate synthase would promote enhanced Al tolerance; (4) The overexpression of the iron-binding protein ferritin has shown to increase up to 1.3-fold higher the iron level in tobacco leaves; (5) The transfer of Zn transporter-ZAT gene from *Thlaspi goesingense* to *Arabidopsis thaliana* resulted in two-fold higher Zn accumulation in its roots; (6) The transfer of *Escherichia coli* ars C and γ -ECS genes to *Arabidopsis* plants resulted in individuals that could transport oxyanion arsenate to aboveground tissues, reduce to arsenite, and sequester it to thiol peptide complexes; (7) The overexpression of glutathione-S-transferase and peroxidase in *Arabidopsis* plants resulted in enhanced Al tolerance; and (8) The increasing phytohormones synthesis can increase biomass of transgenic trees with genetically induced increase in giberellin biosynthesis.

12.5.3 Remediation of Dispersed Radioactive Contaminats

Radioactive pollution may occur due to accumulation of naturally-occurring or artificial radioactive substances, or radionuclides in a site due to handling, storing and accidental release. Radionuclides may be released also from hospitals and industrial and research facilities (Vandenhove et al. 2000). Human activities can cause radioactive pollution in concentrated in areas not normally controlled by regulatory bodies to levels beyond the set limits (IAEA 2003).

The major radionuclide contaminants of concern include those of the naturally occurring uranium and thorium series and man-made radionuclides such as ^{60}Co, ^{137}Cs, ^{90}Sr, ^{239}Pu, ^{241}Am, and others. These radioactive elements exert harmful effects on human health through inhalation, external exposure to gamma radiation, and ingestion of radionuclides through food and water (IAEA 1999). Several remediation methods are applied for soils contaminated with various radionuclides. The planning for remediation technology should be based on (i) potential human health and ecological impacts, (ii) possibility of spread of contaminants, (iii) availability of resource, and (iv) financial capability. An evaluation of the site characteristics including the nature, distribution and extent of radioactive contaminants and their sources; risks associated with human health and environment; and further spread to larger areas, groundwater and through the water courses is necessary. Remediation of soils contaminated with radionuclides (e.g. uranium) could be done through *ex situ* and *in situ* techniques. Remediation methods can be classified into natural attenuation, physical methods, chemical methods, biological methods and Electrokinetic methods (Gavrilescu et al. 2009). A limited number of chemical-physical (Agnew et al. 2011) or biological (Mihalik et al. 2012) methods have been developed for remediation of radioactively contaminated soils. These treatments may be prohibitively costly, time consuming or environmentally unsustainable. Moreover, it is extremely difficult to dispose of separated and contaminated soils if large areas of land are involved (Falciglia et al. 2013).

12.5.3.1 Natural Remediation

The level of contamination starts varying from the initial state due to natural processes. Radioactive decay of radionuclides reduces the contaminant loading. Noticeable decrease in soil contamination due to decay occurs in about 5–10 years after contamination from fission products. However, there are some long lived isotopes such as ^{239}Pu which would not undergo significant decrease in activity even after 500 years. The characteristics of soil also affect natural restoration of contaminated sites. Some soils can naturally retain radionuclides and hold them in place while they decay. The chemical state of the radioactive contaminants can be altered by weathering, and microbial action transform and their solubility and mobility can be reduced (IAEA (1999).

12.5.3.2 Physical Remediation

Among the remediation techniques currently used for radionuclides, soil excavation is the most common treatment. Other physical methods include encapsulation, size separation, and soil washing (Ebbs et al. 1998). According to Entry et al. (1996), the ultimate remediation of radionuclide-contaminated soils might require soil removal from the affected site and to be treated with various dispersing and chelating chemicals. Soil removal may be done only in sites contaminated in small scale with high activity concentrations of radionuclides. The part of the soil that has been contaminated can be removed by front loader, bulldozer, grader, manual digging, turf harvester, lawn mower, etc. Contaminants can be diluted by ordinary plowing, deep plowing, burial plowing, etc. (USEPA 2016). Zhu and Shaw (2000) noted that removing and transporting soil may be very costly. Cement-based Stabilisation/solidification technique was proposed to treat radionuclides polluted soils (Falciglia et al. 2012). In this *in situ* remediation technique, Portland cement, with or without other materials (such as ground blast furnace slag, fly ash cement kiln dust or clay), are mixed with contaminated soils in order to produce high resistance solidified hard masses where contaminants are bounded (Falciglia et al. 2013).

12.5.3.3 Chemical Remediation

Mallampati et al. (2013) developed a nano-Fe/Ca/CaO dispersion mixture based remediation and volume reduction method of real radioactive cesium contaminated soils. After treatment of soil with 10 percent by weight of nano-Fe/Ca/CaO dispersion mixtures, emitting radiation intensity decreased from 4.00 µSv h^{-1} to 0.95 µSv h^{-1} in non-magnetic fraction soils. They suggested that cesium contaminated soil volume can be reduced and simple mixing soil with nano-Fe/Ca/CaO may be a potential technology for the remediation and separation of radioactive Cs contaminated soil in dry conditions. Stojanovic et al. (2013) observed that the use of adsorbents such as zeolite, apatite, diatomite and bentonite individually and in mixtures may be effective for in situ stabilization of uranium ions preventing inclusion of uranium in the food chain. It has been observed that when fast-acting sequestering agents (diatomite, organomodified zeolite and bentonite) are applied along with slow-acting adsorbent (natural phosphate), a synergistic effect leading to permanent solution for the *in situ* stabilization of uranium ions can be achieved. The synergistic effect of a mixture of apatite and organomodified zeolite caused rapid binding of uranium with organomodified zeolite and the formation of stable phase uranium-phosphate-autunite and eliminated the risk of desorption of organomodified zeolite due to changes in soil conditions (Stojanovic et al. 2013).

12.5.3.4 Bioremediation

Radionuclides are found in soils in different oxidation states and may be present as oxide, coprecipitates, inorganic, and organic complexes. Microorganisms cannot degrade them, but can play a major role in the mobilization and immobilization of radionuclides by biochemical actions (Francis 2006). Some anaerobic bacteria can stabilize the radionuclides by reductive precipitation from higher to lower oxidation state. Microbial oxidation-reduction reactions are vital in transforming radionuclides and affecting their solubility, bioavailability and toxicity. For example, reduction of Pu(IV) → Pu(III) increases its solubility, while reduction of U(VI) → U(IV) or Pu(VI) → Pu(IV) decreases their solubility (Francis et al. 2002). The anaerobic bacterium *Clostridium* sp. has the unique metabolic capabilities of solubilizing and precipitating radionuclides (Francis et al. 1994).

Many plant species have been successful in efficiently accumulating the radionuclides in their stems and leaves and hence remediating the contaminated site (Pavel and Gavrilescu 2008). Soils contaminated with radionuclides, particularly ^{137}Cs and ^{90}Sr, pose a long-term radiation hazard to human health through exposure via the food chain and other pathways. Since removal or immobilization of radionuclides from contaminated surface soils through mineral and chemical amendments are often physically difficult and economically impractical, phytoextraction of radionuclides by specific plant species from contaminated sites is a promising bio-remediation method (Zhu and Shaw (2000). Soil pH, redox potential and metal complexation influence the bioavailability of radionuclides and their phytoextraction efficiency. Zhu and Shaw (2000) also suggested that phytoremediation of soils contaminated with radionuclides is still likely to take an excessively long time. For example, Stojanovic et al. (2013) noted that the major disadvantage of the technique is the time requirement - from 18 to 60 months or even decades. To speed up the process of selection of suitable plant taxa, a special plant breeding program assisted by molecular biotechnology may be useful. Some chemical amendments such as addition of chelates may enhance phytoextraction of radionuclides by different plants. Such chelating agents could include EDTA, HEDTA (Nhydroxyethylethylenediamine-N,N',N'-triacetic acid), DTPA (diethylenetrinitrilo pentacetic acid), natural fulvic acid, humic acid and low molecular weight organic acids such as citric, malic, oxalic, and acetic acid. Among these, EDTA has been reported as more effective than other synthetic chelators (Lestan 2006).

Study Questions
1. Define soil pollution. How do soils get polluted? How does soil pollution affect air and water quality? How does it affect human health?
2. What are the point and diffuse sources of soil pollutants? What are the principles of the prevention of soil pollution? Discuss waste management for the prevention of soil pollution.
3. Give an account of persistent organic pollutants (POPs). Discuss the physical and chemical methods of the remediation of soil organic pollutants. Explain why bioremediation of organic pollutants is more environment-friendly.
4. What are the environmentally important heavy metals? What are the effects of heavy metals on soil organisms and plants? Explain metal hyperaccumulation by plants with examples.
5. Write notes on: (a) sewage sludge as organic fertilizer, (b) mining wastes, (c) *exsitu* soil remediation, (d) war and soil pollution, and (e) remediation of radioactive pollution of the soil.

References

Abbaspour A, Golcin A (2011) Immobilization of heavy metals in a contaminated soil in Iran using di-ammonium phosphate, vermicompost and zeolite. Environ Earth Sci 63(5):935–943

Adefila EO, Onwordi CT, Ogunwande IA (2010) Level of heavy metals uptake on vegetables planted on poultry droppings dumpsite. Archives of. Appl Sci Res 2(1):347–335

Adriano DC (2003) Trace elements in terrestrial environments: biogeochemistry, bioavailability and risks of metals. Springer, New York Ahn CK, Kima YM, woo SH, park JM (2008) oil washing using various nonionic surfactants and their recovery by selective adsorption with activated carbon. J Hazard Mater 154:153–160

Agnew K, Cundy AB, Hopkinson L, Croudace IW, Warwick PE, Purdie P (2011) Electrokinetic remediation of plutonium-contaminated nuclear site wastes: results from a pilot-scale on-site trial. J Hazard Mater 186:1405–1414

Ahn WY, Busemeyer JR, Wagenmakers EJ, Stout JC (2008) Comparison of decision learning models using the generalization criterion method. Cogn Sci 32:1376–1402

References

Alam MGM, Allinson G, Stagnitti F, Tanaka A, Westbrooke M (2002) Arsenic contamination in Bangladesh groundwater: a major environmental and social disaster. Int J Environ Health Res 12(3):235–253

Alder A, McArdell CS, Giger W, Golet M, Molnar E, Nipales NS (2000) Determination of antibiotics in Swiss wastewater and in surface water. Presented at Antibiotics in the Environment, February 2, 2000, Cranfield, UK

Alexander EB, Coleman RG, Keeler-Wolf T, Harrison SP (2007) Serpentine geoecology of western North America: geology, soils and vegetation. Oxford Univ Press, New York

Alloway BJ (2012) Heavy metals in soils: trace metals and metalloids in soils and their bioavailability (environmental pollution). Springer, Dodrecht

Amanda L (2010) Determination of lead levels in soil and plant uptake studies. Undergraduate Review: a Journal of Undergraduate Student Research 12:48–56

Anjum NA, Ahmad I, Pacheco M, Duarte AC, Pereira E, Umar S, Ahmad A, Iqbal M (2011) Modulation of glutathione, its redox couple and related enzymes in plants under abiotic stresses. In: Anjum NA, Umar S, Ahmad A (eds) Oxidative stress in plants: causes, consequences and tolerance. International Publishing House, New Delhi

Arwidsson Z, Elgh-Dalgren K, von Kronhelm T, Sjöberg R, Allard B, van Hees P (2010) Remediation of heavy metal contaminated soil washing residues with amino polycarboxylic acids. J Hazard Mater 173:697–704

Aydinalp C, Marinova S (2003) Distribution and forms of heavy metals in some agricultural soils. Pol J Environ Stud 12(5):629–633

Baeder-Bederski O, Kuschk P, Stottmeister U (1999) Phytovolatilization of organic contaminants. In: Heiden S, Erb R, Warrelmann J, Dierstein R (eds) Biotechnologie im Umweltschutz. Erich Schmidt, Berlin, pp 175–183

Bagdatlioglu N, Nergiz C, Ergonul PG (2010) Heavy metal levels in leafy vegetables and some selected fruits. Journal fur Verbraucherschutz ebensmittelsicherheit 5:421–428

Baker AJM (1981) Accumulators and excluders strategies in the response of plants to heavy metals. J Plant Nutr 3:643–654

Balashova NV, Kosheleva IA, Golovchenko NP, Boronin AM (1999) Phenanthrene metabolism by *Pseudomonas* and *Burkholderia* strains. Process Biochem 35:291–296

Baldrian P (2008) Wood-inhabiting ligninolytic basidiomycetes in soils: ecology and constraints for applicability in bioremediation. Fungal Ecol 1:4–12

Banasova V, Horak O (2008) Heavy metal content in *Thlaspi caerulescens* J et C Presl growing on metalliferous and non-metalliferous soils in Central Slovakia. Int J Environ Pollut 33:133–145

Bañuelos GS, Ajwa HA, Mackey B et al (1997) Evaluation of different plant species used for phytoremediation of high soil selenium. J Environ Qual 26(3):639–646

Barakat MA (2011) New trends in removing heavy metals from industrial wastewater. Arab J Chem 4:361–377

Barcelo J, Poschenrieder C, Prasad MNV (2004) Structural and ultrastructural changes in heavy metal exposed plants. In: Heavy metal stress in plants. Springer, Berlin, Heidelberg

Barnswell KD (2005) Phytoremediation potential at an inactive landfill in northwest Ohio. Masters thesis, University of Toledo, Spain

Basta NT, McGowen SL (2004) Evaluation of chemical immobilization treatments for reducing heavy metal transport in a smelter-contaminated soil. Environ Pollut 127:73–82

Batchelor B (2006) Overview of waste stabilization with cement. Waste Manag 26:689–698

Bech J, Duran P, Roca N, Poma W, Sanchez I, Barcelo J, Boluda R, Roca-Perez L, Poschenrieder L (2012) Shoot accumulation of several trace elements in native plant species from contaminated soils in the Peruvian Andes. J Geochem Explor 113:106–111

Becher D, Specht M, Hammer E, Francke W, Schauer F (2000) Cometabolic degradation of dibenzofuran by biphenyl-cultivated *Ralstonia* sp. strain SBUG 290. Appl Environ Microbiol 66:4528–4531

Benisek M, Kukucka P, Mariani G, Suurkuusk G, Gawlik BM, Locoro G, Giesy JP, Blaha L (2015) Dioxins and dioxin-like compounds in composts and digestates from European countries as determined by the in vitro bioassay and chemical analysis. Chemosphere 122:168–175

Bentum JK, Dodoo DK, Kwakye PK (2012) Accumulation of metals and polychlorinated biphenyls (PCBs) in soils around electric transformers in the central region of Ghana. Adv Appl Sci Res 3(2):634–643

Berken A, Mulholland MM, LeDuc DL, Terry N (2002) Genetic engineering of plants to enhance selenium phytoremediation. Crit Rev Plant Sci 21:567–582

Bert V, Bonnin I, Saumitou-Laprade P, de Laguérie P, Petit D (2002) Do Arabidopsis halleri from nonmetallicollous populations accumulate zinc and cadmium more effectively than those from metallicolous populations? New Phytol 155:47–57

Bes C, Mench M (2008) Remediation of copper-contaminated topsoils from a wood treatment facility using in situ stabilisation. Environ Pollut 156:1128–1138

Bezdicek D, Fauci M, Caldwell D, Finch R, Lang J (2001) Persistent herbicides in compost. Biocycle 42(7):25–30

Bidwell SD, Crawford SA, Woodrow IE, Summer-Knudsen J, Marshal AT (2004) Sub-cellular localization of Ni in the hyperaccumulator *Hybanthus floribundus* (Lindley) F. Muell. Plant Cell Environ 27:705–716

Bilgin M, Tulun S (2015) Removal of heavy metal (cu, cd and Zn) from comtaminated soils using EDTA and $FeCl_3$. Global NEST J 18(1):98–107

Boopathy R (2000) Factors limiting bioremediation technologies. Bioresour Technol 74:63–67

Boyd RS, Davis MA, Wall MA, Balkwill K (2002) Nickel defends the south African hyperaccumulator *Senecio coronatus* (Asteraceae) against *Helix aspersa* (Mollusca: Pulmonidae). Chemoecology 12:91–97

Bradl H (2005) Heavy metals in the environment: origin. Interaction and remediation. Elsevier Academic Press, Amsterdam

Bradl H, Xenidis A (2005) Remediation Techniques. In: Bradl HB (ed) Heavy metals in the environment: origin, interaction and remediation. Elsevier, Academic Press, San Diego

Brady KU, Kruckeberg AR, Bradshaw HDJR (2005) Evolutionary ecology of plant adaptation to serpentine soils. Annual Review of Ecology Evolution and Systematics 36:243–266

Bramley-Alves J, Wasley J, King C, Powell S, Robinson SA (2014) Phytoremediation of hydrocarbon contaminants in subantarctic soils: an effective management option. J Environ Manag 142:60–69

Brandli KRC, Bucheli T, Zennegg TD, Huber M, Ortelli S, Muller D, Schaffner J, Iozza C, Schmid S, Berger P, Edder U, Oehme P, Stadelmann M, Tarradellas FX (2007) Organic pollutants in compost and digestate. Part 2. Polychlorinated dibenzo-p-dioxins, and -furans, dioxin-like polychlorinated biphenyls, brominated flame retardants, perfluorinated alkyl substances, pesticides, and other compounds. J Environ Monit 9:465–472

Bressler DC, Fedorak PM (2001) Purification, stability, and mineralization of 3-hydroxy-2- formylbenzothiophene, a metabolite of dibenzothiophene. Appl Environ Microbiol 67:821–826

Broadhurst CL, Chaney RL, Angle JS, Erbe EF, Maugel TK (2004) Nickel localization and response to increasing Ni soil levels in leaves of the Ni hyperaccumulator *Alyssum. murale*. Plant Soil 265:225–242

Brooks RR (1998) Plants that hyperaccumulate heavy metals. CAB Intern, Wallingford

Brooks RR, Lee J, Reeves RD, Jaffré T (1977) Detection of nickeliferous rocks by analysis of herbarium specimens of indicator plants. J Geochem Explo 7:49–57

Bruins MR, Kapil S, Oehme FW (2000) Microbial resistance to metals in the environment. Ecotoxicol and Environ Safety 45:198–207

Bucheli TD, Gustafsson O (2000) Quantification of the soot-water distribution coefficient of PAHs provides mechanistic basis for enhanced sorption observations. Environ Sci Technol 34:5144–5151

Buekers J (2007) Fixation of cadmium, copper, nickel and zinc in soil: kinetics, mechanisms and its effect on metal bioavailability, PhD Thesis, Katholieke Universiteit Lueven, Belgium

Burgess RM, Perron MM, Friedman CL, Suuberg EM, Pennell KG, Cantwell MG, Pelletier MC, Ho KT, Serbst JR, Ryba SA (2009) Evaluation of the effects of coal fly ash amendments on the toxicity of a contaminated marine sediment. Environ Toxicol Chem 28:26–35

Burken JG, Schnoor JL (1998) Predictive relationships for uptake of organic contaminants by hybrid poplar trees. Environ Sci Technol 32:3379–3385

Burrell AM, Hawkins AK, Pepper AE (2012) Genetic analysis of nickel tolerance in a north American serpentine endemic plant, Caulanthus amplexicaulis var. Barbarae (Brassicaceae). Am J Bot 99(11):1875–1883

CAC (2003) Evaluation of certain food additives and contaminants. FAO/WHO, Codex stan. 230–2001, Rev, 1–2003, Codex Alimentarius Commission, Rome

Caldwell RJ, Stegemann JA, Shi C (1999) Effect of curring on field – solidified waste properties, part 1: physical properties. Waste Management and Research 17:37–43

Calheiros CSC, Rangel ADSS, Castro PML (2008) The effects of tannery wastewater on the development of different plant species and chromium accumulation in *Phragmites australis*. Arch Environ Contam Toxicol 55:404–414

Cameselle C, Gouveia S, Akretche DE, Belhadj B (2013) Advances in Electrokinetic remediation for the removal of organic contaminants in soils. In: Rashed MN (ed) Organic pollutants - monitoring, risk and treatment. INTECH Open Access Publishers. http://www.intechopen.com/

Campos VM, Merino I, Casado R, Pacios LF, Gómez L (2008) Review. Phytoremediation of organic pollutants. Span J Agric Res 6(Special issue):38–47

Chakraborti D, Rahman MM, Paul K, Chowdhury UK, Sengupta MK, Lodh D et al (2002) Arsenic calamity in the Indian subcontinent: what lessons have been learned? Talanta 58:3–22

Chandrasekhar K, Kamala CT, Chary NS, Balaram V, Garcia G (2005) Potential of *Hemidesmus indicus* for phytoextraction of lead from industrially contaminated soils. Chemosphere 58:507–514

Chávez FP, Lünsdorf H, Jerez CA (2004) Growth of polychlorinated-biphenyl-degrading bacteria in the presence of biphenyl and chlorobiphenyls generates oxidative stress and massive accumulation of inorganic polyphosphate. Appl Environ Microbiol 70:3064–3072

Chen C, Huang D, Liu J (2009) Functions and toxicity of nickel in plants: recent advances and future prospects. Clean 37:304–313

Chen S, Chao L, Sun L, Sun T (2013a) Effects of bacteria on cadmium bioaccumulation in the cadmium hyperaccumulator plant *Beta vulgaris* var. *Cicla* L. Int J Phytoremediation 15:477–487

Chen J, Xu QX, Su Y, Shi ZQ, Han FX (2013b) Phytoremediation of organic polluted soil. J Bioremed Biodegr 4:e132

Chen YS, Zhang HB, Luo YM et al (2012) Occurrence and assessment of veterinary antibiotics in swine manures: a case study in East China. Chin Sci Bull 57:606–614

Chibuike GU, Obiora SC (2014) Heavy metal polluted soils: effect on plants and bioremediation methods. Appl Environ Soil Sci 2014, Article ID 752708, 12 pages. https://doi.org/10.1155/2014/752708

Cho M, Chardonnens AN, Dietz KJ (2003) Differential heavy metal tolerance of Arabidopsis halleri and Arabidopsis thaliana, a leaf slice test. New Phytol 158:287–293

Choi H, Kim Y, Lim H, Cho J, Kang J, Kim K (2001) Oxidation of polycyclic aromatic hydrocarbons by ozone in the presence of sand. Water Sci Technol 43:349–356

Choi H, Lim H, Kim J (2000) Ozone-enhanced remediation of petroleum hydrocarbon-contaminated soil. In: Wickramanayake GB, Gavaskar AR, Chen AS (eds) Chemical oxidation and reactive barriers: remediation of chlorinated and recalcitrant compounds. Battelle Press, Columbus

Correia JP (2014) Remediation by means of chemical passivation for trace metals in contaminated soils. B. Sc. Project, Worcester Polytechnic Institute, Massachusetts

Cui S, Zhang T, Zhao S, Li P, Zhao Q, Zhang Q, Han Q (2013) Evaluation of three ornamental plants for phytoremediation of Pb-contaminated soil. Int J Phytoremediation 15:299–306

Cummins I, Dixon DP, Freitag-Pohl S, Skipsey M, Edwards R (2011) Multiple roles for plant glutathione transferases in xenobiotic detoxification. Drug Metab Rev 43:266–280

D'Annibale A, Rosetto F, Leonardi V, Federici F, Petruccioli M (2006) Role of autochthonous filamentous fungi in bioremediation of a soil historically contaminated with aromatic hydrocarbons. Appl Environ Microbial 72(1):28–36

Dan T, Hale B, Johnson D, Conard B, Stiebel B et al (2008) Toxicity thresholds for oat (*Avena sativa* L.) grown in Ni-impacted agricultural soils near Port Colborne Ontario Canada. Can J Soil Sci 88:389–398

Danh L, Truong P, Mammucari R, Foster N (2014) A critical review of the arsenic uptake mechanisms and phytoremediation potential of *Pteris vittata*. Int J Phytoremediation 16:429–453

Davarynejad GH, Vatandoost S, Soltész M, Nyéki J, Szabó Z, Nagy PT (2010) Hazardous element content and consumption risk of nine apricot cultivars. Internat. J Hortic Sci 16(4):61–65

de Rosa MS, Zarrilli S, Paesano L, Carbone U, Boggia B, Petretta M et al (2003) Traffic pollutants affect fertility in men. Hum Reprod 18:1055–1061

Dean-Ross D, Cerniglia CE (1996) Degradation of pyrene by *Mycobacterium flavescens*. Appl Microbiol Biotechnol 46:307–312

Dean-Ross D, Moody J, Cerniglia CE (2002) Utilization of mixtures of polycyclic aromatic hydrocarbons by bacteria isolated from contaminated sediment. FEMS Microbiol Ecol 41(1):1–7

Deinlein U, Weber M, Schmidt H, Rensch S, Trampczynska A, Hansen TH, Husted S, Schjoerring JK, Talke IN, Kramer U, Clemens S (2012) Elevated nicotianamine levels in *Arabidopsis halleri* roots play a key role in zinc hyperaccumulation. Plant Cell 24:708–723

Denef VJ, Patrauchan MA, Florizone C, Park J, Tsoi TV, Verstraete W, Tiedje JM, Eltis LD (2005) Growth substrate- and phase-specific expression of biphenyl, benzoate, and C1 metabolic pathways in *Burkholderia xenovorans* LB400. J Bacteriol 187:7996–8005

Dermatas D, Moon DH, Menounou N, Meng X, Hires R (2004) An evaluation of arsenic release from monolithic solids using a modified semi-dynamic leaching test. J Hazard Mater 116:25–38

Diez MC (2010) Biological aspects involved in the degradation of organic pollutants. J Soil Sci Plant Nutr 10(3):244–267

Dikinya O, Areda O (2010) Comparative analysis of heavy metal concentration in secondary treated wastewater irrigated soils cultivated by different crops. Int J Environ Sci Technol 7(2):337–346

Diwan H, Ahmad A, Iqbal M (2008) Genotypic variation in the phytoremediation potential of Indian mustard against chromium. toxicity Environ Manage 41:734–741

Doick KJ, Klingelmann E, Burauel P, Jones KC, Semple KT (2005) Long-term fate of polychlorinated biphenyls and polycyclic aromatic hydrocarbons in agricultural soil. Environ Sci Technol 39:3663–3670

Domingo JL, Agramunt MC, Nadal M, Schuhmacher M, Corbella J (2002) Health risk assessment of PCDD/PCDF exposure for the population living in the vicinity of a municipal waste incinerator. Arch Environ Contam Toxicol 43(4):461–465

Doty SL, Shang TQ, Wilson AM, Moore AL, Newman LA, Strand SE, Gordon MP (2003) Metabolism of the halogenated hydrocarbons, TCE and EDB, by the tropical leguminous tree, Leuceana leucocephala. Water Res 37(2):441–449

Dubey RC (2004) A text book of biotechnology, 3rd edn. S Chand and Company Ltd, New Delhi

Duruibe JO, Ogwuegbu MOC, Egwurugwu JN (2007) Heavy metal pollution and human biotoxic effects. Int J Physical Sci 2(5):112–118

Eapen S, D'Souza SF (2005) Prospects of genetic engineering of plants for for phytoremediation of toxic metals. Biotechnol Adv 23:97–114

Ebbs SD, Brady JD, Kochian VL (1998) Role of uranium speciation in the uptake and translocation of uranium by plant. J Exp Bot 49(324):1183–1190

Ebbs SD, Kochian LV (1997) Toxicity of zinc and copper to *Brassica* species: implications for phytoremediation. J Environ Qual 26(3):776–781

Ellegaard-Jensen L (2012) Fungal degradation of pesticides - construction of microbial consortia for bioremediation. PhD thesis, The PhD School of Science, Faculty of Science, University of Copenhagen, Denmark

Ellis LBM, Hou BK, Kang W, Wackett LP (2003) The University of Minnesota biocatalysis/biodegradation database: postgenomic datamining. Nucleic Acids Res 31:262–265

Entry JA, Vance NA, Hamilton MA, Zabowsky D, Watrud LS, Adriano DC (1996) Phytoremediation of soil contaminated with low concentrations radionuclides. Water, Air Soil Pollution 88:167–176

Environment Agency (2007) Environmental concentrations of polychlorinated dibenzo-p-dioxins and polychlorinated dibenzofurans in UK soil and herbage, UK SHS report no. 10. Environment Agency, Bristol

Etim EE (2012) Phytoremediation and its mechanisms: A review. Int J Environ Bioenergy 2:120–136

EU (2000) Working Document on Sludge, 3rd Draft. Unpublished. Cited from EU (2001) Organic Contaminants in Sewage Sludge for Agricultural Use. Institute for Environment and Sustainability Soil and Waste Unit, European Commission Joint Research Centre

Falciglia PP, Cannata S, Pace F, Romano S, Vagliasindi (2013) Stabilisation/solidification of radionuclides polluted soils: a novel analytical approach for the assessment of the γ-radiation shielding capacity. Chem Eng Trans 32:223–228

Falciglia PP, Cannata S, Romano S, Vagliasindi FGA (2012) Assessment of mechanical resistance, γ-radiation shielding and leachate γ-radiation of stabilised/solidified radionuclides polluted soils: preliminary results. Chem Eng Trans 28:127–132

Favas PJC, Pratas JMS, Prasad MNV (2012) Accumulation of arsenic by aquatic plants in large-scale field conditions: opportunities for phytoremediation and bioindication. Sci Total Environ 433:390–397

Ferreiro PJ, Lu H, Fu S, Mendez A, Gasco G (2014) Use of phytoremediation and biochar to remediate heavy metal polluted soils: a review. Solid Earth 5:65–75

Filip Z (2002) International approach to assessing soil quality by ecologically-related biological parameters. Agric Ecosyst Environ 88(2):169–174

Finzgar N, Kos B, Lestan D (2006) Bioavailability and mobility of Pb after soil treatment with different remediation methods. Plant Soil Environ 52(1):25–34

Fitz WJ, Wenzel WW (2002) Arsenic transformations in the soil rhizosphere - plant system: fundamentals and potential application to phytoremediation. J Biotechnol 99:259–278

Flotron V, Delteil C, Padellec Y, Camel V (2005) Removal of sorbed polycyclic aromatic hydrocarbons from soil, sludge and sediment sample using the Fenton's reagent process. Chemosphere 59:1427–1437

Fodor F, Prasad MNV, Strzalka K (2002) Physiological responses of vascular plants to heavy metals. In: Physiology and biochemistry of metal toxicity and tolerance in plants. Kluwer Academic Publishers, Dordrecht, p 149

Fortnagel P, Harms H, Wittich R-M, Krohn S, Meyer H, Sinnwell V, Wilkes H, Francke W (1990) Metabolism of dibenzofuran by *Pseudomonas* sp. strain HH69 and the mixed culture HH27. Appl Environ Microbiol 56:1148–1156

Francis AJ (2006) Microbial transformations of radionuclides and environmental restoration through bioremediation. Paper presented at the Symposium on Emerging Trends in Separation Science and Technology SESTEC 2006, held at Bhabha Atomic Research Center (BARC), Trombay, Mumbai, India on September 29–October 1, 2006

Francis AJ, Dodge CJ, Lu F, Halada GP, Clayton CR (1994) XPS and XANES studies of uranium reduction by *Clostridium* sp. Environ Sci Technol 28:636–639

Francis AJ, Dodge CJ, Meinken GE (2002) Biotransformation of pertechnetate by clostridia. Radiochmica Acta 90:791–797

Freeman JL, Zhang LH, Marcus MA, Fakra S, McGrath SP, Pilon-Smits EA (2006) Spatial imaging, speciation, and quantification of selenium in the hyperaccumulator plants *Astragalus bisulcatus* and *Stanleya pinnata*. Plant Physiol 142:124–134

Friends of the Earth (2002) The safety of incinerator ash A review of an Environment Agency Report. November 2002. Friends of the Earth, 26–28 Underwood Street, London

Fu D, Teng Y, Luo Y, Tu C, Li S et al (2012) Effects of alfalfa and organic fertilizer on benzo[a] pyrene dissipation in an aged contaminated soil. Environ Sci Pollut Res Int 19:1605–1611

Fu JJ, Zhou QF, Liu JM, Liu W, Wang T, Zhang QH et al (2008) High levels of heavy metals in rice (Oryza sativa L.) from a typical E-waste recycling area in southeast China and its potential risk to human health. Chemosphere 71:1269–1275

Gadd GM (2001) Fungi in Bioremediation. Cambridge University Press, Cambridge, MA

Gai Z, Yu B, Li L, Wang Y, Ma C, Feng J, Deng Z, Xu P (2007) Cometabolic degradation of dibenzofuran and dibenzothiophene by a newly isolated carbazole-degrading *Sphingomonas* sp. strain. Appl Environ Microbiol 73:2832–2838

Gao D, Du L, Yang J, Wu W-M, Liang H (2010) A critical review of the application of white rot fungus to environmental pollution control. Crit Rev Biotechnol 30:70–77

Garbisu C, Alkorta I (2003) Basic concepts on heavy metal soil bioremediation. The European Journal of Mineral Processing and Environmental Protection 3(1):58–66

Garbisu C, Alkorta I, Llama MJ, Serra JL (1998) Aerobic chromate reduction by Bacillus subtilis. Biodegradation 9(2):133–141

Garbisu C, Ishii T, Leighton T, Buchanan BB (1997) Bacterial reduction of selenite to elemental selenium. Chem Geol 132:199–204

Gardea-Torresday JL, De la Rosa G, Peralta-Videa JR, Montes M, Cruz-Jiminez G, Cano-Aguilera I (2005) Differential uptake and transport of trivalent and hexavalent chromium by tumble wed (*Salsola kali*). Arch Environ Contam Toxicol 48:225–232

Garret RG (2002) Natural sources of cadmium. In Morrow H (ed) Sources of Cadmium in the Environment. OECD (Organization For Cooperation And Development) Proceedings (703) 759–7003

Gavrilescu M, Pavel LV, Cretescu I (2009) Characterization and remediation of soils contaminated with uranium. J Hazard Mater 163:475–510

Georgieva SS, McGrath SP, Hooper DJ, Chambers BS (2002) Nematode communities under stress: the long-termeffects of heavy metals in soil treated with sludge. Applied Soil Ecol 20:27–42

Gerhardt KE, Huang X-D, Glick BR, Greenberg BM (2009) Phytoremediation and rhizoremediation of organic soil contaminants: potential and challenges. Plant Sci 176:20–30

Gianfreda L, Bollag JM (2002) Isolated enzymes for the transformation and detoxification of organic pollutants. In: Burns RG, Dick R (eds) Enzymes in the environment: activity, ecology and applications. Marcel Dekker, New York

Gianfreda L, Rao M (2004) Potential of extra cellular enzymes in remediation of polluted soils: a review. Enzym Microb Technol 35:339–354

Gitipour S, Ahmadi S, Madadian E, Ardestani M (2011) Soil washing of chromium-and cadmium-contaminated sludge using acids and ethylenediaminetetra acetic acid chelating agent, Environ Technol, 37(1):1–7

Glazer AN, Nikaido H (2007) Microbial biotechnology: fundamentals of applied microbiology, 2nd edn. Cambridge University Press, Cambridge, MA

GOC (2003) Site remediation technologies: a reference manual. Contaminated Sites Working Group, Government of Canada, Ontario

Goi A, Trapido M, Kulik N (2009) Contaminated soil remediation with hydrogen peroxide oxidation. Int J Chemi Biol Eng 52:185–189

Gomes MP, Cristina T, Lanza L, Marques SM (2013) Cadmium effects on mineral nutrition of the cd-hyperaccumulator *Pfaffa glomerata*. Biologia 68:223–230

Gong Z, Wilke B–M, Alef K, Li P, Zhou Q (2006) Removal of polycyclic aromatic hydrocarbons from manufactured gas plant-contaminated soils using sunflower oil: laboratory column experiments. Chemosphere 62:780–787

Goyal MK, Chauhan A (2015) Environmental pollution remediation through solidification/fixation of heavy metal ions in Portland cement. J Environ Anal Toxicol 5:323

Graffham A (2006) EU legal requirements for imports of fruits and vegetables (a suppliers guide). Fresh Insights no. 1, DFID/IIED/NRI. www.agrifoodstandards.org

Greenberg BM (2006) Development and field tests of a multi-process phytoremediation system for decontamination of soils. Can Reclam 1:27–29

Greenberg BM, Huang X-D, Dixon DG, Glick BR (2005) An integrated multi-process phytoremediation system (MPPS) for removal of persistent organic contaminants from soil, Proceedings of the Eighth International In Situ and On-Site Bioremediation Symposium, Baltimore, Maryland. Batelle Press, Columbus

Greičiūtė K, Vasarevičius S (2007) Decontamination of heavy-metal polluted soil by electrokinetic remediation. Geologija 57:55–62

Grosser RJ, Warshawsky D, Vestal JR (1991) Indigenous and enhanced mineralization of pyrene, benzo[a]pyrene, and carbazole in soils. Appl Environ Microbiol 57:3462–3469

Gulz PA (2002) Arsenic uptake of common crop plants from contaminated soils and interaction with Phosphate. Ph. D. Thesis, Swiss Federal Institute of Technology, Zurich

Guo W, Li D, Tao Y, Gao P, Hu J (2008) Isolation and description of a stable carbazole-degrading microbial consortium consisting of *Chryseobacterium* sp. NCY and *Achromobacter* sp. NCW. Curr Microbiol 57:251–257

Habe H, Chung JS, Kato H, Ayabe Y, Kasuga K, Yoshida T, Nojiri H, Yamane H, Omori T (2004) Characterization of the upper pathway genes for fluorene metabolism in *Terrabacter* sp. strain DBF63. J Bacteriol 186:5938–5944

Habe H, Chung JS, Lee JH, Kasuga K, Yoshida T, Nojiri H, Omori T (2001) Degradation of chlorinated dibenzofurans and dibenzo-p-dioxins by two types of bacteria having angular dioxygenases with different features. Appl Environ Microbiol 67:3610–3617

Habe H, Ide K, Yotsumoto M, Tsuji H, Yoshida T, Nojiri H, Omori T (2002) Degradation characteristics of a dibenzofuran-degrader *Terrabacter* sp. strain DBF63 toward chlorinated dioxins in soil. Chemosphere 48:201–207

Hamscher G, Abu-quare S, Sczesny S, Hoper H, Nau H (2000) Determination of tetracyclines in soil and water samples from agricultural areas in lower saxony. Presented at EuroResidue IV, Veldhoven. 2000

Hansda A, Kumar Anshumali V, Usmani Z (2014) Phytoremediation of heavy metals contaminated soil using plant growth promoting rhizobacteria (PGPR): a current perspective. Recent Res Sci Technol 6(1):131–134

Harrison EZ, Oakes SR, Hysell M, Hay A (2006) Organic chemicals in sewage sludges. Sci Total Environ 367:481–497

Hedlund BP, Geiselbrecht AD, Staley JT (2001) *Marinobacter* strain NCE312 has a *Pseudomonas*-like naphthalene dioxygenase. FEMS Microbiol Lett 201:47–51

Hilber I, Bucheli TD (2010) Activated carbon amendment to remediate contaminated sediments and soils: a review. Global NEST J 12(3):305–317

Hirano T, Tamae K (2011) Earthworms and soil pollutants. Sensors (Basel) 11(12):11157–11167

Hobbelen PHF, Koolhaas JE, Van Gestel CAM (2006) Effects of heavy metals on the litter decomposition by the earthworm *Lumbricus rubellus*. Pedobiologia 50:51–60

Hogarh JN, Fobil JN, Ofosu-Budu GK, Carboo D, Ankrah NA, Nyarko A (2008) Assessment of heavy metal contamination and macronutrient content of composts for environmental pollution control in Ghana. Global Journal of Environmental Research 2(3):133–139

HPA (2008) HPA compendium of chemical hazards: Dioxins (2,3,7,8-Tetrachlorodibenzo-p-dioxin). CHAPD HQ, HPA 2008, Version 1. Chilton: Health Protection Agency. Available at http://www.extension.umn.edu/distribution.horticulture/DG2543html

Hu PJ, Qiu RL, Senthilkumar P, Jiang D, Chen ZW, Tang YT, Liu FJ (2009) Tolerance, accumulation and distribution of zinc and cadmium in hyperaccumulator *Potentilla griffithii*. Environ Exp Bot 66:317–325

Huang D, Xu Q, Cheng J, Lu X, Zhang H (2012) Electrokinetic remediation and its combined technologies for removal of organic pollutants from contaminated soils. Int J Electrochem Sci 7:4528–4544

Huq SMI, Rahman A, Sultana N, Naidu R (2003) Extent and severity of arsenic contamination in soils of Bangladesh. In: Feroze AM, Ashraf AM, Adeel Z (eds) Fate of arsenic in the nvironment. ITN Centre, Dhaka, Bangladesh

Hussein H, Farag S, Kandil K, Moawad H (2005) Resistance and uptake of heavy metals by pseudomonas. Process Biochem 40:955–961

IAEA (1999) Technologies for remediation of radioactively contaminated sites. Waste Technology Section, International Atomic Energy Agency, Vienna

IAEA (2003) Extent of Environmental Contamination by Naturally Occurring Radioactive Material (NORM) and Technological Options for Mitigation, Technical Reports Series No. 419, International Atomic Energy Agency, Vienna

Iannelli R, Masi M, Ceccarini A, Ostuni MB, Lageman R, Muntoni A, Spiga D, Polettini A, Marini A, Pomi R (2015) Electrokinetic remediation of metal-polluted marine sediments: experimental investigation for plant design. Electrochim Acta 181:146–159

Igwe JC, Nnorom IC, Gbaruko BC (2005) Kinetics of radionuclides and heavy metals behaviour in soils: implications for plant growth. Afr J Biotechnol 4(13):1541–1547

International Committee of the Red Cross (2011) Management of medical wastes. ICRC, Geneva

Inui H, Shiota N, Motoi Y, Ido Y, Inoue T, Kodama T, Ohkawa Y, Ohkawa H (2001) Metabolism of herbicides and other chemicals in human cytochrome P450 species and in transgenic potato plants co-expressing human CYP1A1, CYP2B6 and CYP2C19. J Pest Sci 26:28–40

Iwamoto T, Nasu M (2001) Current bioremediation practice and perspective. J Biosci Bioeng 92(1):1–8

IWW (2014) Global occurrence of pharmaceuticals in the environment: results of a global database of Measured Environmental Concentrations (MEC), Presentation by Weber FA on the Geneva conference on Pharmaceuticals in the Environment, 8th–9th April 2014, IWW Water Centre, Germany

Jaffré T, Brooks RR, Lee J, Reeves RD (1976) Sebertia acuminata: a Hyperaccumulator of nickel from New Caledonia. Science 193(4253):579–580

Jhee EM, Boyd RS, Eubanks MD (2005) Nickel hyperaccumulation as an elemental defense of *Streptanthus polygaloides* (Brassicaceae): influence of herbivore feeding mode. New Phytol 168:331–343

Jiang W, Tao T, Liao Z (2011) Removal of heavy metal from contaminated soil with chelating agents. Open Journal of Soil Science 1:70–76

Juhasz AL, Stanley GA, Britz ML (2000) Microbial degradation and detoxification of high molecular weight polycyclic aromatic hydrocarbons by *Stenotrophomonas maltophilia* strain VUN 10,003. Lett Appl Microbiol 30:396–401

Juhasz AL, Stanley GA, Britz ML (2002) Metabolite repression inhibits degradation of benzo[*a*]pyrene and dibenz[*a,h*]anthracene by *Stenotrophomonas maltophilia* VUN 10,003. J Ind Microbiol Biotechnol 28:88–96

Kabata-Pendias A, Mukherjee AB (2007) Trace elements from soil to human. Springer-Verlag, Berlin

Kabata-Pendias A, Pendias H (2000) Trace elements in soil and plants. CRC Press, Boca Raton

Kabata-Pendias A, Pendias H (2001) Trace elements in soils and plants, 3rd edn. CRC Press, Boca Raton

Kanaly RA, Harayama S, Watanabe K (2002) *Rhodanobacter* sp. strain BPC-1 in a benzo[*a*]pyrene-mineralizing bacterial consortium. Appl Environ Microbiol 68:5826–5833

Kang JW (2014) Removing environmental organic pollutants with bioremediation and phytoremediation. Biotechnol Lett 36:1129–1139

Karthika N, Jananee K, Murugaiyan V (2016) Remediation of contaminated soil using soil washing-a review. Int J Eng Res Appl 6(1) (Part - 2): 13–18

Kawahigashi H, Hirose S, Ohkawa H, Ohkawa Y (2007) Herbicide resistance of transgenic rice plants expressing human CYP1A1. Biotechnol Adv 25:75–84

Kay P, Boxsall AB (2000) Environmental risk assessment of veterinary medicines in slurry; SSLRC contract JF 611OZ. Cranfield University, Cranfield, UK

Kazunga C, Aitken MD (2000) Products from the incomplete metabolism of pyrene by polycyclic aromatic hydrocarbon-degrading bacteria. Appl Environ Microbiol 66:1917–1922

Keeling SM, Stewart RB, Anderson CW, Robison BH (2003) Nickel and cobalt phytoextraction by the hyperaccumulator *Berkheya coddii*: implications for polymetallic phytomining and phytoremediation. Int J Phytoremediation 5:235–244

Kelly JJ, Haggblom MM, Tate RL (2003) Effects of heavy metal contamination and remediation on soil microbial communities in the vicinity of a zinc smelter as indicated by analysis of microbial community phospholipids fatty acid profiles. Biol Fertil Soils 38:65–71

Kemper N (2008) Veterinary antibiotics in the aquatic and terrestrial environment. Ecol Indic 8:1–13

Kensa VM (2011) Bioremediation – an overview. J Industrial Pollution Control 27(2):161–168

Kent AD, Triplett EW (2002) Microbial communities and their interactions in soil and rhizosphere ecosystems. Annu Rev Microbiol 56:211–236

Khan S, Cao Q, Zheng YM, Huang YZ, Zhu YG (2008) Health risks of heavy metals in contaminated soils and food crops irrigated with wastewater in Beijing, China. Environ Pollut 152(3):686–692

Kim D, Park MJ, Koh SC, So JS, Kim E (2002) Three separate pathways for the initial oxidation of limonene, biphenyl, and phenol by Rhodococcus sp strain t104. J Microbiol 40:86–89

Kim YH, Freeman JP (2005) Effects of pH on the degradation of phenanthrene and pyrene by *Mycobacterium vanbaalenii* PYR-1. Appl Microbiol Biotechnol 67:275–285

Kim SJ, Kweon O, Freeman JP, Jones RC, Adjei MD, Jhoo JW, Edmondson RD, Cerniglia CE (2006) Molecular cloning and expression of genes encoding a novel dioxygenase involved in low- and high-molecular-weight polycyclic aromatic hydrocarbon degradation in Mycobacterium vanbaalenii PYR-1. Appl Environ Microbiol 72:1045–1054

King-Heiden TC, Mehta V, Xiong KM, Lanham KA, Antkiewicz DS Ganser A, Peterson RE (2012) Reproductive and developmental toxicity of dioxin in fish. Mol Cell Endocrinol 354:121–138

Kirpichtchikova TA, Manceau A, Spadini L, Panfili F, Marcus MA, Jacquet T (2006) Speciation and solubility of heavy metals in contaminated soil using X-ray microfluorescence, EXAFS spectroscopy, chemical extraction, and thermodynamic modeling. Geochim Cosmochim Acta 70(9):2163–2190

Kjøller AH, Struwe S (2002) Fungal communities, succession, enzymes, and decomposition. In: Burns RG, Dick RP (eds) Enzymes in the environment: activity, ecology and applications. Marcel Dekker, New York

Knox AS, Seamans JC, Mench MJ, Vangronseveld J (2000) Remediation of metals and radionuclides: contaminated soil using in situ stabilization techniques. Macmillan Publishers, New York

Koller K, Brown T, Spurgeon A, Levy L (2004) Recent developments in low-level lead exposure and intellectual impairment in children. Environ Health Perspect 112:987–994

Kolpin DW, Riley D, Meyer MT, Weyer P, Thurman EM (2000) Pharm-Chemical contamination: Recconnaissance for antibiotics in Iwoa streams 1999. Proceeding, effects of animal feeding operations on water resources and the environment. Wilde FD, Britton LJ, Miller CV, Kolpin DW (eds) US Geological Survey Open File Report 00-204, US Geological Survey, Reston, Virginia

Kramer U (2010) Metal hyperaccumulation in plants. Annu Re Plant Biol 61:517–534

Kubota H, Takenaka C (2003) *Arabis gemmifera* is a hyperaccumulator of cd and Zn. Int J Phytoremediation 5:197–201

Kuiper I, Lagendijk EL, Bloemberg GV, Lugtenberg BJ (2004) Rhizoremediation: a beneficial plant-microbe interaction. Mol Plant-Microbe Interact 17:6–15

Kupper H, Mijovilovich A, Gotz B, Kupper FC, Meyer-Klaucke W (2009) Complexation and toxicity of copper in higher plants (I): characterisation of copper accumulation, speciation and toxicity in *Crassula helmsii* as a new copper hyperaccumulator. Plant Physiol 151:702–714

Küpper H, Parameswaran A, Leitenmaier B, Trtilek M, Setlik I (2007) Cadmium induced inhibition of photosynthesis and long-term acclimation to cadmium stress in the hyperaccumulator *Thlaspi caerulescens*. New Phytol 175:655–674

Lanno RP, Wells J, Conder JM, Bradham K, Basta N (2004) The bioavailability of chemicals in soil for earthworms. Ecotoxicol Environ Saf 57(1):39–47

Leblanc M, Petit D, Deram A, Robinson B, Brooks RR (1999) The phytomining and environmental significance of hyperaccumulation of thallium by *Iberis intermedia* from southern France. Econ Geol 94:109–113

Lee SE, Seo JS, Keum YS, Lee KJ, Li QX (2007) Fluoranthene metabolism and associated proteins in *Mycobacterium* sp. JS14. Proteomics 7:2059–2069

Lestan D (2006) Enhanced heavy metal phytoextraction. In: Phytoremediation Rhizoremediation. Springer, The Netherlands, pp 115–132

Levine et al. 2006 cited from Mbhele PP (2007) Remediation of soil and water contaminated by heavy metals and hydrocarbons using silica encapsulation. M. S. Thesis, University of the Witwatersrand, Johannesburg

Li L, Xu ZR, Zhang C, Bao J, Dai X (2012) Quantitative evaluation of heavy metals in solid residues from sub- and super-critical water gasification of sewage sludge. Bioresour Technol 121:169–175

Li LY (2007) Remediation treatment technologies: reference guide for developing countries facing persistent organic pollutants. Published by United Nations Industrial Development Organization (UNIDO), Vienna

Li Q, Jiang Y, Liang WJ (2006) Effect of heavy metals on soil nematode communities in the vicinity of a metallurgical factory. J Environ Sci 18:323–328

Liang W, Li Q, Zhang X, Jiang S, Jiang Y (2006) Effect of heavy metals on soil nematode community structure in Shenyang suburbs. American-urasian J Agric & Environ Sci 1(1):14–18

Lombi E, Zhao FJ, Dunham SJ, McGrath SP (2001) Phytoremediation of heavy-metal contaminated soils: natural hyperaccumulation versus chemically enhanced phytoextraction. J Environ Qual 30:1919–1926

Lu L, Tian S, Zhang J, Yang X, Labavitch JM, Webb SM, Latimer M, Brown PH (2013) Efficient xylem transport and phloem remobilization of Zn in the hyperaccumulator plant species *Sedum alfredii*. New Phytol 198:721–731

Lu XC, Chen LH, Xu Q, Bi SP, Zheng Z (2005) China J Environ Sci (Chinese) 25:89–91

Luo YM (2009) Current Research and Development in soil remediation technologies. Progress in. Chemistry 21:558–565

Ma Y, Prasad MNV, Rajkumar M, Freitas H (2011) Plant growth promting rhizobacteria and endophytes accelerate phytoremediation of metalliferous soil. Biotechnol Adv 29:248–258

Madyiwa S, Chimbari M, Nyamangara J, Bangira C (2002) Cumulative effects of sewage sludge and effluent mixture application on soil properties of sandy soil under a mixture of star and kikuyu grasses Zimbabwe. Phys Chem Earth 27:747–753

Mahajan MC, Phale PS, Vaidyanathan CS (1994) Evidence for the involvement of multiple pathways in the biodegradation of 1- and 2-methylnaphthalene by *Pseudomonas putida* CSV86. Arch Microbiol 161:425–433

Mahendran RP (2014) Phytoremediation – insights into plants as remedies. Malaya J Biosci 1(1):41–45

Mallampati SR, Mitoma Y, Okuda T, Sakita S, Kakeda M (2013) Novel approach for the remediation of radioactive cesium contaminated soil with nano-Fe/Ca/CaO dispersion mixture in dry condition. E3S Web of Conferences 1, 08003 (2013), DOI: https://doi.org/10.1051/e3sconf/20130108003., http://www.e3s-conferences.org

Mallick S, Chatterjee S, Dutta TK (2007) A novel degradation pathway in the assimilation of phenanthrene by *Staphylococcus* sp. strain PN/Y via *meta*-cleavage of 2-hydroxy-1-naphthoic acid: formation of trans-2,3-dioxo-5-(2′-hydroxyphenyl)-pent-4-enoic acid. Microbiology 153:2104–2115

Marcel van der Perk M (2013) Soil and water contamination, 2nd edn. CRC Press, Hoboken

Marcel van der Perk M (2006) Soil and water contamination. Taylor and Francis, London

Marco-Urrea E, Reddy CA (2012) Degradation of Chloro-organic pollutants by white rot fungi. In: Singh SN (ed) Microbial degradation of xenobiotics, environmental science and engineering. Springer-Verlag, Berlin Heidelberg

Markus J, McBratney AB (2001) A review of the contamination of soil with lead II. Spatial distribution and risk assessment of soil lead. Environ Int 27(5):399–411

Marques APGC, Rangel AOSS, Castro PML (2009) Remediation of heavy metal contaminated soils: phytoremediation as a potentially promising clean-up technology. Crit Rev Environ Sci Technol 39(8):622–654

Masayuki S, Aya W, Masahiro I, Sakae S, Toshikazu K (2007) Phytoextraction and phytovolatilization of arsenic from as-contaminated soils by Pteris vittata. Proceedings of the Annual International Conference on Soils, Sediments, Water and Energy: Vol. 12, Article 26

Master ER, Lai VWM, Kuipers B, Cullen WR, Mohn WM (2002) Sequential anaerobic-aerobic treatment of soil contaminated with weathered aroclor 1260. Environ Sci Technol 36:100–103

Master ER, Mohn WW (2001) Induction of bphA, encoding biphenyl dioxygenase, in two polychlorinated biphenyl-degrading bacteria, psychrotolerant pseudomonas strain cam-1 and mesophilic Burkholderia strain lb400. Appl Environ Microbiol 67:2669–2676

Mata-Sandoval J, Karns J, Torrent A (2002) Influence of rhamnolipids and triton X-100 on the desorption of pesticides from soils. Environ Sci Technol 36:4669–4675

Matschullat J (2000) Arsenic in the geosphere--a review. Sci Total Environ 249:297–312

References

Matsubara T, Ohshiro T, Nishina Y, Izumi Y (2001) Purification, characterization, and overexpression of flavin reductase involved in dibenzothiophene desulfurization by *Rhodococcus erythropolis* D-1. Appl Environ Microbiol 67:1179–1184

McGrath SP, Zhao FJ, Dunham SJ, Crosland AR, Coleman K (2000) Long-term changes in extractability and bioavailability of zinc and cadmium after sludge application. J Environ Qual 29:875–883

McGrath SP, Zhao FJ (2003) Phytoextraction of metals and metalloids from contaminated soils. Curr Biotechnol 14:277–282

McGowen SL, Basta NT, Brown GO (2001) Use of diammonium phosphate to reduce heavy metal solubility and transport in smelter-contaminated soil. J Environ Qual 30:493–500

Meagher RB, Rugh CL, Kandasamy MK, Gragson G, Wang NJ (2000) Engineered phytoremediation of mercury pollution in soil and water using bacterial genes. In: Terry N, Bañuelos G (eds) Phytoremediation of contaminated soil and water. Lewis Publishers, Boca Raton, pp 201–219

Megharaj M, Ramakrishnan B, Venkateswarlu K, Sethunathan N, Naidu R (2011) Bioremediation approaches for organic pollutants: a critical perspective. Environ Int 37(8):1362–1375

Mengoni A, Gonnelli C, Hakvoort HWJ, Galardi F, Bazzicalupo M, Gabbrielli R, Schat H (2003) Evolution of copper-tolerance and increased xpression of a 2b-type metallothionein gene in *Silene paradoxa* L. populations. Plant Soil 257:451–457

Mihalik J, Henner P, Frelon S, Camilleri V, Fevrier L (2012) Citrate assisted phytoextraction of uranium by sunflowers: study of fluxes in soils and plants and resulting intra-planta distribution of Fe and U. Environ Exp Bot 77:249–258

Miller CD, Hall K, Liang Y-N, Nieman K, Sorensen D, Issa B, Anderson AJ, Sims RC (2004) Isolation and characterization of polycyclic aromatic hydrocarbon-degrading mycobacterium isolates from soil. Microb Ecol 48:230–238

Millward RN, Bridges TS, Ghosh U, Zimmerman JR, Luthy RG (2005) Addition of activated carbon to sediments to reduce PCB bioaccumulation by a polychaete (*Neanthes arenaceodentata*) and an amphipod (*Leptocheirus plumulosus*). Environ Sci Technol 39:2880–2837

Mohammed FA (2009) Pollution caused by vehicle exhausts and oil trash burning in Kirkuk city. Iraqi Journal of Earth Sciences 9(2):39–48

Momani KA, Jiries AG, Jaradat QM (2000) Atmospheric deposition of Pb, Zn, cu, and cd in Amman, Jordan. Turk J Chem 24:231–237

Mongkhonsin B, Nakbanpote W, Nakai I, Hokura A, Jiyaranaikoon N (2011) Distribution and speciation of chromium accumulated in *Gynura pseudochina* (L.) DC. Environ Exp Bot 74:56–64

Montemurro F, Convertini G, Ferri D, Maiorana M (2005) MSW compost application on tomato crops in Mediterranean conditions: effects on agronomic performance and nitrogen utilization. Compost Science & Utilization 13(4):234–242

Moon DH, Park J-W, Koutsospyros A, Cheong KH, Chang Y-Y, Baek K, Jo R, Park J-H (2016) Assessment of soil washing for simultaneous removal of heavy metals and low-level petroleum hydrocarbons using various washing solutions. Environ Earth Sci 75:884

Moradi AB, Swoboda S, Robinson B, Prohaska T, Kaestner A, Oswald SE, Wenzel WW, Schulin R (2010) Mapping of nickel in root cross-sections of the hyperaccumulator plant *Berkheya coddii* using laser ablation ICP–MS. Environ Exp Bot 69:24–31

Moreno JL, Hernandez T, Garcia C (1999) Effects of cadmium - contaminated sewage sludge compost on dynamics of organic matter and microbial activity in an aird soil. Biol Fertil Soils 28:230

Mougin C, Kollmann A, Jolivalt C (2002) Enhanced production of laccase in thefungus Trametes versicolor by the addition of xenobiotics. Biotechnol Lett 24:139–142

Mudgal V, Madaan N, Mudgal A (2010a) Heavy metals in plants: phytoremediation: plants used to remediate heavy metal pollution. Agric Biol J North America Science Huβ. http://www.scihub.org/abjna

Mudgal V, Madaan N, Mudgal A, Singh RB, Mishra S (2010b) Effect of toxic metals on human health. The Open Nutraceuticals Journal 3:94–99

Muegge J (2008) An assessment of zero valence iron permeable reactive barrier projects in California. Document No 1229. Office of Pollution Prevention and Technology Development, California

Mueller JG, Chapman PJ, Blattmann BO, Pritchard PH (1990) Isolation and characterization of a fluoranthene-utilizing strain of *Pseudomonas paucimobilis*. Appl Environ Microbiol 56:1079–1086

Mulligan CN, Yong RN, Gibbs BF (2001) Surfactant-enhanced remediation of contaminated soil: a review. Eng Geol 60:371–380

Nam IH, Kim YM, Schmidt S, Chang YS (2006) Biotransformation of 1,2,3-tri- and 1,2,3,4,7,8-hexachlorodibenzo-p- dioxin by *Sphingomonas wittichii* strain RW1. Appl Environ Microbiol 72:112–116

Naranjo L, Urbina H, De Sisto A, Leon V (2007) Isolation of autochthonous non-white rot fungi with potential for enzymatic degrading of Venezuelan extra-heavy crude oil. Biocatal Biotransformation 25(2–4):341–349

Navarro S, Vela N, Navarro G (2007) Review. An overview on the environmental behaviour of pesticide residues in soils. Span J Agric Res 5(3):357–375

Navas A, Flores-Romero P, Sánchez-Moreno S, Camargo JA, McGawley EC (2010) Effects of heavy metal soil pollution on nematode communities after the Azancollar mining spill. Nematropica 40(1):13–28

Nicell JA (2001) Environmental applications of enzymes. Interdisc. Environ Rev 3:14–41

Nichols GE et al (2014) Phytoremediation of a petroleum-hydrocarbon contaminated Shallow Aquifer in Elizabeth city. Wiley online library, North Carolina

Nicholson FA, Chambers BJ, Williams JR, Unwin RJ (1999) Heavy metal contents of livestock feeds and animal manures in England and Wales. Bioresour Technol 70:23–31

Njoku KL, Akinola MO, Oboh BO (2009) Phytoremediation of crude oil contaminated soil: the effect of growth of *Glycine max* on the physico-chemistry and crude oil contents of soil. Nature and Science 7(10):79–86

Nkansah MA, Amoako CO (2010) Heavy metal content of some common spices available in markets in the Kumasi metropolis of Ghana. Am J Sci Ind Res 1(2):158–163

Nwoko CW (2010) Trends in phytoremediation of toxic elemental and organic pollutants. Afr J Biotechnol 9(37):6010–6016

O'Mahony M, Dobson A, Barnes J, Singleton I (2006) The use of ozone in the remediation of polycyclic aromatic hydrocarbon contaminated soil. Chemosphere 63:307–314

Onianwa PC, Odukoya OO, Alabi HA (2002) Chemical composition of wet precipitation in Madan, Nigeria. Bull Chem Soc Ethiop 16(2):41–147

Orser CS, Salt DE, Pickering IJ, Prince R, Epstein A, Ensley BD (1999) *Brassica* plants to provide enhanced human mineral nutrition: selenium phytoenrichment and metabolic transformation. J Med Food 1:253–261

Osman KT, Kashem MA (2016) Phytoremediation of soil. In: Lal R (ed) Encyclopedia of soil science, 3rd edn. Taylor and Francis. https://doi.org/10.1081/E-ESS3-120053533

Ostermann A, Gao J, Welp G, Siemens J, Roelcke M, Heimann L, Nieder R, Xue Q, Lin X, Sandhage-Hofmann A, Amelung W (2014) Identification of soil contamination hotspots with veterinary antibiotics using heavy metal concentrations and leaching data – a field study in China. Environ Monit Assess 186:1–15

Oustan S, Heidari S, Neyshabouri MR, Reyhanitabar A, Bybordi A (2011) Removal of heavy metals from a contaminated calcareous soil using oxalic and acetic acids as chelating agents. 2011 International Conference on Environment Science and Engineering IPCBEE vol.8 (2011) IACSIT Press, Singapore

Oven M, Grill E, Golan-Goldhirsh A, Kutchan TM, Zenk MH (2002) Increase of free cysteine and citric acid in plant cells exposed to cobalt ions. Phytochemistry 60:467–474

Page V, Schwitzguébel JP (2009) The role of cytochromes P450 and peroxidases in the detoxification of sulphonated anthraquinones by rhubarb and common sorrel plants cultivated under hydroponic conditions. Environ Sci Pollut Res Int 16:805–816

Paulsrud B, Wien A, Nedland KT (2000) A survey of toxc organics in Norwegian sewage sludge, compost and manure. Aquateam, Norwegian Water Technology Centre ASOSLO

Pavel LV, Gavrilescu M (2008) Overview of ex situ decontamination techniques for soil cleanup. Environ Eng Manag J 7(6):815–834

Pawlowska TE, Charvat I (2004) Heavy metal stress and developmental patterns of arbuscular mycorrhizal fungi. Appl Environ Microbiol 70(11):6643–6649

Pedro JS, Valente S, Padilha PM, Florentino AO (2006) Studies in the adsorption and kinetics of photodegradation of a model compound for heterogeneous photocatalysis onto TiO_2. Chemosphere 64:1128–1133

Pensaert S, De Groeve S, Staveley C, De Puydt S (2008) Immobilisation, stabilisation, solidification: a new approach for the treatment of contaminated soils. Case studies: London Olympics & Total Ertvelde. 15de Innovatieforum Geotechniek – 8 oktober 2008, Belgium

Perelo LW (2010) In situ and bioremediation of organic pollutants in aquatic sediments. J Hazard Mater 177(1–3):81–89

Persans MW, Nieman K, Salt DE (2001) Functional activity and role of cation-efflux family members in Ni hyperaccumulation in *Thlaspi geosingense*. Proc Nat Acad Sci USA 98:9995–10000

Petruzzelli G, Gorini F, Pezzarossa B, Pedron F (2010) The fate of pollutants in soil. In: Bianchi F, Cori L, Moretti PF (eds) CNR Environment and Health Inter-departmental Project. Consiglio Nazionale delle Ricerche – Roma

Pierpoint AC, Hapeman CJ, Torrents A (2003) Ozone treatment of soil contaminated with aniline and trifluralin. Chemosphere 50:1025–1034

Pinyakong O, Habe H, Supaka N, Pinpanichkarn P, Juntongjin K, Yoshida T (2000) Identification of novel metabolites in the degradation of phenanthrene by *Sphingomonas* sp. strain P2. FEMS Microbiol Lett 191:115–121

Pointing SB (2001) Feasibility of bioremediation by white rot fungi. Appl Microbiol Biotechnol 57:20–33

Pokhrel D, Viraraghavan T (2005) Municipal solid waste management in Nepal: practices and challenges. Waste Manag 25(5):555–562

Pourrut B, Shahid M, Dumat C, Winterton P, Pinelli E (2011) Lead uptake, toxicity, and detoxification in plants. Rev Environ Contam Toxicol 213:113–136

Prescott LM, Harley JP, Klein DA (2002) Microbiology, 5th edn. McGraw-Hill, New York

Pumphrey GM, Madsen EL (2007) Naphthalene metabolism and growth inhibition by naphthalene in *Polaromonas naphthalenivorans* strain CJ2. Microbiology 153:3730–3738

Rababah A, Matsuzawa S (2002) Treatment system for solid matrix contaminated with fluoranthene. I – modified extraction technique. Chemosphere 46(1):39–47

Ramirez N, Cutright T, Ju LK (2001) Pyrene biodegradation in aqueous solutions and soil slurries by *Mycobacterium* PYR-1 and enriched consortium. Chemosphere 44:1079–1086

Rao MA, Scelza R, Scotti R, Gianfreda L (2010) Role of enzymes in the remediation of polluted environments. J Soil Sci Plant Nutr 10(3):333–353

Rascio N, Navari-Izzo F (2011) Heavy metal hyperaccumulating plants: how and why do they do it? And what makes them so interesting? Plant Sci 180:169–181

Raskin I, Ensley BD (2000) Phytoremediation of toxic metals: using plants to clean up the environment. John Wiley & Sons, New York

Raychoudhury T, Scheytt T (2013) Potential of zerovalent iron nanoparticles for remediation of environmental organic contaminants in water: a review. Water Sci Technol 68(7):1425–1439

Reddy CA, Mathew Z (2001) Bioremediation potential of white rot fungi. In: Gadd GM (ed) Fungi in bioremediation. Cambridge University Press, London

Reddy KR (2010) Nanotechnology for site remediation: Dehalogenation of organic pollutants in soils and groundwater by nanoscale iron particles. 6th International Congress on Environmental Geotechnics, New Delhi

Reddy KR (2013) Electrokinetic remediation of soils at complex contaminated sites: technology status, challenges, and opportunities. In: Manassero et al (eds) Coupled phenomena in environmental geotechnics. Taylor & Francis Group, London

Reeves RD (2006) Hyperaccumulation of trace elements by plants. In: Morel JL, Echevarria G, Goncharova N (eds), Phytoremediation of metal-contaminated soils, NATO Science Series: IV: Earth and environmental sciences, Springer, New York

Reeves RD, Adiguzel N, Baker AJM (2009) Nickel hyperaccumulation in *Bornmuellera kiyakii* and associated plants of the Brassicaceae from Kizildaay Derebucak (Konya), Turkey. Turk J Bot 33:33–40

Reeves RD, Baker AJM (2000) Metal-accumulating plants. In: Raskin I, Ensley BD (eds) Phytoremediation of toxic metals: using plants to clean up the environment. John Wiley, New York

Rentz JA, Alvarez PJJ, Schnoor JL (2008) Benzo[a]pyrene degradation by *Sphingomonas yanoikuyae* JAR02. Environ Pollut 151:669–677

Resnick SM, Gibson DT (1996) Regio- and stereospecific oxidation of fluorene, dibenzofuran, and dibenzothiophene by naphthalene dioxygenase from *Pseudomonas* sp. strain NCIB 9816-4. Appl. Environ Microbiol 62:4073–4080

Romero MC, Cazau MC, Giorgieri S, Arambarri AM (1998) Phenanthrene degradation by microorganisms isolated from a contaminated stream. Environ Pollut 101:355–359

Roper JC, Dec J, Bollag J (1996) Using minced horseradish roots for the treatment of polluted waters. J Environ Qual 25:1242–1247

Rubilar O, Feijoo G, Diez MC, LuChau TA, Moreira MT, Lema JM (2007) Biodegradation of pentachlorophenol in soil slurry cultures by Bjerkandera adusta and Anthracophyllum discolor. Indust Engin Chem Res 46:744–751

Sahi SV, Bryant NL, Sharma NC, Singh SR (2002) Characterization of a lead hyperaccumulator shrub, *Sesbania drummondii*. Environ Sci Technol 36:4676–4680

Saichek RE, Reddy KR (2005) Electrokinetically enhanced remediation of hydrophobic organic compounds in soils: a review. Crit Rev Environ Sci Technol 35:115–192

Santos EE, Lauria DC, Porto da Silveira CL (2004) Assessment of daily intake of trace elements due to consumption of foodstuffs by adult inhabitants of Rio de Janeiro city. Sci Total Environ 327:69–79

Saruhan V, Gul I, Aydin I (2010) The effects of sewage sludge used as fertilizer on agronomic and chemical features of bird's foot trefoil *(Lotus corniculatus* L.) and soil pollution. Sci Res Essays 5(17):2567–2573

Saveyn H, Eder P (2014) End-of-waste criteria for biodegradable waste subjected to biological treatment (compost & digestate): Technical proposal. EC JRC Scientific and Policy Reports (EUR 26425 EN)

Schneider J, Grosser R, Jayasimhulu K, Xue W, Warshawsky D (1996) Degradation of pyrene, benz[a]anthracene, and benzo[a]pyrene by *Mycobacterium* sp. strain RJGII-135, isolated from a former coal gasification site. Appl Environ Microbiol 62:13–19

Schnoor JL, Licht LA, McCutcheon SC, Wolfe NL, Carreira LH (1995) Phytoremediation of organic and nutrient contaminants. Environ Sci Technol 29:318–323

Schoof RA (2003) Guide for incorporating bioavailability adjustments into human health and ecological risk assessments part 1: overview of metals bioavailability, Tri-Service Ecological Risk Assessment, NFESC, AFCEE, AEC, USA

Schroder P, Harvey PJ, Scwitzguebel JP (2002) Prospects for the phytoremediation of organic pollutants in Europe. Environ Sci Pollut Res 9(1):1–3

Schwitzguébel JP, van der Lelie D, Baker A, Glass DJ, Vangronsveld J (2002) Phytoremediation: European and American trends. J Soils Sediments 2(2):91–99

Semple KT, Morriss AWJ, Paton GI (2003) Bioavailability of hydrophobic organic contaminants in soils: fundamental concepts and techniques for analysis. Eur J Soil Sci 54:809–818

Seo JS, Keum YS, Li QX (2009) Bacterial degradation of aromatic compounds. Int J Environ Res Public Health 6(1):278–309

Sepic E, Leskovsek H (1999) Isolation and identification of fluoranthene biodegradation products. Analyst 124:1765–1769

Shah K, Nongkynrih JM (2007) Metal hyperaccumulation and bioremediation. Biol Plant 51(4):618–634

Sharma HD, Reddy KR (2004) Geoenvironmental engineering: site remediation, waste containment, and emerging waste management technologies. John Wiley, Hoboken

Shi C, Fernandez-Jimenez A (2006) Stabilization/solidification of hazardous and radioactive wastes with alkali-activated cements. J Hazard Mater B137:1656–1663

Shukla KP, Singh NK, Sharma S (2010) Bioremediation: developments, current practices and perspectives. Gen Eng Biotechnol J 3:1–20

Siegel FR (2002) Environmental geochemistry of potentially toxic metals. Springer-Verlag, Heidelberg
Sikka SC, Wang R (2008) Endocrine disruptors and estrogenic effects on male reproductive axis. Asian J Androl 10:134
Singer AC, Bell T, Heywood CA, Smith JAC, Thompson IP (2007) Phytoremediation of mixed contaminated soil using the hyperaccumulator plant *Alyssum lesbiacum*: evidence of histidine as a measure of phytoextractable nickel. Environ Pollut 147:74–82
Singer AC, Crowley DE, Thompson IP (2003) Secondary plant metabolites in phytoremediation and biotransformation. Trends Biotechnol 21:123–130
Singh H (2006) Mycoremediation: fungal bioremediation. John Wiley & Sons, London
Singh S, Jawaid SMA, Deep S (2014) Heavy metal removal from contaminated soil by soil washing – a review. GJESR Review Paper 1(8):11–15
Sinha RK, Valani D, Sinha S, Singh S, Heart S (2009) Bioremediation of contaminated sites: a low cost nature's biotechnology for environmental clean up by versatile microbes, plants and earthworms. In: Faerber T, Herzog J (eds) Solid waste management and environmental remediation. Nova Science Publishers Inc, Hauppauge, N.Y
Siuta J (1999) Sposoby przyrodniczego uzytkowania osadow sciekowych. In: Siuta J, Miluniec R (eds) Przyrodnicze uzytkowanie osadow sciekowych, Swinoujscie
Sivaci A, Elmas E, Gumu F, Sivaci ER (2008) Removal of cadmium by *Myrophyllum heterophyllum* Michx and *Potamogeton crispus* L. and its effect on pigments and total phenolic compounds. Arch Environ Contam Toxicol 54:612–618
Smol M, Włodarczyk-Makuła M, Włóka D (2014) The effectiveness adsorption of carcinogenic PAHs on mineral and on organic sorbents. Zeszyty Naukowe Wyższej Szkoły Zarządzania Ochrona Pracy W Katowicach 1(10):5–16
Smolders E (2001) Cadmium uptake by plants. Int J Occup Med Environ Health 14(2):177–183
Stegmann R, Brunner G, Calmano W, Matz G (2001) Treatment of contaminated soil: fundamentals, analysis, applications. Springer, New York
Stojanovic M, Lopicic Z, Mihajlovic M, Petrovic M, Radulovic D, Milojkovic J (2013) New uranium remediation approach based on mineral row materials and phytoaccumulator. Acta Technica Coviniensis – Bulletin of Engineering, Fascicule 3:31–35
Sun HQ, Liu GX, Wei ZH, Yang WW (2014) Removal of heavy metals from contaminated soils by washing with citric acid and subsequent treatment of soil-washing solutions. Adv Mater Res 937:646–651
Sun R, Jin C, Zhou Q (2010) Characteristics of cadmium accumulation and tolerance in *Rorippa globosa* (Turcz.) Thell., a species with some characteristicsof cadmium hyperaccumulation. Plant Growth Regul 61:67–74
Sun Y, Zhou Q, Wang L, Liu W (2009) Cadmium tolerance and accumulation characteristics of *Bidens pilosa* L. as a potential cd-hyperaccumulator. J Hazard Mater 161:808–814
Suresh B, Ravishankar GA (2004) Phytoremediation – a novel and promising approach for environmental clean-up. Crit Rev Biotechnol 24:97–124
Susarla S, Medina VF, McCutcheon SC (2002) Phytoremediation: an ecological solution to organic chemical contamination. Ecol Eng 18:647–658
Szymański K, Janowska B, Jastrzębski P (2011) Heavy metal compounds in wastewater and sewage sludge. Annu set. Environ Prot 13:83–100
Taiz L, Zeiger E (2002) Plant physiology, 3rd edn. Sinauer Associates Inc, Mass, Sunderland
Tamaoki M, Freeman JL, Pilon-Smits EAH (2008) Cooperative ethylene and jasmonic acid signaling regulates selenite resistance in *Arabidopsis*. Plant Physiol 146:1219–1230
Tian S, Lu L, Labavitch J, Yang X, He Z, Hu H, Sarangi R, Newville M, Commisso J, Brown P (2011) Cellular sequestration of cadmium in the hyperaccumulator plant species *Sedum alfredii*. Plant Physiol 157:1914–1925
Tolls J (2001) Sorption of veterinary pharmaceuticals in soils: a review. Environ Sci Technol 35:3397–3406
Torre DL, Iglesias A, Carballo I, Ramírez P, Muñoz MJ (2012) An approach for mapping the vulnerability of European Union soils to antibiotic contamination. Sci Total Environ 414:672–679

Tortella G, Diez MC, Durán N (2005) Fungal diversity and use in decomposition of environmental pollutants. Crit Rev Microbiol 31:197–212

Tortella G, Rubilar O, Valenzuela E, Gianfreda L, Diez MC (2008) Enzymatic characterization of Chilean native wood-rotting fungi for potential use in the bioremediation of polluted environments with chlorophenols. World J Microb Biot 24:2805–2818

Trakal L, Neuberg M, Tlustoš P, Száková J, Tejnecký V, Drábek O (2011) Dolomite limestone application as a chemical immobilization of metal-contaminated soil. Plant Soil Environ 57(4):173–179

Trapp S, Karlson U (2001) Aspects of phytoremediation of organic pollutants. J Soils & Sediments 1:1–7

Traunfeld JH, Clement DL (2001) Lead in garden soils. Home and Garden. Maryland Cooperative Extention, University of Maryland, 2001. http://www.hgic.umd.edu/_media/documents/hg18.pdf

Truu J, Truu M, Espenberg M, Nõlvaka H, Juhanson J (2015) Phytoremediation and plant-assisted bioremediation in soil and treatment wetlands: a review. The Open Biotechnology Journal 9(Suppl 1-M9):85–92

Tsai YA, Mao F, Chi KH, Chang MB, Feng CC, Lin CH, Hung PC, Chen ML (2014) Health risk from exposure to PCDD/fs from a Waelz Plant in Central Taiwan. Aerosol Air Qual Res 14:1310–1319

Turgut C, Pepe MK, Cutright TJ (2004) The effect ofEDTA and citric acid on phytoremediation of Pb, Cr,and Ni from soil using Helianthus annuus. Environ Pollut 131:147–154

Umlauf G, Christoph EH, Lanzini L, Savolainen R, Skejo H, Bidoglio G, Clemens J, Goldbach H, Scherer H (2011) PCDD/F and dioxin-like PCB profiles in soils amended with sewage sludge, compost, farmyard manure, and mineral fertilizer since 1962. Environ Sci Pollut Res Int 18:461–470

USDA (2003) Zinc in foods-draft for comments. Foreign Agricultural Service (GAIN Report) # CH3043, Peoples Republic of China

USEPA (2000) Introduction to phytoremediation. EPA 600/R-99/107, United States Environmental Protection Agency, Office of Research and Development, Cincinnati, Ohio, USA

USEPA (2001) A citizen's guide to vitrification. United States Office of Solid Waste and EPA 542-F-01-017 Environmental Protection Emergency Response December 2001 Agency (5102G)

USEPA (2016) Current and emerging Post-Fukushima technologies, and techniques, and practices for wide area radiological survey, remediation, and waste management. National Homeland Security Research Center, Office of Research and Development, United States Environmental Protection Agency, Research Triangle Park, NC

Valentın L, Nousiainen A, Mikkonen A (2013) Introduction to organic contaminants in soil: concepts and risks. In: Vincent T et al. (eds) Emerging organic contaminants in sludges: analysis, fate and biological treatment. Hdb Env Chem,24:1–30

Van Ginneken L, Meers E, Guisson R et al (2007) Phytoremediation for heavy metal-contaminated soils combined with bioenergy production. J Environ Eng Landsc Manag 15(4):227–236

van Herwijnen R, Springael D, Slot P, Govers HAJ, Parsons JR (2003) Degradation of anthracene by *Mycobacterium* sp. strain LB501T proceeds via a novel pathway, through o-phthalic acid. Appl Environ Microbiol 69:186–190

Van Hoof NALM, Hassinen VH, Hakvoort HWJ, Ballintijn KF, Schat H, Verkleij JAC, Earnst WHO, Karenlampi SO, Tervahauta AI (2001) Enhanced copper tolerance in *Silene vulgaris* (Moench) Garcke populations from copper mines is associated with increased transcript levels of a 2b-type metallothionein gene. Plant Physiol 126:1519–1526

Vandenhove H, et al. (2000) Investigation of a possible basis for a common approach with regard to the restoration of areas affected by lasting radiation exposure as a result of past or old practice or work activity — CARE, Radiation Protection 115, Final Report to the European Commission. http://europa.eu.int/comm/energy/nuclear/radioprotection/

Vasilyeva GK, Strijakova ER, Shea PJ (2006) Use of activated carbon for soil remediation. In: Twardowska et al (eds) Water pollution monitoring, protection and remediation. Springer, Dodrecht

References

Vidali M (2001) Bioremediation. An overview. Pure Appl Chem 73(7):1163–1172

Viehweger K, Geipel G (2010) Uranium accumulation and tolerance in *Arabidopsis halleri* under native versus hydroponic conditions. Environ Exp Bot 69:39–46

Visioli G, Vincenzi S, Marmiroli M, Marmiroli N (2012) Correlation between phenotype and proteome in the Ni hyperaccumulator *Noccaea caerulescens* subsp. *caerulescens*. Environ Exp Bot 77:156–164

Vogel-Mikuš K, Pongrac P, Kump P, Necˇemer M, Regvar M (2006) Colonisation of a Zn, cd and Pb hyperaccumulator *Thlaspi praecox* Wulfen with indigenous arbuscular mycorrhizal fungal mixture induces changes in heavy metal and nutrient uptake. Environ Pollut 139:362–371

Vos JG, Dybing E, Greim HA, Ladefoged O, Lambre´ C, Tarazona JV, Brandt I, Vethaak AD (2000) Health effects of endocrine-disrupting chemicals on wildlife, with special reference to the European situation. Crit Rev Toxicol 30:71

Walliwalagedara C, Atkinson I, van Keulen H, Cutright T, Wei R (2010) Differential expression of proteins induced by lead in the dwarf sunflower *Helianthus annuus*. Phytochemistry 71:1460–1465

Wang B, Lai Q, Cui Z, Tan T, Shao Z (2008) A pyrene-degrading consortium from deep-sea sediment of the west pacific and its key member *Cycloclasticus* sp. P1. Environ Microbiol 10:1948–1963

Wang H, Zhong G (2011) Effect of organic ligands on accumulation of copper in hyperaccumulator *Commelina communis*. Biol Trace Elem Res 143:489–499

Wang YC, Qu GZ, Li HY, Wu YJ, Wang C, Liu GF, Yang CP (2010) Enhanced salt tolerance of transgenic poplar plants expressing a manganese superoxide dismutase from *Tamarix androssowii*. Mol Biol Rep 37:1119–1124

Ward O, Singh A, Hamme JV (2003) Accelerated biodegradation of petroleum hydrocarbon waste. J Ind Microbiol Biotechnol 30:260–270

Watanabe K, Kodoma Y, Stutsubo K, Harayama S (2001) Molecular characterization of bacterial populations in petroleum-contaminated ground water discharge from undergoing crude oil storage cavities. Appl Environ Microbiol 66:4803–4809

Watts RJ, Stanton PC, Howsawkeng J, Teel AL (2002) Mineralization of a sorbed polycyclic aromatic hydrocarbon in two soils using catalyzed hydrogen peroxide. Water Res 36:4283–4292

Wei CY, Chen TB, Huang ZC, Zhang XQ (2002) Cretan brake-an arsenic-accumulating plant. Acta Ecol Sin 22:777–782

Wei L, Luo C, Li X, Shen Z (2008) Copper accumulation and tolerance in *Chrysanthemum coronarium* L. and *Sorghum sudanens* L. Arch Environ Contam Toxicol 55:238–246

Weissenfels WD, Beyer M, Klein J (1990) Degradation of phenanthrene, fluorene, and fluoranthene by pure bacterial cultures. Appl Microbiol Biotechnol 32:479–484

Werle S, Dudziak M (2014) Analysis of organic and inorganic contaminants in dried sewage sludge and by-products of dried sewage sludge gasification. Energies 7:462–476

Wesp HF, Tang X, Edenharder R (2000) The influence of automobile exhausts on mutagenicity of soil: contamination with, fractionation, separation, and preliminary identification of mutagens in the salmonella/reversion assay and effects of solvent fractions on the sister-chromatid exchanges in human lymphocyte cultures and in the in vivo mouse bone marrow micronucleus assay. Mutat Res 472:1–21

West L (2014) Atrazine exposure has serious health consequences for animals and humans. Free Environmental Issues Newsletter http://environment.about.com/od/healthenvironment/a/How-Dangerous-Is-Atrazine.htm

White C, Sayer JA, Gadd GM (1997) Microbial solubilization and immobilization of toxic metals: key biogeochemical process for treatment of contamination. FEMS Microbiol Rev 20:503–516

Whiteley CG, Lee DJ (2006) Enzyme technology and biological remediation. Enzym Microb Technol 38:291–316

Wick AF, Haus NW, Sukkariyah BF, Haering KC, Daniels WL (2011) Remediation of PAH-contaminated soils and sediments: a literature review. Virginia Polytechnic Institute and State University Department of Crop and Soil Environmental Sciences Blacksburg, VA 24061, 102 pp, http://landrehab.org/

Wiren-Lehr S, Scheunert I, Dorfler U (2002) Mineralization of plant-incorporated residues of 14C-isoproturon in arable soils originating from different farming systems. Geoderma 105:351–366

Wuana RA, Okieimen FE (2011) Heavy metals in contaminated soils: a review of sources, chemistry, risks and best available atrategies for remediation. ISRN Ecology doi:https://doi.org/10.5402/2011/402647

Xi XY, Liu MY, Huang Y, Chen Y, Zhang Y (2010) Response of flue-cured tobacco plants to different concentration of lead or cadmium. 4th International Conference on Bioinformatics and Biomedical Engineering (iCBBE). June, Chengdu, China

Yamazoe A, Yagi O, Oyaizu H (2004) Degradation of polycyclic aromatic hydrocarbons by a newly isolated dibenzofuran-utilizing *Janibacter* sp. strain YY-1. Appl Microbiol Biotechnol 65:211–218

Yanqun Z, Yuan L, Jianjun C, Haiyan C, Li Q, Schvartz C (2005) Hyperaccumulation of Pb, Zn and cd in herbaceous grown on lead-zinc mining area in Yunnan,China. Environ Int 31:755–762

Yap CL, Gan S, Ng HK (2011) Fenton based remediation of polycyclic aromatic hydrocarbons-contaminated soils. Chemosphere 83:1414–1430

Yu MH (2005) Environmental toxicology- biological and health effects of pollutants, 2nd edn. CRC Press, Boca Raton

Yu XY, Ying GG, Kookana RS (2009) Reduced plant uptake of pesticides with biochar additions to soil. Chemosphere 76:665–671

Yun SW, Yu C (2015) Immobilization of Cd, Zn, and Pb from soil treated by limestone with variation of pH using a column test. Journal of Chemistry Article ID 641415, 8 pages, https://doi.org/10.1155/2015/641415

Zand E, Baghestani MA, Nezamabadi N, Shimi P (2010) Application guide of registered herbicides in Iran. Jihade-e-Daneshgahi Press, Mashhad, Iran

Zeng X-W, Qiu R-L, Ying R-R, Tang Y-T, Tang L, Fang X-H (2011) The differentially-expressed proteome in Zn/cd hyperaccumulator *Arabis paniculata* Franch. In response to Zn and cd. Chemosphere 82:321–328

Zevenhoven R, Kilpinen P (2001) Control of pollutants in flue gases and fuel gases. Picaset Oy, Espoo. Chapter 8. http://users.abo.fi/rzevenho/gasbook.html

Zhang XH, Liu J, Huang HT, Chen J, Zhu YN, Wang DQ (2007) Chromium accumulation by the hyperaccumulator plant *Leersia hexandra* Swartz. Chemosphere 67:1138–1143

Zhao L, Sun Y-L, Cui S-X, Chen M, Yang H-M, Liu H-M, Chai T-Y, Huang F (2011) Cd-induced changes in leaf proteome of the hyperaccumulator plant *Phytolacca americana*. Chemosphere 85:56–66

Zhou YY, Chen DZ, Zhu RY, Chen JM (2011) Substrate interactions during the biodegradation of BTEX and THF mixtures by pseudomonas oleovorans DT4. Bioresour Technol 102(12):6644–6649

Zhu YG, Shaw G (2000) Soil contamination with radionuclides and potential remediation. Chemosphere 41:121–128

Zvinowanda CM, Okonkwo JO, Shabalala PN, Agyei NM (2009) A novel adsorbent for heavy metal remediation in aqueous environments. Int J Environ Sci Tech 6(3):425–434

Chapter 13
Degraded Soils

Abstract Soils are used in many different ways including agriculture, horticulture, forestry, agro-forestry, pasturing, foundation engineering, mining, ceramics, pottery, medicine, and in many other purposes. Agricultural soils are used for cropping, livestock rearing, and other types of biomass production. In whatever way the soils are used they need some sort of management; and we often mismanage soils while doing so. Misuse and mismanagement of the soil include cultivation of marginal soils, inappropriate tillage, use of heavy machinery for cultivation and harvesting, faulty irrigation and drainage systems, improper use of agrochemicals including fertilizers, lime and pesticides, removal of crop residues, continuous cropping, deforestation, overgrazing, etc. Misuse and mismanagement of soils lead to soil degradation and desertification. Although soils have some capacity or resilience to return naturally to their original state if some minor changes have occurred, soils suffering from human induced degradation cannot usually regain their productive capacity without some sort of conservation and rehabilitation efforts. So, soil degradation needs to be prevented well ahead of its reaching to an irreversible state. Where degradation is moderate, rehabilitation of the soil can be technically and economically feasible. When irreversible degradation takes place, it is almost impossible to return it to a meaningful productive state. About 2 billion hectares of agricultural land have been abandoned due to soil degradation since farming began; these lands are now bare wastelands. Over exploitation and mismanagement have converted many drylands in arid, semi-arid and dry subhumid areas into desert-like lands. The process is popularly known as desertification. Desertification affects 3.6 billion hectares of rain-fed croplands, rangelands, and irrigated lands. However, some efforts of desrtified land rehabilitation are encouraging.

Keywords Soil degradation · Soil mismanagement · Soil sealing · Soil compaction · Soil erosion · Conservation tillage · Cropping systems · Cover crops · Vegetative barriers · Windbreak

13.1 Soil Use and Misuse

The soil is a multi-functional medium. It provides about 90 per cent of all human food, livestock feed, fiber, wood and fuel, and also gives services beyond productive functions. For example, the soil supplies water, minerals and construction materials, and it may be used to meet economic and social necessity. Some types of soil use other than plant production are shown in Figs. 13.1, 13.2, 13.3, 13.4, 13.5, 13.6, 13.7, and 13.8.

The use of soil, however, must depend on its characteristics; the physical, chemical and biological properties of soil that make it capable of providing utility materials, ecosystem functions and life support services for the world of today and tomorrow. According to European Soil Charter (resolution 72(19), soil is a thin layer that covers part of the earth's surface and that forms slowly by physical, physico-chemical, and biological processes over a long time. It is said that formation of one centimeter fertile surface soil may need over one thousand years, but a soil can be destroyed

Fig. 13.1 Soil use in pottery

Fig. 13.2 Soil use in clay jewelerry

13.1 Soil Use and Misuse

Fig. 13.3 Soil in face musking

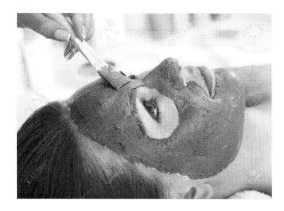

Fig. 13.4 Soil for making idols

Fig. 13.5 Soil for constructing house

within a few years by careless human action. Careful management can improve productive capacity of soil over years or decades but once soil quality is diminished or destroyed its reconstitution may take centuries (Council of Europe 2003).

Soil uses targeted at high economic returns such as high yield of crops or amount of biomass production per unit area on short term basis may degrade the soil eventually. Modern machineries, fertilizers, manures, pesticides, soil modifiers, irri-

Fig. 13.6 Soil in making bricks

Fig. 13.7 Clay tiles for roofs of houses

Fig. 13.8 Rail roads on soil

gation and drainage may cause considerable increases in yields, but their indiscriminate use may create nutrient and water imbalances, and deteriorate physical, chemical and biological properties of the soil. As a result, soil organic matter and soil fertility can decline alarmingly. Additionally, aggregation, porosity, infiltration and aeration are adversely affected, and the productivity of agricultural soils may decline to a considerable extent The ways in which the soil is used also have some impacts, positive or negative, on microclimate, hydrological and biogeochemical cycles, gas emission, groundwater quality, etc. Land use changes and land use conflicts including conversion of forests and grasslands into croplands, draining of wetlands for agriculture, plantation of palm in tropical peatlands and peat extraction in North America have been associated with major environmental problems around the world since long ago. The largest terrestrial biome of the world at present is constituted by croplands and pastures (Foley et al. 2005), and this area will increase in the immediate future. Then more areas would suffer from continued deforestation (Alcamo et al. 2005), which has occurred at an estimated global rate of 130,000 km^2 per year over the period 2000–2005 (FAO 2006).

Soil degradation occurs due to one or more of the following causes (Mbagwu and Obi 2003): (i) deforestation, (ii) over-stocking and overgrazing in rangeland, (iii) overuse of marginal land, (iv) inappropriate tillage in croplands, (v) the use of heavy machinery for cultivation and harvesting, (vi) faulty irrigation and drainage, (vii) unbalanced fertilizer application, (viii) use of contaminated biosolids, (ix) overliming, (x) burning and removal of crop residues, and (xi) continuous cropping and inappropriate cropping patterns. Human actions for soil degradation may include (i) unsustainable land use that cannot be continued into the future and that leads to irreversible soil degradation, and (ii) inappropriate soil management which includes techniques that cause soil degradation usually to such an extent that the degraded soils can be rehabilitated by appropriate management techniques at the expense of permissible energy and cost. Extremely bare, devegetated and eroded surfaces, popularly known as badlands, have in reality undergone irreversible soil degradation, and efforts of their rehabilitation often become time consuming, uneconomic and usually unsuccessful. The loss of soil organic matter, biodiversity and soil quality is often associated with unsustainable practices such as deep plowing on fragile soils and cultivation of erosion-facilitating crops. Continuous use of heavy machinery destroys soil structure and leads to soil compaction. Soil erosion, which often takes the form of the most severe soil degradation, may be linked to agricultural mismanagement practices and deforestation (van Lynden 2000). Soil degradation, when uncontrolled, may adversely affect many important ecological services of the soil, including the regulation of water storage and quality, biogeochemical cycling and carbon sequestration. Inappropriate soil management practices such as over-tillage, faulty irrigation and drainage, and injudicious use of agrochemicals may result in salinization, alkalization, and pollution of the soil.

13.2 Soil Degradation

Soil degradation may be defined as the progressive deterioration in soil quality in a way that the soil cannot regain the capacity to perform its normal ecological functions without the aid of some soil conservation and sustainable management practices. Soil degradation reduces soil fertility and productivity, and hence yields of crops. Soil degradation may occur slowly and cumulatively and may have long lasting impacts on ecosystems (Muchena 2008). Degraded soils cannot produce desirable yields even if large inputs are given, and many agricultural soils had to be abandoned for severe soil degradation (Benayas et al. 2007). According to Lal (2004), important global hotspots of soil degradation are sub-Saharan Africa, South Asia, the Himalayan-Tibetan ecoregion, the Andean region, Central America and the Caribbean. There are several indices to describe the severity of soil degradation. These are (i) soil degradation status – the current situation of soil degradation, (ii) soil degradation rate – the relative decrease or increase of degradation over last 5–10 years leading to the present status and (iii) soil degradation risk – the probability of loss in soil quality when the external conditions (climate or soil management) are changed (van Lynden 2000). Bai et al. (2008) stated that soils of more than 20% of all cultivated areas, 30% of forests and 10% of grasslands were undergoing degradation. Some soils have high degree of resilience – the capacity of a soil to recover from some degree of deterioration naturally by themselves. In most cases, however, the level of human-induced soil degradation far exceeds that capacity. In some of the degraded soils, the productivity and normal ecosystem functioning can be restored with sustainable management practices. However, the loss of soil quality is irreversible in many instances.

13.2.1 Causes of Soil Degradation

The causes of soil degradation have been mentioned in Sect. 13.1. They may differ, however, with regions, soil types and farming systems. For example, mining of soil nutrients declines average yields of crops in much of Africa and in most non-irrigated dry lands in Asia and Latin America. Low-input agricultural practices are being done in cleared forest land to produce more food and in drier and more vulnerable pasture lands. Often soil quality deteriorates for shortening of fallow periods to get more food. Excessive irrigation without facilitating drainage in the semi-arid and arid regions has led to water-logging and secondary salinization. Oldeman (2000) suggested that excessive use of fertilizers and pesticides, and inadequate animal waste containment resulted in soil acidification in many areas. Mining is common throughout North America to extract minerals, coal, or oil and gas; mining operations can deteriorate soil functions and can lead to soil pollution.

A brief description of the major causes of soil degradation is given below.

13.2.1.1 Mining

Mining is responsible for large-scale soil degradation. Mining and milling operations, including grinding, concentrating ores and disposal of tailings, along with mine and mill waste water, provide obvious sources of contamination in soil and water. Soils in abandoned mine areas generally contain high concentrations of the metals Cd, Cu, Ni, Pb and Zn and the metalloid As (Navarro et al. 2008; Zhiyuan Li et al. 2014). Details of heavy metal pollution of soil and its prevention as well as soil quality restoration have been described in Chap. 12.

13.2.1.2 Deforestation

The estimated total forest area of the world is slightly over four billion hectares, and more than 60 per cent is concentrated in seven countries, like Russia, Brazil, Canada, the USA, China, Indonesia, and the Democratic Republic of Congo. Ten countries have no forest at all, and an additional 54 countries have forest on less than 10% of their total land area (FAO 2010). World Resources Institute (1997) estimated that only about 22% of the world's original forest cover was intact. These areas are distributed in three regions: the Canadian and Alaskan boreal forest, the boreal forest of Russia, and the tropical forest of the northwestern Amazon Basin and the Guyana Shield (Guyana, Suriname, Venezuela, Columbia, etc.).

Forests of the world have suffered bitterly for millennia from human interference, and its impact has been enormous. Deforestation has been mainly caused by controlled or uncontrolled felling of forest trees, extraction of forest products for food, wood, building materials, encroachment of farm settlements onto forest land and conversion of forest land into non-forest use. Non-forest uses include farmland, ranches, pasture, industrial complexes, and urban settlements. About one-half of the forests that originally covered the earth have been cleared (Kapos 2000). Human demands on forests are likely to increase further in future because of increased population growth. So, the degree and extent of deforestation are feared to increase further in the future. Most deforestation occurred in Europe, North Africa, and the Middle East in recent time. The vast majority of deforestation has occurred in the tropics in last few decades with the greatest total area in Indonesia, Sudan, Myanmar, and the Democratic Republic of Congo.

Removal of the forest cover immediately exposes the soil surface to the scorching heat of the sun, the beating action of rain and the sweeping action of the wind. As a result of these processes, soil aggregates are broken down, soil particles are detached from the aggregates, soil pores are clogged, infiltration rate is reduced, and runoff and erosion increase. Accelerated erosion caused by deforestation has been found to be the primary factor limiting sustainable utilization of soil resources on the Loess Plateau of Northwestern China (An et al. 2008). The physical, chemical, and microbiological processes of soil along a chronosequence of deforestation in this area indicated that soil wet aggregate stability and mean aggregate diameter decreased with years following deforestation. Accelerated erosion due to deforestation resulted in notable losses of organic matter and total N content of soil. Drastic reduction of soil organic matter content occurs due to the low input of organic residues, increased rate of decomposition and accelerated erosion after deforestation (Pulleman et al. 2000). Hajabbasi et al. (1997) observed that deforestation and clear cutting of the forests in the central Zagrous mountain of Iran resulted in lowering of soil quality and decreased the productivity of the natural soil. Their results revealed a significant decrease in organic matter, total nitrogen and soluble nutrients, and an increase in bulk density of soil. The tilth index coefficient of the forest site was significantly higher than the cultivated forest and the deforested sites. According to Asghari et al. (2016), land use change from natural to managed ecosystems causes serious soil degradation.

13.2.1.3 Overgrazing

The carrying capacity of a pasture or range is the number of a specific type of animals that can subsist on a unit area and produce at a required rate over a specified period. If the pasture can no longer carry as many animals as before, or that its productivity has declined so that the performance of the animal in growth has worsened, it is said to have suffered from overgrazing. Overgrazing also occurs when animals are kept in a paddock too long or brought back too soon, the latter means that a plant is grazed before it has recovered from a previous grazing (Pratt 2002). Again, marginal lands are usually used for grazing; as many marginal lands are being brought under cropping for food, the area under grazing is shrinking in many places. It is one cause of overgrazing. In addition, when cattle ranching becomes an economically lucrative business, farmers increase their herd size. Overgrazing is one of the major causes of soil degradation worldwide. The soil is compacted by animal loads, and soil structure is broken by strokes of cattle hooves. Soil particles are detached from aggregates and the soil becomes susceptible to wind erosion in arid and semi-arid regions. However, Warren and Khogali (1992) suggested that soil degradation caused by overgrazing is especially widespread in Australia and Africa, where it accounts for 80.6% and 49.2% respectively of all soil degradation. Kairis et al. (2015) compared water runoff, sediment loss, soil moisture, air and soil temperature in two sites experiencing overgrazing and sustainable grazing. The study identified overgrazing as a driver of land degradation in southern Europe. Results of this study revealed that sustainable grazing was associated with lower water runoff, reduced sediment loss and lower soil temperature than overgrazing. Villamil et al. (2001) observed that overgrazing in semi-arid regions caused increased bulk density and penetration resistance, higher soil loss from water and wind erosion, changes in pore space distribution, and a decrease in soil aggregate stability and infiltration capacity. Azarnivand et al. (2011) reported from a study in Iran that soil bulk density increased and soil moisture, soil porosity, and aggregate stability index decreased as the density of grazing increased.

13.2.1.4 Over-Cultivation

The need for more food and the lust for more economic returns from unit land area have led farmers to go for over cultivation. As a result, lands are fragmented, fallow periods are reduced, and cultivation is expanded into marginal lands. However, marginal lands that are brought into cropping can give low yields for a few initial years, and cannot retain enough soil fertility afterwards to support sustainable plant growth. Such inherently low fertility soils become further degraded within a few years and are eventually abandoned (Reenberg 2001). For example, agricultural expansion in the Sahelian part of Senegal has taken place at the expense of the decline of fallow lands and savannah vegetation (Van Vliet et al. 2013). Ruelland et al. (2011) reported that a steady increase in crop lands and eroded bare soils followed a drastic decline in woody vegetation cover in the Sahelian region of Mali.

13.2.1.5 Inappropriate Tillage

Tillage, or the pre-planting physical manipulation of the soil, is done to obtain smoothness of soil surface, fineness of tilth, enhancement of aeration, optimization of moisture content and controlling weeds. It facilitates sowing and covering of seeds, germination of seeds and emergence of seedlings and extension of roots. These are very important factors for a healthy crop stand and profitable yield. Manures and fertilizers are mixed with the soil, and movement of water and nutrients within the soil and from the soil to plant roots is improved by tillage. However, inappropriate tillage has been found to be a major cause of soil degradation. Inappropriate tillage includes such operations as (i) tilling the soil in inappropriate time (tilling in dry and wet soil conditions which makes the soil surface cloddy, powdery or puddled), (ii) more than enough and a large number of serial passes, (iii) deep inversion and bringing the relatively less fertile subsoil to the surface, (iv) leaving the tilled soil bare and, (v) using inappropriate tools. Inappropriate tillage causes aggregate instability, poor soil structure, soil sealing, soil compaction, reduced infiltration and poor drainage. It also increases mineralization and loss of soil organic matter and nutrients. Inappropriate tillage causes subsoil compaction (Zhang et al. 2006) and erosion (Wells et al. 2013).

13.2.1.6 Use of Heavy Machinery

Heavy wheeled farm machines such as tillers, spreaders, harvesters, and combines are used in many modern mechanized agricultural systems. These heavy machines often adversely affect soil properties involving (a) increased soil resistance to penetration; (b) reducing conductivity of water and gas flow in soil through damage of pores, and (c) reduced number, size, and stability of aggregates. Eventually, the soil compacted. Soil compaction is a function of the ground pressure and total load (ground pressure x contact area of the tire or track), and soil characteristics such as texture and structure and moisture conditions at the time of operation. Tillage and traffic using heavy machines can induce subsoil compaction in different soil types and climatic conditions in cropped systems (Mosaddeghi et al. 2000). Increased use of heavy machinery in European soils increases the risk of their compaction (Van den Akker et al. 2003).

13.2.1.7 Faulty Irrigation and Drainage

The scope of stable crop production without irrigation is limited in areas of rainfall uncertainty and marked variability in rainfall distribution. However, faulty irrigation (over-irrigation, under-irrigation, wrong irrigation method, use of saline water for irrigation, etc.) is also a major cause of soil degradation. Applying water in excess of irrigation requirements often creates waterlogging, poor aeration and root suffocation. It also reduces mineralization, nitrification and nutrient uptake. The groundwater table rises when excess water is applied, and it draws salts in the root zone. As a result, plant roots may be damaged and nutrient uptake is reduced. Soils may also be salinized by irrigating with saline water. The salinization process eventually renders the soil unusable; many soils had to be abandoned due to such secondary salinization in semi-arid regions. Irrigation induced salinization or sodification can cause flocculation, deflocculation, dispersion and swelling of soil colloids. These processes and the fluctuation of ESR-SAR relationships are responsible for changes in hydro-physical behavior of soils (Chaudhari et al. 2006). Severe damage to growing crops can be caused if excess water is not allowed to drain away from the soil. Therefore, balanced irrigation and drainage are needed for sustainable crop production.

13.2.1.8 Burning and Removal of Crop Residues

Crop residues help to enhance and protect soil quality. The effects of crop residues on soil functions include: protection from erosive forces, increased or maintained soil organic matter, addition of nutrients to the available pool of soil, enhanced biological activity, improved soil structure and improved crop yields (USDA-NRCS 2006). However, the availability of crop residues for application on or incorporation into the soil is reduced in the rural areas of poverty stricken countries of Asia and Africa where crop residues are usually harvested for use as fuel for cooking and for making thatched huts. On the other hand, a large proportion of crop residues are used for ethanol production in the industrial countries (Blanco-Canqui and Lal 2009). Crop residues would be harvested for biofuel in large scale as technologies for the transformation of the high-cellulose biomass into biofuel (i.e., ethanol) develop and demands for ethanol intensify. Corn stover is mainly preferred for ethanol production at present in the United States because corn stover represents nearly 80% (245 million Mg^{-1} yr^{-1}) of the total crop residue production (Kadam and McMillan 2003). Crops that produce large amounts of residues with high content of cellulose include corn, wheat, sorghum and rice. In the Great Plains, wheat straw and sorghum stover are potential biofuel feedstocks (Sarath et al. 2008).

Crop residues are often burnt in the field to clean agricultural land after crop harvest to facilitate tillage (Stan et al. 2014). Burning crop residues may effectively control insects, plant diseases, and the emergence of invasive weed species (Gonçalves et al. 2011). However, burning of crop residues in field is not desired for several reasons, such as: economic loss (Kludze et al. 2013), environmental degradation (Viana et al. 2013), adverse health impact (Agarwal et al. 2012) and the loss of soil organic matter (Granged et al. 2011). Additionally, the removal of crop residues accelerates the rain drop impact on soil aggregates and increases the detachment of fine soil particles that tend to seal the soil surface and lead to crust formation. Surface crust reduces infiltration and promotes runoff and erosion (van Donk and Klocke 2012). Removal of crop residues increases soil temperature and decreases soil moisture storage. Harvesting of crop residues from fields may lead to a decrease in soil organic matter levels, soil fertility and ultimately productivity (Johnson et al. 2006), unless other mitigating management practices are implemented (Laird and Chang 2013).

13.2.1.9 Unbalanced Fertilizer Application

Fertilizers increase crop yields and produce crops of better quality (Savci 2012). Inorganic fertilizers usually contain phosphate, nitrate, ammonium and potassium salts. Micronutrient fertilizers may also be used. However, application of fertilizers for agricultural and horticultural purposes at rates higher than needed has become a growing environmental concern. As nitrogen fertilizers usually show immediate effects on crop plants such as the appearance of dark green leaves and vigorous growth a few days after their application (FAO-IFIA 2000), farmers generally have a tendency of using excess nitrogen and relatively less phosphate and potassium fertilizers. Unbalanced and excess nitrogen application to cereals may result in their lodging, greater weed competition and pest attacks. Moreover, the whole amount of applied nitrogen fertilizer is not absorbed by plants; the left over nitrogen is likely to be transferred to other environmental components, including water and air. Nitrate leaching and groundwater pollution is a common phenomenon associated with nitrogen fertilizer application and nitrification in soil. Fertilizing fields with nitrogen releases nitrous oxide, which is a gas that is 310 times more detrimental to the climate than carbon dioxide (Kotschi 2013). In spite of the fact that soil can act as a filter and buffer for N, and can protect water and atmosphere against N pollution, this capacity of soil is frequently exceeded by excess of N applied to agricultural soils. The application of N fertilizers alone and in excess may result in a decline of soil organic matter. Acidification of soil may result from excess and prolonged use of nitrogen fertilizers; lime can be used to counteract acidification. However, over-liming may cause micronutrient deficiency (Velthof et al. 2011). Low input of potassium in excess of nitrogen has been found to be responsible for negative K balances in soils of India, China and Egypt (Magen 2008). Raw materials of fertilizers often contain important soil contaminants including heavy metals such as Hg, Cd, Pb, Cu, Ni, and natural radionuclides such as ^{238}U, and ^{232}Th (FAO 2009). Exponential increase in fertilizer use in the recent years has created serious risks of environmental problems. Savci (2012) suggested that fertilizer application may result in the accumulation of heavy metals in soil and plant systems.

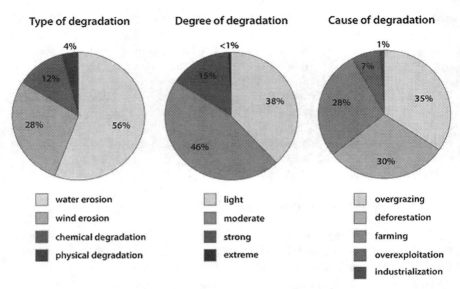

Fig. 13.9 Types, degree and causes of global soil degradation. (Image courtesy of Nature Education)

13.2.2 Types of Soil Degradation

There are three major types of soil degradation: physical degradation, chemical degradation and biological degradation. The subtypes within these types are given below.

Physical degradation Reduction in aggregation, decline in soil structure, sealing, crusting, compaction, lowering of infiltration capacity, waterlogging and drought, and soil erosion.

Chemical degradation Acidification, sodification, salinization, nutrient depletion, nutrient imbalance and disruption in elemental cycles.

Biological degradation Reduction in activity and species diversity of soil fauna, decline in biomass C and depletion of soil organic carbon pool.

Gruver (2013) identifies four types of soil degradation including water erosion, wind erosion, chemical degradation and physical degradation; four degrees of soil degradation such as light, moderate, strong and extreme; and five dominant causes of soil degradation such as overgrazing, deforestation, farming, overexploitation and industrialization. The shares of each category are shown in Fig. 13.9.

Dominant chemical degradation processes such as nutrient depletion, salinization, acidification and pollution have been discussed in Chaps. 9, 10, 11, and 12 respectively. Some physical degradation processes such as soil sealing and crusting, soil compaction, and water erosion and wind erosion in agricultural soils are discussed briefly here.

13.2.2.1 Soil Sealing and Crusting

Crusting in the surface soil usually follows soil sealing which refers to the change in the physical conditions of the soil leading to the reduction in porosity, infiltration capacity and permeability. Soil sealing takes place due to the clogging of the pores by fine particles detached from soil aggregates. Particles are detached from soil aggregates due to physical pressures, such as raindrop impact, improper tillage operations and hoof-strokes of grazing animals. Prolonged and repeated cultivation of soils during cropping regimes also destroys natural aggregates. Sandy loams, sandy clay loams, and sodic soils tend to be the most aggregate deteriorating soils in the dry climates and coastal regions. Sodic soils are also susceptible to slaking and dispersion. Low organic matter containing soils cannot develop stable peds that could resist slacking. Tilling the soil in dry condition grinds soil aggregates in to fine powders which fill the macropores of the surface soil. Particles may be dispersed from the peds due to the presence of excess exchangeable sodium. Again, unstable peds may slack when wetted. On the other hand, tilling in the wet condition puddles the soil; drying the soil afterwards leaves the surface soil sealed and crusted. As a result, the surface soil becomes impervious to water and plant roots. The ecological soil functions are severely impaired or even prevented due to soil sealing as well; and when sealed, the soils become unsuitable for many uses including agriculture and forestry. Soil sealing increases bulk density (Eynard et al. 2004) and slows down solute transport (Assouline 2006) and root growth. The risk of soil erosion and flooding are substantially increased due to the loss of soil water storage and infiltration capacities. Surface sealing, which is nearly irreversible, gives rise to surface crusts on further consolidation and drying.

According to Zejun et al. (2002), crust is a thin layer at the soil surface, and is characterized by a greater density, higher shear strength, and lower hydraulic conductivity than the underlying soil. There are two types of surface crusts: structural crust and depositional crust. A structural crust is a surface layer of the soil, a few millimeters to a few centimeters thick, more compact than the material beneath. The import of external materials is not involved in the formation of the structural crust. Structural crusts are developed also due to trampling by livestock or through traffic by agricultural machinery. Structural crusts may be hardsetting crusts and traffic crusts.

Sealing and crusting are very common in soils worldwide; more of their occurrences are found in soils of arid and semiarid regions. Crusting results from a series of physical processes associated with interactions between water and soils: disintegration, detachment, entrainment, deposition, and compression. The formation and the characteristics of seals and crusts are influenced by several factors, including soil texture and stability of aggregates, intensity and energy of rainfall, kind of tillage and farm machinery, gradients and length of slope, and electrolyte concentration of the soil solution and rainwater.

Hardsetting is a process of compaction of the soil along with increased bulk density occurring without the application of an external load. Hardsetting develops a surface crust by the collapse of most or all of the aggregates due to wetting of a previously loosened topsoil. According to Daniells (2012), hardsetting soils are hard, structureless mass of soil developed in the surface due to rapid increase in soil strength during drying of a previously wetted soil. These soils are difficult or impossible to cultivate until the profile is rewetted. The aggregates are disrupted and dispersed during wetting which leads to coalescence. When soil dries from this coalesced state a massive soil structure is left. There are two types of hardsetting processes – slumping and uniaxial shrinkage. Slumping occurs on wetting a soil that contains water-unstable aggregates. Slumping causes the aggregates to soften and swell, and some or all of the finer particles (silt and clay) become suspended. Aggregates disintegrate due to stresses built up by rapid water uptake, rapid release of heat on wetting, trapped air, or by differential swelling. On the other hand, uniaxial shrinkage occurs due to realignment of the remnants of the disrupted aggregates and/or the internal fabric of the soil. Cracking may not occur in uniaxial shrinkage if forces holding the soil together are long range and non-specific. Probably, matric potential provides the force for uniaxial shrinkage (Lal and Stewart 1990).

The principal cause of traffic crusts is the external pressure of farm machinery and farm animals. These crusts can cause a serious reduction in the penetration of water and the emergence of seedlings. Grazing can induce crust formation by surface compaction of wet or moist soils and mechanical destruction of the surface soil aggregates. Many soils in the tropical areas are naturally and genetically hardsetting. These soils are impervious, compacted as well as cemented. Cementing agents like amorphous silica and imogolite-like aluminosilicates are produced in tropical and sub-tropical soils due to intensive weathering, oxidation and leaching in older parent materials under warm humid conditions. Often cemented layers are found in surface soil or a little beneath the surface. Again, the presence of high exchangeable sodium in sodic soils may lead to slacking and dispersion of fine particles of soil which on drying may be hardest. However, a depositional crust develops due to deposition of soil particles, suspended in water, on the soil surface during infiltration or evaporation. Development of a depositional crust depends on externally derived materials or suspended fine particles, the sources of which can include flood and furrow irrigation water, raindrop splash, overflow and sheet erosion. Clay particles can remain either in dispersed or flocculated condition depending on the electrolyte concentration in the turbid suspension. When the electrolyte concentration exceeds the flocculation threshold, the clay particles flocculate. On the other hand, clay particles disperse if electrolyte concentration in suspension is below the flocculation threshold. Depositional crusts can develop from the settling of both flocculated and dispersed particles. Crusts formed from flocculated particles have an open structure and high permeability but those originating from the deposition of dispersed particles have very low hydraulic conductivity. Depositional crusts can develop both in cultivated and uncultivated soils.

13.2.2.2 Soil Compaction

Soil compaction, or the physical consolidation of soil, is characterized by destruction of soil structure, compression of soil volume, increased soil strength and bulk density, reduced porosity, and limited movement of water and air within the soil.

Soil compaction is caused predominantly by such human actions as overuse of heavy machinery, intensive cropping, short crop rotations, intensive grazing and inappropriate soil management. It is exacerbated by low organic matter content in soil and by tilling and/or grazing when the soil is wet. Soil compaction occurs due to the compression of soil particles together in wet condition, which leads to the development of a more dense soil layer that is relatively hard for crop roots and water to penetrate. Such agricultural practices as increased number of field operations and larger equipment have made soil compaction a great concern in regions of mechanized farming. Soil compaction occurs in a wide range of soils and climates. Excessive grazing cattle on range and farmlands can also cause soil compaction, but such compaction is usually short-lived, and the total weight of grazing animals is often not sufficient to initiate deeper soil compaction (Baumhardt et al. 2011). However, overgrazing can cause puddling due to trampling of soil by animals under very wet conditions breaks down surface soil structure and subsequently results in crust formation upon drying out of the soil.

Soil compaction can happen at any layer in the soil profile depending on soil properties, soil moisture conditions and the nature of external pressure. Typically, there are two types of soil compaction: surface and subsurface soil compaction. When the dense soil layer occurs at the surface of the soil, it is called surface soil compaction. Surface soil compaction happens due to the disruption of the surface soil aggregates through the impact of falling raindrops, runoff, standing water during irrigation, or tillage. Sealing, crusting, and hardsetting are some soil compaction processes that occur in surface soil by the slaking of soil aggregates, and detachment and dispersion of fine soil particles. Subsurface compaction results in a compacted layer somewhere in the soil below the surface. This type of compaction may arise from natural or human-induced causes. Natural compaction develops through the soil-forming processes and is normally found in the subsoil. Subsurface compaction can also occur as a result of use of modern agricultural practices such as tillage, use of heavy farm equipment, and farm traffic. For example, plowing the soil at the same depth for several years may cause a compacted layer just beneath the plow layer, which is called a plow pan.

There may be shallow and deep soil compaction; shallow compaction occurs near the soil surface and deep compaction may extend as far as 75 cm below the surface. Surface crusts and hardest layers are examples of shallow soil compaction. Compacted layers near the surface can be broken up by normal tillage operations. Deep soil compaction is caused mainly by axle load. Van den Akker and Schjonning (2004) suggested that extensive areas have undergone subsoil compaction due to the use of heavy farm machineries. As the deep compact soil layers remain well below the normal tillage zone (Fig. 13.10), it is very difficult to correct them without much labor and cost.

Deep compaction reduces water and air storage in the deeper part of the soil profile and hamper the development of desirable root systems of most deep-rooted crop plants.

Soil compaction is regarded as the most serious environmental problem caused by conventional agriculture, but it is the most difficult type of soil degradation because it shows little evidence on the soil surface (McGarry 2001). However,

Fig. 13.10 A compacted layer at a depth in the soil (http://aces.nmsu.edu/pubs/_circulars/CR672.pdfP)

stunted plant growth, shallow and malformed plant roots, standing water, formation of large clods after tillage, and physically dense soil may exhibit that the soil has been compacted. Oldeman et al. (1991) reported that there were about 68.3 M ha compacted soils globally, which accounts for 4% of anthropogenic soil degradation. It was estimated that 37.5% of the 54 M hectares of total agricultural land in seven countries in Eastern Europe were affected by human-induced soil compaction (Birkas 2008).

Physical characteristics of the soil, weight and design of farm machines, distribution of the load over the number of axles and tires, and number of passes across the field affect the extent of compaction of a soil. Traffic over the fields with large and heavy farm machineries such as tractors, combines, and other equipment is the major cause of compaction in agricultural soils in mechanized farming systems. Figure 13.11 shows the signs of soil compaction by the wheels of a tractor weighing about 20 tons. The most important soil factors that determine the susceptibility of the soil to compaction include texture, structure, organic matter, and water. Soils composed of particles of about the same size are compacted less than soils with a variety of particle sizes. When compressed, smaller particles fill the pores between larger particles and the rearrangement makes the soil denser. A sandy loam soil is more easily compacted than a sandy or clayey soil. Soils rich in organic matter form stable aggregates and are less susceptible to slacking. Again, the higher water content in soil during tilling makes the soil more susceptible to compaction. However, a soil saturated with water has a little risk of soil compaction since water occupies

13.2 Soil Degradation

Fig. 13.11 Soil compaction by wheels of heavy tillage equipment (http://aces.nmsu.edu/pubs/_circulars/CR672.pdf)

the pores and water is not compressible. The risk of soil compaction is the highest in soil moisture contents slightly higher than field capacity when tillage is done (NCASI 2004).

Soil compaction decreases soil physical fertility through reduced supply of water and nutrients, which leads to higher fertilizer requirement and increasing production cost, reduced plant growth and crop yield, lower production of crop residues, reduced biological activities, and reduced nutrient recycling and mineralization (Hamza and Anderson 2005). Soil compaction reduces infiltration, increases surface runoff and accelerates soil erosion by water. It adversely modifies soil properties and soil processes so that seed germination, seedling emergence, root growth, use efficiency of water and nutrients, and growth as well as yield of crops are reduced (Fageria et al. 2006). Reduced root volume and unfavorable root distribution may mainly be responsible for inhibited plant growth because acquisition of water and nutrients is primarily determined by the dimension of root zone and the distribution of roots.

13.2.2.3 Soil Erosion

As discussed in Chap. 8, soil erosion is a severe problem in soils on steep slopes; erosion may be very serious in agricultural soils as well. Although most agricultural soils are situated on level, nearly level flat and gently sloping lands, tillage and other farming practices often cause soil erosion to a great extent. Such human induced soil erosion is popularly known as accelerated erosion. The extent and intensity of accelerated erosion has become tremendous in crop lands predominantly due to human actions aimed at exploiting land, water, vegetation, and soil resources. Of the six billion hectares of crop and pasture lands in the world, nearly all need some degree of protection from soil erosion by water and wind (Sharaiha and Ziadat 2007). Oldeman et al. (1991) estimated that eroded soils account for 84% of the total global area of degraded soils. The remaining 16% soil degradation occurs due to physical degradation (sealing, crusting and compaction) and chemical degradation (acification, salinization and pollution). However, accelerated soil erosion, which can be caused either by water or wind, removes tons of nutrients along with thousands of tons of soil materials from agricultural soils annually. About 2 billion ha of cropland has been abandoned due to loss of productivity caused by human actions since farming began (Lal 1990). Soil erosion reduces the productive capacity of soils and the diversity of plants, animals, and microbes. Pimentel (2006) reported that an estimated 10 million ha of cropland are abandoned each year worldwide because of the loss in productivity due to soil erosion. Worldwide, soil erosion losses are the highest in agro-ecosystems of Asia, Africa, and South America, averaging 30–40 t ha^{-1} yr^{-1}. Soil erosion can occur silently and little signs or scars can be noticed only when the light colored subsoil has been exposed, or tiny channels or rills are apparent. At this state, however, soil erosion has far advanced and soil quality has substantially declined. Soil erosion reduces soil productivity and crop yield on-site, and damages land, water, and installations off-site.

On-site effects The on-site effects of soil erosion include the loss of soil materials, plant residues, soil organic matter, and nutrients, and the reduction in soil fertility, productivity, biodiversity as well as biological activity. Finer soil particles and soil organic matter are physically and chemically the most active components of the soil. During the initiation of the erosion processes soil particles are detached from soil aggregates and are removed by run-off water or by wind. Crop residues and soil organic matter are light-weight substances which are carried away at the very onset of the run-off. The falling rain drops on bare or sparsely vegetated soils destruct unstable soil aggregates, and more of the detached particles are transported off the field. Detached soil particle also clog the pores and make the soil impervious and crusted. Often, the entire top soil is removed and the relatively light colored subsoil is exposed due to erosion. Erosion may completely remove the soil profile in some shallow soils on steep slopes. The loss of soil often exceeds the soil loss tolerance value or the T-value which is the maximum amount of soil loss due to erosion that does not cause significant loss in productivity. The maximum tolerable limits of erosion vary with soil type (Blanco-Canqui and Lal 2008). Foster et al. (2006) described results of a survey of soil loss at 70 sites throughout Western Australia and observed that soil loss values ranged from less than 1 Mg ha^{-1} yr^{-1} to more than 20 Mg ha^{-1} yr^{-1}. Soil erosion also removes huge amounts of nutrients; 10 mm top soil loss may equal to 350 kg ha^{-1} N, 90 kg ha^{-1} P and 1000 kg ha^{-1} K (Hicks and Anthony 2001). Soil fertility and productivity decline with concomitant reduction in crop growth and yield as soil erosion proceeds. However, soil erosion by water may be benignant in some cases, and malignant in some others. For example, in well vegetated forestlands, pasturelands, and in level and mulched croplands, natural erosion is low, gradual and harmless.

13.2 Soil Degradation

Off-site effects Soil erosion can affect land, water, vegetation, and installations far away from the site where erosion is taking place. Transported soil materials can bury crops, damage roads and demolish houses. Runoff water loaded with sediments often washes away seeds, seedlings standing crops. Sediments can be deposited in streams and other reservoirs and can reduce reservoir capacity. Drainage ditches and stream channels are filled with sediments. It causes floods to be frequent and intense. Water quality of surface reservoirs can decline due to the deposition of transported soil particles, nutrients, and agrochemicals. As a result, surface water is eutrophicated and polluted, and biodiversity of aquatic ecosystems is drastically reduced.

Soil erosion by water Falling raindrops in humid areas have usually sufficient energy to detach soil particles from already weakened aggregates. In previous discussions on the causes of soil erosion it was pointed out how soil aggregates become unstable and the ways the soil particles are detached from aggregates. As these particles clog soil pores and reduce infiltration, rain water gradually accumulates on soil surface, and at one stage it starts flowing down the slope if there is any or cutting channels into the soil if there is a chance. Soil particles suspended in water are transported away with runoff water. Suspended particles in runoff water that flows across the surface cause further detachment of soil particles. The amount of soil loss due to erosion by water may depend on rainfall characteristics (amount, distribution, intensity), soil characteristics (texture, structure, porosity), slope characteristics (degree of slope, length of slope), management characteristics (tillage, cropping systems, soil conservation measures), etc. A well aggregated porous and deep soil on a nearly level flat topography under fairly well distributed rainfall regimes suffers less from erosion due to water. However, the following four types of erosion by water are generally recognized: splash erosion, sheet erosion, rill erosion and gully erosion. Gully formation, gully erosion and gully control have been discussed in relation to management of soils on steep slopes (Chapter 8). Other types mainly occur in agricultural soils on nearly level to gently sloping topography.

Splash erosion Splash erosion is caused by the impact of falling rain drops. The impact velocity of falling raindrops, however, depends on drop size and drop velocity, and the detaching power of raindrops depends on their velocity, crop cover and soil aggregate stability. After rainfall proceeded some time a thin film of water covers the soil surface. Further rain drops beat the water and splash the suspended soil particles away. Usually, splash erosion is the beginning of sheet erosion.

Sheet erosion Sheet erosion is the process of more or less uniform removal of soil from the whole soil surface. It occurs in almost level to nearly level smooth slopes by the shallow overland flow of water. In this type of soil erosion, soil particles are detached mainly by raindrops and in some cases by frost, hooves of farm animals, tillage, and by the mechanical activities of farm machines. Detached soil particles are transported by runoff water. Sheet erosion also removes a lot of organic matter and nutrients rendering the soil unproductive. Sheet erosion gradually removes the deeper soil layers if allowed to proceed unhindered. The less fertile subsoil is exposed eventually over a large area. Although less noticed until the light colored subsoil is exposed, it is the most dangerous type of soil erosion. However, soil surface is often not so uniform over the whole area, and water accumulates in tiny channels, so that the surface is criss-crossed by discontinuous rillets. It is then known as the interrill erosion.

Rill erosion Rill erosion occurs by the concentration and movement of runoff water along the slope in tiny channels due to irregularities in the smoothness of soil surface, between rows of standing crops, and along tillage lines, etc. Rills are channels less than 30 cm deep created by the cutting action of flowing water. Soil particles may also be detached by mechanical forces of tillage implements and animal hooves, and the soil materials are transported away by moving water along tiny channels. These channels form a dense network of rills across the whole soil surface. Rill erosion may remove high amount of soil, but the small channels may not interfere with tillage implements. Usually, the rills are leveled by normal tillage operations. However, rills may deepen and widen to form gullies eventually if they are allowed to continue for a long time, and no soil conservation measure is taken. Rill erosion occurs mainly on relatively steep slopes with impermeable soil material consisting of younger sediments.

Soil erosion by wind Soil erosion can occur by the sweeping action of blowing wind which carries the detached soil particles far away from the soil surface. It is a serious problem in the arid and semi-arid regions, where vegetation is sparse and over exploited, rainfall is low and erratic and temperature is high. Here, potential evaporation usually exceeds precipitation, and there is severe deficit in soil moisture most of the year. Organic matter is depleted and soil structure is deteriorated. Soil particles are detached from soil aggregates by the mechanical action of tillage equipment during tilling the land in dry conditions and by the strokes of animal hooves. Wind blows soil particles into the air, and transports the dust particles to variable distances depending on particle size, and velocity and direction of wind and vegetation barriers. Dust in the air can affect visibility of light and quality of air due to the presence of sediments and agrochemicals. Natural processes may be associated with the blowing of dust by the wind, but according to Zhibao et al. (2000), agricultural activities can disturb the soil, and can greatly increase the frequency and amount of airborne dust. Human actions responsible for acceleration of wind erosion include overgrazing of pastures, overstocking in ranches, and agricultural operations such as plowing, leveling beds, planting, weeding, seeding, fertilizing, mowing, cutting, baling, spreading compost or herbicides and burning fields (Nordstrom and Hotta 2004). Wind erosion is a global problem, particularly in Europe (Goossens et al. 2001), Africa (Bielders et al. 2000), Asia (Zhibao et al. 2000), Australia (Gillieson et al. 1996), and South America (Buschiazzo et al. 1999).

Wind erosion removes the fertile top soil. Deposition of wind borne sediments damages crops, buildings, fences, and highways. Wind erosion causes loss of soil fertility and productivity. According to estimates of UNEP and ISRIC (1990), the area degraded by wind erosion amounts to 5.05×10^6 km², which accounts for 46.4% of the global degraded land.

Three typical processes of particle transport operate simultaneously during soil erosion by wind. These processes are saltation, suspension, and surface creep. Saltation, which is the movement of soil particles by successive jumps, is the primary process that causes other particles to move in suspension and surface creep. Without saltation, neither creep nor suspension can occur.

Saltation Saltation refers to the movement of soil particles over the soil surface through successive jumps by wind pressure. Direct wind pressure causes rolling of fine soil particles of 0.1–0.5 mm in diameter to some distance over the soil surface and then abruptly jumping up vertically to a height of 20–30 cm. After gaining in velocity these particles then descend at an angle of 5–12° from the horizontal in an almost straight line to a distance of 50–100 cm. The particles, on striking the ground, rebound into the air and descend again in the same fashion. Moreover, the falling particles may knock other soil particles into the air, and these particles may also start saltation in the similar manner. Saltation is responsible for 50–75 percent of the soil loss due to wind erosion.

Suspension The staying of soil particles, uplifted by wind, as dust load in the air and their horizontal transport to some meters or hundreds of kilometers downwind is the process of suspension in wind erosion. The particles range in size from about 2 to 100 μm with a median diameter of about 50 μm. However, particles <20 μm in diameter predominate in long-distance transport because the larger particles have settled at shorter distances earlier.

Surface creep Soil particles or aggregates of 0.5–1.0 mm in diameter are not usually lifted up by the forces of normal erosive winds. Spinning particles in saltation push, roll and drive them over the soil surface. When the velocity of wind is high the whole soil surface appears like creeping slowly forward. Creeping is not very effective in roughened soil surfaces.

13.3 Prevention of Soil Degradation

13.3.1 *Principles of Soil Degradation Prevention*

- **Minimum disturbance of soil:** Agricultural soils are mainly disturbed by tillage and intercultural operations. Tillage, if excess or more than optimum, favors soil sealing, compaction and erosion. Tillage loosens soil, and soil particles tend to be detached by rains and transported by runoff water. Some soils are more susceptible to erosion than others, and in those soils conservation tillage systems,

including no-tillage, minimum tillage and residue management are some of the options that can reduce soil disturbance.
- Maintaining soil organic matter: Soil organic matter improves soil structure and stabilizes soil aggregates so that they become less susceptible to detaching and slacking. Organic residues and manures are the principle sources of organic matter inputs in soil. Stubles and other crop residues should be left on soil instead of their harvest, removal and burning.
- **Reducing raindrop impact:** Raindrops are strong agents of soil aggregate destruction and soil particle detachment, especially in bare soils. Raindrop impact can efficiently be reduced by providing a cover on the soil during the rainy season. Mulching, cover crops, close-growing crops, dense forest canopy, etc. can offer necessary protection against raindrop impact.
- **Stabilizing soil aggregates:** Soil particles from unstable soil aggregates are easily detached and dispersed. These particles clog soil pores and create sealing, crusting, and compaction. Aggregates in poorly aggregated soils may need to be stabilized by amendments of natural and/or synthetic soil aggregating agents.
- **Increasing infiltration capacity:** Physically degraded soils have usually low infiltration capacity which enhances runoff and erosion. Infiltration capacity can be increased by manuring, mulching and modifying the slope.
- **Reducing amount and velocity of runoff:** As the amount and velocity of runoff increase the amount and rate of soil loss as well as soil fertility deterioration due to erosion increase. The velocity of runoff is higher in steeper slopes of the land. Modifying the degree and length of slope and vegetative barriers can reduce the velocity of runoff. Contour cropping, strip cropping, and grass waterways effectively reduce run off velocity and also trap detached particles. Close growing crops may prevent concentration of water in narrow channels and formation of rills and gullies.
- **Reducing velocity of wind:** Vegetative barriers, stubbles and soil ridging reduce wind velocity and trap wind-blown soil particles.
- **Selecting an appropriate cropping system:** Cropping systems have great impacts on soil compaction and erosion. For example, growing shallow rooted crops continuously in a land over a long period may cause soil crusting and nutrient depletion of the surface soil. Alternating sod type of crops with deep rooted crops are known to enhance infiltration and favor better soil nutrient utilization. Crop rotations help to conserve soil.
- **Integrating management practices:** Several soil conservation methods need to be integrated simultaneously because no single method alone is effective enough to prevent degradation of agricultural soils.

13.3.2 Practices for the Prevention of Soil Degradation

13.3.2.1 Amendments and Soil Stabilization

Several natural and synthetic materials are used to improve soil conditions especially related to the prevention of detachment of soil particles and enhancement of porosity and infiltration. These soil amendments are popularly known as soil conditioners which are defined by the Soil Science Society of America as materials which measurably improve specific soil physical characteristics or physical processes for a given use (SSSA 2008). Soil conditioners increase aggregation and stability of aggregates, and prevent soil particle detachment and dispersion. Natural products used as soil amendments include organic residues, manures, composts, lime and gypsum. Some synthetic polymers are also used as soil amendments for increasing aggregate stability. Soil particles, detached by physical disintegration or physicochemical dispersion, block soil pores, form surface crust, decrease infiltration and increase runoff as well as erosion (Yu et al. 2003). Soil amendments, particularly organic colloids, are efficient in binding soil particles, and forming and stabilizing soil aggregates. Addition of manures and composts favors structure formation, increases aggregate stability, porosity and infiltration and thus reduces runoff and erosion.

Among synthetic organic polymers, polyacrylamide (PAM), which is an industrial flocculant, has been used with success as an aggregate former and soil stabilizer. Polyacrylamide is a water-soluble, very long chain, high molecular weight organic polymer that has characteristics to be suited as a useful soil amendment for increasing aggregate stability, reducing the release and transport of sediments, increasing infiltration, and decreasing runoff as well as soil loss (Lentz et al. 2001; Flanagan et al. 2002). The PAM prevents detachment and dispersion of particles and flocculates dispersed clay colloids (Santos and Serralheiro 2000). Bjorneberg et al. (2000) reported that the application of 2–4 kg ha^{-1} of PAM could reduce soil erosion by 70–90% in some soils. Polyacrylamide, applied in agricultural soils with irrigation water at the rate of 10 kg ha^{-1}, was found to improve infiltration from 7 to 8 times as compared with the control (Chavez et al. 2010). Application of PAM at a rate of 20 kg ha^{-1} in granular form has proven to be effective in controlling soil erosion (Lentz and Sojka 2000; Chavez et al. 2009). However, some investigators observed that application of high molecular weight PAM dissolved in irrigation water in low concentration gives better results (Lentz and Sojka 2000; Bjorneberg et al. 2003; Chavez 2007; Chavez et al. 2009). The use of PAM has been increasing consistently in irrigated fields for preventing and controlling soil erosion (Sojka 2006). As observed by Lee et al. (2010), the application of PAM and gypsum individually and in combination decreases runoff and erosion considerably. Through their effects on aggregate stability, soil conditioners and modifiers may effectively reduce wind erosion as well.

Biochar can also be used as an effective soil amendment for the improvement of soil structure and prevention of soil degradation (Barrow 2012). It is a charcoal-like product that can be prepared on farm by incomplete combustion (pyrolysis) of biomass under limited oxygen supply in a closed furnace at temperatures ≤ 700 °C (Lehmann and Joseph 2009) or can be purchased from the market. It is a stable product that can remain unaltered but active for a very long time in the soil. Biochar is mainly used for improving soil fertility, moisture availability and productivity, and can be an efficient agent of carbon sequestration in soil (Verheijen et al. 2010). Wu et al. (2014) observed that biochar amendment could improve properties of saline and sodic soils. Biochar greatly reduced pH, increased soil organic carbon content and cation exchange capacity. Biochar also reduced exchangeable sodium percentage. Sodic soils have inherently poor soil structure. Slacking, clay dispersion and tunnel erosion are the general problems of these soils. Soil structure in such soils can be improved by gypsum amendment. Gypsum treatment over large areas can, however, be very expensive and, its effect can only be temporary (Morgan 2005). In addition to all these substances, Lal (2007) pointed out the scope of the application of nanotechnology for soil fertility improvement and soil remediation.. According to Padidar et al. (2014), montmorillonite nanoclays were able to stabilize soil structure, increase aggregation and decrease soil erosion by wind.

13.3.2.2 Conservation Tillage

Although tillage is an essential element of pre-planting operations for most crop production systems, inappropriate tillage – excess tillage, tillage during dry or wet soil conditions, intensive tillage, or use of heavy tillage equipment – is a major cause of soil degradation (Sect. 13.2.1.5). Therefore, some sort of tillage method that could ensure the maintenance of sustainable soil productivity with minimum soil disturbance and with little risk of soil compaction and erosion has to be adopted. Such tillage systems are popularly known as conservation tillage systems which usually involve fewer number of passes of plows, localized placement of tillage lines, maintenance of crop residues, cover crops and mulches, and no tillage but making only some small holes to sow seeds. However, the appropriate tillage method may depend on soil conditions, crop type and farm facilities. Here, appropriate tillage practices are those that avoid the degradation of soil but sustainably maintain desired crop yields as well as ecosystem stability. Compared with conventional tillage, conservation tillage systems reduce the number of tillage operations and the cost of tillage (Upadhyaya et al. 2001), the amount of dust that is generated (Madden et al. 2008), and the volume of soil that is disturbed (Mitchell et al. 2004; Reicosky and Allmaras 2003).

There are several conservation tillage systems including no tillage, minimum tillage, reduced tillage, strip tillage and mulch tillage, all essentially having the common principle of maintaining at least 30 per cent soil surface covered with organic residues. As there is no or little soil disturbance except making a small slit for seed sowing and retaining most of the residues of the past crop on the soil surface in no tillage system, it is often excluded from conservation tillage methods (Mitchell et al. 2009). Conservation tillage systems involve more soil disturbance than no tillage and less soil work than conventional tillage. However, residue management, which is an important component of no tillage and other conservation tillage systems, provides such benefits as reducing rain drop and wind impact, decreasing runoff and erosion, increasing infiltration and soil moisture storage, building soil organic matter, and improving soil fertility as well as crop productivity (Blanco-Canqui and Lal 2008).

The term 'zero tillage' implies a planting system that involves no preparatory tillage – plowing and harrowing – before sowing seeds in the field. In the strict sense it is a surface seeding system in which seeds are simply broadcast on a moist soil without any soil work. This traditional system of growing cereals and pulses is practiced in many active floodplain regions of the world including Asia and Africa. Seeds are broadcast on the surface soil immediately after the floodwater recedes. The fine silts deposited by floodwater are moist, soft and fertile enough for the germination of seeds and establishment of seedlings. The advantages of this system are its simplicity, making it accessible to even the poorest farmers, and it's enabling of timely sowing in areas where planting machinery is not available. However, there is a risk of high weed infestation at the later stages of crop growth. No-till or zero till may also involve sowing seeds directly into the residues of the previous crop without any prior topsoil loosening. The most practiced zero till system, however, uses seed-fertilizer drills with either 'inverted-T' or chisel type openers to create narrow slits of 2–3 cm width and 5–10 cm depth for sowing of seeds and incorporating fertilizer while minimizing soil disturbance. The fertilizer-seed drills can work well in presence of anchored crop residues. Herbicides are often needed to control weeds before sowing seeds or planting seedlings. Use of crop rotations and cover crops may reduce weed problems and the dependence on herbicide in no-till systems. If farmers do not have ready access to fertilizer-seed drills and/or face excessive weed problems, other reduced tillage options may be favored rather than zero till. Figure 13.12 shows a field with cotton planted into a cover crop in no tillage system.

Reduced tillage is a practice of minimizing soil disturbance and allowing crop residues or stubbles to remain on the ground instead of being harvested or incorporated into the soil. Reduced tillage involves fewer numbers of passes, tilling only in lines of sowing and leaving the gaps between strips undisturbed, mulching, cover cropping and keeping the land fallow for variable periods. The major benefits of these technologies include (i) reduced costs due to savings in fuel and labor, (ii) timely planting of kharif and winter season crops, resulting in higher yields, (iii) reduced erosion problem, (iv) less need of irrigation, and (vi) better water and nutrient use efficiency.

Strip tillage is a reduced tillage system in which the field is mechanically tilled in narrow alternate strips, 5–10 cm wide, for sowing of seeds or planting of seedlings and incorporating fertilizers. The gaps between strips are left undisturbed and pro-

Fig. 13.12 No-till cotton planted into a cover crop. (Image courtesy of USDA-NRCS)

tected by mulch (Fig. 13.13). A rotary strip tiller or a shank-coulter tiller can be used for tilling the soil in strips. These machinery need to be set up in a way that the planter runs precisely where the strips are placed. Strip tillage has the following advantages: (i) equal or greater crop yields compared to conventional tillage, (ii) higher profit due to saving labor, fuel, and fertilizer costs, (iv) reduced runoff, erosion and compaction (v) reduced leaching of nutrients, and (vi) conservation of water (Foley et al. 2012; Luna and Staben 2002).

In the ridge tillage system, land preparation is done by establishing 15–20 cm high permanent ridges during the second cultivation or after harvest for the next crop (Blanco-Canqui and Lal 2008). The ridge beds are established and maintained through the use of special cultivators and planters. Shallow tillage is applied only on the ridge tops where the crops are grown often leaving the space between the rows covered with crop residues. It is actually a controlled traffic tillage system; implements move through the rows between the ridges. For such confinement of traffic, ridge tillage does not compact soil as the conventional systems do. This system has become very popular for maize and soybean production system in the USA. Ridge tillage reduces costs of tillage, improves crop yields, and reduces losses of runoff and soil.

Ridging and surface roughening are also done for effective control of wind erosion in dry areas. The ridges and clods reduce the wind velocity and trap drifting soils. Ridging cultivated soils also reduces the severity of drifting soil particles by wind. Figure 13.14 shows a view of a ridged field containing sufficient residues.

Mulch tillage is mulching and tillage combined. Mulch is a cover formed by natural organic residues or synthetic materials over the soil surface for the protection and improvement of soil (also see Chap. 3; Sect. 3.6.7). In mulch tillage systems the maximum crop residues, native or imported, are retained on the soil surface

13.3 Prevention of Soil Degradation 435

Fig. 13.13 Strip tillage. (Photo courtesy of USDA-NRCS)

Fig. 13.14 Residue management in a ridged field for the prevention of wind erosion. (Photo courtesy of USDA-NRCS)

and the minimum soil disturbance is done. For mulch tillage, a chisel plough is often used to break hard crusts or hard pans in the soil if there is any. *In situ* mulch, formed from the residue of a dead or chemically killed cover crop left in place, is generally becoming an integral component of mulch tillage techniques. Mulch tillage generally involve (a) uniformly spreading the residue on the soil surface to

Fig. 13.15 Soybean mulched with corn straw. (Photo courtesy of USDA-NRCS)

accommodate the following crop, (b) using non-inversion tillage tools that only partially incorporate surface organic material, (c) planning the number, sequence, and timing of tillage operations to achieve the prescribed amount of surface residue needed to accomplish the objectives of the practice, (d) using planting equipment designed to operate in high residue situations, and (e) minimizing removal of organic residue by burning, baling or grazing. Mulch tillage reduces raindrop or wind impact and soil particle detachment, improves infiltration and soil moisture storage, increases soil organic matter and aggregation, and decreases runoff and erosion. It provides a quick seed germination, and adequate stand and a satisfactory yield. Mulching with organic residues gradually improves soil fertility and productivity through enhanced nutrient supply by continued organic matter decomposition and improved tilth. Figure 13.15 shows a field of soybean mulched with corn straw.

Stubble mulch tillage is another efficient mulch tillage system for the control of erosion, particularly wind erosion. In this system, minimum soil disturbance is done leaving most stubbles of the past crop intact on soil surface during the entire period of the next crop. Stubble mulching is a crop residue management system using tillage, generally without soil inversion and usually with blades or V-shaped sweeps with the objective of keeping desirable quantity of plant residue on the surface of the soil at all times (Fig. 13.16). Plastic mulch can also be useful for soil protection.

13.3 Prevention of Soil Degradation

Fig. 13.16 Stubbles effectively reduce water and wind erosion. (Photo courtesy of USDA NRCS)

Cover Crops

Cover crops are close-growing crops that provide soil protection between periods of normal crop production and favor soil improvement in crop lands, orchards and vineyards (SSSA 2008). Cover crops can be used along with zero tillage, reduced tillage and mulch tillage to reduce soil erosion and improve soil quality. The benefits of cover crops include: (i) protection against soil erosion, (ii) improvement of soil structure, porosity and infiltration capacity (iii) enhancing soil fertility, (iv) suppressing weeds, (v) increasing soil organic matter content, (vi) increasing crop yields, (vii) recycling nutrients, (viii) preventing leaching of nutrients, and (ix) improving water quality (Blanco-Canqui and Lal 2008). Cover crops effectively control soil erosion through reducing raindrop impacts, spreading runoff water throughout the field rather than concentrating in channels, increasing infiltration, preventing detachment of soil particles by increasing aggregate stability and supplying organic matter. Many species of grasses and legumes can be grown closely to develop cover crops. When legumes are grown as cover crops, biological nitrogen fixation may benefit the succeeding crop. Cover crops also control weeds, a major constraint in reduced and no-till systems.

Cover crops can be used as green manures. For green manuring a legume crop is incorporated into the soil when it reaches the flowering stage, and when the plants are still soft, succulent and slender. When a leguminous cover crop is turned under for green manuring it favors decomposition, increases biological activity, causes a rapid nutrient release, and improves the supply of nitrogen for the next crop. However, retaining cover crops as mulch is more benefitting than being turned

Table 13.1 Plants suitable for growing as cover crops

Scientific name	Common name	Scientific name	Common name
Arachis glabrata	Rhizoma peanut	*Melilotus officinalis*	Sweetclover
Arachis hypogaea	Peanut	*Melilotus alba*	White sweet clover
Avena sativa	Common oat	*Mucuna pruriens*	Velvet bean
Avena strigosa	Black oats	*Panicum miliaceum*	Proso millet
Brassica juncea	Brown mustard	*Pennisetum glaucum*	Pearl millet
Brassica napus	Rape	*Pisum sativum*	Garden pea
Brassica nigra	Black mustard	*Poa pratensis*	Kentucky bluegrass
Brassica rapa	Field mustard	*Raphanus sativus*	Cultivated radish
Bromus ordeaceus	Soft brome	*Secale cereale*	Cereal rye
Cajanus cajan	Pigeonpea	*Sesbania aculeata*	Dhaincha
Canavalia ensiformis	Jack bean	*Sesbania herbacea*	Bigpod sesbania
Crotalaria juncea	Sunn hemp	*Sesbania exaltata*	Coffeeweed
Echinochloa crus-galli	Barnyardgrass	*Sesbania sesban*	Egyptian riverhemp
Eruca vesicaria.	Rocketsalad	*Setaria italica*	Foxtail millet
Fagopyrum sculentum	Buckwheat	*Sinapis alba*	White mustard
Glycine max	Soybean	*Sorghum bicolor*	Sorghum
Helianthus annuus	Common sunflower	*Trifolium fragiferum*	Strawberry clover
Hordeum	Barley	*T. lexandrinum*	Egyptian clover
Hordeum pusillum	Little barley	*T. hirtum*	Rose clover
Indigofera hirsuta	Hairy indigo	*T. hybridum*	Alsike clover
Lablab purpureus	Hyacinthbean	*T. incarnatum*	Crimson clover
Lathyrus sativus	White pea	*Trifolium pratense*	Red clover
Lathyrus sylvestris	Flat pea	*Trifolium repens*	White clover
Lens culinaris	Lentil	*T. subterraneum*	Subterranean clover
Lolium perenne	Italian ryegrass	*T. vesiculosum*	Arrowleaf clover
Lolium rigidum	Wimmera ryegrass	×*Triticosecale*	Triticale
Lolium temulentum	Darnel ryegrass	*Triticum aestivum*	Common wheat
Lotus corniculatus	Bird's-foot trefoil	*Urochloa ramosa*	Browntop millet
Lupinus	Lupine	*Brachiaria ramosa*	
Lupinus albus	White lupine	*Vicia benghalensis*	Purple vetch
L. angustifolius	Narrowleaf lupine	*Vicia faba*	Fava bean
Medicago littoralis	Water medick	*Vicia grandiflora*	Large yellow vetch
M. lupulina	Black medick	*Vicia sativa*	Garden vetch
M. polymorpha	Burclover	*Vicia villosa*	Winter vetch
M. rugosa	Gama medic	*Vigna unguiculata*	Cowpea
M. scutellata	Snail medick	*Vulpia myuros*	Annual fescue
M. truncatula	Barrelclover		
Melilotus indicus	Annual yellow sweet clover		

Source: USDA NRCS. http://plants.usda.gov/java/covercrops

under in soils in which the erosion rate is likely to be high. Cover crop mulch on the soil surface increases soil organic matter content and suppresses weeds in addition to providing protection against erosion and loss of soil moisture through evaporation. Cover mulches also keep soil warmer in the cool season and colder in the warm season. Some suitable plants for use as cover crops are listed in Table 13.1.

13.3.2.3 Cropping Systems

Cropping patterns and cropping systems are two important terms associated with crop production systems. The kind of crops that are grown in a field over space and time constitute the cropping patterns. For example, corn-soybean, corn-oats-wheat, and corn-oat-wheat-clover are three different cropping patterns. In some places, only one crop is grown in a year; it is called mono-cropping. Crops can be grown one after harvesting another, or a fallow period can be there between two crops. Growing crops one after another with or without any fallow is known as sequential cropping. Two crops can even be mixed together haphazardly in different proportions or in alternate rows and can be grown in the same field at the same time. This is mixed cropping; the latter type being known as intercropping. Seeds of a crop can be sown in a field of another crop some days before its harvest to match sowing dates and enhance crop maturity of the second crop. It is known as relay cropping. Sequential cropping, mixed cropping, intercropping, and relay cropping are called multiple cropping because they involve more than one crop in a field on an annual basis. These are examples of cropping systems which can be defined as the way the crops are grown over space and time. Cropping systems have important impacts on soil conservation because the management of cropping systems involves tillage, crop residues, fertilizers, manures, irrigation and drainage.

Multiple cropping is a cropping system where (i) a single crop species is grown more than once in a year, or (ii) two or more crops are grown sequentially (sowing one after harvesting another), or (iii) seeds of one crop are sown shortly before harvesting another crop (relay), or (iv) different crops are simultaneously planted on the same field during the same season (mixed or intercrop). Multiple cropping is a popular practice among small farmers in developing regions because they can integrate diverse crops, more crops, farm animal energy, livestock feed, etc. Planting several crops extends the harvest season either with earlier or later ripening crops while providing greater vegetative surface cover over a long period of time (Blanco-Canqui and Lal 2008). If soil, climate and farm facilities permit multiple cropping, it may be a source of grains, fruits, and vegetables all the year round. Major advantages of multiple cropping are: (i) production of diverse food crops, (ii) better soil erosion control due to continuous cover, variable biomass production and rooting systems, (iii) reduced risk of the total loss of crops from adverse climate conditions or diseases, (iv) diversified farm products from a small piece of land, (v) reducing annual production costs, (vi) improving soil fertility and reducing soil erodibility by planting grass, grain crops, and legumes, and (vii) opportunity of planting crops in different seasons, spreading the harvest and supply of produce.

Intercropping is also a multiple cropping system. It involves the cultivation of two or more crops in the same field at the same time in a definite fashion, either in alternate rows or in alternate strips. The component crops of an intercropping system do not necessarily have to be sown at the same time, nor do they have to be harvested at the same time, but they should be grown simultaneously for a great part of their growth periods (Lithourgidis et al. 2011). In intercropping, there is normally one main crop and one or more companion crop(s). It is an old and commonly used cropping practice aiming at efficiently matching crop demands to the available growth resources and labor. Farmers grow most of their beans with maize, potatoes, and other crops in Latin America, whereas maize is intercropped on many of the maize-growing areas of the region. It has been reported that the majority of cowpeas in Africa and most of beans in Colombia are intercropped. In the tropical regions, intercropping is mostly associated with food grain production, whereas in the temperate regions it is receiving much attention as a means of efficient forage production (Lithourgidis et al. 2006). Strip cropping is a variant of intercropping where multiple crops are grown in narrow, adjacent strips that allow the interaction between the different species on the one hand, and management with modern equipment on the other. The strips are sufficiently wide so that each can be managed independently, yet are narrow enough so that the crops, which are rotated annually, can influence the microclimate and yield potential of adjacent crops. Strip intercropping can provide important agronomic and environmental benefits. A well-managed strip-intercropping system could result in higher profitability and greater soil and water conserving potential than most monocrops. Crop selection, strip width, planting direction, plant population, and crop strip orientation are important considerations; they all influence crop growth and soil fertility use efficiency. The most common advantages of the intercropping system include (i) the production of greater yield on a given piece of land by making more efficient use of the available growth resources using a mixture of crops of different rooting ability, canopy structure, height, and nutrient requirements, (ii) improving soil fertility through biological nitrogen fixation with the use of legumes, (iii) conserving soil through greater ground cover than sole cropping, (iv) providing better lodging resistance for crops susceptible to lodging than when grown in monoculture, reducing pests and diseases, and (v) providing insurance against crop failure in areas subject to extreme weather conditions such as frost, drought, and flood (Lithourgidis et al. 2011). Zougmore et al. (2000) reported that planting sorghum with cowpeas is an effective technique against runoff and soil erosion. Intercropping of cereals with legumes is of particular interest for several specific benefits: (i) reducing runoff and soil loss, (ii) providing nitrogen to associated crops, (iii) suppression of weeds, (iv) increasing soil organic matter, and (4) suppressing crop diseases. Intercropping of soybean, groundnut, cowpea etc. with corn, sorghum, pearl millet, etc. is a common practice in India. This practice gives a better cover on the land, good protection to the soil from the beating of rain and soil erosion, by binding the soil particles (Sharma and Singh (2013). Small grain strips in intercrops slow down surface runoff, and reduce sediment and chemical loads in runoff water which may improve water quality. Intercropping maize and cassava is a popular practice at Ibadan, southwest Nigeria. This system provides better cover than growing cassava alone. Intercropping of maize and cassava produces 50% canopy cover 50 days after planting in April, compared with 63 days for cassava as a monoculture (Morgan 2005).

13.3 Prevention of Soil Degradation

Crop rotation is a systematic sequence of crops grown in combination with other crops or with grasses and legumes. A set of crops repeats after successive stipulated periods. There are fewer problems with weeds, insects, parasitic nematodes, diseases caused by bacteria, fungi, and viruses when using rotations compared to monocultures. When legumes are part of the rotation, nitrogen is supplied to the succeeding crop. With forage rotations, soil organic matters increase as a result of longer rotations. Rotations can be simple, corn followed by soybeans, or very complex, tobacco with a cover crop for two years followed by corn-double cropped wheat and soybeans using conservation tillage. Crop yields in rotation are often higher than those grown in monoculture. Practices such as conservation tillage in combination with rotations benefit soil quality by maintaining or increasing soil organic matters. There are three main types of rotations based on the duration: monoculture (a single crop with no diversity), short rotation (basically a 2-yr rotation, e.g., corn-soybean), and extended rotation (>2-yr rotations, e.g., corn-oat-wheat-clover-timothy). In planning a rotation, crop habits such as leaf characteristics, root characteristics, nutrient demands, water requirements, nitrogen fixing capacity, etc. are carefully considered. Usually, a shallow rooted crop is alternated with a deep rooted crop, a high nutrient demanding crop is alternated with a low demanding crop, cereals can be alternated with legumes, root crops can be alternated with grains, etc. Rotating different crops is an ecologically viable alternative to monocropping and is relevant to addressing agricultural and environmental concerns. Long rotations are preferred over monocropping and short-rotations. Benefits of crop rotations include: reduction in soil erosion, improving soil properties, increasing soil organic matter, improving soil fertility, increase crop yields, reducing pests and diseases of crops, increasing net profits, improving wildlife habitat, reducing the use of chemicals, and reducing water pollution (Blanco-Canqui and Lal 2008). Crop rotations that utilize the land more intensively such as corn, wheat and soybeans grown in two years produce larger amounts of biomass during the rotation and are more effective in reducing erosion than a continuous cropping sequence. Increased cover from grass and or legume rotations or high residue crops combined with other conservation practices such as conservation tillage will reduce upland erosion which, in turn, reduces sediment from surface runoff. Crop rotation systems that promote an increase in organic matter and aggregate stability can maintain or improve the presence of pores for infiltration. Decaying roots, especially those of deep rooted crops like alfalfa and safflower, may leave channels for improved infiltration. Other conservation practices may be needed in crop rotations such as crop residue management to ensure surface protection and improve infiltration. Crop rotation improves soil structure, enhances permeability, increases biological activity, increases water and nutrient storage capacity, and reduces soil erosion. These benefits can be further improved by combining crop rotations with cover crops and reduced or no-tillage practices (Zentner et al. 2002).

13.3.2.4 Vegetative Barriers and Windbreaks

Vegetative barriers are used typically in two ways: narrow filter-strips (or grass barriers) and filter-strips. The grass barriers are narrow strips, approximately 1.2 m wide, of tall, erect, stiff-stemmed, native perennial grasses planted on the contours to reduce sediment yield, retard and disperse the runoff and facilitate benching of the slopes. However, the vegetative filter-strips are much wider, more than 5 m, established between field borders and water ways (Blanco-Canqui et al. 2004). Windbreaks are vegetative barriers composed of trees, shrubs, tall grasses or even crop plants that obstruct wind, reduce field width and wind velocity, change the direction of wind impact, and traps saltated soil particles. Windbreaks can be used as field wind barriers or around agricultural fields, farm houses and as shelterbelts in shorelines. Windbreaks composed of linear plantings of one or more rows of trees or shrubs and placed perpendicular to the direction of wind to reduce its velocity and impact on the crop and the soil can significantly decrease wind erosion. Many species of trees and shrubs can be found suitable to build windbreaks. However, annual crops like small grains, corn, sorghum, sudangrass, sunflowers and tall wheatgrass can be used in field barrier systems to prevent wind erosion. Figure 13.17 shows a field wind barrier system.

Fig. 13.17 Field windbreak. (Photo courtesy of USDA-NRCS)

13.4 Restoration and Rehabilitation of Degraded Soils

Rehabilitation refers to bringing degraded soils into productive use again. It is aimed at optimizing the production of useful biomass of a degraded and/ or abandoned site. The main purpose is the economic utilization of a soil that lost its fertility and its capacity of sustaining productivity due to human use, misuse and mismanagement. The goals of rehabilitation of degraded soils are: (a) to halt the depletion, deterioration and degradation of the soil, (b) to regain and maintain productivity, and (c) to ensure sustainability and proper land use. On the other hand, restoration implies the reversion of degradation and regaining, as close as possible, the pre-disturbance biodiversity and ecosystem functions (Ward 2006). Restoration is a more complex task than rehabilitation. Returning to the natural state by artificial processes is too ambitious. Rehabilitation could convert degraded soils into completely a new use if that appears to give the best results for humans. It is relevant to the immediate needs of people because it emphasizes that biomass production can ameliorate hunger, fuel shortages and poverty. Thus, rehabilitation can be achieved within a relatively short period of time if degradation has not gone too far.

Degraded soils are generally denuded and dry. They lose their moisture and nutrient retaining capacity, and become unproductive gradually. Soil degradation may be slight and almost unnoticeable in the beginning, but at the later stages deterioration accelerates and becomes irreversible in the long run. The degree of degradation depends on soil characteristics, climate and management. Soil degradation can be distinguished into three categories depending on its intensity: (a) early stages, (b) moderate stages, and (c) critical stages. The symptoms of degradation that start appearing in the early stage include the thinning of vegetative cover, decline in growth and yield of crop plants, lowering in soil organic matter, deceasing soil moisture, deterioration of soil structure, surface crusting, reduction in infiltration rate, and slight disturbance in surface soil. The pace of degradation is slow at this stage, and rehabilitation can be easy, cheap, rapid and effective. In the moderate stage, the top soil is usually lost and the poorly fertile compact subsoil is exposed. Still, some plant growth is possible, though not probably profitable, and rehabilitation may be economically feasible at the expense of moderate to large inputs. In the critical stage, the entire top soil is lost; sometimes the bed rock is exposed due to severe erosion, and the land is desertifed in arid regions. Cropping is not possible if soil degradation has reached the critical stage. At this stage, the soil, water, vegetation, and biodiversity are lost, and the changes are mostly irreversible. The cost, effort and time needed for rehabilitation are quite high and the success is very uncertain. Engineering and bioengineering methods of rehabilitation can be undertaken after doing a proper environmental and economic analysis. Many dryland soils are severely degraded. Rehabilitation of degraded drylands can restore essential ecosystem services such as biodiversity and sustainable supplies of fodder, food and renewable energy (Mussery et al. 2013). Rehabilitation of degraded soils can be achieved by long term conservation, halting over-grazing, addition of organic residues or manure, application of fertilizer based on soil test and crop response, crop rotations, cover crops, alley cropping and agroforestry. It requires varying time spans and financial resources.

Rehabilitation of degraded soils need re-carbonization of the depleted soil organic carbon pool, which is essential to numerous functions, through regular input of biomass-C and essential elements, for example, N, P, and S (Lal 2014). There are three basic strategies of restoring soil quality: (i) minimizing losses from the pedosphere or soil solum; (ii) creating a positive soil C budget, while enhancing biodiversity; and (iii) strengthening water and elemental cycling (Lal 2015). The steps for soil quality restoration include: (i) reducing erosion and eliminating compaction, (ii) improving soil /agro-biodiversity, and (iii) soil restorative farming/cropping system.

(i) **Minimizing losses from the pedosphere:** Loss from the pedosphere occurs mainly through erosion and leaching. Soil erosion must be limited by all means within the soil loss tolerance value (T-value) which is usually less than 12.5 Mg ha^{-1} (Lal 2015). But, accelerated soil erosion caused by misuse and mismanagement of the soil often exceeds the T-value. For example, a survey of soil loss at 70 sites throughout Western Australia indicated that about 10% of the sites had soil losses in excess of 20 Mg ha^{-1} yr^{-1} and about 30% of the sites had soil losses of 5–20 Mg ha^{-1} yr^{-1} (Foster et al. 2006). In addition to soil loss, erosion causes huge loss of organic matter and nutrients, for example 350 kg ha^{-1} N, 90 kg ha^{-1} P, and 1000 kg ha^{-1} K (Hicks and Anthony 2001). Reduction of soil erosion can be effectively achieved by the building up of organic matter, covering the soil with plant cover or mulch, afforestation/reforestation, slope modification, conservation tillage, conservation farming, vegetative barriers and crop-livestock mixed farming. Burning of the native vegetation or crop residues aggravates soil erosion. Experiments conducted in Spain indicated high post-fire soil degradation risks (Badia and Marti 2008). Leaching is the downward transport of soluble materials including soluble organic matter and nutrient ions below the solum. The leaching risk for a nutrient increases with its mobility in the soil. Among nutrient anions, nitrate is particularly easily leached because of its high mobility (Lehmann and Schroth 2003). To reduce nitrate leaching, Kirchmann et al. (2002) proposed the use of a range of counter-measures including catch crops, minimum tillage, control of biological processes, etc. depending on how sensitive the farming system, soil and climate are to the risk of nitrate leaching. Phosphate ions are also leachable to a considerable extent (Svanback et al. 2014). Basic cations such as Ca^{2+}, Mg^{2+} and K^+ are also leached from soils of the humid region. Soils with high water infiltration and percolation and low nutrient retention capacity, such as sandy soils and soils with low activity clays and low organic matter content are very susceptible to nutrient leaching. For minimizing leaching loss, Chen and Neill (2006) suggested (i) re-cropping rather than fallowing, (ii) following conservation tillage, (iii) diversifying crops including perennial and/or deep-rooted annual crops, (iv) inclusion of legumes in cropping patterns, (v) optimizing crop density, (vi) replacing flood irrigation by sprinkler irrigation, and (vii) optimizing irrigation need and avoiding over-irrigation. Several amendments including manures, biochar, calcium compounds may be used to reduce

13.4 Restoration and Rehabilitation of Degraded Soils

leaching loss. The use of $CaCl_2$, $CaCO_3$, or their combination can significantly reduce P leaching from sandy soils (Yang et al. 2007).

(ii) **Enhancing soil/agro-biodiversity and creating positive soil carbon budget:** Soil biota perform critical roles in key ecosystem functions, including biomass decomposition, geochemical cycling, nutrient transformations, carbon sequestration, degradation of toxins, moderating CO_2 in the atmosphere, remediation of pollutants and suppressing diseases. Improving activity and species diversity of soil fauna and flora is, therefore, essential to restoring and improving soil quality and reducing risks of soil degradation. Macro-organisms, such as earthworms and termites, also have significant roles in restoring soil quality. Organic mulching, crop residues and cover cropping can increase biotic activity, build-up soil organic matter and improve structural properties. Therefore, risks of soil degradation can be mitigated through the adoption of land use and management systems which improve soil biological processes, and introduction of beneficial organisms into soils by selective inoculation. For these reasons, the presence of earthworms, termites and other soil biota are often identified as important indicators of quality in tropical soils (Ayuke et al. 2012). Blanco-Canqui et al. (2014) observed soil carbon accumulation under switchgrass barriers.

(iii) **Soil restorative farming/cropping systems:** Farming/cropping systems affect the type, rate and severity of soil degradation. Cropping systems can significantly impact soil organic carbon (SOC) pool and associated soil properties. Similar to arable lands, managing the quality of rangeland soils is also essential for reducing risks of degradation. Sustainable management of rangeland soils is especially challenging because of high variability, harsh environments, and the temptation for over-grazing. A reduction in the proportion of palatable perennials increases soil compaction and declines SOC (Snyman and du Preez 2005). Establishment and management of forage trees such as *Acacia fadherbia* (Garrity et al. 2010) and grass-legume mixtures (Muir et al. 2011) can also improve the quality of rangeland soils.

If degradation can be arrested in the early stage, the soil may, in many cases, recover itself by natural processes. It is known as soil resilience which is the soil's ability to recover after disturbance through natural soil processes. According to Blancoo-Canqui (2008), soil resilience refers to the intrinsic ability of a soil to recover from degradation and return to a new equilibrium similar to the antecedent state. Lal (2015) defines soil resilience as the ability of the soil to recover its quality in response to any natural or anthropogenic perturbations. Soil resilience is not the same as soil resistance, because resilience refers to the elastic attributes that enable a soil to regain its quality upon alleviation of any perturbation or destabilizing influence. The processes which make a soil effectively resilient include swelling and shrinking, freezing and thawing, chemical and biochemical reactions associated with organic matter and mineral transformations, and biological activity. These processes tend to restore soil properties to pre-disturbance conditions. There are three criteria necessary for soil resilience to be effective: (1) the soil must be sensitive to the process, (2) the climate must produce the temperature and moisture regimes necessary for the process to occur, and (3) the cycles or processes must occur with sufficient frequency and duration. Some soils are more resilient to physical impacts than are other soils, and mitigation measures or corrective treatments are not always needed. The time required for natural recovery from compaction varies with soil physical characteristics, chemical characteristics, climate, and the severity of compaction. Blanco-Canqui et al. (2015) reported that cattle manure application reduces the soil's susceptibility to compaction and increases water retention.

Rehabilitation of compacted soil can be attempted by mechanical manipulation, such as disk harrowing, bedding or mounding, subsoiling or ripping, and spot cultivation. Deep ripping or deep cultivation, is an important practice for eliminating soil compaction, destroying hard pans and ameliorating hard setting soils (Torella et al. 2001; Laker 2001). It has become a common management technique, used to shatter dense subsurface soil horizons that limit percolation of water and penetration of roots (Bateman and Chanasyk 2001). If properly applied, these tillage practices can favorably alter soil properties and enhance seedling survival and growth. Tillage under non-optimum conditions (e.g., wet soil), however, can cause additional soil compaction and/or puddling, and create further risk to long-term productivity (NCASI 2004). A ripper can be used to loosen the light textured soils where the hardpan is near the surface. A ripper is a chisel-shaped implement pulled by animals or a tractor. It breaks up surface crusts and opens a narrow slot or furrow in the soil about 5–10 cm deep. Unlike a mouldboard plow, a ripper does not turn the soil over. Ripping can be done during the dry season, or at planting time. Seeds can be sown in the slot by hand, or using a planter attached to the ripper. Ripping can be shallow or deep. Deep ripping is applied on sandy soils with compacted layer at more than 30 cm depth, and the subsoil is not highly acidic. Performing tillage operations below the normal tillage depth to modify the physical or chemical properties of a soil is known as deep tillage. The objectives of deep tillage are (i) fracturing restrictive soil layers, and (ii) mixing soil deposits. Deep tillage operations are performed when the soil moisture is less than 30 percent of the field capacity at the maximum depth to which the tillage will be done. The effects of deep tillage can be enhanced by including deep rooted crops in rotation so that the roots are able to extend through the fractures of the restrictive layers. For example, rapeseed (canola), cotton and turnip produce large taproots that can protrude a depth up to 2 m (Clark 2007). Tillage equipments include chisels, subsoilers, bent- and rippers that have the ability to reach the required soil depth. The depth of tillage should be deeper than the depth of the restrictive layer. Complete fracturing of the restrictive layer is not required. The fractured zone, as a minimum, shall be sufficient to permit root penetration below the restrictive soil layer. Chiseling can be used to break up or fracture pans or compacted layers that are within 40 cm of the surface. The implement would penetrate through the compact layer and completely shatter it between implement penetration points for close grown crops such as alfalfa and small grains. For row crops, the area just below or to the side of the plant needs shattering only. The excessively dense hardpans at shallow depths impede root growth and result in crop yield reduction because plant roots cannot exploit soil moisture and nutrient reserves in the lower soil strata. Mechanical loosening has been widely reported to give beneficial responses at such situations (Drewry and Paton 2000; Burgess et al. 2000). Farmers apply subsoiling either annually or biennially to mechanically disrupt the hardpan layers and improve the rooting environment for optimal crop growth (Busscher and Bauer 2003; Raper et al. 2005). The application of this energy-intensive subsoiling operation is based on the assumption that the compacted layers are located at a constant depth across the field. The relative strength and depth to the hardpans, however, vary from field to field and within fields (Goodson et al. 2000; Isaac et al. 2002). With uniform depth subsoiling, tillage may be applied in areas of the field where there is no soil compaction problem or at depths that do not necessarily correspond to the hardpan depth. Figure 4.5 shows a subsoiler to break the hardpans. There is a great amount of variability in the depth and thickness of hardpan layers; moreover, it does not exist in some parts of the field (Raper et al. 2007). Optimum tillage depth may be deeper or shallower than what is conventionally applied, making uniform-depth tillage costly. Therefore, deciding the tillage depth based on the thickness and depth of the compacted layer is important (Keskin et al. 2011).

13.5 Desertification and Desert Reclamation

If processes of soil degradation continue for a long time at any place, they can create irreversible changes to soil and vegetation of the ecosystems and can ultimately lead to desertification. According to UNCCD (2000), land degradation due to climate variability and human activities in drylands has led to desertification. Desertification can be defined as the resultant of ecological changes that lead to degradation of land in arid, semi-arid and dry sub-humid areas with persistent loss of soil quality, biodiversity and human health. Some areas in arid and semi-arid regions are being desertified by human over exploitation and mismanagement of land and vegetation. At the same time, there are some efforts of desert reclamation. Desertification converts drylands into desert-like lands and diminishes ecosystem services that are fundamental to sustaining life. Desertification affects large dryland areas around the world and is a major cause of stress in human societies (D'Odorico et al. 2013). Overgrazing is the major cause of desertification worldwide. Other factors that cause desertification include urbanization, climate change, overdrafting of groundwater, deforestation, natural disasters and tillage practices in agriculture that place soils more vulnerable to wind. The United Nations Conference on Environment and Development concluded that desertification affects 3.6 billion hectares of rain-fed croplands, rangelands, and irrigated lands, the equivalent of one-quarter of the world's total land surface and 70 percent of all drylands (UNCED 1992).

Desertification occurs globally but the highest intensity is found in the developing countries. For example, developing countries in Asia, Africa, and South America have larger populations living in drylands (GEF and GM 2006). An estimated 40% of people in Africa and Asia live in areas constantly threatened by desertification

(Stather 2006). Achieving the core objectives of sustainable development will remain an impossible mission for nearly two billion people living in the world's drylands, whose biological productivity is under serious threat from the intensifying trend of desertification (FFO 2007). Due to desertification, the annual loss of income is estimated at US $ 65 billion, and this does not include the costs incurred in social and environmental aspects (Kannan 2012). The costs of desertification are most often measured in terms of lost productivity, which includes the reduced crop yields, grazing intensities, etc. Secondary costs are the loss of ecosystem services and ecological functions that affect the very sustainability of the planet (de Sherbinin 2002).

To control desertification, the Chinese government implemented a series of large-scale mitigation programs, including the Three Norths Shelter Forest Program and the Combating of Desertification Program (Runnstrom 2000). These projects focus on increasing the vegetation cover by prohibiting grazing, planting trees and grasses, and constructing shelter forests to protect farmland against blowing sand (Wang et al. 2007). Desert reclamation is a priority strategy in China to convert desertified lands into productive rangelands and croplands. Coordinated efforts have been undertaken by the Institute of Desert Research of Academia Sinica (IDRAS) at Lanzhou Province. The Institute operates nine field stations in desertified areas and evaluates various techniques of desert reclamation to recommend appropriate strategies in reclamation programs. Some of the initial pioneering work was carried out at Shapotou Research Station and Yanchi Experimental Station for Cooperative Desertification Control Research, both in Ningxia Autonomous Region. Desertified land in both of these regions has been reclaimed by various techniques over different time periods. The Shapotou station was established in 1957 to find ways of protecting 40 km of the Lanzhou to Baotou railway line from sand burial. Yanchi station is located in the Shabianzi Region, part of the transitional area between the Ordos and loess plateaus. IDRAS uses a number of stabilization and reclamation techniques at Shapotou, Yanchi and other field stations. These techniques aim at halting the desert advance and decreasing aeolian damage to rangelands and croplands downwind. These include the use of windbreaks, the establishment of straw or clay checkerboards, managed successions of xerophytes, irrigation, land enclosure, extracting palaeosols and chemical treatment.

Planting trees in a desert might sound foolish, but UNCCD (2000) reported that Xinjiang poplars and several species of willow can be successfully raised in the Kubqi Desert which is one of the wettest deserts in the world and, just 20 cm below the dusty surface, the sand is relatively moist. Planted in the spring or autumn, the saplings are protected by wooden frames, which are sunk deep into the sand to prevent movement. They give the young trees stability and the time to take root. Planted correctly, they grow quickly and their spidery roots help to stop the migration of sand and thereby stabilize mobile sand dunes. An example of successful rehabilitation of deserted land is offered by the Badia rangelands in Syria. An IFAD-supported project has restored vegetation in about one third of the Badia rangelands (http://www.ifad.org/pub/factsheet/desert/e.pdf). The success in this regard was due to the involvement of the local people's participation in decision making, full ownership of the rehabilitation, and management of the rangelands. Bedouin herders, with their extensive local knowledge, worked with project experts to draft and implement management plans, determining how many animals should graze in a given area at a given time and taking seasonal conditions into account. The project took three strategies of rehabilitation: resting, reseeding and planting. Where possible the land was simply rested for up to 2 years. Native plants that had long since disappeared returned, and the full range of vegetative cover has come back to life. Where degradation was too advanced, the project introduced reseeding, using native rangeland forage plants or plants acclimatized to local conditions. Soils were first furrowed to encourage rainwater infiltration. More than 930,000 hectares of the Badia have been regenerated by resting, a further 225,000 have been reseeded, and about 94,000 hectares have been planted with nursery shrubs, each plant surrounded by a small handmade soil bank to protect the plant and collect rain. In this way, the shrubs are watered just once when they are planted, and afterwards rely on this simple irrigation method. Regular cropping by livestock keeps the shrubs from becoming woody and prolongs their life. Eventually, they reseed themselves.

Study Questions
1. Explain soil use, misuse and mismanagement. Discuss how mismanagement causes soil degradation.
2. Define soil quality and soil degradation. Briefly describe the causes of soil degradation.
3. What are the types of soil degradation? Discuss soil sealing, crusting and compaction with emphasis on the associated problems and their mitigation.
4. What are the advantages and disadvantages of tillage? Distinguish between reduced tillage and no-tillage. Describe different conservation tillage systems in brief.
5. Differentiate restoration from rehabilitation. Explain how you can rehabilitate degraded soils. What do you mean by irreversible soil degradation?

References

Agarwal R, Awasthi A, Singh N, Gupta PK, Mittal SK (2012) Effects of exposure to rice-crop residue burning smoke on pulmonary functions and oxygen saturation level of human beings in Patiala (India). Sci Total Environ 429:161–166

Alcamo J, van Vuuren D, Ringler C, Cramer W, Masui T, Alder J et al (2005) Changes in nature's balance sheet: model-based estimates of future worldwide ecosystem services. Ecol Soc 10:19

An S, Zheng F, Zhang F, Van Pelt S, Hamer U, Makeschin F (2008) Soil quality degradation processes along a deforestation chronosequence in the Ziwuling area, China. Catena 75:248–256

Asghari S, Ahmadnejad S, Keivan Behjou F (2016) Deforestation effects on soil quality and water retention curve parameters in eastern Ardabil, Iran. Eurasian Soil Sc 49:338. https://doi.org/10.1134/S1064229316030029

Assouline S (2006) Modeling the relationship between soil bulk density and the hydraulic conductivity function. Vadose Zone J 5:697–705

Ayuke F, Karanja N, Okello J, Wachira P, Mutua G, Lelei D, Gachene C, Hester R, Harrison R (2012) Agrobiodiversity and potential use for enhancing soil health in tropical Soils of Africa. Soils Food Secur 35:94–134

Azarnivand H, Farajollahi A, Bandak E, Pouzesh H (2011) Assessment of the effects of overgrazing on the soil physical characteristic and vegetation cover changes in rangelands of Hosainabad in Kurdistan Province, Iran. J Rangeland Sci 1(2 H):95–102

Badia D, Marti C (2008) Fire and rainfall energy effects on soil erosion and runoff generation in semi-arid forested lands. Arid Land Res Manage 22:93–108

Bai ZG, Dent DL, Olsson L, Schaepman ME (2008) Global assessment of land degradation and improvement 1. Identification by remote sensing. Report 2008/01, ISRIC, Wageningen

Barrow CJ (2012) Biochar: potential for countering land degradation and for improving agriculture. Appl Geogr 34:21–28

Bateman JC, Chanasyk DS (2001) Effects of deep ripping and organic matter amendments on Ap horizons of soil reconstructed after coal strip-mining. Can J Soil Sci 8:113–120

Baumhardt RL, Schwartz RC, MacDonald JC, Tolk JA (2011) Tillage and cattle grazing effects on soil properties and grain yields in a dryland wheat– sorghum–fallow rotation. Agron J 103:914–922

Benayas JMR, Martins A, Nicolau JM, Schulz JJ (2007) Abandonment of agricultural land: an overview of drivers and consequences. CAB Rev Perspect Agric Vet Sci Nutr Nat Resour 2(057):1–14. http://www.cababstractsplus.org/cabreviews

Bielders CL, Michels K, Rajot JL (2000) On-farm evaluation of ridging and residue management practices to reduce wind erosion in Niger. Soil Sci Soc Am J 64:1776–1785

Birkas M (2008) Environmentally sound adaptable tillage. Akademia Kiado, Budapest

Bjorneberg DL, Aase JK, Westermann DT (2000) Controlling sprinkler irrigation runoff, erosion, and phosphorus loss with straw and polyacrylamide. Trans ASAE 43:1545–1551

Bjorneberg DL, Santos FL, Castanheira NS, Martins OC, Reis JL, Aase JK, Sojka RE (2003) Using polyacrylamide with sprinkler irrigation to improve infiltration. J Soil Water Conserv 58:283–289

Blanco-Canqui H, Lal R (2008) Principles of soil conservation and management. Springer, New York

Blanco-Canqui H, Lal R (2009) Crop residue removal impacts on soil productivity and environmental quality. Crit Rev Plant Sci 28(3):139–163

Blanco-Canqui H, Gantzer CJ, Anderson SH, Alberts EE (2004) Grass barriers for reduced concentrated flow induced soil and nutrient loss. Soil Sci Soc Am J 68:1963–1972

Blanco-Canqui H, Gilley J, Eisenhauer D, Boldt A (2014) Soil carbon accumulation under switchgrass barriers. Agron J 106:2185–2192

Blanco-Canqui H, Hergert GW, Nielsen RA (2015) Cattle manure application reduces soil's susceptibility to compaction and increases water retention after 71 years. Soil Sci Soc Am J 79:212–223

Blancoo-Canqui (2008) Soil resilience and conservation. Springer, Dodrecht

Burgess CP, Chapman R, Singleton PL, Thom ER (2000) Shallow mechanical loosening of a soil under dairy cattle grazing: effects on soil and pasture. N Z J Agric Res 43:279–290

Buschiazzo DE, Zobeck TM, Aimar SB (1999) Wind erosion in loess soils of the semiarid Argentinian pampas. Soil Sci 164:133–138

Busscher WJ, Bauer PJ (2003) Soil strength, cotton root growth and lint yield in a southeastern USA coastal loamy sand. Soil Till Res 74:151–159

Chaudhari SK, Singh R, Singandhupe RB, Mahadkar UV, Kumar A (2006) Irrigation induced soil degradation in Mula Command Area of Maharashtra. Publication No 33, Water Technology Centre for Eastern Region, Indian Council of Agricultural Research, Bhubaneswar, India

Chavez C (2007) Erosion control in furrow irrigation system and water use efficiency with application of gypsum and PAM (in Spanish). Thesis. M.Sc. Universidad Autónoma de Querétaro, México

Chavez C, Ventura-Ramos E, Fuentes C (2009) Erosion control in furrow irrigation using Polyacrylamide (in Spanish). Hydraulic Eng Mexico 24(4):135–144

Chavez C, Fuentes C, Ventura-Ramos E (2010) Water use efficiency in furrow irrigation with application of gypsum and polyacrylamide (in Spanish). Terra Latinoamericana 28(3):231–238

References

Chen C, Neill K (2006) Response of spring wheat yield and protein to row spacing, plant density, and nitrogen application in central Montana. Fertilizer Fact No. 37. http://landresources.montana.edu/FertilizerFacts/pdf/FF37.pdf

Clark A (2007) Managing cover crops profitably, 3rd ed. National SARE Outreach Hardbook Series Book 9. Natl Agric Lab, Beltsville

Council of Europe (2003) Revised European Charter for the Protection and Sustainable Management of Soil (CO-DBP/documents/codbp2003/10e). Committee for the activities of the Council of Europe in the field of biological and landscape diversity, Strasbourg, France

D'Odorico P, Bhattachan A, Davis KF, Ravi S, Runyan CW (2013) Global desertification: drivers and feedbacks. Adv Water Resour 51:326–344

Daniells IG (2012) Hardsetting soils: a review. Soil Res 50(5):349–359

van Donk SJ, Klocke NL (2012) Tillage and Crop Residue Removal Effects on Evaporation, Irrigation Requirements, and Yield. In: Proceedings of the 24st Annual Central Plains Irrigation Conference, February 21–22, 2012, Colby, Kansas

Drewry JJ, Paton RJ (2000) Effect of subsoiling on soil physical properties and dry matter production on a brown soil in Southland, New Zealand. N Z J Agric Res 43:259–268

Eynard A, Schumacher TE, Lindstrom MJ, Malo DD (2004) Porosity and pore-size distribution in cultivated Ustolls and Usterts. Soil Sci Soc Am J 68:1927–1934

Fageria NK, Balingar VC, Clark RB (2006) Physiology of crop production. The Haworth Press Inc., New York

FAO (2006) The Global Forest Resources Assessment 2005. FAO, Rome, Italy

FAO (2009) ResourceSTAT-Fertilizer. Food and Agriculture Organization of the United Nations. [Online]. Available: http://faostat.fao.org/site/575/DesktopDefault.aspx?PageID=575#ancor, 12.03.2009

FAO (2010) Global forest resources assessment 2010. FAO, Rome

FAO-IFIA (2000) Fertilizers and their uses, a pocket guide for extension officers, 4th edn. Food and Agriculture Organization of the United Nations and International Fertilizer Industry Association, Rome

FFO Federal Foreign Office, Government of Germany (2007) Desertification: a security threat? Analysis of risks and challenges. Technische Zusammenarbeit, Berlin

Flanagan D, Chaudhari K, Norton LD (2002) Polyacrylamide soil amendment effects on runoff and sediment yield on steep slopes: part II. Rainfall conditions. Trans ASAE 45(5):1339–1351

Foley JA, DeFries R, Asner GP, Barford C, Bonan G, Carpenter SR et al (2005) Global consequences of land use. Science 309:570–574

Foley KM, Shock CC, Norberg OS, Welch TK (2012) Making strip tillage work for you: a Grower's Guide, Oregon State University, Department of Crop and Soil Science Ext/CrS 140

Foster G, Romkens MJ, Dabney SM (2006) Soil erosion predictions from upland areas – a discussion of selected RUSLE2 advances and needs. In: Proceedings of the Sino-American workshop on advanced computational modeling in hydroscience and engineering, 25–26 Nov, 2006, Beijing

Garrity D, Akinnifesi F, Ajayi O, Weldesemayat S, Mowo J, Kalinganire A, Larwanou M, Bayala J (2010) Evergreen agriculture: a robust approach to sustainable food security in Africa. Food Secur 2:197–214

GEF and GM (2006) Resource mobilization and the status of funding of activities related to land degradation. Report prepared for the Global Environment Facility (GEF) and the Global Mechanism of the United Nations Convention to Combat Desertification (UNCCD), Washington, DC

Gillieson D, Wallbrink P, Cochrane E (1996) Vegetation change, erosion risk and land management on the Nullarbor Plain, Australia. Environ Geol 28:145–153

Gonçalves C, Evtyugina M, Alves C, Monteiro C, Pio C, Tomé M (2011) Organic particulate emissions from field burning of garden and agriculture residues. Atmos Res 101:666–680

Goodson R, Letlow R, Rester D, Stevens J (2000) Use of precision agriculture technology to evaluate soil compaction. In Proc 23rd Annual Southern Conservation Tillage Conference for Sustainable Agriculture, 19–21 June, Monroe, Louisiana

Goossens D, Gross J, Spaan W (2001) Aeolian dust dynamics in agricultural land areas in Lower Saxony, Germany. Earth Surf Proc Land 26:701–720

Granged AJP, Zavala LM, Jordán A, Bárcenas-Moreno G (2011) Post-fire evolution of soil properties and vegetation cover in a Mediterranean heathland after experimental burning: a 3- year study. Geoderma 164:85–94

Gruver JB (2013) Prediction, prevention and remediation of soil degradation by water erosion. Nature Educat Knowl 4(12):2

Hajabbasi MA, Jalalian A, Karimzadeh HR (1997) Deforestation effects on soil physical and chemical properties, Lordegan, Iran. Plant Soil 190:301

Hamza MA, Anderson WK (2005) Soil compaction in cropping systems a review of the nature, causes and possible solutions. Soil Tillage Res 82:121–145

Hicks DH, Anthony T (2001) Soil conservation technical handbook. The Ministry for the Environment, New Zealand

Isaac NE, Taylor RK, Staggenborg SA, Schrock MD, Leikam DF (2002) Using cone index data to explain yield variation within a field. Agricultural Eng Inter, The CIGR Journal of Scientific Research and Development. Manuscript PM 02 004: IV

Johnson JMF, Allmaras RR, Reicosky DC (2006) Estimating source carbon from crop residues, roots and rhizodeposits using the national grain-yield database. Agron J 98:622–636

Kadam KL, McMillan JD (2003) Availability of corn stover as a sustainable feedstock for bioethanol production. Bioresour Technol 88:17–25

Kairis O, Karavitis C, Salvati L, Kounalaki A, Kosmas K (2015) Exploring the impact of overgrazing on soil erosion and land degradation in a dry mediterranean agro-forest landscape (Crete, Greece). Arid Land Res Manag 29(3):360–374

Kannan A (2012) Global environmental governance and desertification: a study of Gulf Cooperation Council countries. The Concept Publishers, New Delhi

Kapos V (2000) Original forest cover map. UNEP-WCMC, Cambridge

Keskin SG, Khalilian A, Han YJ, Dodd RB (2011) Variable-depth Tillage based on Geo-referenced Soil Compaction Data in Coastal Plain Soils. Int J Appl Sci Technol 1(2):22–32

Kirchmann H, Johnston AE, Bergstrom LF (2002) Possibilities for reducing nitrate leaching from agricultural land. Ambio 31(5):404–408

Kludze H, Deen B, Weersink A, Van Acker R, Janovicek K, De Laporte A, McDonald I (2013) Estimating sustainable crop residue removal rates and costs based on soil organic matter dynamics and rotational complexity. Biomass Bioenergy 56:607–618

Kotschi J (2013) A soiled reputation: adverse impacts of mineral fertilizers in tropical agriculture. Heinrich Böll Stiftung (Heinrich Böll Foundation), WWF, Germany

Laird DA, Chang C-W (2013) Long-term impacts of residue harvesting on soil quality. Soil Tillage Res 134:33–40

Laker MC (2001) Soil compaction: effects and amelioration. In: Proceedings of the 75th Annual Congress of the South African Sugar Technologists' Association, 31 July 3 August 2001, Durban, South Africa

Lal R (1990) Soil erosion and land degradation: the global risks. In: Lal R, Stewart BA (eds) Soil degradation. Springer, New York

Lal R (2004) Review: soil carbon sequestration to mitigate climate change. Geoderma 123:1–22

Lal R (2007) Soil science and the carbon civilization. Soil Sci Soc Am J 71:1425–1437

Lal R (2014) Societal value of soil carbon. J Soil Water Conserv 69:186A–192A

Lal R (2015) Restoring soil quality to mitigate soil degradation. Sustainability 7(5):5875–5895

Lal R, Stewart BA (1990) Soil degradation, Advances in soil science, Vol II. Springer, New York

Lee SS, Gantzer CJ, Thompson AL, Anderson SH (2010) Polyacrylamide and gypsum amendments for erosion and runoff control on two soil series. J Soil Water Conserv 65(4):233–242

Lehmann J, Joseph S (2009) Biochar for environmental management: an introduction. In: Lehmann J, Joseph S (eds) Biochar for environmental management, Science and technology. Earthscan, UK, pp 1–2

Lehmann J, Schroth G (2003) Nutrient leaching. In: Schroth G, Sinclair FL (eds) Trees, crops and soil fertility. CAB International, Wallingford, UK

Lentz RD, Sojka RE (2000) Applying polymers to irrigation water: evaluating strategies for furrow erosion control. Trans ASAE 43:1561–1568

Lentz RD, Sojka RE, Robbins CW, Kincaid DC, Westermann DT (2001) Polyacrylamide for surface irrigation to increase nutrient-use efficiency and protect water quality. Comm Soil Sci Plant Anal 32:1203–1220

Li Z, Ma Z, van der Kuijp TJ, Yuan Z, Huang L (2014) A review of soil heavy metal pollution from mines in China: pollution and health risk assessment. Sci Total Environ 843–853:843–853

Lithourgidis AS, Vasilakoglou IB, Dhima KV, Dordas CA, Yiakoulaki MD (2006) Silage yield and quality of common vetch mixtures with oat and triticale in two seeding ratios. Field Crops Res 99:106–113

Lithourgidis AS, Dordas CA, Damalas CA, Vlachostergios DN (2011) Annual intercrops: an alternative pathway for sustainable agriculture. Aust J Crop Sci 5(4):396–410

Luna JM, Staben ML (2002) Strip tillage for sweet corn production: yield and economic return. Hortscience 37(7):1040–1044

van Lynden GWJ (2000) Soil degradation in central and eastern Europe. The assessment of the status of human-induced degradation. FAO Report 2000/05. FAO and ISRIC, Rome, Italy and Wageningen, The Netherlands

Madden NM, Southard RJ, Mitchell JP (2008) Conservation tillage reduces PM10 emissions in dairy forage rotations. Atmos Environ 42:3795–3808

Magen H (2008) Balanced crop nutrition: fertilizing for crop and food quality. Turk J Agric For 32:183–193

Mbagwu JSC, Obi ME (2003) Land degradation agricultural productivity and rural poverty. Environmental implications. In: Proceedings of the 28th annual conference of soil Science society of Nigeria, 7 Nov 2003, National Pool Crop Research Institute, Umudike Umuahia, Nigeria

McGarry D (2001) Tillage and soil compaction. In: Garcia-Torres L, Benites J, Martinez-Vilela A (eds) First World Congress on Conservation Agriculture, Natural Resource Sciences, 1–5 October 2001, Madrid, Spain

Mitchell JP, Jackson L, Miyao G (2004) Minimum tillage vegetable crop production in California. Oakland: University of California Agriculture and Natural Resources Publication 8132. UC ANR CS Web site. http://anrcatalog.ucdavis.edu/VegetableCropProductionin California/8132.aspx

Mitchell JP, Pettygrove, GS, Upadhyaya DS, Shrestha A, Fry R, Roy R, Hogan P, Vargas R, Hembree K (2009) Classification of conservation tillage practices in California Irrigated Row Crop Systems. Publication 8364, University of California, Division of Agriculture and Natural Resources, http://anrcatalog.ucdavis.edu

Morgan RPC (2005) Soil erosion and conservatiom, 3rd edn. Blackwell Publishing, Malden

Mosaddeghi MR, Hajabbasi MA, Hemmat A, Afyuni M (2000) Soil compactibility as affected by soil moisture content and farmyard manure in central Iran. Soil Tillage Res 55:87–97

Muchena FN (2008) Indicators for sustainable land management in Kenya's context. GEF Land Degradation Focal Area Indicators, ETC-East Africa. Nairobi, Kenya

Muir J, Pitman W, Foster J (2011) Sustainable, low-input, warm-season, grass-legume grassland mixtures: mission (nearly) impossible? Grass Forage Sci 66:301–315

Mussery A, Leu S, Lensky I, Budovsky A (2013) The effect of planting techniques on arid ecosystems in the Northern Negev. Arid Land Res Manag 27:90–100

Navarro MC, Pérez-Sirvent C, Martínez-Sánchez MJ, Vidal J, Tovar PJ, Bech J (2008) Abandoned mine sites as a source of contamination by heavy metals: a case study in a semi-arid zone. J Geochem Explor 96(2–3):183–193

NCASI (2004) Effects of heavy equipment on physical properties of soils and on long-term productivity: a review of literature and current research. Technical Bulletin No 887, National Council for Air and Stream Improvement, Inc, Research Triangle Park, NC

Nordstrom KF, Hotta S (2004) Wind erosion from cropland in the USA: a review of problems, solutions and prospects. Geoderma 121:157–167

Oldeman LR (2000) Impact of soil degradation: a global scenario. International Soil Reference and Information Centre, Wageningen

Oldeman LR, Hakkeling RTA, Sombroek WG (1991) World map of the status of human-induced soil degradation: an explanatory note. ISRIC, Wageningen

Padidar M, Jalalian A, Abdouss M, Najafi P, Honarjoo N, Fallahzade J (2014) Effect of nanoclay on soil erosion control. NANOCON 2014, Nov 5th – 7 th, Brno, Czech Republic, EU

Pimentel D (2006) Soil erosion: A food and environmental threat. Environ Dev Sustain 8:119–137

Pratt D (2002) Stop Overgrazing. Beef. Minneapolis 38:12. 22

Pulleman MM, Bouma J, van Essen EA, Meijles EW (2000) Soil organic matter content as a function of different land use history. Soil Sci Soc Am J 64:689–693

Raper RL, Schwab EB, Balkcom KS, Burmester CH, Reeves DW (2005) Effect of annual, biennial, and triennial in-row subsoiling on soil compaction and cotton yield in Southeastern U.S. silt loam soils. Appl Eng Agric 21(3):337–343

Raper RL, Reeves DW, Shaw JN, Van Santen E, Mask PL (2007) Benefits of site-specific subsoiling for cotton production in Coastal Plain soils. Soil Tillage Res 96:174–181

Reenberg A (2001) Agricultural land use pattern dynamics in the Sudan-Sahel—towards an event-driven framework. Land Use Policy 18(4):309–319

Reicosky DC, Allmaras RR (2003) Advances in tillage research in North American cropping systems. In: Shrestha A (ed) Cropping systems: trends and advances. Haworth Press, New York

Ruelland D, Tribotte A, Puech C, Dieulin C (2011) Comparison of methods for LUCC monitoring over 50 years from aerial photographs and satellite images in a Sahelian catchment. Int J Remote Sens 32(6):1747–1777

Runnstrom MC (2000) Is northern China winning the battle against desertification? Satellite remote sensing as a tool to study biomass trends on the Ordos plateau in semiarid China. Ambio 29:468–476

Santos FL, Serralheiro RP (2000) Improving infiltration of irrigated Mediterranean soils with polyacrylamide. J Agric Engng Res 76:83–90

Sarath G, Mitchell RB, Sattler SE, Funnell D, Pedersen JF, Graybosch RA, Vogel KP (2008) Opportunities and roadblocks in utilizing forages and small grains for liquid fuels. J Ind Microbiol Biotechnol 35:343–354

Savci S (2012) An agricultural pollutant: chemical fertilizer. Int J Environ Sci Develop 3(1):77–80

Sharaiha RK, Ziadat FM (2007) Alternative cropping systems to control soil erosion in arid to semi-arid areas of Jordan. African Crop Sci Conf Proc 8:1559–1565

Sharma NK, Singh RJ (2013) Agronomic practices for erosion control. Popular Kheti 1(3):57–60

de Sherbinin A (2002), A CIESIN thematic guide to land-use and land-cover change, Centre for International Earth Science Information Network (CIESIN), Columbia University, New York

Snyman H, du Preez C (2005) Rangeland degradation in a semi-arid South Africa—II: Influence on soil quality. J Arid Environ 60:483–507

Sojka RE (2006) PAM research project. Cited from Blanco H, Lal R (2008). Principles of soconservation and management. Springer, New York

SSSA (2008) Glossary of soil science terms. (Soil Science Society of America). http://www.soils.org/sssagloss/

Stan V, Fintineru G, Mihalache M (2014) Multicriteria analysis of the effects of field burning crop residues. Not Bot Horti Agrobo 42(1):255–262

Stather E (2006) Germany's engagement. In: FMECD, The Role of Governance in Combating Desertification, GTZ Haus, Berlin

Svanback A, Ulen B, Etana A (2014) Mitigation of phosphorus leaching losses via subsurface drains from a cracking marine clay soil. Agric Ecosyst Environ 184:124–134

Torella JL, Ceriani JC, Introcaso RM, Guecaimburu J, Wasinger E (2001) Tillage, liming and sunflowers. Agrochimica 45:14–23

UNCCD (2000) An introduction to the United Nations Convention to Combat Desertification. http://www.unccd.int

UNCED (1992) Managing fragile ecosystems: combating desertification and drought. Agenda 21, Chapter 12, Doc. A/CONF.151/26, Vol. II, United Nations Conference on Environment and Development
UNEP (United Nations Environment Program) and ISRIC (International Soil Research Information Center) (1990) World map of the status of human induced soil degradation. Wageningen, The Netherlands
Upadhyaya SK, Lancas KP, Santos-Filho AG, Raghuwanshi NS (2001) One-pass tillage equipment outstrips conventional tillage method. Calif Agric 55(5):44–47
USDA-NRCS (2006) Crop residue removal for biomass energy production: effects on soils and recommendations. Soil Quality National Technology Development Team, USDA-Natural Resource Conservation Service, Washington, DC
Van den Akker JJH, Schjonning P (2004) Subsoil compaction and ways to prevent it. In: Schjønning P, Elmholt S, Christensen BT (eds) Managing soil quality: challenges in modern agriculture. CAB International, Wallingford, Oxon
Van den Akker JJH, Arvidsson J, Horn R (2003) Introduction to the special issue on experiences with the impact and prevention of subsoil compaction in the European Union. Soil Tillage Res 73:1–8
Van Vliet N, Reenberg A, Rasmussen LV (2013) Scientific documentation of crop land changes in the Sahel: a half empty box of knowledge to support policy? J Arid Environ 95:1–13
Velthof G, Barot S, Bloem J, Butterbatch-Bahl K, de Vries W, Kros J, Lavelle P, Olesen JE, Oenema O (2011) Nitrogen as a threat to European soil quality. In: Sutton MA, Howard CM, Erisman JW, Billen G, Bleeker A, Grennfelt P, van Grinsven H, Grizzetti B (eds) The European nitrogen assessment. Cambridge University Press, Cambridge
Verheijen F, Jeffery S, Bastos AC, van der Velde M, Diafas I (2010) Biochar applications to soils: a critical scientific review of effects on soil properties, processes and functions. JRC Scientific and Technical Reports EUR240.99EN, JRC European Commission & Institute for Environmental Sustainability. Office for the Official Publications of the European Communities, Luxembourg
Viana M, Reche C, Amato F, Alastuey A, Querol X, Moreno T, Lucarelli F, Nava S, Calzolai G, Chiari M, Rico M (2013) Evidence of biomass burning aerosols in the Barcelona urban environment during winter time. Atmos Environ 72:81–88
Villamil MB, Amiotti NM, Peinemann N (2001) Soil degradation related to overgrazing in the semi-arid Southern Caldenal area of Argentina. Soil Sci 166:441–452
Wang GY, Innes JL, Lei JF, Dai SY, Wu SW (2007) China's forestry reforms. Science 318:1556–1557
Ward SC (2006) Restoration: success and completion criteria. In: Lal R (ed) Encyclopedia of soil science. CRC Press, Boca Raton
Warren A, Khogali M (1992) Assessment of desertifcation and drought in the Sudano-Sahelian Region IW-/99/. United Nations Sudano-Sahelian Office
Wells MS, Reberg-Horton SC, Smith AN, Grossman JM (2013) The Reduction of Plant-Available Nitrogen by Cover Crop Mulches and Subsequent Effects on Soybean Performance and Weed Interference. Agronomy Journal 105(2):539–545
WRI (1997) World resources 1996–1997. World Resources Institute, Washington, DC
Wu Y, Xu G, Shao HB (2014) Furfural and its biochar improve the general properties of a saline soil. Solid Earth 5:665–671
Yang J, He Z, Yang Y, Stoffella P, Yang X, Banks D, Mishra S (2007) Use of amendments to reduce leaching loss of phosphorus and other nutrients from a sandy soil in Florida. Environ Sci Pollut Res Int 14(4):266–269
Yu J, Lei T, Shainberg I, Mamedov AI, Levy GJ (2003) Infiltration and erosion in soils treated with dry PAM and gypsum. Trans ASAE 43:1561–1568
Zejun T, Tingwu L, Qingwen Z, Jun Z (2002) The sealing process and crust formation at soil surface under the impacts of raindrops and polyacrylamide. In: Proceedings of the 12th ISCO conference, Beijing

Zentner RP, Wall DD, Nagy CN, Smith EG, Young DL, Miller PR, Campbell CA, Campbell BG, Brandt SA, Lafond GP, Johnston AM, Derksen DA (2002) Economics of crop diversification and soil tillage opportunities in the Canadian prairies. Agron J 94:216–230

Zhang S, Grip H, Lövdahl L (2006) Effect of soil compaction on hydraulic properties of two loess soils in China. Soil Tillage Res 90:117–125

Zhibao D, Zunming W, Lianyou L (2000) Wind erosion in arid and semiarid China: an overview. J Soil Water Conserv 55:439–444

Zougmore R, Kambou FN, Outtara K, Guillobez S (2000) Sorghum-cowpea intercropping: an effective technique against runoff and soil erosion in the Sahel (Saria, Burkina Faso). Arid Soil Res Rehabil 14:329–342

Index

A
Abelmoscus esculantus, 203
Abney level, 188
Abrus precatorius, 203
Absorption, 75, 92, 228, 319, 378
Acacia albida, 26
Acacia raddiana, 58
Acacia senegal, 58, 417
Acacia tortilis, 26
Ace pseudoplantanus, 161
Acer campestre, 101
Acer ginnala, 101
Acer platanoides, 101, 161
Acer rubrum, 101
Acetaldehyde, 91
Acetic acid, 91, 337, 386, 390
Achillea sp., 43
Acidification, 2, 4, 305, 306, 308, 309, 317, 414, 419, 420
Acidithiobacillus ferooxidans, 312
Acidity, 2, 4, 9–11, 13, 40, 41, 49, 112, 146, 148, 152, 165, 220, 231, 247, 299, 300, 303–317, 319, 321, 323, 326, 328, 355
Acid mine drainage, 344
Acid soil, 10, 42, 90, 99, 168, 231, 299–328, 351
Acid sulfate soils, 2–4, 9, 90, 94, 168, 299
Acorus calamus, 175
Acquired soil fertility, 220
Activated carbon, 363, 378
Adsorption, 156, 320
Afforestation, 152, 161, 444
Agave fourcroydes, 26
Agave sisalana, 26
Agricultural lime, 322, 325

Agrochemicals, 52, 108, 110, 162, 163, 292, 334, 339, 343, 413, 427, 428
Agroforestry, 1, 203–210, 215, 443
Alcaligenes, 368, 369
Alchornia cordifolia, 204
Alder, 112, 161, 175, 206, 337
Aldrin, 340
Alfisols, 12, 22, 37, 133, 302, 303, 326
Alginite, 45
Alkaline soils, 4, 9, 90, 99, 314, 315, 317, 351
Alkylbenzene sulfonates, 335
Alley cropping, 46, 203–206, 215, 443
Allium cepa, 268, 269, 272
Allium sativum, 269
Almond, 24, 268, 271, 272, 359
Alnus cordata, 161
Alnus glutinosa, 161
Alnus spp., 161
Aluminium toxicity, 4, 5, 7, 8, 10, 301, 318, 319, 321–323, 326
Alyssum sp., 77
Amendments, 27, 45–50, 60, 75, 78, 79, 138–139, 230, 235, 247, 260, 270, 275, 276, 283, 285–290, 292, 293, 313, 340, 379, 381, 383, 389, 390, 430, 431
American cranberry, 112, 273
American hazelnut, 206
American plum, 206
Americium-241, 355
Ammonia (NH_3), 50, 92, 100, 126, 131, 237, 239, 292, 307, 315, 345
Ammonification, 92, 93, 315
Amur maple, 101
Anacardium occidentale, 25
Anaerobic composting, 232, 336, 360, 361

Anaerobiosis, 88
Andes, 151
Andisols, 12
Apatite, 313, 389
Apium graveolens, 272, 284
Aporrectodea caliginosa, 356
Apple, 24, 26, 323, 354, 359
Apricot, 24, 268, 269, 271, 272, 359
Aquerts, 119
Arabidopsis halleri, 351, 385, 386
Arachis glabrata, 438
Arachis hypogaea, 244, 245, 269, 271, 438
Araucaria cunninghamii, 48
Arenosols, 9, 37, 38, 303
Aridisols, 22
Aristida pungens, 58
Aronia arbutifolia, 273
Aronia melanocarpa, 210
Arrowwood, 206, 273
Arsenic (As), 282, 339, 350, 352, 355, 357, 358, 376, 385
Arsenopyrite, 352
Artificial drainage, 81, 97, 98, 102–108, 112, 147, 162, 170, 275
AsFeS, 352
Ash, 78, 137, 161–163, 233, 290, 360, 361, 364, 379, 381, 388
Asparagus, 165, 167, 284
Asparagus officinalis, 284
Aspen, 161
AsS, 352
Atmospheric deposition, 225, 256, 260, 307, 334, 345, 350, 353
Atrazine, 342, 344, 365, 368, 374
Atriplex halimus, 58
Atriplex nummularia, 58
Austrian pine, 44, 273
Avena sativa, 284, 438
Avena strigosa, 438
Avocado, 269
Azadirachta indica, 48, 240
Azotobacter, 317
Azotobacterplaque test, 222

B

Bajra, 284
Balanites aegyptiaca, 125
Bald cypress, 101, 273
Balsam fir, 323
Bamboo fences, 212
Band placement, 242
Barley, 24, 26, 27, 30, 38, 48, 130, 137, 205, 229, 269, 271, 281, 284, 322, 323, 375, 438

Barnyardgrass, 438
Barrelclover, 438
Barringtonia acutangula, 101
Basal dressing, 232, 241
Basella alba, 203
Bayberry, 42, 273
Beach plum, 42, 273
Bean, 24–26, 52, 54, 76, 110, 130, 131, 158, 165, 167, 175, 205, 235, 236, 242, 244, 245, 268, 269, 271, 272, 281, 319, 354, 357, 358, 438
Bearing capacity, 7, 155, 159, 162, 166, 195, 248
Bedrock, 68, 76, 80, 191, 192
Beets, 26, 165
Benincasa hispida, 203
Bentonite, 45, 59, 231, 378, 389
Bentonite amendments, 47–48
Benzene, 338, 340, 346, 347, 368, 374
Berberis, 42
Berberis thunbergii, 42
Bergenia, 77
Bergenia sp., 77
Berkheya coddii, 385
Bermuda grass, 271, 284
Berseem, 245, 284, 322
Beta vulgaris, 271, 272, 284, 385
Betula nigra, 101
Betula pendula, 161
Betula pubescens, 161
Big bluestem, 206
Big tooth aspen, 58
Bioaccumulation, 349, 356, 384
Bioaugmentation, 366, 368
Bioavailability, 351, 363, 375, 381, 384, 389, 390
Biochar, 6, 45, 49–50, 230, 235, 364, 432
Bioconcentration, 349
Biodegradation, 364, 366, 371, 372
Biodiversity conservation, 145, 150, 157, 163, 164, 173, 175
Biomagnification, 349
Biomass, 20, 21, 27, 47, 49, 54, 75, 112, 172, 175, 225, 230, 232, 235, 243, 247, 352, 384, 386, 411, 418, 420, 432, 439, 441, 443–445
Bioremediation, 363, 366–368, 370–373, 375, 376, 381, 389, 390
Biosolids, 138, 413
Biosparging, 368
Biostimulation, 366, 368
Bioventing, 368
Birch, 58, 323
Bird's-foot trefoil, 438

Index 459

Blackberry, 206, 268, 269, 271, 272
Black chokeberry, 206, 209, 210
Black Clays, 118
Black Cotton Soils, 118, 127
Black Cracking Clays, 118
Blanket bog, 148, 149
Blanket flower, 42, 78
Black gum, 273
Blackhaw, 206
Black medick, 438
Black mustard, 438
Black oats, 438
Black soils, 133
Black spruce, 111, 161
Bluegrass, 205, 284, 438
Blue-green algae, 317
Bog asphodel, 152
Bog bean, 175
Bog cotton, 152
Bog rosemary, 152
Bogs, 9, 87, 94, 95, 111, 148, 150, 155, 157, 158, 303
Bone meal, 231–233
Boreal, 161, 169
Boreal forests, 111, 415
Boreal tundra, 110, 153
Bornmuellera sp., 385
Boulders, 2, 70, 191, 202, 212
Bouteloua curtipendula, 209
Boysenberry, 268, 272
Brachiaria mutica, 284
Branch packing, 208
Brassica juncea, 284, 383–387, 438
Brassica napus, 383, 384, 438
Brassica nigra, 438
Brassica oleracea, 272, 284, 386
Brassica rapa, 165, 268, 272, 384, 438
Bridging, 73
Broad-based terraces, 199, 200
Broad-bed, 132–136
Broadcasting, 136, 169, 240, 241, 289
Broadleaves, 161
Broccoli, 165, 243, 244, 272
Bromus ordeaceus, 438
Brown mustard, 438
Brushlayer, 208
Brush mattress, 208
Brussels sprouts, 243
BTEX, 346, 347, 349, 368, 374, 375
Buckwheat, 27, 54, 438
Buddleia davidii, 77, 102
Buffalo grass, 175
Buffering capacity, 11, 39, 309, 323, 326

Bulk density, 39, 73, 99, 121, 122, 125, 126, 136, 147, 153, 154, 162, 171, 192, 200, 287, 290, 416, 421, 422
Burclover, 438
Burning bush, 210
Bush clover, 42
Bush honeysuckle, 210
Butterfly bush, 77, 102
Buttonbush, 206
Butyrospermum paradoxum, 25

C

Cabbage, 76, 165–167, 169, 243, 244, 272, 284
Ca-chlorapatite, 313
Cadmium, 282, 323, 336, 339, 350, 352, 353, 355, 357, 358, 376, 382, 385
Caesium–137, 355
Ca-fluorapatite, 313
Ca-hydroxyapatite, 313
Cajanus cajan, 25, 135, 236, 244, 438
Calcisols, 22
Calcite, 121, 291, 312, 323, 324
Calcium bentonite, 47
Calcium sulfate, 285
California poppy, 42
Callery Pear, 101
Calycanthus floridus, 102
Campsis radicans, 44
Canada wildrye, 209
Canavalia ensiformis, 236, 245, 438
Canyons, 185
Capricum frutescens, 203
Capsicum annuum, 246, 268, 272
Carbamates, 339–341
Carbaryl, 340, 368
Carbohydrates, 91, 246
Carbon dioxide (CO2), 49, 89, 152, 175, 290, 305, 309, 324, 360, 361, 419
Carbon sequestration, 50, 145, 162, 169, 175, 413, 432, 445
Carbonate rocks, 71
Carica papaya, 203
Carissa carandus, 25
Carpinus betulus, 101
Carrot, 76, 165–167, 243, 244, 268, 271, 272
Cash crops, 26, 52, 55, 207, 266
Cashew, 25
Cassava, 25, 158, 204, 440
Casuarina equisetifolia, 58
Cation exchange capacity (CEC), 4, 5, 9, 10, 39, 46, 47, 121, 127, 128, 164, 219, 323, 326, 351, 353, 432
Cauliflower, 165, 167, 243, 244, 284, 321

Celastrus scandens, 44
Celery, 76, 165, 167, 272, 284
Celtis occidentalis, 101
Cereals, 17, 20, 24–27, 53, 130, 145, 167, 169, 205, 206, 231, 240, 246, 248, 258, 267, 307, 322, 359, 419, 433, 438, 440, 441
Chaenomeles speciosa, 43
Chalk, 71, 76
Charcoal, 49, 50, 130, 235, 379, 432
Chard, 358
Chaya, 25
Chelation, 380
Chemical wastes, 350
Chemolithotrophic bacteria, 312
Cherry, 77, 78, 269, 273, 358
Chickpea, 24, 26, 130, 135, 245, 322
Chinese water chestnut, 158
Chinese witchhazel, 101
Chiseling, 80, 433, 435, 446
Chlorinated aromatic compounds, 346
Chloris gayana, 284
Chlorite, 128
Chlorpyrifos, 339
Chokeberry, 206, 273
Chromium (Cr), 282, 336, 350, 354, 356–358, 376, 385
Cicer arietinum, 131, 135, 244, 245, 269
Cinquefoil, 77
Citrullus lanatus, 319
Citrus limon, 269
Citrus sinensis, 268, 269, 272
Clay, 2, 21, 37, 71, 117, 146, 168, 192, 220, 258, 300, 344
Cleome, 42
Cleome hassleriana, 42
Clethra alnifolia, 102, 273
Cliffs, 185
Climate, 4, 8, 11, 21, 26, 37, 38, 51, 76, 78, 96, 102, 111, 119, 121, 125, 150–152, 158, 161, 163–165, 172, 175, 196, 203, 228, 271, 301, 303, 305, 414, 419, 421, 423, 439, 443, 444, 446, 447
Cnidoscolus chayamansa, 25
Coarse texture, 2, 11, 21–23, 31, 37, 40, 57, 81, 100, 165, 210, 223, 282, 302, 326, 327
Coastal dunes, 58
Cobalt-60, 355
Colcasia esculanta, 203
Cold soils, 2, 110–112
Combined drilling, 242
Combustion, 129, 171, 235, 306, 346, 360, 361, 432
Common oat, 438
Common sunflower, 438

Compacted soils, 3, 7, 208, 417, 424, 446
Compaction, 2, 4, 11, 21, 28, 39, 41, 76, 99, 103, 106, 121, 124, 125, 128, 166, 171, 228, 413, 417, 420–422, 426, 429, 430, 432, 434, 444, 446, 449
Composite waste, 334
Compost, 6, 25, 29, 40, 45, 46, 55, 78, 131, 230, 232, 233, 247, 290, 336, 360, 361, 379, 381, 428, 431
Composting, 249, 336, 360, 361
Concave slopes, 187, 188
Conifer, 70, 76, 161, 302
Conservation tillage, 13, 27–32, 137–138, 230, 247, 429, 432, 441, 449
Contaminants, 324, 335, 336, 344, 345, 347, 350, 359, 360, 362–369, 371–378, 380, 387, 388, 419
Contour bunds, 196–197, 204
Contour cropping, 430
Contour crops, 204
Contour strip cropping, 204–206
Controlled traffic tillage, 434
Conventional tillage, 27, 41, 137, 432–434
Convex slope, 187
Copper (Cu), 50, 71, 157, 165, 220, 237, 282, 315, 323, 328, 336, 339, 343, 350, 355, 358, 376, 385
Coralberry, 206
Coriandrum sativum, 294
Coriaria sinica, 205
Corn, 24, 26, 31, 32, 51–53, 110, 165, 167, 168, 205, 207, 228, 229, 231, 242, 271, 272, 281, 284, 354, 418, 436, 439–442
Corn stover, 49, 233, 418
Cornus sericea, 273
Corsican pine, 161
Cosmos, 77
Cosmos bipinnatus, 42
Cotton, 8, 19, 24, 26, 28, 52, 109, 111, 130, 138, 266, 269, 271, 281, 284, 322, 433, 434, 446
Cottonwood, 58, 198, 208
Coulters, 434
Cover crops, 6, 13, 47, 53–55, 79, 80, 110, 230, 236, 244, 245, 247–248, 270, 430, 432–435, 437, 438, 441, 443
Cowpea, 26, 54, 76, 158, 204, 236, 245, 246, 269, 271, 284, 322, 438, 440
Crabapple, 78, 101, 206
Crape myrtle, 43, 101, 102
Crataegus phaenopyrum, 43
Crateagus crusgalli, 101
Creep, 42, 58, 191, 193, 210, 285, 429
Creeping juniper, 210, 273

Crimson clover, 236, 245, 438
Crop management, 1, 23–33, 41, 130, 132–136, 165, 176, 230, 244, 269, 283
Crop removal, 225, 228–229
Crop residue, 6, 23, 25, 28, 29, 31, 46, 54, 57, 131, 136–138, 198, 228, 243, 247, 287, 360, 413, 418, 425, 426, 430, 432–434, 436, 439, 441, 445
Cropping patterns, 31, 125, 206, 236, 242, 249, 413, 439
Cropping systems, 24, 26–27, 108, 133, 134, 136, 203–206, 242–246, 249, 290, 427, 430, 439–441, 444, 445
Crotalaria cobalticola, 385
Crotalaria juncea, 438
Crusting, 10–12, 28, 57, 79, 166, 265, 420–423, 426, 430, 431, 435, 443, 449
Cryerts, 119
Cucumber, 24, 55, 81, 110, 165, 243, 244, 272
Cucumis sativus, 272
Cucurbita maxima, 203
Cucurbita pepo, 203, 272, 284, 372
Cunninghamellaplaque test, 222
Curcuma longa, 203
Cynodon dactylon, 109, 284
Cyperus papyrus, 164
Cytotoxic wastes, 337

D

Dark clay soils, 118
Darnel ryegrass, 438
Date palm, 24, 272, 359
Daucus carota, 268, 272
DCE, 346
DDT, 339, 340, 343, 347, 374
Dead nettle, 77, 209
Debris avalanches, 191
Debris flows, 191, 195
Debris torrents, 191
Decompaction, 29, 80, 270
Deep placement, 242
Deep soils, 69–70, 75, 160, 423, 427
Deep-water rice, 99
Deforestation, 152, 163, 190, 195, 203, 413, 415, 420, 447
Denitrification, 13, 91, 92, 99, 168, 225, 227, 239–241, 316
Depositional crust, 421, 422
Desalinization, 263
Desertification, 16, 20, 23, 32, 447, 448
Deserts, 2, 16, 17, 19, 21, 57, 58, 448
   ecosystem, 59
   reclamation, 448

Desmodium, 205, 206, 245
Destabilization, 328
Diammonium phosphate (DAP), 237, 238
Dianthus, 77
Diazinon, 339, 368
Dibrom, 339
Dichloroethylene, 346, 365
2,4-Dichlorophenoxy acetic acid (2, 4-D), 337, 340, 344, 368, 375
Dieldrin, 340, 343
Diervilla lonicera, 210
Diethylhexyl phthalate, 335
Diplachne fusca, 284
Disinfectants, 337
Disk harrow, 446
Dispersion, 11, 125, 126, 138, 258–260, 268, 289, 389, 418, 421–423, 431, 432
Diversion terraces, 199, 200
Dogwoods, 58, 206, 273
Dolichos lablab, 25, 244
Dolomite, 168, 323, 324, 327, 339, 350, 381
Dove plum, 25
Downy birch, 161
Drainage, 2, 41, 83, 85–91, 93–106, 108–111, 130, 157, 188, 260, 301, 335, 412
Drip irrigation, 52, 242, 270, 278
Dry soils, 2, 3, 304
Dryland
   farming, 20, 25, 28–30
   salinity, 2, 22, 262, 283
   soils, 13, 15–32, 443
Dryness, 2, 3, 11, 25, 260, 280

E

Earthflows, 191, 192
Earthquakes, 192, 195
Earthworms, 2, 110, 232, 321, 356
Eastern Gama grass, 206
Ecosystems, 16, 41, 50, 58, 76, 86, 94, 96–98, 110, 145, 146, 150–152, 157, 162–164, 170, 172–175, 220, 292, 304, 315, 328, 410, 413, 416, 426, 427, 432, 443, 445, 447, 448
Effective Neutralizing Value, 309
Elaeagnus angustifolia, 44
Elderberry, 78, 206, 273
Electrokinetic remediation, 365, 380
Elettaria cardamomum, 44
Elevation, 68, 78, 150, 186, 206, 210, 213
Elymus canadensis, 209
Embankments, 129, 155, 196, 199, 202
Empetrum, 175
Encapsulation, 377–380, 388

Endive, 323
Entisols, 12, 22, 38, 70
Eroded soils, 3, 7, 203, 208, 212, 228, 426
Erodibility, 21, 201, 439
Erosion, 2, 20, 70, 96, 124, 153, 187, 256, 314, 413
Erosion hazard, 2, 4, 8, 78, 230–231
Eruca vesicaria, 438
Eschscholzia californica, 42
Estuarine areas, 99, 261, 306, 314
Ethanol, 91, 363, 418
Ethylbenzene, 346, 347, 368
Eucalyptus citriodora, 48
Eucalyptus grandis, 48
Eucommia ulmoides, 101
Euonymous fortunei, 44
Euonymus alatus, 210
Euphorbia balsamifera, 58
European beech, 161
European hornbeam, 101
Eutrophic peat, 153, 157, 168
Evaporation, 16, 25, 28, 29, 41, 52, 54, 55, 123, 137, 247, 261, 265, 270, 274, 275, 278, 283, 422, 428, 438
Evapotranspiration, 10, 15, 16, 21, 22, 52, 53, 74, 109, 149, 208, 256, 261, 262, 276, 375
Exchangeable acidity, 300
Exchangeable Al, 311, 314, 326
Exchangeable sodium, 3, 9, 124–126, 130, 256, 258, 260, 262, 268, 269, 283, 284, 287, 289, 292, 421, 422, 432
Exchangeable sodium percentage (ESP), 124–126, 138, 256, 258–260, 283, 287, 289–291, 432
Expansive soils, 8, 139
Extended dryness, 11
Extended wetness, 11
Extreme textures, 2

F

Faba bean, 24, 52, 130, 131, 244, 245
Fagopyrum sculentum, 438
Fagus sylvatica, 78, 161
Fallow, 20, 24, 27, 31, 47, 53, 75, 125, 127, 131, 132, 203, 205, 236, 247, 248, 414, 416, 433, 439, 444
Feed legumes, 26
Fens, 83, 94, 95, 110–112, 148–150, 155, 167, 303
Ferrihydrite, 311, 364
Ferrous sulfide (FeS), 311, 315
Fertigation, 52, 239, 242

Fertility depletion, 4, 248
Fertilizer efficiency, 242
Fertilizer placement, 133, 241
Fescue, 78, 205, 209, 284, 322, 438
Festuca ovina, 209
Festuca rubra, 209
Fiber plants, 26
Fibrists, 147, 153–155
Field mustard, 438
Field pea, 54, 130, 131, 245
Fieldstone, 202
Fig, 24
Firethorn, 42
Firki, 118
Flat pea, 438
Flocculation, 125, 258, 259, 418, 422
Flood hazard, 4, 8
Flooding, 2, 52, 86, 88–91, 99, 160, 188, 195, 260, 261, 275, 278, 327, 421
Floods, 4, 8, 10, 104, 278, 326, 422, 427, 440
Florida Anise, 101
Fluid lime, 324
Fluvisols, 303, 304
Folists, 88, 90, 94, 146, 147
Food and Agriculture Organization (FAO), 2, 4, 7–10, 16–18, 21, 22, 24, 37, 38, 47, 68, 70, 94, 98, 100, 105, 109, 118, 119, 134, 137, 146, 147, 150, 151, 162, 198, 200, 202, 204, 220, 225, 229, 236, 248, 263–266, 270, 279, 303, 413, 415, 419
Forage, 6, 26, 54, 75, 80, 109, 203, 205, 221, 236, 258, 271, 284, 440, 441, 445, 449
Forest fires, 192
Forestlands, 1, 426
Forest peat, 153
Formation of gullies, 210–211
Fragaria sp., 268, 272
Fragipan formation, 73
Fragrant sumac, 43, 210
Fragaria sp, 268, 269, 272
Fraxinus excelsior, 161
Fraxinus pennsylvanicum, 101
Free calcium carbonate, 291
Freezing and thawing, 445
Frigid temperature regime, 12, 119
Frost, 88, 110, 165, 172, 189, 427, 440
Fuchsia, 77, 78
Fungicide, 339, 342–344, 358
Furrow irrigation, 52, 278–280, 422

G

Gabions, 211, 213–214
Gama medic, 438

Gardening, 1, 70
Gazania sp., 42
Gelisols, 12, 87, 146, 148
Genotoxic wastes, 337
Geomorphological properties, 97
Geomorphology, 95, 96, 102, 151, 152
Geranium, 78
Germination, 2, 27, 41, 48, 49, 99, 102, 103, 109–111, 166, 236, 247, 266, 278, 357, 417, 425, 433, 436
Gilgai, 118, 119, 122–124
Ginger, 207
Ginkgo, 101
Ginkgo biloba, 101
Gleditsia triacanthos, 101
Gleysols, 304
Gliding, 192, 194
Gliricidia sepium, 204, 206
Global warming, 152, 153
Glycine max, 135, 244, 245, 271, 372, 438
Glycophytic plants, 266–267
Glyphosate, 341, 342, 344
Goethite, 311, 312, 315, 364
Golden plover, 152
Gold Moss, 42
Gossypium harbaceum, 203
Grading, 161, 196, 198–202, 359
Grain legumes, 25, 244
Grape, 24, 269, 272, 323, 358
Grass clippings, 29, 55
Grassed waterways, 133, 137
Grass pea, 131, 137
Grasslands, 1, 17, 18, 71, 119, 130, 301, 304, 413, 414
Gravels, 2, 5, 29, 70
Gravity, 104, 155, 189–191
Grazing, 8, 19, 20, 38, 68, 70, 75, 79, 96, 123, 130, 131, 145, 152, 171, 205, 225, 416, 421–423, 436, 448
Green ash, 101
Green gram, 158, 245
Greenhouse gas, 90, 95, 98, 146, 152, 163
Green manuring, 47, 136, 235–236, 244, 245, 437
Grevillea robusta, 48
Groundnut, 19, 24, 26, 130, 158, 242, 245, 269, 271, 317, 440
Ground gypsum, 289
Groundwater table (GWT), 2, 11, 68, 69, 74, 80–81, 87–89, 100, 102, 103, 105, 108, 147, 167, 175, 259, 261, 270, 275, 282, 292, 418
Guizotia abyssinica, 131
Gullies, 31, 57, 195, 208, 210–215, 428, 430

Gypsisols, 22
Gypsum, 6, 21–23, 79, 126, 139 231, 257, 260, 261, 276, 285–292, 339, 431, 432

H

Hackberry, 101, 206
Hairy indigo, 438
Hamamelis virginiana, 101, 372
Hardpans, 21, 23, 69, 73, 75, 76, 81, 121, 289, 446, 447
Hard setting, 29, 79, 446
Harrier, 152
Haumaniastrum robertii, 385
Hazardous wastes, 359, 360, 373, 379
Hazelnut, 24, 206
Heather, 323
Heavy cracking clays, 3
Heavy metals, 9, 156, 317, 323, 335–338, 343, 345, 350–362, 365, 372, 375–384, 386, 387, 390, 415, 419
Hedge, 205, 206
Hedge maple, 101
Hedgerows, 196, 203–206
Helianthus annuus, 269, 385, 438
Hematite, 311, 315
Hemerocallis spp, 209
Hemists, 147
Hemlock, 323
Henequen, 26
Heptachlor, 340, 343
Herbicides, 27, 31, 138, 337, 339–344, 363, 370, 371, 428, 433
Hexachlorobenzene (HCB), 343, 346, 347
Hickory, 206
Hierochloe odorata, 175
High Al, 10, 12, 318
High altitude, 70, 79
High mountains, 70, 151
High phosphorus fixation, 4, 5, 10
Highbush blueberry, 112, 273
Himalaya, 151
Histels, 146, 148
Histosols, 3, 12, 88, 90, 94, 111, 146–150, 302–304, 320, 326
Holly, 78, 273
Hordeum, 438
Hordeum pusillum, 438
Hordeum vulgare, 131, 269, 271, 284
Horn meal, 233
Horse beans, 26
Horseradish tree, 25
Hospital wastes, 334, 337
Hosta spp., 209

Hudson Bay Lowland, 151
Humus, 8, 37, 49, 309
Hyacinth bean, 438
Hybrid larch, 161
Hydraulic conductivity, 99, 125, 126, 136,
    139, 154, 166, 259, 270, 289, 290, 378,
    421, 422
Hydric organic soils, 94
Hydric soils, 81, 83–100, 110, 112
Hydroabsorbents, 45, 48
Hydrogel, 48
Hydrology, 96, 108, 125, 145, 151, 152, 158,
    174, 192
Hydrolysis, 92, 258, 259, 292, 310
Hydromorphy, 4, 5, 7, 9
Hydrophytic plants, 93, 100, 101
Hyperaccumulator, 351, 383, 384, 386

I
Ilex crenata, 273
Ilex glabra, 101, 273
Ilex verticillata, 101
Ilex vomitoria, 101
Illicium floridanum, 101, 273
Ilmenite, 311
Immobilization, 168, 225, 316, 351, 377, 380,
    381, 389
Impermeability, 11
Inceptisols, 70, 100, 302, 303, 326
Incineration, 361, 362
Inclination, 185, 193
Indian grass, 206
Indian hawthorn, 273
Indian rhubarb, 323
Indigofera hirsuta, 438
Industrial wastes, 338
Inert wastes, 334
Infectious wastes, 360
Inkberry holly, 101
Inorganic mulches, 55
Insecticide, 336, 339–342
Interceptor drainage, 104
Inter-cropping, 27, 133, 135, 204, 206, 242,
    245, 246, 439, 440
International Centre for Integrated Mountain
    Development (ICIMOD), 200–202
The International Crops Research Institute for
    the Semi-Arid Tropics (ICRISAT),
    132, 133
Interrill erosion, 428
Ion exchanger, 156, 377, 380
Ion uptake, 351
Ipomoea aquatica, 164

Ipomoea batatas, 203, 268, 269, 272, 372
Irish hare, 152
Irish yew, 161
Irrigated rice, 99
Italian alder, 161
Italian clover, 236
Italian ryegrass, 438

J
Jack bean, 236, 245, 438
Jack pine, 58
Japanese barberry, 42, 77
Japanese holly, 273
Japanese kerria, 43
Japanese larch, 161
Japanese pagoda tree, 43, 101
Japanese rose, 43
Japanese yew, 210, 273
Jarosite, 311, 326
Jerusalem thorn, 26
Juniperus chinensis, 273
Juniperus horizontalis, 210, 273
Juniperus sabina, 210
Juniperus virginiana, 43, 272

K
Kale, 165
Karanda, 25
Karnal grass, 284
Kerria japonica, 43

L
Lablab bean, 25, 236, 244
Lablab purpureus, 203, 236, 438
Lacebark elm, 101
Lactuca sativa, 268, 272
Lageneria siceraria, 203
Lagerstroemia indica, 101, 102
Lagerstroemia speciosa, 101
Lamium spp, 209
Land degradation, 11, 23, 136, 416, 447
Land farming, 367
Landslides, 8, 11, 70, 78, 191–195, 211
Land subsidence, 98, 157, 171, 174
Land use, 1, 3, 8, 9, 17–21, 23, 71, 84,
    95–100, 102, 124, 126, 145, 153,
    160–162, 164, 166, 172, 173, 185, 188,
    189, 198, 200, 201, 210, 211, 248, 283,
    305, 413, 416, 443, 445
Lapland rhododendron, 317
Larix kaempferi, 161

Index 465

Larix laricina, 111, 161
Laterite, 70, 75
Lathyrus sativus, 131, 438
Lathyrus sylvestris, 438
Lava flow, 192, 195
Lavandula sp., 43
Lavender, 43
Lead (Pb), 90, 96, 131, 137, 157, 164, 171, 175, 266, 282, 308, 314, 327, 336, 345, 346, 350, 353–355, 357, 358, 376, 378, 383, 385, 414, 419, 422, 447
Leafy vegetables, 25, 165, 359
Legume, 6, 24, 25, 27, 46, 47, 110, 132, 169, 204, 206, 208, 231, 235, 243–246, 248, 266, 284, 307, 317, 321, 323, 359, 437, 439–441
Lemon, 269, 359
Lens culinaris, 131, 244, 245, 269, 284, 438
Lentil, 24, 26, 27, 130, 131, 136, 229, 244, 245, 269, 322, 438
Lepidocrocite, 311
Leptadenia pyrotechnica, 58
Leptosols, 7, 22, 37, 68, 70, 71, 303
Leptospirillum ferrooxidans, 312
Lespedeza thunbergii, 42
Lettuce, 76, 81, 165–167, 268, 272, 281, 358
Leucaena, 25
Leucaena leucocephala, 25, 204–206, 317, 372
Lignin, 6, 46, 155, 371
Lime, 21–23, 59, 129, 214, 230, 238, 247, 286, 290, 308, 317, 318, 321, 322, 324–326, 328, 353, 381, 419, 431
Lime requirement, 168, 322, 325, 326
Limestone, 71, 120, 121, 168, 287, 308, 323–325, 339, 381
Liming, 2, 6, 69, 165, 168, 171, 188, 240, 292, 309, 315, 317, 318, 320–324, 327, 328, 339
Liming material, 6, 168, 306, 322–325, 381
Lindane, 340, 343
Linear slope, 187
Lineseed, 131
Ling heather, 152
Linseed, 26, 130
Linum usitaissium, 131
Liquid fertilizers, 239
Liquidambar styraciflua, 101, 273
Lithosols, 7, 70, 71, 81
Little bluestem, 209
Live facine, 208
Live stake, 208

Livestock, 16, 17, 19–21, 23, 26, 70, 131, 191, 198, 205, 210, 228, 232, 248, 334, 336–338, 410, 421, 439, 449
Localized placement, 242, 432
Lodgepole pine, 161
Logs, 211
Lolium perenne, 438
Lolium rigidum, 438
Lolium temulentum, 438
Lonicera tatarica, 273
Lotus corniculatus, 438
Low BSP, 11
Low buffering capacity, 11, 306
Low CEC, 7, 9, 11, 23, 75, 231, 306
Low fertility, 4, 10, 41, 153, 162, 417
Low infiltration, 11, 126, 137, 430
Lowland rice, 99, 319
Low organic matter, 5, 6, 9–13, 21, 23, 29, 37, 90, 100, 127, 292, 327, 421, 423, 444
Low P availability, 4
Low soil strength, 4, 7
Low soil temperature, 11, 110
Lucerne, 47, 76, 236, 322
Luffa acutangula, 203
Lumbricus rubellus, 232, 356
Lupine, 245, 269, 438
Lupinus albus, 236, 244, 245, 438
Lupinus albusi, 319
Lupinus angustifolius, 438

M

Macedonian pine, 161
Maesopsis eminii, 48
Maghemite, 311
Magnetite, 311, 364
Magnum terrace, 199
Maize, 19, 24, 26, 30, 49, 130, 133, 135, 139, 158, 204, 205, 227, 232, 242, 269, 271, 272, 290, 352, 434, 440
Makande, 118
Malathion, 339
Malus spp., 78, 101
Manganese, 10, 71, 86, 89, 91, 92, 100, 118, 119, 156, 157, 220, 282, 292, 312, 313, 315, 317–319, 321, 328, 339, 350, 356, 371
Manganite, 312
Mangrove swamps, 306
Manihot esculenta, 25, 46, 203
Manure, 6, 25, 29, 40, 41, 46, 47, 50, 54, 55, 79, 108, 131, 138, 204, 205, 225, 230–233, 235, 241, 243, 248, 276, 287, 290, 321, 336, 338, 339, 350, 379, 411, 417, 430, 431, 437, 439, 443, 446

Marama bean, 25
Margalite, 118
Marl, 121, 123, 127, 324
Marshes, 9, 83, 87, 94, 95, 110, 111, 148, 185, 301
Mass movement, 2, 4, 8, 188–192, 196, 215
Mass wasting, 189
Mat bean, 25
Matric potential, 422
Mbuga, 118
Medicago rugosa, 438
Medicago littoralis, 438
Medicago lupulina, 438
Medicago polymorpha, 438
Medicago scutellata, 438
Medicago truncatula, 438
Melia volkensii, 48
Melilotus indicus, 438
Melissa officinalis, 245, 438
Melon, 24, 81, 165, 207, 243, 244
Mercury, 350, 354, 355, 357, 358, 376, 383
Merlin, 152
Mesa, 185
Mesquite, 26
Metal hyperaccumulator plants, 384–386
Metal sequestration, 351, 384
Metasequoia, 101
Metasequoia glyptostroboides, 101
Methane, 89, 102, 152, 360, 368
Metroxylon rumphii, 164
Metroxylon sagus, 164
Microbial remediation, 366, 367, 371, 381
Microbial respiration, 291, 305
Micronutrient, 10, 22, 31, 39, 100, 102, 108, 131, 168, 169, 220, 238, 239, 246, 247, 267, 292, 310, 312, 315, 319, 350, 351, 419
Micronutrient deficiency, 11, 419
Millet, 19, 24–26, 38, 54, 130, 290, 438, 440
Mineralization, 13, 92, 93, 138, 161, 163, 171, 175, 225, 227, 308, 313, 315, 318, 371, 373, 417, 418, 425
The Mindanao Baptist Rural Life Center (MBRLC), 206
Minimum tillage, 28, 45, 136, 137, 340, 430, 432, 444
Mints, 165, 167
Mires, 94, 148, 150, 152, 155, 172, 174, 175
Mirex, 343
Mock orange, 77, 273
Mole drainage, 106, 167
Mollisols, 12, 22, 88
Molybdenum, 157, 220, 268, 282, 315, 317, 321, 328
Momordica charantia, 203

Monoammonium phosphate (MAP), 237, 238
Monoaromatic hydrocarbons, 346
Monocropping, 24, 135, 441
Montmorillonite, 118, 125, 128, 289, 432
Moringa oleifera, 25
Moss, 42, 94, 112, 148, 150, 152, 153, 168
Mottling, 86, 87, 89
Muck, 94, 146–150, 164, 171, 303
Mucuna pruriens, 236, 245, 438
Mulch tillage, 24, 28, 29, 40, 45, 432, 434, 436, 437
Mulching, 13, 28–30, 55–57, 59, 60, 80, 230, 430, 433, 434, 436, 445
Multiple cropping, 27, 242, 439
Mung bean, 130, 357
Municipal solid wastes (MSW), 360, 361
Municipal wastes, 334, 335, 337, 360, 361
Muriate of potash, 237, 238
Mushroom, 175, 205, 359
Mustard, 130, 241, 284, 386, 438
Mycobacterium, 368, 369
Mycorrhiza inoculation, 246
Mycorrhizae, 6, 246
Myrica pensylvanica, 42

N
Nannyberry ninebark, 206
Narrowleaf lupine, 438
Native vegetation, 39, 71, 125
Natural gas, 237–238
Natural Resource Conservation Service (NRCS), 32, 54, 55, 87, 94, 106, 110, 206, 211, 214, 268, 275, 284, 300, 325, 418, 435–438, 442
Natural soil fertility, 21, 219
Nematodes, 2, 246, 356, 441
Neutralizing power, 324, 325
Nichols terrace, 199
Nickel, 31, 157, 220, 282, 323, 336, 345, 350–352, 355–359, 361, 376, 384–386, 414, 419
Nitraria retusa, 58
Nitrate, 11, 50, 89, 91, 92, 108, 156, 168, 220, 237, 239, 240, 257, 307, 308, 316, 320, 353, 419, 444
Nitrate leaching, 11, 419, 444
Nitrification, 50, 92, 227, 240, 308, 315–316, 318
Nitrobacter, 92, 316
Nitrogen fertilizers, 31, 49–51, 229, 236, 237, 239–241, 307, 419
Nitrogen fixation, 47, 54, 235, 245, 307, 316, 317, 322, 323, 356, 437, 440

Nitrosococcus, 92, 316
Nitrosomonas, 92, 316
Nitrous oxide, 49, 419
Non-recycleable wastes, 360
Nonylphenole (NP), 138, 139, 335, 347
Northern red oak, 206
Norway maple, 101, 161
Norway spruce, 101, 161
No-tillage, 20, 24, 28, 45, 137, 430, 432, 433, 441, 449
Nutrient availability, 2, 4, 9, 10, 49, 220, 228, 310
Nutrient toxicity, 10, 11, 219, 223, 256, 293, 321
Nyssa sylvatica, 273

O

Oats, 26, 49, 130, 205, 229, 284, 322
Oenanthe javanica, 164
Oil plants, 25
Oilseeds, 24, 26, 27, 130
Oligotrophic peat, 153, 156, 157, 165, 169
Olive, 24, 49, 77
Ombrogenous peat, 148
Onion, 81, 165, 167, 168, 243, 244, 268, 269, 271, 272, 287, 358
Open dumping, 152
Opuntia sps., 25
Orange, 90, 209, 268, 269, 271, 272, 320, 341, 359
Orchard, 52, 205, 241, 242, 437
Orchard grass, 205
Ordinary super phosphate (OSP), 237, 238
Organic fertilizer, 46, 169, 206, 207, 230–233, 237, 241, 249, 335, 361, 390, 419
Organic matter, 3, 21, 37, 70, 86, 125, 146, 147, 152, 154, 156, 158, 162, 163, 166, 167, 172, 175, 220, 258, 300, 335, 412
Organic mulches, 29, 55
Organic pollutants, 3, 335, 336, 343, 345–347, 349, 362–376
Organic soil, 93–95, 111, 146–150, 155–157, 159, 165, 171, 176, 376
Organic wastes, 138, 231, 360, 361, 366
Organochlorine, 336, 337, 339–341, 343, 347
Organophosphate, 339
Organophosphorus herbicides, 341
Origanum spp., 77
Orpiment, 352
Oryza sativa, 81, 98, 165, 203, 284, 319, 372
Otter, 152
Overhead sprinkler, 53
Overland flow, 186, 189, 427
oxaloacetic acid, 91

Oxisols, 12, 68, 302, 303, 306, 319, 320
Oxychlordane, 343
Ozone treatment, 364, 365

P

Palygorskite, 128
Pan, 99, 258, 423
Pan formation, 127
Panicum turgidum, 58
Panicum virgatum, 209, 215
Papyrus, 164
Para grass, 284
Parent material, 4, 21, 38, 69, 71, 92, 120–121, 123, 127, 154, 256, 259, 303, 305, 308, 309, 311, 350, 353, 422
Parkinsonia aculeate, 26
Parsley, 167, 284, 323, 358
Parsnip, 165, 167, 243, 244
Pasturing, 1
Pathogens, 2, 108, 246, 335
Pea, 24–27, 130, 229, 236, 244, 267, 269, 387, 438
Peach, 24, 269, 271, 272
Peanut, 52, 207, 271, 438
Pearl millet, 24–26, 242, 246, 438, 440
Peat extraction, 152, 153, 157, 169–170, 172, 174, 176, 413
Peat moss, 172, 233
Peat reclamation, 159, 171
Peat soils, 3, 7, 94, 98, 145–176
Peatland drainage, 98, 110, 152, 158, 160, 162, 163, 165, 167, 170, 171, 412
Peatland rehabilitation, 173
Peatland restoration, 173
Peatlands, 86, 145, 150–153, 169, 172, 413
Pebbles, 29
Pedunculate oak, 161
Pellet placement, 242
Pennesitum americanum, 25
Pennisetum typhoides, 284
Penstemon sp., 43
Pentachlorophenol, 346, 363, 365, 368, 374
Pepper, 165, 243, 244, 246, 268, 272, 323, 375
Percolation, 21, 22, 40, 52, 99, 102, 108, 261, 276, 444, 446
Peregrine falcon, 152
Perennial grasses, 19, 27, 32, 442
Perennial vegetation, 18, 304
Permafrost, 38, 88, 110, 146–148, 151, 152
Permeability, 4, 30, 71, 88, 102, 104, 106, 119, 130–132, 155, 192, 201, 258, 260, 268, 274–276, 282, 289, 291, 351, 354, 364, 366, 378, 421, 422, 441

Persea Americana, 269
Persica salvadora, 58
Persimmon, 206
Persistent organic pollutants (POPs), 336, 343, 345, 347, 371, 390
Pest, 20, 24, 26, 27, 110, 135, 236, 243, 248, 361, 440, 441
Petroselinum crispum, 284
Phaseolus acutifolius, 25, 244
Philadelphus coronarius, 102
Phoenix dactylifera, 25, 272
Phosphate fixation, 2, 314
Phosphogypsum, 293
Phosphoric acid, 237, 238, 308
Phosphorus fixation, 4, 5, 10
Photosynthesis, 266, 357
Physical problems, 11, 131
Phytodegradation, 372–374
Phytoextraction, 352, 372–375, 383, 384, 386, 387, 389
Phytoremediation, 290–291, 293, 352, 366, 367, 372–376, 381, 382, 386, 390
Phytovolatilization, 372–374, 376, 383
Picea abies, 101, 161
Picea glauca, 44, 101
Picea mariana, 161
Picea sitchensis, 161
Pigeon pea, 24–27, 130, 133, 135, 158, 236, 244, 245, 322
Pin oak, 273
Pinus caribaea, 48
Pinus contorta, 161
Pinus nigra, 78, 161, 273
Pinus peuce, 161
Pinus rigida, 44
Pinus strobus, 44, 58
Pinus sylvestris, 161
Pipe drainage, 106
Pistachio, 24
Pisum sativum, 131, 245, 269, 438
Pitch pine, 44
Pithecellobium dulce, 203
Plough sole placement, 135, 435
Plum, 25, 206, 268, 269, 271, 272
Pocosin, 148
Podzol, 37, 303
Polluted soils, 3, 7, 333–390
Pollution, 2, 4, 11, 50, 96, 109, 220, 225, 227, 240, 333, 334, 337, 339, 343, 346, 354, 368, 378, 387, 413, 414, 419, 420, 426, 441
Polyacrylamide (PAM), 49, 59, 79, 198, 431
Polychlorinated biphenyl (PCB), 335, 336, 345–348, 363, 365, 367, 368, 371, 373–375

Polychlorinated di-benzodioxins (PCDD), 336, 345, 347
Polychlorinated di-benzofurans (PCDFs), 336, 345
Polychloroethylene (PCE), 346
Polycyclic aromatic hydrocarbons (PAHs), 335, 336, 345–348, 363, 364, 366–368, 374, 375
Polymers, 32, 48, 79, 237, 287, 339, 381, 431
Polynuclear aromatic hydrocarbons (PAHs), 346
Polyphenol, 46
Polysacharides, 48
Poor drainage, 3, 11, 73, 75, 86, 102, 265, 417
Poor fertility, 11, 247
Poorly drained soils, 2, 74, 83–88, 98, 101–102, 104, 110, 112, 227, 282
Poorly fertile soils, 2, 3, 23, 39, 219–249
Poor soil structure, 11, 259, 260, 292, 417, 432
Popilus tremula, 161
Poplar, 161, 205, 273, 448
Populus beaupre, 161
Populus spp., 161, 372
Porosity, 38, 40, 74, 76, 103, 106, 111, 122, 125, 126, 154, 155, 158, 162, 169, 194, 200, 230, 282, 290, 351, 378, 412, 416, 421, 422, 427, 431, 437
Potassium fertilizer, 229, 419
Potato, 26, 49, 51, 53, 76, 81, 138, 165, 205, 207, 229, 242–244, 268, 272, 281, 359, 373, 440
Potentilla, 43
Potentilla fruticosa, 43, 77, 273
Poultry manure, 46, 232, 233, 339
Precipitation, 3, 15, 16, 21, 22, 30, 31, 87, 102, 111, 148, 193, 256, 260, 261, 265, 275, 283, 291, 306, 320, 345, 351, 377, 380, 381, 383, 389, 428
Preservation, 173
Pressmud, 290
Prickly pear, 25
Proso millet, 24, 438
Prosopis juliflora, 58
Prosopsis sps., 26
Prunus armeniaca, 268, 269, 272
Prunus avium, 269
Prunus domestica, 268, 269, 272
Prunus dulcis, 268, 272
Prunus maritima, 42
Prunus persica, 268, 269, 272
Psamments, 37, 38
Pseudomonas, 368, 369, 382
Pulses, 17, 20, 24, 26, 27, 130, 194, 244, 246, 433
Pumpkins, 243, 244
Punica granatum, 25

Index

Pyracantha coccinea, 42, 273
Pyrite, 9, 94, 285–288, 301, 304, 306, 309, 311, 312, 314, 326, 327
Pyrolusite, 312
Pyrolysis, 49, 50, 235, 362, 432
Pyrus calleryana, 101
Pyruvic acid, 91

Q

Quartz, 38, 39, 73, 120, 121, 128, 309, 363
Quercus acutissima, 101
Quercus alba, 58, 273
Quercus palustris, 273
Quercus petracea, 161
Quercus phellos, 273
Quercus robur, 161, 273
Quercus schumardii, 101
Quince, 24, 43, 359

R

Radioactive materials, 337, 355, 387
Radionuclides, 345, 355, 372, 380, 387–389, 419
Radish, 76, 165, 167, 268, 272, 323, 373, 375, 438
Rainfall intensity, 22, 51, 193, 200, 204, 427
Rainfed agriculture, 17, 20–21, 24, 76
Raised bed, 78, 104, 110, 137, 166
Ramsar Convention, 95
Rangelands, 17, 19, 23, 413, 445, 447, 448
Rape, 375, 438
Raphanus sativus, 268, 272, 373, 438
Realgar, 352
Reclamation, 11, 136, 158–159, 170, 266, 269–293, 306, 327, 328, 448
Recyclable materials, 334, 377
Recyclable wastes, 360
Redbud, 206
Red cedar, 206, 272
Red clover, 47, 236, 245, 317, 322, 438
Red fescue, 209
Red maple, 101, 206
Red-osier dogwood, 206, 273
Redoximorphic features, 88–90, 100, 119
Redox potential, 89, 91, 92, 99, 156, 171, 312, 313, 351, 390
Red pine, 58
Reduced tillage, 6, 20, 24, 28, 45, 56, 110, 136–138, 432, 433, 437, 449
Reed, 95, 101, 150, 153, 175
Regur, 117, 118
Relief, 118, 122, 124

Rendzina, 7, 71, 72
Reshaping, 198, 214
Residue management, 23, 25, 28, 31, 131, 136, 137, 231, 247, 430, 433, 435, 436, 441
Restoration, 31, 95–98, 173, 174, 214, 248, 371, 388, 415, 443–447, 449
Retaining walls, 196, 202
Revegetation, 59, 96, 97, 170, 173, 174, 196, 208, 383
Rhapiolepis indica, 273
Rhizobium, 245, 317, 321
Rhizobium galegae, 316, 317
Rhizobium leguminosarum, 317
Rhizodegradation, 372–375
Rhizoma peanut, 438
Rhizoremediation, 396, 399
Rhizosphere, 52, 100, 222, 267, 322, 373–375, 386
Rhodes grass, 284
Rhodochrosite, 312
Rhodococcus, 368, 370
Rhododendron, 317, 323
Rhubarb, 323
Rhus aromatica, 43, 210
Ridging, 96, 159, 166, 170, 430, 434
Rill erosion, 427, 428
Ripping, 80, 446
River birch, 101
Rock check dams, 212
Rockfalls, 191
Rock dams, 212, 213
Rock outcrop, 2, 4, 6, 68, 71, 188
Rock phosphate, 163, 237, 238, 307, 313
Rocky mountains, 191
Root crops, 25, 441
Root restrictive layers, 7, 11, 22, 23, 29, 68, 71–74, 79–80, 106, 258, 270
Rosa rugosa, 43, 273
Rosemary, 77, 210
Rosmarinus officinalis, 210
Rotations, 6, 20, 24, 27, 47, 80, 111, 132, 136, 192, 203, 205, 207, 230, 231, 235, 242–244, 290, 323, 423, 430, 433, 441, 443, 446
Rubus ursinus, 268, 272
Runoff, 8, 57, 79, 88, 92, 98, 102, 104, 111, 132, 137, 149, 171, 186–188, 195–200, 202, 204, 205, 208, 210, 212–214, 227, 236, 240, 277, 335, 415, 416, 419, 423, 425, 427–431, 433, 434, 436, 437, 440–442
Russian Olive, 44
Rye, 109, 110, 229, 233, 271, 322, 438
Ryegrass, 205, 271, 352

S

Saccharum officinarum, 271, 284
Safflower, 26, 130, 441
Sage, 77
Sago, 164
Sago palm, 175
Sahel, 10, 38, 126
Saline-sodic soils, 257, 258, 283, 284, 287, 289–291, 293
Salinity, 2–6, 9, 11, 12, 22, 26, 102, 103, 108, 109, 136, 220, 228, 231, 256, 258–261, 263–270, 275–282, 292, 293, 353
Salinization, 4, 52, 109, 261, 265, 413, 414, 418, 420, 426
Salix dasyclados, 161
Salix viminalis, 161, 372
Salt flushing, 274–275
Salt scraping, 273–274
Salt stress, 26, 31, 246, 266
Saltation, 59, 429
Sand dropseed, 209
Sambucus canadensis, 273
Sand dropseed, 209
Sand dunes, 21, 38, 40, 57–60, 227, 448
Sandy soils, 9, 10, 13, 38–60, 209, 219, 227, 230
Saprists, 147, 153–155
Savannas, 304
Savin juniper, 210
Saw dust, 28, 49, 55, 231, 233, 371
Sawtooth Oak, 101
Schizachyrium scoparium, 209
Schumard oak, 101
Scots pine, 161
Sedge, 95, 101, 112, 150, 152, 153, 164
Sedge peat, 153
Sedimentation, 96, 186, 204, 212, 290, 345
Sedum sp., 42, 385, 386
Seed husks, 55
Selenium, 282, 358, 382, 383, 386
Semi-desert, 2, 17
Senecio sp., 374, 375, 385
Serviceberry, 206
Sesame, 26, 130, 269
Sesamum indicum, 203, 269
Sesbania bispinosa, 136
Sesquioxides, 9, 303, 306, 309, 328
Sessile oak, 161
Settlement, 11, 17, 130, 155, 190, 195, 415
Sewage sludge, 335, 336, 350, 361, 390
Sewage wastes, 49
Shallow soils, 4, 7, 10, 12, 13, 67, 68, 70–76, 78–81, 227, 426
Shea butter, 25

Shear strength, 155, 194, 421
Sheeps fescue, 209
Sheet erosion, 189, 422, 427
Shelterbelts, 442
Shifting cultivation, 8, 70, 75, 203–204, 206, 228, 230
Shifting sands, 37, 58
Shorelines, 442
Shredded leaves, 29
Shrink-swell potential, 4, 124, 128, 129
Shrink-swell soils, 8, 118, 129, 139
Shrubby cinquefoil, 273
Shrubs, 17, 19, 32, 42, 68, 77–78, 101, 108, 111, 153, 204–206, 208, 214, 215, 272, 442, 449
Shrub willow, 58
Sidcoat grama, 209
Silica, 21, 23, 73, 422
Silky dogwood, 206
Silt fences, 196, 197
Silver, 205, 209
Silver birch, 161
Sinorhizobium meliloti, 316, 317
Sisal, 26
Slacking, 421
Slaking, 22, 126, 423
Slickensides, 118, 119, 121–123
Slide, 123, 189, 191, 193, 194
Sliding, 191, 192, 194
Slippage, 129, 189
Slipping, 194
Slit tillage, 80
Slope, 8, 13, 31, 51, 104, 106, 133, 215
Slope disturbance, 185
Slope gradient, 57, 185–189, 194, 215
Slope modification, 192, 196, 198, 444
Slope reshaping, 185
Slumps, 8, 191, 192, 202, 422
Smectite, 47, 117, 120, 121, 123–125, 128, 310
Snail medick, 438
Sod crop, 6
Sodic soils, 3, 231, 240, 256
Sodicity, 2, 4, 5, 9, 11, 22, 131, 138, 220, 231, 256, 259–263, 268–269, 281, 284, 287, 292, 293
Sodium adsorption ratio (SAR), 256, 258, 287, 289, 418
Soft brome, 438
Soil acidity, 292, 301, 305, 308, 310, 314–323
Soil alkalinity, 11, 364
Soil banking, 213, 449
Soil bioengineering, 207–210
Soil biotechnology, 207
Soil carbon loss, 2

Soil colloids, 221, 228, 262, 325, 344, 353, 356, 380, 418
Soil compaction, 111, 413, 417, 420, 422–425, 430, 432, 445–447
Soil contamination, 2, 367, 378, 388
Soilcrete, 213
Soil degradation, 11, 20, 27, 32, 165, 203, 244, 321, 413–443, 445, 449
Soil erodibility, 439
Soil flushing, 269, 274, 378
Soil limitations, 2, 4, 11, 22–23
Soil mining, 334
Soil moisture, 12, 22–25, 27–32, 40, 41, 46, 53–55, 76, 83, 88, 89, 102, 106, 110, 119, 123, 126, 130–132, 165, 236, 247, 268, 275, 312, 373, 416, 419, 423, 425, 428, 433, 436, 438, 443, 446–447
Soil organisms, 2, 225, 349, 356, 390
Soil pollution, 11, 333, 334, 337, 339, 343, 345, 356, 359–362, 367, 379, 390, 414
Soil productivity, 45, 426, 432
Soil protection, 436, 437
Soil quality, 2, 16, 24, 28, 130, 133, 137, 138, 164, 221, 287, 411, 413–415, 418, 426, 437, 441, 444, 445, 447, 449
Soil resources, 1, 2, 7, 68, 70, 118, 146, 147, 415, 426
Soil stabilizers, 431
Soil strength, 4, 7, 155, 422
Soil Survey Staff, 22, 37, 38, 68, 73, 83, 86, 94, 118, 146, 257, 302, 304, 306
Soil Taxonomy, 7, 22, 37, 38, 68, 70, 71, 87, 90, 118, 124, 146, 302, 304
Solanum melongena, 203
Solanum tuberosum, 138, 268, 272, 372, 373
Solidification, 377, 378
Solum, 37, 67, 119, 444
Solvents, 337, 338, 346, 363, 368, 372, 374, 375, 378
Sophora japonica, 101, 273
Sorghum, 19, 24, 26, 30–32, 49, 54, 76, 130, 133, 139, 158, 204, 205, 207, 246, 271, 284, 385, 418, 440, 442
Sorption, 4, 10, 93, 156, 163, 235, 314, 344, 363, 378, 379, 383
Soybean, 47, 49, 54, 76, 110, 130, 135–137, 165, 205, 242, 244–246, 271, 317, 322, 323, 434, 436, 438, 441
Sphagnum moss, 111, 112, 148, 152, 153, 174
Sphenostylis stenocarpa, 25
Sphingomonas, 368, 370
Splash erosion, 427
Spodosols, 12, 37, 302
Sporobolus cryptadrus, 209

Sprinkler irrigation, 51, 53, 54, 169, 239, 241, 242, 270, 278
Stabilization, 10, 32, 57–60, 188, 198, 199, 207–211, 213, 215, 377–379, 383, 389, 431, 448
Staghorn sumac, 206, 273
Steepness, 2, 202, 204
Steep slopes, 2, 4, 8, 10, 11, 26, 68, 70, 71, 75, 78–79, 215, 303, 425–428
Steppe, 17, 304
Steppe soils, 22
Stones, 5, 155, 196, 197
Stoniness, 4, 5, 11, 70, 78
Straw, 28, 29, 32, 49, 55, 57, 59, 137, 198, 231, 235, 290, 307, 418, 436, 448
Strawberry, 56, 57, 78, 268, 269, 271, 272, 281, 284, 322, 323, 359, 438
Straw checkerboard, 59
Straw mulching, 56
Streptanthus polygaloides, 385
Strip cropping, 203, 205, 206, 430, 440
Strip dunes, 58
Strip tillage, 6, 28, 29, 432–435
Strontium–90, 345
Structural crust, 421
Stubble mulches, 29, 56, 436
Submergence, 86, 89, 90, 92, 96, 292
Subsoiling, 75, 80, 81, 214, 276, 446, 447
Subsurface drainage, 6, 81, 103–106, 131, 132, 161
Succinic acid, 91
Sudangrass, 32, 54, 442
Sugar beet, 26, 109, 165, 229, 271, 281, 284, 375
Sugarcane, 76, 130, 136, 227, 271, 284
Sulfaquent, 94, 306, 319
Sulfaquept, 304, 306, 319
Sulfidic material, 3, 90, 301, 304, 327
Sundew, 152
Sunflowers, 26, 32, 42, 76, 130, 158, 229, 269, 322, 352, 386, 442
Sunken bed, 135
Sunnhemp, 245
Superphosphates, 238
Supplemental irrigation, 25–27, 30–32, 38, 132
Surface creep, 429
Surface crusting, 10, 22, 56, 79, 174, 443
Surface drainage, 88, 104, 112, 133, 196, 282
Surface drains, 70
Surface mats, 198
Surface roughening, 434
Surface sealing, 137, 421
Surfactants, 363, 366, 377, 378
Suspension, 239, 299, 304, 309, 324, 422, 429

Sustainable management, 75, 270, 292, 413, 445
Swamps, 83, 85, 87, 94, 95, 101, 110, 148, 149, 151, 153, 162, 164, 166, 172, 175, 303
Sweeps, 436
Sweet corn, 110, 165, 243, 244
Sweet gum, 101
Sweet potato, 26, 158, 165, 207, 243, 244, 268, 269, 272, 323
Sweet shrub, 102
Switchgrass, 209, 215
Sycamore, 161
Sylvite, 238

T

Table beet, 165
Tamarack, 111, 112, 161, 323
Tamarisk, 43, 273
Tamarix aphylla, 58
Tamarix parviflora, 43
Tamarix senegalensis, 58
Tatarian honeysuckle, 273
Taxodium distichum, 101, 273
Taxus baccata, 161, 273
Taxus cuspidata, 210
TCE, 346, 368, 374, 376
Tea, 322
Teen Suda, 118
Temephos, 339
Tensiometers, 53
Tepary bean, 25, 244
Terbufos, 339
Terminalia superba, 48
Terra rossa, 71, 72
Terracing, 31, 163, 196, 198–202
Thlaspi caerulescens, 385, 386
Thlaspi praecox, 385, 386
Thorium, 387
Thornless honey locust, 101
Threshold ECe, 271, 277, 293
Thuja plicata, 77, 161
Thyme, 42
Thymus spp., 78
Thymus vulgaris, 42
Tidal wetland rice, 99
Tile drainage, 106, 107, 109, 167, 227, 282
Tillage erosion, 20, 24, 27–30
Timber plant, 26
Timothy, 205, 441
Tin, 343
Tirs, 118
Toluene, 346, 347, 363, 374

Top dressing, 241
Topography, 4, 57, 69, 75, 104, 106, 125, 204, 305, 427
Torrerts, 119
Tourist resorts, 198
Trichloroethylene, 346, 368, 376
2,4,5-Trichlorophenoxyacetic acid (2,4,5-T), 340
Trichosanthes anguina, 203
Trichosanthes diocia, 203
Trifolium alexandrinum, 284
Trifolium incarnatum, 236, 245
Trifolium repens, 236, 245
Trigonella foenumgraecum, 131
Triple super phosphate (TSP), 237, 238, 308
Triticale wheat, 26
Trumpet Vine, 44
T values, 426, 444
Tylosema esculentum, 25

U

Uderts, 119
Ulmus parvifolia, 101
Ultisols, 12, 37, 302, 303, 306, 320, 326
Umbrella thorn, 26
Understory, 76
UNDP, 175, 176
UNESCO, 38, 150, 263, 265
Uranium, 339, 387, 389
Uranium-235, 355
Uranium-238, 355, 419
USDA, 7, 32, 54, 55, 86, 87, 94, 106, 110, 118, 123, 149, 163, 210–214, 247, 259, 268, 275, 284, 300, 325, 418, 435–438, 442
USEPA, 352, 363, 375, 376, 379, 383, 388
Usterts, 119

V

Vaccinium, 175
Vaccinium corymbosum, 112
Valleys, 8, 9, 69, 70, 79, 87, 185, 188, 210
Vanadium, 220, 282, 339
VC, 346
Vegetables, 26, 52, 55, 110, 130, 158, 165, 168, 169, 206, 236, 355, 358, 439
Vegetated geogrid, 208
Vegetative barriers, 211, 214–215, 430, 442, 444
Velvet bean, 54, 158, 236, 245, 438
Vermiculite, 40, 121
Vertic properties, 4, 8, 11

Index 473

Vertical drains, 108
Vertisols, 7, 8, 12, 22, 100, 118–133, 135–139
Vetch, 55, 236, 244, 245, 248
Vetiveria zizanioides, 205, 215
Vicia faba, 131, 244, 245, 268, 271, 438
Vigna aconitifolius, 25
Vigna radiata, 244, 245
Vigna unguiculata, 236, 244, 245, 269, 271, 284, 438
Vineyard, 78, 437
Vinyl chloride, 346, 365, 374
Viola spp, 210
Violet, 210, 323
Virginia Creeper, 44
Vitamin, 246, 347, 368
Vitex negundo, 205
Vitis spp., 44
Vitrification, 363, 377, 379
Volatilization, 93, 100, 126, 131, 168, 225, 239, 240, 292, 316, 349, 367, 382
Volcanic eruption, 192

W

Waste management, 334, 360–362, 390
Waste water, 335, 414
Water celery, 164
Water conservation, 25, 30, 134, 158, 166, 207, 228
Water erosion, 2, 4, 7, 20, 26, 131, 202, 210, 228, 420
Water holding capacity, 4, 10, 11, 40, 41, 46, 47, 49, 50, 54, 75, 125, 130, 154, 158, 163, 169, 232, 256
Waterlogging, 2, 4, 40, 69, 73, 83, 84, 88–92, 99–104, 132, 133, 136, 137, 230, 319, 327, 418, 420
Water medick, 438
Watermelon, 319
Water potential, 39, 266
Water regulation, 145
Water repellency, 40
Water spinach, 164, 165
Water stress, 4, 22, 25, 26, 30, 132, 256, 266
Wave attack, 4
Weeds, 20, 24, 25, 27, 31, 41, 54–56, 96, 99, 132, 135, 137, 153, 198, 236, 241, 244, 247, 340–343, 374, 417–419, 433, 437, 438, 440, 441
Weigela, 43, 77
Weigela florida, 43

Western red cedar, 77, 161
Wetland ecosystems, 83, 95–98
Wetland plants, 354
Wetland rehabilitation, 95–97, 101
Wetland soils, 83–87, 93, 94, 98
Wetness, 2, 9, 11, 87, 88, 91, 112, 265
Wheat, 20, 24, 26, 30, 32, 38, 49, 99, 130, 136, 137, 205, 227, 229, 230, 242, 269, 271, 284, 290, 319, 322, 323, 358, 375, 418, 438, 441
Wheatgrass, 109, 284, 285, 442
White clover, 47, 236, 245, 322, 438
White forsythia, 77
White fronted goose, 152
White lupine, 236, 244, 319, 438
White oak, 206, 273
White pea, 438
White pine, 58
White spruce, 101
Wild blackberry, 206
Wildlife habitat, 196, 441
Wild raspberry, 206
Wild rice, 164, 175
Willow, 161, 175, 198, 205, 208, 273, 448
Willow oak, 273
Wimmera ryegrass, 438
Windbreaks, 442–443, 448
Wind erosion, 2, 3, 7, 20, 24, 29, 31, 32, 40, 41, 56, 59, 60, 108, 138, 416, 420, 428, 429, 431, 434–437, 442
Wind machine, 417, 421, 433, 434
Windrow composting, 47, 232, 336, 360, 361
Winterberry holly, 101
Wintercreeper, 44
Winter wheat, 20, 24, 229
Witch hazel, 206
Wood chips, 28, 49, 55, 235, 338
Woodlands, 18, 19, 78, 126, 173, 304
Wood stakes, 211
World Reference Base (WRB), 7, 22, 37, 38, 68, 70, 71, 118, 146, 147, 303, 304

X

Xererts, 119
Xylenes, 346, 347, 368, 374

Y

Yarrow, 43
Yaupon holly, 101, 273
Yunnan camellia, 317

Z

Zeolite, 45, 379, 381, 389
Zero tillage, 28, 433, 437
Zinc, 22, 71, 100, 157, 220, 237, 282, 292,
    315, 323, 328, 336, 339, 353, 355, 356,
    358, 376, 382, 386

Zingiber officinale, 203
Zizania aquatica, 164, 175
Zizania palustris, 164
Zygophyllum spp., 58